INJURIES IN ATHLETICS: CAUSES AND CONSEQUENCES

INJURIES IN ATHLETICS: CAUSES AND CONSEQUENCES

by

Semyon Slobounov
The Pennsylvania State University
University Park, PA, USA

 Springer

Editor
Semyon Slobounov
The Pennsylvania State University
University Park, PA 16802

ISBN: 978-0-387-72576-5   e-ISBN: 978-0-387-72577-2

Library of Congress Control Number: 2008921945

Printed on acid-free paper

9 8 7 6 5 4 3 2 1

springer.com

DEDICATION

To elite athletes all over the world who sustain both physical and psychological traumas throughout their athletic careers - it is for their love of sports, their achievements, and their pain and suffering that I am most indebted. Nothing more could influence my inspiration and effort in preparation of this book.

Semyon Slobounov

ABOUT THE AUTHOR

 **Semyon Slobounov, Ph.D.,** is a Professor in the Department of Kinesiology College of Health of Human Development, and Adjunct Professor of Orthopaedics and Medical Rehabilitation with Hershey Medical College at the Pennsylvania State University, with primary responsibilities to teach undergraduate and graduate courses in the areas of psychology of injury, neural basis of motor behavior, and psychophysiology. His coaching background and clinical work with numerous injured athletes for more than 25 years was instrumental for development of ideas and topics elaborated in this book. His research focused on the neural basis of human movements with special emphasis on rehabilitation medicine, psychology and neurophysiology, including traumatic brain injuries. Dr. Slobounov is an adjunct investigator with the National Institute of Health, National Institute of Neurological Disorders and Stroke. He also is an adjunct Professor of the Neuroscience and an affiliate Professor of Gerontology Center at Penn State. He received his first Ph.D. from the University of Leningrad, Department of Psychology, USSR in 1978 and his second Ph.D. from the University of Illinois at Urbana-Champaign, Department of Kinesiology in 1994.

CONTENTS

Dedication   v

About the Author   vii

Preface   xi

Foreword   xvii

Acknowledgements   xix

Part I   **Foundations of Injury in Athletics**   **1**

Chapter 1.   Classification of Injuries in Athletics   3

Chapter 2.   Science of Training and Injury in Athletics   25

Chapter 3.   Balance as a Risk Factor for Athletic Injuries   45

Chapter 4.   Fatigue-Related Injuries in Athletes   77

Chapter 5.   Nutrition as a Risk Factor for Injury in Elite Athletes   97

Part II   **Coaches and Athletes' Perspectives of Injury**   **111**

Chapter 6.   Injury in Athletics: Coaches' Point of View   113

Chapter 7.   Injury from Athletes' Perspectives   147

Chapter 8.   Interviews with Injured Athletes   161

Chapter 9.   Overuse Injuries: Students' Points of View   197

Chapter 10   Fitness Assessment in Athletes   217

Part III   **Psychological Traumas in Athletes**   **241**

Chapter 11.   Psychological Trauma: Unfortunate Experience   243
                    in Athletics

Chapter 12.   Fear as Adaptive or Maladaptive Form of Emotional   269
                    Response

Chapter 13.   Fear of Injury, Kinesiophobia & Perceived Risk   289

Chapter 14.   Multiple Facets of Pain due to Injury   311

Chapter 15.   Psychological Trauma: Case Studies   331

Chapter 16.   Psychological Trauma: Age & Gender Factors   357

Part IV   **Concussion in Athletes**   **375**

Chapter 17.   Concussion: Why Bother?   377

Chapter 18.   Concussion Classification: Historical Perspectives   399
                       and Current Trends

Chapter 19.   Evaluation of Concussion: Signs and Symptoms   415

Chapter 20.   Traumatic Brain Injuries in Children   447

Part V   **Injury Rehabilitation**   **467**

Chapter 21.   Integrated Injury Rehabilitation   469

Chapter 22.   EEG & Neurofeedback in Rehabilitation   493

Chapter 23.   Virtual Reality in Injury Rehabilitation   515

Index   **541**

PREFACE

As the mud kicked up from the moist earth clung within his cleats, the pulse of his breath setting the rate of his stride, the distance between him and the goal narrowed. Hugging the sideline he kept pace looking towards the center to find aid, he stepped back and released. The rotation of his body contorting his spine sent radiating pains pulsing throughout his limbs. Collapsing, cushioned only by the soft earth, he fell to his knees reeling in pain. The game, though far from over, was now finished for him. Making his way to the sideline, the physical pain dulled by the psychological wounds, he braved through the torment. Unfortunately, the events of this day will haunt the memories of a once fearless athlete for a lifetime...

Injury is an unfortunate risk that is still an unavoidable part of athletics today. Over the past decade, the scientific information on athletic injury in general, and integrative models of injury rehabilitation in particular, has increased considerably. As an example, a database search of peer-review articles from Medline, *SportDiscus* and *PsycInfo* between 1970 and 2006, using a variety of search items and combinations of terms (e.g., "return to sport," "psychology of athletic injury," "sport injury") returned more than 2,000 sport injury articles. Using the search engine *PubMed* (National Library of Medicine) for the term *"psychology of injury"* there were 1,990 articles available between the years 1994-2005, compared to 930 for the years 1966-1993. In recent years, a number of models, theories and hypotheses describing the physical, biological, behavioral, cognitive and affective aspects interactively influencing the healing process have been developed. The feasibility of these models, although in most cases contradictory, has been overall justified in a clinical setting. That said, despite dramatic advances in the physical education of coaches, the fields of medicine, athletic training and physical therapy, sport-related traumatic injuries is our major concern. It is a matter of fact that athletic injuries, both single and multiple, have a tendency to grow dramatically. Accordingly, the prediction, prevention, and, if possible, reduction of sport-related injuries are among the major challenges facing the sports medicine world, research and clinical community to-date.

The purpose of this book is to accumulate the latest developments in science of athletes' training from "injury-free" perspectives, along with psychological analyses, evaluation, and management of sport-related injuries, including traumatic brain injuries. It is this author's attempt to classify athletic injury with respect to its underlying causes and consequences. Clearly we are still far from a complete understanding of the major causes and multimodal consequences of sport-related injuries. The *clinical significance* of research into sport-related injures stems from the fact that the number and severity of injuries in athletes have a tendency to grow exponentially, despite advances in coaching techniques, and technological advances in sports equipment and protective devices. For example, it have has been estimated that just in high school football alone, there are more than 250,000 incidents of mild traumatic brain injury each season, which translates into approximately 20% of all boys who participate in this sport. The incidence of injury for men's basketball games is 9.9 injuries per 1,000 athlete-exposures. The injury rate in gymnastics is about 15.19 injuries per 1,000 athlete-exposures. These are really "scary" statistics clearly indicating that modern sport is far from safe. Some details of injury epidemiology in athletics are depicted in this book.

Currently, there is no consensus among medical practitioners in terms of a generic definition and classification of injuries in athletics. The existing diversity in definitions of the term "sports injury" is apparent in the relevant literature and most likely accounts for disagreements in reported research findings and clinical practices when dealing with injured individuals. The classification of injuries, such as acute, chronic, etc., should be defined in conjunction with the severity of the injury (mild, moderate, major, sport disabling, catastrophic) to be recognized and fully accepted by coaches, athletes and medical professionals. Obviously, psychological risk factors, athletes' personalities, fear-related issues, adherence toward rehabilitation protocols and numerous other attributes of injury have never been considered within the scope of epidemiological research on the prevalence of certain injuries in certain sports. However, neither proper assessment nor appropriate treatment protocols could be developed unless multiple physical, biological, psychological and sociodemographic substrates are interactively considered when dealing with injured athletes.

It is also important to stress the ***conceptual significance*** of basic science and clinical research on various perspectives of injury. This issue has been addressed in a number of chapters of this book. For example, the effects of improper balance as a fundamental skill, progressive muscle fatigue, fear of injury and pain issues from both basic neuroscience and clinical research viewpoints will be discussed within the scope of this book. The need for an advanced conceptualization of injury in athletics stems from the fact that no two traumatic injuries are alike in terms of mechanism, symptoms, or symptoms resolution.

There is still confusion among coaches and medical personnel in terms the criteria for injured athletes' *readiness for sport participation* versus *readiness for competition*. It is important to note that: *"physical symptoms resolution of an injury is not an indication of injury resolution per se."* Although, the reality of athletics is that return-to-sport participation criteria are defined by presence and/or absence of symptoms of injury. Specifically, physical symptoms resolution (i.e., no evidence of residual tissue damage, restored anatomical integrity of joint, etc.) and functional symptoms resolution (i.e., ROM, strength, stamina) are two major criteria of return-to-play. Regarding traumatic brain injury, athletes are allowed to return to play when common symptoms of concussion (i.e., headache, fatigue, light or sound sensitivity, etc) are resolved. Is this really the cornerstone for clearance of the athlete for sport participation? In fact, residual dysfunctions and structural damage may still be present, but not observed due to numerous factors, including both extrinsic (i.e., lack of sensitivity of the assessment tools) and intrinsic (i.e., the athlete's desire to quickly return to sport participation because "… an injured athlete is worthless.")

Now, in terms of concussion in modern sport, that indeed should be treated as a "silent epidemic." The need for further understanding of concussion stems from the fact that, according to Dr. Robert Cantu, injury to the brain is the most common cause of death in athletes. It is conventional wisdom that athletes with uncomplicated and single mild traumatic brain injuries (mTBI) experience rapid resolution of symptoms within one to six weeks after the incident with minimal prolonged consequences. However, there is a growing body of knowledge indicating long-term disabilities that may persist up to ten years post injury. Recent brain imaging studies have clearly demonstrated the signs of cellular damage and diffuse axonal injury, not previously recognized by conventional imaging and neuropsychological examinations, in subjects suffering from concussion. It is a most striking fact that progressive neuronal loss in concussed subjects, as evidenced by abnormal brain metabolites, may persist up to thirty-five days post-injury. Note that current clinical practice is that athletes suffering from mild to moderate forms of TBI are usually cleared for sport participation within ten days post-injury. As a result, athletes who prematurely return to play based upon subjective symptoms resolution may be highly susceptible to future and often more severe brain injuries. In fact, concussed athletes often experience a second TBI or even multiple concussions within one year post initial brain injury. Moreover, every athlete with a history of a single mild TBI who returns to competition upon symptoms resolution still has a risk of developing a post-concussive syndrome with potentially fatal consequences.

Humans, in general, and athletes in particular, are able to compensate for mild or even severe physical and functional deficits because of redundancy in human neural, motor and cognitive systems. This in turn, allows for the reallocation of existing resources such that undamaged pathways and functions are used to perform cognitive and motor tasks. This functional reserve and overall capability to accomplish the testing protocols gives the appearance that an athlete has returned to pre-injury health status, while in actuality the injury is still present and hidden from the observer. As a result, premature return to sport participation based upon physical symptoms resolution may put athletes at high risk for recurrent injuries and the development of permanent psychological trauma. In fact, there is still no agreement upon a psychological diagnosis and definition of psychological trauma, and there is no known comprehensive treatment of psychological trauma in athletes. It is a growing concern among medical practitioners and coaches that athletes with an initial injury are prone to suffering from recurrent and more severe injuries. It is feasible to suggest that one of the major factors of recurrent injuries in athletes is premature return to sport participation based upon questionable assessment of symptoms resolution.

OUTLINE OF THE BOOK

We will now provide a few more details on the organization of this book's content. There are **5** main parts, which provide analysis of the most recent basic science and clinical research on sport-related injuries. This book is focused on both applied and conceptual issues regarding the classification of injuries, common coaches' errors leading to injury, coaches' and athletes' viewpoints on injury, the development of psychological trauma in athletes, traumatic brain injuries and basic principles of rehabilitation.

Currently accepted in clinical practice and research classification of injury, prevalence of injuries in different sports, athletic injuries from coaches' and athletes' perspectives constitute **Part 1**. Several chapters will discuss basic principles of elite athletes' preparation and common coaches' errors including problems associated with:

- Confusing classifications of injury;
- Improper planning and training periodization;
- Whole body postural control and balance;
- Progressive muscle fatigue and overloading;
- Nutritional aspects

Coaches and athletes' viewpoints on injury, including psychological responses to injury, constitute **Part 2**. Numerous interviews with collegiate and professional coaches and athletes, and descriptions of psychological methods and diagnostic procedures well-accepted in clinical practice, case studies, current practices dealing with injured athletes and future challenges are the heart of this section. In addition, a discussion of overuse "abuse" injuries in athletics is included in this part of the book. It is important to note that one of the major coaching errors in modern sport is the lack of appreciation for the proper assessment of physical fitness. This issue will be also discussed within the scope of Part 2.

Current psychological research within the conceptual framework of "psychological trauma" in athletes constitutes **Part 3**, which includes a number of chapters summarizing experimental research on fear of injury and different forms of pain resulting from sport-related injuries. Special emphasis will be given to the aspect of the development of fear of re-injury and fear of movement due to anticipated pain, i.e., "*kinesiophobia*," as a predisposing factor for long-term psychological trauma in athletes. Behavioral indexes of fear/generalized anxiety and the development of *bracing behaviors* as a result of injury will be discussed as well. Moreover, factors of age and gender as a predisposition for athletes' individuated responses to injury will be discussed in a special chapter of Part 3.

Part 4 of the book constitutes current information on traumatic brain injuries in athletes. Assessment scales and return-to-play guidelines that are well accepted and currently debated in clinical practice will be discussed within the scope of this section. Pediatric concussion, which is a major concern among medical practitioners today, will be also discussed within the scope of Part 4.

Finally, basic concepts and principles of integrated rehabilitation aimed at a timely return to sport participation will be discussed in **Part 5**. The special chapters of this section will be focused on the specialized treatment and rehabilitation of injured athletes, including the feasibility and applicability of virtual reality (VR), goal-setting and neurofeedback protocols in a clinical setting.

To my knowledge, multimodal perspectives of injury in athletics, including a discussion of the major causes and consequences of sport-related injuries with a special emphasis on coaches and athletes' viewpoints, have never been accumulated in a single source. Since the topic of sport-related injuries is included in most of the Kinesiology, Sports Psychology, Exercise and Sport Science, Athletic Training, Physical Therapy and Neuroscience curricula, it is anticipated that this book will be considered for adoption as a valuable asset and/or supplementary reading source within kinesiology, exercise and medical sciences programs.

Professor
Semyon Slobounov

FOREWORD

 Sport-related injury is one of the major factors preventing the realization of full potentials in high performance athletes. It is still an ongoing debate among sport medicine professionals and coaches whether athletic injuries are an inevitable part of athletics, or whether they may be predicted or even prevented by utilization of advanced coaching strategies and technologically safe equipment. To fully elaborate on the issues of injury among elite athletes it is necessary to consider multiple factors, both the external and internal causes and consequences of injury. Among external factors is the nature of specific sports, creating differential demands on athletes. For example, according to a recent report provided by FINA after the Olympic Games in Greece, injuries during competition are predominant in sport games (i.e., water polo), while injury during practices most often happened in complex coordination sports (i.e., springboard diving). An internal factor that is important to note is the athlete's gender. Male athletes are more prone to injury than their female counterparts, partly due to hormonal differences. In addition, psychological factors, such as attitude toward risky activity, personality, level of fear and perception of pain are just a few things that contribute to injury in elite athletes. Consequences of injury in athletes are another tremendous concern among coaches and medical professionals. Of course, this depends on type of injury, musculo-skeletal, nerves, brain injury, etc. Nevertheless, it is no longer a secret that athletes who are not treated properly and who participate in sports again before their injury is fully healed, put themselves at tremendous risk for recurrent injury and often career-ending injuries. Most importantly, multiple improperly-treated injuries may develop fear syndrome with huge psychological/psychiatric consequences. Of course, dealing with injury in elite athletes requires proper attention, highly professional knowledge and active involvement of coaches, athletes and medical staff. This book by Professor Semyon Slobounov covering various issues of sport-related injuries from the perspective of athletes, coaches, medical practitioners and scientific researchers. It is a very timely addition and very much welcomed.

Professor Eide Lübs
Federation Internationale de Natation (**FINA**)
Sport Medicine Committee Honorary Secretary
China, Nan Jing FINA Diving World Series, 2007

ACKNOWLEDGEMENTS

This book would not have been possible without the author's inspiration from observing and working with numerous athletes, who, despite constant injuries and suffering throughout their athletic careers, still amaze us by their dedication and commitment to their loving sports. It is because of their commitment to sport, incredible achievements and suffering both physically and psychologically, that our knowledge of causes and consequences of sport-related injuries has advanced so far in recent years. I would like to thank numerous students at The Pennsylvania State University whom I had the privilege to teach and share my professional growth and maturation with. It is because of their urgent need to understand, not injury per se, but rather every single injured athlete's response to injury, and the underlying causes and factors thereof, that this book was written. I would like to thank my numerous colleagues' valuable suggestions and their insight on the problems associated with sport-related injuries during preparation of this book. My special thanks go to Dr. Karl Newell and Dr. Wayne Sebastianelli who have helped me mature both professionally and personally.

In addition, I would like to thank all of the Penn State student athletes, coaching and medical staff that have given me the privilege of collaborating and taking part in their programs. I would like to acknowledge my academic departments of Kinesiology for allowing me to pursue the area of sports-related injuries. I am especially grateful to my growing family (Elena Slobounov, Vera & Craig Anderson, Kate, Stanley and Dalton Preschutti, and Anton Slobounov) for their unconditional love and support. My special thanks extended to Kate Contacos. This book wouldn't be readable without her editorial effort, enthusiasm and commitment in 'fine-tuning' the chapters. Finally, I would like to thank the staff at Springer Publishing Company for their eagerness to see this book completed and their help in making this book possible.

PART I: FOUNDATIONS OF INJURY IN ATHLETICS

CHAPTER 1

CLASSIFICATION OF INJURIES IN ATHLETICS

1. INTRODUCTION

Injury is an unfortunate risk that, according to most coaches, athletes, and medical practitioners, is an unavoidable part of athletics. Most athletes that participate in high level sports experience some type of injury during their athletic careers. Ironically, despite significant advances in science of coaching, improvement in coaching techniques, technological progress in the design of athletic equipment, protective devices and facilities, the incidence of sport injury has actually increased during the past 15-20 years (Orchard & Powell, 2003). Clear conceptualization of the term "injury" is needed in order to fully appreciate existing research and clinical practice dealing with injured athletes. Accordingly, in the following text a number of approaches to define athletic injuries will be outline and discussed.

It should be noted that diversity in definitions of the term "sport injury" is apparent in the literature and probably accounts for some disagreements in reported research findings (Pargman, 2007). The most common keyword searches used for research in popular Ovid MEDLINE and Pub Med among others are: "athletic injury", "athletic injury reporting", "injury definition", "injury risk", "injury rate", "sports injury", "injury surveillance", "injury patterns", etc., may come up with pretty diverse and controversial definitions of injury, and as a result, may imply confusing treatment procedures.

Examination of various sources, including the National Electronic Injury Surveillance System, clearly shows that there is no single consensus on the definition of a reportable injury. For example, according to Pediatric Orthopedic Society of North America (2003) early sport injury surveillance studies primarily focused on traumatic brain (TBI) and catastrophic injuries. Unlike the definitions used in more recent studies that have been more inclusive as "any tissue damage", including even minor bruises (Junge et al., 2002), any physical damage caused by a sport-related incident, where or not it is, results in any incapability to the participant (Finch, 1997), or "any time an athlete sought medical help" (Orchard, 1995; Beachy et al., 1997).

Traditionally, injuries in athletics were classified based on events and associated symptoms, including: (a) *acute* traumatic injuries (i.e. contusions, sprains tears etc.), (b) *chronic* injuries (i.e. jumper' knee, tennis' elbow, thrower's shoulder, etc.), and (c) *overuse* injuries (i.e. low back pain, spondylolisis, etc.). *Acute* injuries occur as a result of a single, sudden impact that creates tissue damage. Most often, the athlete becomes aware of the injury soon after it has occurred. This awareness, in fact, does not mean that the athlete and his/her coach have an accurate and complete comprehension of the impact on injury at the time of the accident. The full understanding of the initial injury may be achieved while monitoring various symptoms resolution (both physical/physiological and psychological) over the course of recovery. Interestingly, almost 60% of injured elite skiers indicated that they immediately realized that something was going wrong after the injury. The other 40% of injured skiers noted that they were not initially aware of the extent and severity of injury (Udry, 1999). Therefore it is important to understand that the process of mental awareness and cognitive appraisal of the injury may or may not follow the objective clinical assessment of injury.

On the other hand, *chronic* and often referred *overuse* injuries are the accumulation of repeated and most often under-threshold injuries, due to exposure to small forces over time which ultimately result in serious damage. This type of injury is frequently manifested in conditions such as tendonitis and/or stress fractures. While acute injuries may be more recognizable because of their sudden impact and often associated obvious functional abnormalities and physical symptoms, chronic injuries often gradually develop, frequently not observable and their influence of athletes may be more insidious. *Chronic* injuries can play havoc with an athletes' motivation and most often are major cause of burnout.

It should be mentioned that physicians' currently have a strong concern for non-traumatic injuries. For example the common non-traumatic cycling-related injuries include the knee, neck/shoulder, hands, buttock and perineum (Dettori & Norvell, 2006). This type of injury is often not severe enough to cause individuals to seek medical assistance. According to American College of Sport Medicine (1974), 117 cases of "handlebar palsy" were reported in athletes observed by a single physician over a four-year period; interestingly, most were not associated with tour riding. Two categories of non-traumatic injuries in cycling sport that may have the greatest impact on disability comparable with traumatic injuries include ulnar nerve palsy, and erectile dysfunctions. Moreover, according to a recent report by Dettori & Norvell (2006) the prevalence of non-traumatic cycling injuries can be as high as 85%. Overall, there are not a lot of records of non-traumatic injuries available to researchers. Accordingly, there is a lack of well-controlled reports regarding non-traumatic injuries in other than cycling sports.

From intervention perspectives, at least 5 categories of injury severity were proposed (see also Hail, 1993)

1) *Mild* – an injury requiring treatment without interruption of training and participating in competitions with low risk of development of physical and/or psychological consequences. It should be noted that mild injuries sometimes are non-recognizable by athletes, coaches and even medical practitioners. This may put the mild injuries

athletes at risk for re-current more severe injuries. It should be noted that multiple recurrent mild injuries may cause the development of "psychological trauma". Therefore, medical practitioners should consider assessment of signs for psychological trauma in athletes with multiple minor injuries and referring these athletes to a qualified psychologist for further evaluation and treatment, if necessary.

2) *Moderate* – a relatively more severe injury that interferes with ongoing practices and potentially limiting participation in practices and competitions. This type of injury definitely requires referral to medical professional for comprehensive evaluation and treatment if necessary. Monitoring for any sings of psychological trauma is highly recommended.

3) *Major* – an injury requiring a long duration of inability to practice, often associated with surgery and/or hospitalization and may potentially lead to chronic and long term physical and/or psychological deficits. It is highly recommended that referral to a sport psychologist with expertise dealing with injured athletes should be made by medical practitioners. The athletes with even a single episode of major injury are highly susceptible for development of psychological trauma.

4) *Sport disabling* - an injury which, because of severity or timing, prevent an injured athlete from returning to prior level of functioning both physical and psychological. Involvement of a clinical psychologist with knowledge of sport-related injuries is critical for this category of injured athletes to predict and prevent potentials for post-traumatic psychological/psychiatric problems (i.e. post-traumatic stress disorder).

5) *Catastrophic* – an injury that causes permanent functional and/or psychological impairment and/or disability, typically from damage to the head and spinal cord, and other injuries of comparable severity. It should be noted that the occurrence of permanent functional disability due to catastrophic injury is infrequent relative to all sports. Care about psychological well-being of this category of athletes is mandatory.

More recently, some consistency has emerged for four broadly acceptable classification/definitions based on:

(A)   *Knowledge* based approach to define injury, including the functional anatomy, an accurate patient history, diligent observation and a thorough examination.

(B)   *Medical treatment*, which incorporates injuries requiring any treatment by a team physician, whether or not they result in loss-of-time from competition or training.

(C)   ***Loss-of-time***, fully inclusive, which incorporates injuries that result in loss-of-time from competition and/or practices (Brooks & Fuller, 2006).

(D)   ***Types of sports*** differentially dominating most common injuries in athletes

2. CLASSIFICATION OF INJURIES

A correct diagnosis and proper classification of injury in athletes depends on knowledge of functional anatomy, an accurate patient history, diligent observation, thorough examination and standardized injury reports. It is critically important that the same research methodology is implemented to classify the injury in athletics. It should be noted that several reports attempt to analyze the incidence of injuries in a given sport but the results found in these reports cannot be compared with one another because of heterogeneous injury definitions, methods and data collection, observation format and design, and sample characteristics. Thus, the risk for injury in different sports can be compared because standardized research methodology is implemented.

From this perspective, it is worth mentioning to review one study conducted by a team of experts who conducted comparative study of exposure-related incidences and characteristics of injuries of elite athletes during 2004 Olympic Games (see for details: Junge et al., 2006). Specifically, all team sport (n=14) tournaments (soccer, handball, volleyball and beach volleyball, basketball, field hockey, baseball, softball and water polo) were analyzed using standardized injury reports obtained from physicians. It should be noted that response rate was 93%. According to this comprehensive report, a total of 377 injuries were reported from 456 matches. The overall injury incidence of the team sport tournaments was almost 1 injury per game. Half of the injuries affected the lower extremities, specifically ankle sprain followed by knee injury. Interestingly, 24 % of injuries involved the head and neck; accordingly, the most prevalent diagnoses were head contusion.

This report noted that the severity of head injuries were not always obvious. On average, almost 80% of injuries were caused by contact with another player. Half of these contact injuries were caused by foul play, as rated by the team physicians. However, a higher % of non-contact injuries prevented the players from participating in his or her sports.

In terms of severity of tournament-related injuries, there were 16 fractures, 17 ligament ruptures, 4 meniscal lesions, and 2 dislocations. Not surprisingly, there was a high risk of non-contact ASL injuries in female athletes. However, injuries in male and female players were similar in location and circumstances but differed significantly in terms of type of injury. Moreover, there was the higher incidence of ligamentous knee injuries in female players, which is in agreement with numerous recent reports. In addition, concussions were diagnosed more frequently among female players, although, it should be noted that some concussions may be missed in male's games and/or recorded as contusions.

Classifying Injuries is multi-factorial by nature. Clinicians take into account the patient's previous medical history which includes information related to the site of injury and general heath and well being. Through observation, clinicians are able to get an overall view of structural abnormalities, influences of posture, and determine the various signs associated with inflammation. Performing palpations provides information related to temperature changes, spasms, and most importantly painful areas. Integrity of the joints can be assessed through range of motion testing and various structural and functional testing. Often times, diagnostic images (i.e. x-ray, MRI CT scans) are used to determine location and sometimes severity of injuries.

How we classify injuries depends largely on the anatomical structures and/or the specific areas involved.

1. Soft Tissue Injury Classification
2. Neurological Injury Classification
3. Low Back Injury Classification

2.1. Soft Tissue Injury Classification

Soft tissue injuries involve structures such as muscles, tendons, and ligaments. Injuries that involve these structures typically are classified as 1st, 2nd, or 3rd degree in nature. Table 1 lists the difference associated with each grade. (Magee, 2008) One recent change is the classification of tendon. The term tendonitis has now been replaced with tendonopath; mainly due to the lack of inflammation present in the tendon.

Table 1. Classification of Muscle Strains, Ligament Sprains and Tendonopathy.

	1° Strain/Sprain	2° Strain/Sprain	3° Strain/Sprain	Tendonopathy
Definition	Few torn fibers	½ of fibers torn	All fibers torn	Inflammation and/or degeneration of tissue
MOI	Overstretch/Overload	Overstretch/Overload	Overstretch/Overload	Overuse
Onset	Acute	Acute	Acute	Chronic, acute
Weakness	Minor	Moderate to major (reflex inhibition)/Minor to moderate	Moderate to major/Minor to moderate	Minor to moderate
Disability	Minor	Moderate	Major /Moderate to major	Minor to major
Spasm	Minor	Moderate to major/Minor	Moderate/Minor	Minor
Edema	Minor	Moderate to major/Moderate	Moderate to major	None
LOF	Minor	Moderate to major	Major (reflex inhibition)/Moderate to major (instability)	Minor to major
Pain with contraction	Minor/No	Moderate to major/No	No to minor/No	Minor to major
Pain on stretch	Yes	Yes	No	Yes
Joint play	Normal	Normal	Normal/Normal to excessive	Normal
Defect	No	No	Yes	Possible
Crepitus	No	No	No	Possible
ROM	↓	↓	May ↑ or ↓ dependent upon edema/May ↑ or ↓ dependent upon edema, dislocation or subluxation possible	↓

2.2. Neurological Injury Classification

Nerves can be challenging to evaluate from an injury standpoint. Typically trauma to nerves lends to radicular symptoms including numbness, tingling, and/or pain. Table 2 outlines some distinct differences associated with neurological injury.

Table 2. Classification of Nerve Injuries According to Seddon (1943).

Grade of Injury	Definition	Signs and Symptoms
Neuropraxia (Sunderland 1°)	A transient physiological block caused by ischemia from pressure or stretch of the nerve with no wallerian degeneration	Pain No or minimal muscle wasting Muscle weakness Numbness Proprioception affected Recovery time: minutes to days
Axonotmesis (Sunderland 2° and 3°)	Internal architecture of nerve preserved, but axons so badly damaged that wallerian degeneration occurs	Pain Muscle wasting evident Complete motor, sensory and sympathetic functions lost Recovery time: months (axon regenerates at rate of 1 inch/month, or 1mm/day)
Neurotmesis (Sunderland 4° and 5°)	Structure of nerve is destroyed by cutting, severe scarring, or prolonged severe compression	No pain (anesthesia) Muscle wasting Complete motor, sensory and sympathetic functions lost Recovery time: months and only with surgery

2.3. Low Back Injury Classification

Low back injury or pain is very common among the general and athletic population. It ranks #1 as the reason patients visit either an orthopedist or neurologist. It falls to #2 for reason that patients visit their primary care physician (Heck & Sparaon, 2000; Wipf & Deyo, 1995). Recently, it has been suggested that low back pain be classified in relation to the type of symptoms that a patient has instead of an actual pathology (Heck, 2000). This is mainly due to the fact that much pathology related to low back pain are not diagnosed or properly identified in over 2/3 of the patient population (Heck & Sparaon, 2000; Spratt et al., 1990).

Table 3. Low Back Injury Classification (Heck, 2000).

Level	Description
1	Local low back pain only
2	Radiating pain
3	Neurological deficits
4	Serious conditions

3. COMMON PROBLEMS WITH EXISTING CLASSIFICATIONS

After reviewing numerous papers on the relevant topic, a conclusion can be made that the definition of injury cannot be expressed in a few simple words, nor is just one definition agreeable to all experts in the field. Analysis of published literature shows that the definition of injury is very broad and diverse due to different criteria used to evaluate it such as activity, severity of injury, and circumstances resulting in injury. However, at least one common agreement is that the injury has to have happened during a sporting event whether it is a practice of performance or a game or competition. According to NEISS-AIP the definition of injury relates to "bodily harm resulting from exposure to an external force or substance". This is exclusive needs to be further elaborated on definition of injury with respect to both physical and psychological factors associated with it. Indeed, perception of injury and its severity are multi-factorial and age, gender, fitness level etc. dependent phenomenon.

It is more or less common opinion among sport injury professionals that a definition of injury needs to include at least one, if not more, of the following: severity of the injury, time loss due to the injury, injury type, and whether a professional examined the injury. A very detailed description defined injury as "any unintentional or intentional damage to the body caused by acute exposure to physical agents such as mechanical energy, heat, electricity, chemicals, and ionizing radiation interacting with the body in amounts or at rates that exceed the threshold of human tolerance" (Baker et al., 1992). It is also agreed that injury is "an event requiring medical attention in which the time absent from participation is directly correlated to the severity of the injury." However, an addition should be made that injury is "any incident that occurs during an athletic event which involves a blow or damage to the head or spine as well as any physical ailment ranging anywhere from minor to severe causes that result in an athlete being unable to perform at 100%."

Besides the physical aspects which define an injury, psychological factors also aide to describe what injury is. Injury is an individual experience that only the injured person truly knows how it feels and what needs to be done to get better. A doctor can determine exactly what is wrong and help diagnose the issue and physical therapists can suggest exercises to perform to get back to normal. However, ultimately a positive attitude and good outlook on the situation at hand can greatly affect the recovery period along with age and gender. Some researchers mention that psychology plays a huge part in the entire injury process. Specifically, injured athletes direction of thoughts, and content of the healing imagery are important factors to consider in order to facilitating the recovery from the injury as the injury and during the recovery and healing process.

Clearly, after experiencing a traumatic injury, it is much more than just going to rehabilitation; it involves mental injury as well. Mental injury can be defined as a psychological trauma. One injured athlete reported that she felt both physical and psychological strains of the injury during her rehabilitation process. She declared that her mind wanted to be at a level that her body was not and felt less stressful when both mind and body were at the same recovering level. She also claimed that when she recovered, she was mentally stressed due to fear of re-injury.

Generally, there are immense pressures to compete early in life. Added stress and overload is often a common predisposing risk factor of injury. Once injured, negative thoughts of becoming re-injured, the amount of stress on an injured athlete and reflecting on whether or not your body is strong enough to compete in a sport again can take over the mind of an athlete. With the rate of injury increasing, coaches and trainers really need to evaluate not only the injury spot itself, but the athletes mental state a lot more than they currently are. Considering the importance of psychological trauma and its consequences on athletes' recuperation process, this topic will be specifically discussed elsewhere in this text.

The National Collegiate Athletic Association (NCAA) Injury Surveillance System (ISS) defines reportable injury as "an event that occurs during participation in intercollegiate athletics, requiring medical attention from the CAT or team physician and results in restriction of the athlete's participation for greater than or equal to one day beyond the day of injury" (Goldberg et al, 2007). In addition, many concur that including time lost from participation in the definition effectively reduces the bias associated with the incidence estimate (Meeuwisse, et al, 2003). Prager, considering *High School Football* injuries, agrees that the definition should include both time factor and a severity component used by the research committee of the American Orthopedic Society for Sports Medicine and by the US National Athletic Injury Reporting System, be adopted by all sports injury surveillance studies (Prager et al, 1989). Some argue that this point leaves room for argument.

However, there are disagreements with both of these definitions stating that one cannot base the definition of injury on time lost from activity because many athletes play through injury and in some cases, the injury is minor enough that it is all right to play through. Some athletes are more willing to fight through pain, recommended or not, than other athletes. One article claims that "70%-90% of all injuries sustained fall in the transient category—that is, by only recording injuries that result in missed matches, the majority of injuries are missed and therefore injury rates are underreported" (Hodgson, 2007). This brings up an interesting point about the frequencies and severity of injuries. Concussions, for example, occur less often than ankle injuries in basketball, yet should these two injuries have the same weight in the definition of injury? One is severe and could potentially result in death, while the other occurs often and is easily looked after.

Some researchers document that it should be considered only if it was an emergency room visit, a complaint requiring the attention of an athletic trainer, time loss that restricted participation or if a consultation with a physician was required. According to the NCAA's definition, an athlete would need to have time loss from play greater than one day (Godlberg, 2007). While other state only injuries including the head and a time loss of more than 48 hours if it occurred during a game or competition would count (Prager, 1989). The severity of the injury also relates to the amount of time a player is away from functioning properly at a game and/or practice. Interestingly, it is not clear how to classify the athletes who function properly and who do not using existing observational formats.

One's definition on injury is correlated with the amount of knowledge and exposure one has had with sports-related injuries. Coaches that are unaware about the importance of the recovery and rehabilitation process of the physical, functional and emotional aspects could allow an athlete back to play prematurely, which may result in more damage to the athlete. Anecdotal observation in general, and an injured student-athlete reported specifically, that while playing soccer, he was knocked unconscious and the coach put him back in not knowing that he had a mild concussion. This proves that a more experienced professional, like a trainer, would be knowledgeable about the injury and not allow the athlete to go back into the game knowing that he could have hurt himself more. The bottom line is that researchers need to be knowledgeable and come to a conclusion on a meaningful definition that allows for severity, time lost, and exposure of a sport.

One researcher used bones as an example to explore the "science" of injury. "Bones in youth generate new tissue at growth plates positioned near the end of most bones. During puberty, a protective band of tissue that supports the growth plate begins to deteriorate so the bone can harden in preparation for adulthood. Without the protective band, the plate is particularly vulnerable to being abnormally compressed or possibly pulled apart. Research suggests that girls are up to eight times as likely to injure their ACL as males are. Knee joints tend to be looser in girls and their quadriceps are often stronger than their hamstrings, destabilizing the knee" (Haycock, 1976). This is a very good example with valid points. It suggests that definition/classification of injuries in athletics must take into consideration the developmental aspects to properly assess the impact of injuries and to develop rehabilitation protocol.

There are various levels of injury perception and attitude because the level of pain/perception of pain associated with this specific injury varies from person to person. One cannot set an exact timeline for the recovery process because everyone's bodies are on different schedules and of different abilities. Many people feel that you cannot gauge severity by the time lost because some athletes heal faster than others do, while still more return to their sport prematurely. The knowledge of professionals is also responsible for treating and diagnosing those athletes who actually report that an injury has occurred. A sideline coach may have little to no knowledge of injury while specific sports teams need trainers on the sidelines because it is a high impact sport. Such a high number of athletes do not report injury in order to keep participating in the game, which may cause more damage to injured athlete. It is important to be able to identify an injury and its mechanism in order to classify it under specific definition and perform effective injury surveillance, which in turn could lead to better preventive ideas and methods.

4. EPIDEMIOLOGY OF INJURIES IN ATHLETICS

It has long been know that sports and injuries go hand in hand. The sports medicine team consisting of physicians, athletic trainers, and coaches all need to be aware of sudden injuries, whether they are minor or catastrophic. Each sport is unique to injuries depending on the nature of the sport. This most current literature review section was elaborated with an intention to classify the sports into contact or non-contact, breakdown the epidemiology of injury, and provide some sport specific prevention strategies.

Both **baseball** and **softball** had relatively low injury rates as compared with the other sports examined. According to a 16 year study done by the NCAA, Dick et al. (2007) reported 5.78 baseball injuries per 1,000 athlete-exposures during games whereas Marshall et al. (2007) reported 4.3 softball injuries per 1,000 athlete-exposures during games. Those same studies reported 1.85 baseball injuries per 1,000 athlete-exposures for practices and 2.7 softball injuries per 1,000 athlete-exposures for practices, respectively. Of those injuries baseball games produced the most upper leg muscle-tendon strains at 11% followed by ankle ligament sprains at 7.4% (Dick, 2007). Baseball practices produced shoulder muscle-tendon strains at10% of total injuries followed by ankle ligament sprains at 8.5% (Dick, 2007). Softball game injuries varied slightly by producing the most ankle ligament sprains at 10.3% of total injuries and practices resulted in the 9.5% of ankle ligament sprains making them the most common injury (Marshall et al., 2007).

One common prevention strategy for both baseball and softball is to use breakaway bases. Many ankle ligament sprains occur from sliding feet first into the base. Throughout the 16 year study, Dick et al. (2007) found there were 439 injuries resulting from sliding into stationary bases and only 40 injuries from breakaway bases. Another similar prevention strategy is proper off season conditioning to acclimatize the athlete to a new training regiment. Ankle ligament sprains are seen in all sports; however, the sport most accountable for this type of injury is basketball. According to Dick et al. (2007) their data collection resulted in 26.2% of all game injuries resulting in an ankle ligament sprain. The same can be said about practices with 26.8% of all injuries being ankle ligament sprains. Agel et al. (2007) reported women's basketball had 24.6% of all game injuries as ankle ligament sprains and 23.6% of practice injuries consisted of a sprained ankle.

The incidence of injury for **men's basketball** games is 9.9 injuries per 1,000 athlete-exposures and for practices the incidence drops to 4.3 injuries per 1,000 athlete-exposures (Dick et al., 2007). The same can be said for women's basketball which had an incidence of injury of 7.68 per 1,000 athlete-exposures for games and 3.99 injuries per 1,000 athlete-exposures for practices (Agel et al., 2007). Some common prevention strategies focus directly on the ankle. Further research is warranted on proprioception training to control the ankle upon landing. Also, the efficacy of prophylactic taping and bracing needs to be discussed. Does one work better than the other to prevent recurrent ankle sprains? Improving off season conditioning to prevent fatigued play during preseason may also help prevent injuries.

Fencing is considered to be a low risk sport when it comes to injury. There are, however, several injuries that can occur from this sport. Murgu (2006) reported the most common lower extremity acute injuries are ligament sprains, meniscus tears, Achilles' tendon ruptures, muscle strains, and nail contusions. The most common upper extremity acute injuries are finger, wrist, and elbow sprains and strains and muscle cramps. Overuse injuries can also happen to a fencer. Prolonged periods of time in the fencing stance can lead to patellofemoral pain syndrome, IT Band friction syndrome, and low back pain. A few suggested prevention strategies are adequate warm up, the use of orthotics, properly checked equipment, and a health care professional who can properly assess injury to prevent recurrent injury.

Golf is closely related to fencing as it is considered a low risk sport in terms of incidence of injury. Theriault and Lachance (1998) reported there to be between 1.19 and 1.31 injuries per golfer. Depending on the swing and level of play, the most commonly injured body part varies from study to study, although the several top injuries remains constant. Parziale and Mallon (2006) reported low back pain is the most common ailment of golfers. Theriault and Lachance (1998) reported that professional golfers most often injure their wrist and amateur golfers most often injure their thoracic spine. Gosheger et al. (2003) found the most common injury to professional golfers was lumbar spine (21% of injuries) followed by wrist (20%) and the most common injury to amateurs is elbow (24.9% of injuries) followed by lumbar spine (15.2%). Prevention for golf injuries has a commonality between all the studies. Proper technique of the golf swing is reiterated throughout these studies (Theriault and Lachance, 1998; Parziale and Mallon, 2006; Gosheger et al. 2003). Quality equipment, proper warm up, flexibility, good aerobic training, and not carrying your golf clubs can all decrease the overuse injuries experienced. Gosheger et al. (2003) also performed a study with one group performing a ten minute warm up as compared to no warm up. The incidence of injury dropped from 1.02 injuries per player (no warm up) to 0.41 injuries per player for those who performed a ten minute warm up.

Another non contact sport growing in popularity is **gymnastics.** The gymnastics injury rate is substantially higher than previous non-contact sports reported. Marshall et al. (2007) found 15.19 injuries per 1,000 athlete-exposures during competition and 6.07 injuries per 1,000 athlete-exposures in practice. Kilt and Kirkby (1999) reported 5.45 injuries per gymnast or 3.31 injuries per 1,000 hours of training. It is important to note that sub-elite gymnasts were also included in these studies, therefore increasing the injury rates by trying more advanced skills. Of the high incidence of injury Marshall et al (2007) reported, during competition knee internal derangement consisted of 20.6% of injuries followed by ankle ligament sprains at 16.4%. During practice Marshall et al. (2007) found 15.2% of injuries consisted of ankle ligament sprains, followed by knee internal derangement at 8.7%. Kolt and Kirkby (1999) studied 64 gymnasts reported 349 injuries. A majority of the injuries (31.2%) were located in the foot and ankle followed by the low back (14.9%), and the knee (13.5%).

Gymnastics injuries can be decreased with some specific prevention strategies. Ensure amateur gymnast only try more difficult routines after many hours of practice. A sting mat can be used to decrease the amount of force generated upon landing. Alter rules for competition to discourage harder routines with a higher risk of injury. Also, neuromuscular training, core stability, constant mental focus throughout practice, and condition so fatigue does not alter routines during enduring practices.

Rowing injuries are often due to overuse, repetitive motions. Rumball et al. (2005) found the most common site of injury in male rowers was the lumbar spine, followed by the forearm and wrist, then the knee. In females, Rumball et al. (2005) found the most common site of injury was the chest, lumbar spine, followed by the forearm and wrist. In this study, spine injuries consisted of 15-25% of all injuries. The reason for the spine injuries is the force generation caused by hyper flexion and twisting motion causing a 4.6-fold increase in body weight. McNally, Wilson, and Seiler (2005) also reported low back pain is the most common site of injury among rowers. This could possibly be due to the new rowing equipment which was made stiffer. McNally, Wilson, and Seiler (2005) reported one in eight rowers will have a rib stress fracture. Rib stress fractures are more common in sweep rowers and women. Rumball et al. (2005) found rib stress fractures account for 6.1-22.6% of all rowing injuries. Rib stress fractures could be caused by muscle weakness leading to less shock absorption and muscle imbalance. Some prevention strategies used in rowers are to strengthen core stabilizers, stretch, ease into strength training and provide the rowers with adequate rest and recovery after a training bout.

Another sport causing a majority of overuse injuries is **running**. Common injuries as noted by Strakowski and Jamil (2006) and Cosca and Navazio (2007) are shin splints, Achilles' tendonitis, plantar fasciitis, patellofemoral pain syndrome, IT band friction syndrome, and patellar tendonitis. These overuse injuries can be caused by numerous things such as: improper footwear, a sudden increase in mileage or intensity, pes cavus or pes planus, tibial varum, or leg length discrepancy (Strakowski and Jamil, 2006). There are many prevention strategies for runners to use to decrease the amount of overuse injuries or recurrent injuries. Stretching, taping, strengthening, and cross training are some easy ways to help prevent overuse injuries. Also, a runner can take NSAIDS (non-steroidal anti-inflammatory drugs). Rest, ice, properly prescribed orthotics, proper footwear, and heel lifts will also help.

Swimming is a non-contact sport involving a lot of upper body strength. Muscle imbalance can lead to injury, specifically at the shoulder. Johnson (2003) reported that shoulder injuries are about 30% of all injuries, followed by back at 20% and knee injuries comprise 10% of all swimming injuries. With up to 1 million shoulder revolutions per year, it is easy to understand why shoulder injuries are so common in swimmers (Johnson, 2003). Weldon and Richardson (2001) reported that 90% of complaints from swimmers are about the shoulder. Overuse in swimming causes a muscle fatigue which can lead to decreased rotator cuff effectiveness and decreased scapular positioning during each stroke (Weldon and Richardson, 2001). Another big risk factor for swimmers and shoulder injury is muscle imbalance which leads to shoulder instability which results from ligamentous laxity and increased range of motion (Weldon and Richardson, 2001). There are ways to prevent so many overuse injuries in swimmers. Increasing shoulder stability and muscle imbalances is a key factor to decreasing the incidence of shoulder injuries. General rotator cuff exercises and scapular positioning will attribute to a decrease in injury. Attention to proper technique and, if injured, cross training exercises will be most beneficial.

As with many non contact sports, **tennis** comes with a lot of overuse injuries. Perkins and Davis (2006) reported on the most common types of injuries among tennis players. Of them, muscle strains and ligament sprains were the most frequent. They found the most common shoulder injury to be rotator cuff tendonitis. Lateral epicondylitis of the elbow can occur frequently. De Quervain's tenosynovitis is one of the most common tendon problems facing tennis players. The most common wrist problem as described by Perkins and Davis (2006) is ulnar wrist pain secondary to extensor carpi ulnar tendonitis. Jumper's knee and patellofemoral pain syndrome are the two most common knee ailments in tennis players. Overuse treatment and prevention strategies are very similar no matter what sport causes the injury. Conservative treatments work well such as rest, ice, and NSAIDS. Strengthening and stretching the injured area can also be effective to relieve some discomfort after an injury.

One more non contact sport to include is **volleyball**. The incidence of injury in volleyball players as reported by Agel et al (2007) is 4.58 injuries per 1,000 athlete-exposures in a game and 4.1 injuries per 1,000 exposures in practice. The incidence of injury does not change much from practices to games because of the activities in both. Ankle ligament sprains make up almost half of all game injuries and almost 30% of al practice injuries. Most ankle sprains occur from landing on another player's foot. Proprioceptive training and strength training should be of utmost importance to volleyball players especially if he/she has suffered an ankle sprain previously. Agel et al. (2007) did report that the rate of ankle sprains over the 16 year period decreased by 1.8% in games and 3% in practices. Prophylactic taping and bracing should be considered once further studies are conducted to support the efficacy for it.

The number of injuries sustained in a contact sport rapidly increases compared with most non contact sports previously mentioned. One example of this is **ice hockey** where Flik, Lyman, and Marx (2005) reported the incidence of injury for men's ice hockey to be 13 injuries per 1,000 athlete-exposures in games but only 2.2 injuries per, 1,000 exposures for practices. This is very similar to Agel et al. (2007) who reported 16.27 injuries per 1,000 exposures for games and 1.96 injuries per 1,000 athlete exposures for practices.

Flik, Lyman, and Marx (2005) found concussions to be the most common injury happening 18.6% of the time, while Agel et al. (2007) found concussion to take only 9% of total injuries during games and 5.3% of all injuries during practices. The most common injury reported in the 16 year study was knee internal derangements consisting of 13.5% of all game injuries and in practice, 13.1% of the injuries were pelvis and hip muscle tendon strains (Agel et al., 2007). Agel et al. (2007) also reported on women's ice hockey showing slightly different injury patterns. Here game injuries consisted mostly of concussions happening 21.6% of the time. In practice, concussions lead all injuries in women's ice hockey taking up 13.2% of all injuries. The incidence of injury for women's ice hockey was as similar as men's ice hockey. Agel et al. (2007) reported 12.6 injuries per 1,000 athlete-exposures in games and 2.5 injuries per 1,000 athlete-exposures in practices.

Women's ice hockey likely causes more concussions because of the no checking policy making players unprepared for accidental contact. Also, some women grow up playing on a men's team so they are used to the physical contact, however, the women who are not used to it don't know how to absorb a hit and sustain a concussion. Prevention strategies should focus on decreasing the amount of brain injury in ice hockey. Making rule changes to increase the size of the ice may decrease the amount of contact a player has with another player and may help decrease the concussion rate (Agel et al, 2007). Increasing the penalty time for elbows and illegal hits to the head may also help decrease the concussion rate (Flik et al, 2005).

Men's lacrosse is very similar in the incidence of injury as ice hockey. Dick et al. (2007) reported an injury rate of 12.58 injuries per 1,000 athlete-exposures during games and an incidence of injury of 3.24 injuries per 1,000 exposures. Concussions also ranked evenly with ice hockey concussion rates comprising 8% of all injuries, the third most common following ankle ligament sprains (11.3%) and knee internal derangement (9.1%). The most common practice injury Dick et al. (2007) reported ankle ligament sprains, accounting for 16.4% of all injuries.

Dick et al. (2007) also showed that **women's lacrosse** had similar concussion injury rates as men's lacrosse, but a substantially greater number of ankle ligament sprains in games, totaling 22.6% of all injuries. Ankle ligament sprains comprised 15.5% of all injuries in practice. Another study on women's lacrosse by Matz and Nibbelink (2004) show almost identical injury rates for games and practices for ankle ligament sprain being the most common injury. An interesting note Dick et al. (2007) reported was that a new kayaking style helmet was introduced in the 1996-97 season for the women's lacrosse team to wear, and it actually increased concussion rates most likely because the helmet was meant to withstand one blow and not multiple blows throughout the course of a competition. Women's lacrosse just implemented a mandatory goggle rule in hopes of decreasing the amount of eye injuries, but no head protection is yet required as is the case in men's lacrosse.

A similar sport and incidence of injury to women's lacrosse is **women's field hockey**. Dick et al. (2007) reported 7.87 injuries per 1,000 athlete-exposures in games and 3.7 injuries per 1,000 exposures in practices. The most common injury in games was ankle ligament sprains (13.7% of injuries) followed by knee internal derangement (10.2%) and concussions was not far behind comprising 9.4% of all game injuries. The study also reported upper leg muscle strains made up over one fourth of the total number of injuries in practice. A few rule changes half way through this study decreased the amount of injuries. The sticks were made of different materials, incorporated some protective equipment, increased play on artificial turf, eliminated off sides, and moved the corner shot to the sideline (Dick et al, 2007). Some more steps can be taken to prevent more injury such as mandatory helmet and padded glove wear, change rules to decrease congestion around the goal area, and prophylactic taping and bracing to decrease ankle ligament sprains.

Soccer is an endurance type sports, however, soccer athletes sustain a high incidence of injury. Manning and Levy (2006) reported 10-15 injuries per 1,000 athlete-exposures. Dick et al. (2007) reported an incidence of injury for women's soccer at 16.4 injuries per 1,000 athlete-exposures for games and 5.2 injuries per 1,000 exposures for practices. Agel et al. (2007) reported an incidence rate for men's soccer slightly higher at 18.75 injuries per 1,000 athlete-exposures for games and 4.34 injuries per 1,000 exposures for practice. Ankle ligament sprains are the most common injury to men and women soccer player. In a men's soccer game they make up 17% of the total injuries in games (Agel, 2007) and in a women's soccer game ankle sprains make up 18.3% of the total injuries (Dick et al., 2007). Manning and Levy (2006) reported a higher incidence of injury playing indoors as compared to outdoors. Prevention of soccer injuries is primarily concentrated on ankle ligament sprains such as strength and neuromuscular training. Properly fitted shoes are also important. A rule change for tackling should be considered to prevent tackling type injuries.

Water polo is a sport more popular overseas but one that is growing in popularity in the United States. Franic, Ivkovic, and Rudic (2007) reported the most common water polo injury is shoulder pain. Shoulder pain is directly correlated with the level of competition in the athlete and the number of years competing. Hand and wrist injuries, facial lacerations, and swimmer's ear are all common because of the nature of water polo. Early identification and treatment of these injuries is paramount to prevent recurrent injuries.

Wrestling is seen by many as a dirty sport consisting of only skin diseases. Agel et al. (2007) reported the injury rate to be 26.4 injuries per 1,000 athlete-exposures for a match and 5.7 injuries per 1,000 exposures for practice. Knee internal derangement was the most common injury at 22.9% of all injuries for matches. In practice, skin infections consisted of 17.2% of all injuries followed closely by knee internal derangements at 14.8% of the total number of injuries. Weekly, wrestlers deprive their bodies of proper nutrition to make weight for a match. One prevention strategy for wrestlers is proper hygiene and nutrition. Because of the high rate of skin infections during practice, a clean and sterile environment to wrestle on is a key element in decreasing the likelihood of spreading skin infections. A rule change for specific submission holds should be considered to decrease match injuries.

Rugby is a violent contact sport consisting of anywhere form 10.6 injuries to 19.8 injuries per 1,000 athlete-exposures (McIntosh, 2005). The University of New South Wales (McIntosh, 2005) reported 35% of match injuries were to the lower extremity, 23% were to the upper extremity, 23%

to the shoulder, and 20% to the head. Concussions comprised 14-17% of all injuries sustained in this study. Carson, Roberts, and White (1999) studied a women's rugby team for one season and reported the incidence of injury to be 23.2 injuries per 1,000 athlete-exposures for matches and 2.5 injuries per 1,000 exposures for practices. The most common injury was ankle sprains. Four out of the 35 total injuries were concussions. Almost half of all injuries occur at the lower extremity. Rugby is a tough sport for injury prevention because of the nature of the game. No helmets, or pads are worn by the athletes so everyone involved needs to be clear on the procedures for injury, especially catastrophic injury.   An extensive injury surveillance system needs to be set up for rugby teams so more injury prevention programs can be specific to the nature of the sports.

Along the same lines as rugby is **football** where injury rate is 35.9 injuries per 1,000 athlete-exposures in games and 3.8 injuries per 1,000 exposures in practice (Dick et al., 2007). Knee internal derangements were the most common injury for fall games, fall practices, and spring practices at 17.8% of total injuries, 12% of total injuries, and 16.4% pf total injuries, respectively (Dick et al., 2007). Proper techniques are the most important injury prevention programs for football.   Heat acclimatization before preseason is also important. The athletes need to come into preseason practices in condition to train in the heat and endure grueling practices. Prophylactic knee bracing is also a prevention strategy to decrease the number of knee internal derangements.

CONCLUSION

Proper classification, assessment and definition of injury in athletes are important components to the development of a treatment plan. Unfortunately, the diversity in definitions of the term "sport injury" is apparent in the literature and probably accounts for some disagreements in reported research findings and clinical practices dealing with injured athletes. Classification of injuries, such as acute, chronic, etc., should be defined in conjunction with severity of injury (mild, moderate, major, sport disabling, catastrophic). This combination makes the definition of injuries extremely complex and confusing not only for athletes but also for medical professionals and coaches.

In addition, rarely psychological status of injured athletes, athletes' self-perception of injury and attitude towards injury are within the scope of definition of traumatic injuries. It is important to note that there are basically NO definitions of psychological trauma that can or cannot be directly related with physical injury in athletes. Moreover, psychological risk factor, and athletes' personality have never been considered within the scope of epidemiological research on prevalence of certain injuries in certain sports. Neither proper assessment nor appropriate treatment protocols can be elaborated unless both physical and psychological impacts are considered when dealing with injured athletes.

Acknowledgments

This chapter was prepared with great help of numerous Penn State undergraduate students who took my "Injury in Athletics, KINES 497 class". My special thanks to Dusty Lang, AT graduate assistant in Fall 2007. A also appreciate contribution of Dr. Nicole McBrier to this chapter.

REFERENCES

Orchard, J.W., & Powell, J.W. (2003). Risk of knee sprains under warious weather conditions in American football. *Medicine and Science in Sports and Exercise, 35,* 1118-1123.

Pargman, D. *Psychological Bases of Sport Injuries.* 3rd Edition. Morgantown: Fitness Infomation Technology, 2007.

Junge, A., Rosch, D., Peterson, L., et al., (2002). Prevention of soccer injuries: A prospective intervention study in youth amateur players. *American Journal of Sports Medicine, 30*(5), 652-659.

Finch, C.F. (1997). An overview of some definitional issues for sports inury surveillance. *Sports Medicine, 24,* 157-163.

Orchard, J. (1995). Orchard sports injury classification system (OSICS). Sports Health, 11, 39-41.

Beachy, G., Akau, C.K., Martinson, M., et al., (1997). High school sports injuries: A longitudinal study in Punahou School: 1988 to 1996. *American Journal of Sports Medicine, 25*(5), 675-681.

Udry, E. (1999). The paradox of injurie: Unexpeted posititive consequences. In D. Pargman (Ed.), *Psycholgoical bases of sport injuries.* (pp.79-88). Morgantown, WV: FIT.

Detorri, N., & Norvell, D. (2006). Non-traumatic bicycle injries.: A review of the literature. *Sports Medicine, 36*(1), 7-18.

Hail, J. *Psychology of Sport Injury.* Champaign: Human Kinetics, 1993.

Brooks, J., & Fuller, C. (2006). The influence of methodological issues on the results and conclusions from epidemiological studies of sports injuries. *Sports Medicine, 36*(6), 459-472.

Junge, A., Langevoort, G., Pipe, A., et al. (2006). Injuries in team sport tournaments during the 2004 Olympic Games. *American Journal of Sports Medicine, 34*(4), 565-576.

Magee, D. *Orthopedic Physical Assessment.* 5th ed. Philadelphia, Elsevier, 2008.

Seddon, H.J. (1943). Three types of nerve injury. *Brain 66,* 17-28.

Heck, J.F., & Sparaon, J.M. (2000). A classification system for the assessment of lumbar pain in athletes. *Journal of Athletic Training, 35*(2), 204-211.

Wipf, J.E., & Deyo, R.A. (1995). Low back pain. *Medicine Clinical in North America, 79*(23), 1-246.

Spratt, K.F., Lehmann, T.R., Weinstein, J.N., Sayre, H.A. (1990). A new approach to the low-back physical examination.: behavioral assessment of mechanical signs. *Spine, 15,* 96-102.

Baker, S.P, O'Neill, B., Ginsburg, M.J., et al. The Injury Fact Book, 2nd ed. New York: Oxford University Press, 1992.

Goldberg, A.S., et al., (2007). Injury Surveillance in Young Athletes, A Clinician's Guide to Sports Injury Literature, Philadelphia, PA. *Sports Medicine, 37* (3): 265-278.

Meeuwisse, W.H., et al. (2003). Rates and Risks of Injury During Intercollegiate Basketball. *American Journal of Sports Medicine, 31*(3), 379-385.

Prager, B.I., et al. (1989). High School Football Injuries: A Prospective Study and Pitfalls of Data Collection" *American Journal of Sports Medicine, 17*(5), 681-865.

Hodgson et al. (2007). For Debate; Consensus Injury Definitions in Team Sports Should Focus on Encompassing All Injuries. *Clinical Journal of Sports Medicine, 17*, 188-191.

Haycock, C.D., & Gillette, J.V. (1976). Susceptibility of women athletes to injury: myths vs. reality. *JAMA, 236*, 163-165.

Dick, R., Sauers , E.L., Agel , J., Keuter, G., Marshall, S.W., McCarty, K., et al. (2007). Descriptive epidemiology of collegiate men's baseball injuries: National collegiate athletic association injury surveillance system, 1988-1989 through 2003-2004. *Journal of Athletic Training. 42*, 183-193.

Marshall, S.W., Hamstra-Wright, K.L., Dick, R., Grove, K.A., & Agel, J. (2007). Descriptive epidemiology of collegiate women's softball injuries: National collegiate athletic association injury surveillance system, 1988-1989 through 2003-2004. *Journal of Athletic Training. 42*, 286-294.

Dick, R., Hertel, J., Agel, J., Grossman, J., & Marshall, S.W. (2007). Descriptive epidemiology of collegiate men's basketball injuries: National collegiate athletic association injury surveillance system, 1988-1989 through 2003-2004. *Journal of Athletic Training. 42*, 194-201.

Agel, J., Olson, D.E., Dick, R., Arendt, E.A., Marshall, S.W., & Sikka, R.S. (2007). Descriptive epidemiology of collegiate women's basketball injuries: National collegiate athletic association injury surveillance system, 1988-1989 through 2003-2004. *Journal of Athletic Training. 42*, 202-210.

Murgu, A.I. (2006). Fencing. *Physical Medicine and Rehabilitation Clinics of North America. 17*, 725-36.

Thériault, G., & LaChance, P. (1998). Golf injuries. *Sports Medicine. 26*, 43-57.

Parziale, J.R., & Mallon, W.J. (2006). Golf injuries and rehabilitation. *Physical Medicine and Rehabilitation Clinics of North America. 17*, 589-607.

Gosheger, G., Liem, D., Ludwig, K., Greshake, O., & Winkelmann, W. (2003). Injuries and overuse syndromes in golf. *American Journal of Sports Medicine. 31*, 438-443.

Marshall, S.W., Covassin, T., Dick, R., Nassar, L.G., & Agel, J. (2007). Descriptive epidemiology of collegiate women's gymnastics injuries: National collegiate athletic association injury surveillance system, 1988-1989 through 2003-2004. *Journal of Athletic Training. 42*, 234-240.

Kolt, G.S., & Kirkby, R.J. (1999). Epidemiology of injury in elite and subelite female gymnasts: a comparison of retrospective and prospective findings. *British Journal of Sports Medicine. 33*, 312-318.

Rumball, J.S., Lebrun, C.M., Di Ciacca, S.R., & Orlando, K. (2005). Rowing injuries. *Sports Medicine, 35*, 537-555.

McNally, E., Wilson, D., & Seiler, S. (2005). Rowing injuries. *Seminars in Musculoskeletal Radiology. 9*, 379-396.

Strakowski, J.A., & Jamil, T. (2006). Management of common running injuries. *Physical Medicine and Rehabilitation Clinics of North America. 17*, 537-552.

Cosca, D..D., & Navazio, F. (2007). Common problems in endurance athletes. *American Family Physician. 76*, 237-244.

Johnson, J.N. (2003). Competitive swimming illness and injury: common conditions limiting participation. *Current Sports Medicine Reports. 2*, 267-271.

Weldon, E.J., & Richardson, A.B. (2001). Upper extremity overuse injuries in swimming: A discussion of swimmer's shoulder. *Clinics in Sports Medicine. 20*, 423-38.

Perkins, R.H., & Davis, D. (2006). Musculoskeletal Injuries in Tennis. *Physical Medicine and Rehabilitation Clinics of North America. 17*, 609-631.

Agel, J., Palmieri-Smith, R.M., Dick, R., Wojtys, E.M., & Marhsall, S.W. (2007). Descriptive epidemiology of collegiate women's volleyball injuries: National collegiate athletic association injury surveillance system, 1988-1989 through 2003-2004. *Journal of Athletic Training. 42*, 295-302.

Flik, K., Lyman, S., & Marx, R.G. (2005). American collegiate men's ice hockey: An analysis of injuries. *American Journal of Sports Medicine. 33*, 183-187.

Agel, J., Dick, R., Nelson, B., Marhsall, S.W., & Dompier, T.P. (2007). Descriptive epidemiology of collegiate women's ice hockey injuries: National collegiate athletic association injury surveillance system, 2000-2001 through 2003-2004. *Journal of Athletic Training. 42*, 249-54.

Agel, J., Dompier, T.P., Dick, R., & Marhsall, S.W. (2007). Descriptive epidemiology of collegiate men's ice hockey injuries: National collegiate athletic association injury surveillance system, 1988-1989 through 2003-2004. *Journal of Athletic Training, 42*, 241-248.

Dick, R., Romani, W.A., Agel, J., Case, J.G., & Marshall, S.W. (2007). Descriptive epidemiology of collegiate men's lacrosse injuries: National collegiate athletic association injury surveillance system, 1988-1989 through 2003-2004. *Journal of Athletic Training. 42*, 255-61.

Dick, R., Lincoln, A.E., Agel, J., Carter, E.A., Marshall, S.W., & Hinton, R.Y. (2007). Descriptive epidemiology of collegiate women's lacrosse injuries: National collegiate athletic association injury surveillance system, 1988-1989 through 2003-2004. *Journal of Athletic Training. 42*, 262-269.

Matz, S.O., & Nibbelink, G. (2007). Injuries in intercollegiate women's lacrosse. *American Journal of Sports Medicine. 32*, 608-11.

Dick, R., Hootman, J.M., Agel, J., Vela, L., Marshall, S.W., & Messina, R. (2007). Descriptive epidemiology of collegiate women's field hockey injuries: National collegiate athletic association injury surveillance system, 1988-1989 through 2002-2003. *Journal of Athletic Training. 42*, 211-220.

Manning, M.R., & Levy, R.S. (2006). Soccer. *Physical Medicine and Rehabilitation Clinics of North America. 17*, 677-95.

Agel, J., Evans, T.A., Dick, R., Putukian, M., & Marshall, S.W. (2007). Descriptive epidemiology of collegiate men's soccer injuries: National collegiate athletic association injury surveillance system, 1988-1989 through 2002-2003. *Journal of Athletic Training. 42*, 270-277.

Dick, R., Putukian, M., Agel, J., Evans , T.A., & Marshall, S.W. (2007). Descriptive epidemiology of collegiate women's soccer injuries: National collegiate athletic association injury surveillance system, 1988-1989 through 2002-2003. *Journal of Athletic Training. 42*, 278-285.

Franić, M., Ivković, A., & Rudić, R. (2007). Injuries in water polo. *Croatian Medical Journal. 48*, 281-8.

Agel, J., Ransone, J., Dick, R., Oppliger, R., & Marshall, S.W. (2007). Descriptive epidemiology of collegiate men's wrestling injuries: National collegiate athletic association injury surveillance system, 1988-1989 through 2003-2004. *Journal of Athletic Training. 42*, 303-310.

McIntosh, A.S. (2005). Rugby injuries. *Medicine and Sports Science. 49*, 120-139.

Carson, J.D., Roberts, M.A., & White, A.L. (1999). The epidemiology of women's rugby injuries. *Clinical Journal of Sports Medicine. 9*, 75-78.

Dick, R., Ferrara, M.S., Agel, J., Courson, R., Marshall, S.W., Hanley, M.J. et al. (2007). Descriptive epidemiology of collegiate men's football injuries: National collegiate athletic association injury surveillance system, 1988-1989 through 2003-2004. *Journal of Athletic Training. 42*, 221-233.

CHAPTER 2

SCIENCE OF TRAINING AND INJURY
IN ATHLETICS

1. INTRODUCTION

Over the past 45 years or so, we have achieved significant scientific understanding of many physical factors involved in the development of various aspects of training, including specific strength and conditioning training. This has allowed more effective programs to be used for athletes' safety and preparation for competitions. Specifically, several components of training, such as skills, speed, strength, stamina and psychological skill training have been a focus of numerous text and research. The current conceptualization of science of training, basic principles of training theories as well as specific safe methods of strength and conditioning for athletes, have been summarized in *Science and Practice of Strength Training* (Zatsiorsky, 1995). The major theme of this book aims to provide scientific basis for the concept of *adaptation* as a *law of training*. Indeed, proper exercise, sport-specific drills and/or regular physical and psychological load is a very powerful stimulus for adaptation (i.e., organisms' adjustment in its environment). Accordingly, the major objective of athletes' preparation should be inducing specific *adaptations* in order to improve sport performance via: (a) carefully planned; (b) skillfully executed; and (c) goal-oriented training programs. From practical perspectives, at least four important features of the adaptation process should be considered by a coach in order to make training programs effective and most importantly safe for the athletes. Otherwise, athletes may experience and express various forms of *maladaptive* responses to training and associated performance saturation/deterioration with high risk for sport-related traumatic injuries. Athletes' adaptive responses are usually characterized by an increase in both physical properties, such as strength, speed, etc., and associated psychological indices, including emotional stability, proper level of motivation and vigor. According to Zatsiorsky (1995) there are four essential features of adaptation process as outlined below:

(1) **Overload.** The most challenging issue that coaches face daily is to provide an opportunity for maximal performance enhancement and secure a safe and injury free coaching environment. There is always a possibility of injury due to the nature of athletic activity that coaches should constantly be aware of. Due to coaches' primary responsibilities, which are an achievement of maximal performance and secure winning, positive (but not negative) training effect should be their major goal. However, positive

training effect may take place only if training load is above the habitual level. In other words, if training load in terms of the volume and intensity is the same over an extensive period of time, there will be no additional adaptation resulted in physical fitness saturation. If the training load and intensity are too low, detraining may occur, meaning that an athlete may not improve his/her physical status despite continuous training. On the other hand, if the training load and intensity are too high, an athlete may experience *maladaptive* responses to training and an increase in risk of injury. Athletes' individual responses (both adaptive and maladaptive) should be carefully monitored by the coaches in order to achieve progressive improvement and most importantly, to prevent overload-related injuries. Specific signs and symptoms of athletes' overtraining will be discussed elsewhere in this book. Overall, training load can be roughly classified according to three important principles:

(a) *progressive stimulating*: when the training volume and intensity are above individually defined neutral zone allowing for adaptation to occur;

(b) *retaining*: when the magnitude of the load remains the same in the individually defined neutral zone, so the level of fitness may maintain for a long time;

(c*) *detraining*: when the magnitude of the load tends to decrease and associated performance deterioration and/or functional capacities of the athlete may be observed.

It should be noted however, that the aforementioned principles should be considered with regards to hierarchy and duration of the periodical training units (i.e., general preparation period, competitive preparation period and transition period). In addition, it is important to note that these principles are also athletic fitness/skill level dependent. The aspect of individualization in terms of novice versus elite athletes' responses to training load will be discussed in more details in the following text.

(2) **Accommodation.** Positive training effect and associated positive psychological responses to the training load may take place if accommodation is prevented via proper training programs. In essence, accommodation refers to the training program when the same training program and type of exercise remains constant over a prolong period of time. For example, a diver that just performed optional dives (regardless of degree of difficulty) and ignores fundamental dry land, gym and conditioning training, a decrease in performance level will ultimately be observed. This is kind of a manifestation of biological law of accommodation. According to this law, an organisms' response to a given constant stimulus saturate or even decrease over time. Not surprisingly, experienced coaches always vary their exercise programs by (a) constantly replacing exercise routine; (b) switching from aerobic to anaerobic types of activity; and (c) balancing specificity and generalization of training sessions. It is also advisable for coaches to schedule flexibility and relaxation exercises between heavy resistance

strength drills to speed up recovery, prevent loss of flexibility and overall to avoid accommodation. According to Zatsiorsky (1995), training programs should satisfy at least two demands to avoid accommodation and to preserve specificity via: (a) quantitative modification (changing training loads in terms of the volume and intensity of exercise); and qualitative modification (replacing the exercises aimed at developing the athletes' specific functions such as strength, coordination/flexibility, and endurance).

(3) **Specificity.** Training adaptation is highly specific in nature. Success and injury free in particular sports require that the athletes posses specific qualities. What would be essential for a long distant runner could de detrimental or even harmful for a long jumper. Well developed upper body for a gymnast may not be beneficial for a springboard diver. Even among divers, depending on the event (springboard versus platform diver) current practices tend to provide differential training in order to develop sport diving specific qualities. As an illustration, excessive muscular development of the lower body compared to the upper body in springboard divers is an obvious necessity that needs to be achieved via specific strength training (see also Figure 1 below).

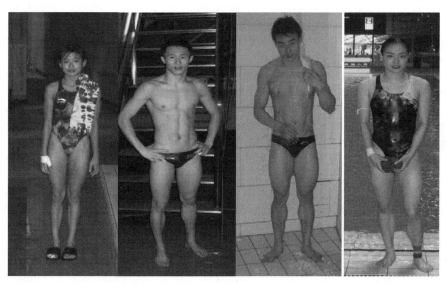

Figure 1. World Best Divers body compositions most likely influenced/selected for the platform (left, both female and male) and springboard (right, both male and female) events.

Current trend in diving is to achieve excessive body mass and explosive power of the lower body allowing the improvement of the jumping skill and height of the dives. As can be seen from this picture, top world springboard divers (right Pictures) are "more developed" and have larger leg muscles compared to the platform divers (left Pictures). Both males and females are most likely to encounter differential and special training programs, even within the same sport of diving. [Pictures were taken during FINA 2007 Diving World Series, Nan Jing, China, with permission from divers].

Another way to consider specificity of training program is to select sport-demand-specific routines. Clearly, fish can swim because they swim, birds can fly because they fly, frogs can jump because they jump, divers can dive because they dive, and so on. Thus, strength, flexibility and endurance training are highly specific in various sports. Unfortunately, this important principle of specificity of training programs is often ignored by the coaches. For example, it is a common practice in collegiate athletics that divers and swimmers utilize similar heavy resistance workouts for upper body, particularly during preparation period. This is inconsistent, at least, with the principle of specificity. Coaches should be aware that "what is honey for a swimmer could be poison for a diver".

Similarly, in the field of athletic training dealing with injured athletes, at an early stage following acute injury, it is important to control inflammation and regain the pre-injury range of motion. Accordingly, a specific exercise rehabilitation program should be utilized for this purpose to reduce probability of slow recovery and/or risk for re-injury. At the later stage of acute injury recovery, the muscle strength should be a major target for rehabilitation, thus, specific strength training drills must be utilized at this stage of recovery. Finally, preparation for the execution of specific drills should be a focus of rehabilitation. Accordingly, more sport-specific rather than general conditioning, strength and flexibility exercise should be incorporated into rehabilitation sessions.

Another aspect of *specificity* may be considered from perspectives of *identical-elements theory* (see also theories of transfer initially developed by Thorndike back in 1914 and further elaborated within the scope of current motor control and learning research). In essence, in order to achieve positive transfer of learning between various skills and exercise routines, the main elements underlying different skills or situations surrounding performance must be identical and similar in nature. In other words, a major assumption of this theory is that positive transfer between skills is not based upon any general and unrelated performance, but rather very specific in nature. Similarities between stimuli (type of exercise) and responses (developed skills) are complementary in nature. The use of dry-land and gymnastic training aimed at practicing complex exercise maneuvers complement (positively transfer) to the springboard diving. Conversely, as the degree of similarity between stimuli and responses are declined, conflicting consequences may be experienced. For example, because of the dissimilarity between diving and gymnastic somersaulting techniques, athletes' transition from gymnastic to diving may not likely foster any positive transfer. Similarly, because of the dissimilarity between the two sports, tackle techniques in football may not be applicable (but rather difficult to transfer) for rugby. In fact, a vast majority of concussive injury in rugby is due to tackle techniques that the rugby players adopted from their past experience playing football. Coaches, who understand basic principles of specificity, may avoid numerous problems and most importantly, may provide an optimal injury free training environment for their athletes.

(4) **Individualization.** Due to genetically predisposed and environmentally influenced individual differences among people, the same exercise routines and training program may elicit differential effect among athletes. Indeed, people are different in terms of anthropometric dimensions (larger/smaller; stronger/weaker; more or less flexible; more or less fatigable; emotionally stable/unstable; risk taker/risk avoider; etc.). Therefore, any attempts to mimic performance style and or techniques of world best athletes have proven to be useless or even harmful. For example, numerous attempts to "copy" Greg Louganis' diving style by novice divers led to significant deterioration of their own styles and overall performance. Similarly, mimicking the best Chinese divers' clean entry and/or fast somersaulting techniques (which was a tendency a few years ago among USA diving coaches), has proven to be devastating. However, the acquisition of fundamental skills and coordination patterns should be essential regardless of aforementioned individual differences among athletes. Fundamentally correct posture and basic skills should be trained regardless of sports, whether it be complex coordination, games and/or cyclic in nature. Not surprisingly, apparatus gymnastics is called the "mother of all sports" and required as an essential training method for youngsters. With coaches' creativity based on solid fundamental skills and qualities, injury controlled training methods proved to be successful. No average methods exist for exceptional athletes. *"Only average athletes, those who are far from excellent, prepare with average methods. A champion is not average, but exceptional"* (Zatsiorsky, 1995).

2. TRAINING PERIODIZATION

One of the common errors leading to athletic injury is improper planning of both athletes training sessions and competitive activities. In essence, "…Failure to plan is planning to fail" (Balui, 1995). Poor planning and inadequate duration of preparatory season and lack of general conditioning prior to competitive season are the major causes of injury in athletics. Both lack of proper planning aimed to reach peak performance at proper time and lack of flexibility in planning aimed at correcting training programs, if necessary, may have severe consequences not only from performance enhancement but also from injury prevention perspectives. For example, adding another challenging competition in prior planned competitive calendar forces athletes and coaches to reconsider not only preparation for this specific event, but also modify the whole competitive season. Premature transition from general preparatory to specific preparatory phases is another coaching error leading to injury early in the season. The most common challenges facing coaches and other sport practitioners may be summarized as: (a) how to design a rational plan for a sufficiently long as well as for short-time training period; (b) how to skillfully execute a well-designed plan in an optimal manner in order to satisfy the general law of training (e.g., adaptaion), reduce the risk of overtraining and potential injuries; and (c) how to reach the optimal peak of athletes best abilities precisely at the time of major events/competitions of the season, not before/and or after. The whole concept of training *periodization* is to address these challenges. It should be noted that there is still a lack of clarity, complete with controversy, in terms of how to define the concept of periodization, and most importantly, how to properly plan a training load.

Generally speaking, *periodization* is "…a sensible and well planned approach to training, which maximizes training gain and performance enhancement." (Dawson, 1996). Training *periodization* can also be defined as "…the purposeful sequencing of different training units (long-, medium- and short-term training cycles and sessions) for the attainment of the athlete's desired state and planned results" (Issurin, 2003). This definition is similar to Nadori and Granek (1989) suggesting that "…periodization is the predetermined sequence of training sessions and competitions."

The following text contains a summary of the most general current notions and ideas regarding the training *periodization* with respect to the *classic approach* (Matveev et al., 1977). This issue is presented based on materials and documents kindly offered by Dr. Issurin (with permission from the author). It should be noted that a "classic approach" has been predominant for decades, particularly with regard to the *block composition* design.

2.1. Training Periodization: Looking Back and Current Trends

Training *periodization* as a sport scientific concept and theory of athletes' preparation was elaborated during the 1950s-1970s in the former USSR by Russian scientists Matveyev (1977), Ozolin (1970) and many other prominent leaders in the field at that time. This theory was adopted and propagated in Eastern Europe and more recently in Western countries (see Bompa, 1984; 1999; Dick, 1980) and has developed as a core foundation of planning in high-performance Olympic sports. In general, the training *periodizaion* theory exploits the periodical changes of all biological systems and social activities typical to human beings. Specific to sport reality, at least four rationales should be considered as the factors that determine the periodical changes in the context of athletes' training:

a) **Repetitive patterns of nature:** Exogenous (external) and endogenous (internal, or circadian) rhythms are one of the fundamentals of biological systems, including humans. The seasonal changes as well as the daily changes experienced by living systems predetermine all biological activities both in terms of volume and intensity. The months and weeks naturally divide social and economic life into historically and traditionally consolidated cycles, which are incorporated into general adaptation: the weekly resting rhythm. Clearly, all biological, social, industrial and other activities are subordinate to exogenous rhythms of nature; it would be strange if sport and athletic activities were an exception to these patterns of nature.

b) **Adaptation as a general law:** As mentioned in the previous text of this chapter, the law of adaptation dictates and determines the athletes' training and preparation for competitions. To reiterate, athletes should avoid excessive accommodation to habitual loads in order to improve desired qualities (i.e., general conditioning, specific strength, flexibility and stamina). Accustomed (habituated) stimuli, such as a constant training load and intensity, cannot continue to be effective. In order to regenerate the adaptability of the athletes, their training program and exercise repertory must be periodically changed and renewed according to demands and individual goals of an athlete. In other words, an excessively stabilized and fixed training program leads the athlete to an adaptation barrier, where he/she is forced to dramatically increase the magnitude of habitual workloads in order to increase the positive body response. From this point of view, periodic changes of the training program should be considered, and carefully planned within the scope of *adaptation law*.

c) **The sequencing of different training aims:** Training in any sport is characterized by complexity, diversity, and variety. General and sport specific motor abilities, both technical and tactical skills/drills, cannot be developed simultaneously and maintained throughout the entire season. A more specific technical skill, for example, should be based on the appropriate level of motor fitness functional and psychological readiness of the athletes. In fact, a great number of injuries that an athlete suffers are from improper techniques and movement forms, which may be a result of improper physical fitness, or forcing the athletes to sacrifice movement fundamentals for the sake of performing the required drills. Fundamental skills and techniques (i.e., proper balance/posture and movement basic forms) must be acquired prior to acquisition of more specific skills. Similarly, excessive range of motion and joint stability/flexibility must precede the learning of advanced skills. Repetitive sequence of various training properties, individually defined and goal-oriented in nature, should be designed within both short and long-term training programs.

d) **Competition schedule:** A vast number of injuries in athletics occur due to *multiple peak performance* problems. The number of "most important events" dramatically increased over the past decade pushing the athletes to force their readiness for each event, skip fundamental training and ignore under-recovery symptoms. Having said that, overall there is established events (competitions) that take place periodically both nationally and internationally. Specifically, various bodies such as the International Olympic Committee and a number of international, national, and domestic sport associations control the frequency and timing of competitions. Thus, the competition calendar established by these bodies determines and dictates preparation, competition and recuperation cycles of athletes' training programs. The quadrennial cycle of Olympic preparation provides an excellent example of periodic changes which affect and dictate the activities of top world athletes. Traditional specification of periodic cycles of training/competition/recuperation phases in world class athletes' preparation is presented in Table 1 below.

Table 1. An Example of Hierarchy and Duration of the Periodical Training Units in Olympic Athletes

Training units	Time duration	Mode of planning
Quadrennial (Olympic) cycle	Four years – period between Olympic Games	Long-term
Macrocycle , may be annual cycle	One year or a number of months	
Training period	A number of months as part of the macrocycle	Medium-term
Mesocycle	A number of weeks	
Microcycle	One week or a number of days	
Workout or training session	A number of hours (usually not more then three)	Short-term
Training exercise	A number of minutes	

Training cycles of medium duration, called mesocycles, were traditionally proposed by classic theory, although there are several authors who do not consider these training units in their research (see, e.g., Bompa, 1999; Letzelter, 1978). The modes, specific aims, and content of mesocycles have been considered by many sport researchers who have suggested up to ten types and sub-types with more or less convincing argument (e.g., Harre, 1982; Matveyev, 1977; Platonov, 1997). The microcyles, as the small training cycles, are the most comprehensive, commonly accepted and least disputable terms, and are mostly defined as weekly cycles.

It should be noted that there is a current tendency to reconsider the major assumption of the training periodization outline in Table 1, due to various practical reasons, such as:

- increased number and level of competition throughout the entire season;
- increased complexity of routines, especially in complex coordination sports;
- earlier maturation of athletes, requiring to consider developmental aspect of athletes preparation;
- increased total volume and intensity of training load the whole year around;
- current practices and attempts to simultaneously develop motor abilities and functions such as strength, flexibility and stamina;
- increased number of alternative views on the nature of training periodization;
- progress in training methods and sport technologies, and protective devices, such as helmets, mouth guards, pads, braces, etc.;
- increase in the financial and other extrinsic sources for motivation to compete constantly at their peak level.

Periodical training units afford a great deal of freedom for creativity for coaches and athletes. However, the competitions calendar in most cases dictates the selection of the appropriate sequences, content, and duration of the training cycles. Specifically, the competition calendar dictates different modes of planning in order to reach the peak performance at the right time of the most important event of the season. Thus, this is the direct responsibility of coaches to design the plan of the training program focusing on principal features such as timing, peaking and training load distribution.

The general conceptual framework for *periodization* of training programs, including preparatory, competitive and transition periods, was initially proposed by Matveev (1977) and Harre (1982) and more recently modified by Issurin (1985-2004). It is important to note that within the preparatory period, there is: (a) an initial stage aimed at developing and enhancing the general motor abilities and functions, and (b) a later stage aimed at developing and enhancing more specific motor abilities and functions. Similarly, within the competitive period, there is also: (a) a general preparatory stage and (b) an acute and most immediate pre-competition stage characterized by different aims and should be achieved by various properties/features of training loads. Finally, the most common coaching error associated with training load and planning is the underestimation of the transition period aimed at the restoring of physical, psychological and functional resources. Some additional details for coaches to fully appreciate the modern approach to *periodization* of training program can be found in Table 2 below.

Table 2. Goal-Oriented Stages of Training Periodization

Period	Stage	Aims	Workloads' features
Preparatory	General preparatory	Enhancing the level of general motor abilities. Enlarging the potential for various motor skills	Relatively large volume and reduced intensity of main exercises; high variety of training tools
	Special preparatory	Development of the special training level; enhancement of more specialized motor and technical abilities	The loads' volume reaches their maximum; the intensity increases selectively
Competitive	Competitive preparation	Enhancing event-specific motor fitness, technical and tactical skills; formation of the model of competitive performance	Stabilization and reduction of the volume; increase of intensity in event-specific drills
	Immediate pre-competitive training	Accomplishing event-specific fitness and reaching readiness for main competition	Low volume, high intensity; the fullest modeling of forthcoming competition
Transitory	Transitory	Recovery and recuperation of physical, functional, and psychological spheres properties	Active rest; use of pleasant, attractive, and variable activities

3. TRAINING EFFECTS

A Coaches' knowledge about differential training effect is fundamental in terms of proper planning of workout load, selection of exercise, its duration and intensity, timing of administration and timing of recuperation. Currently well-accepted in sport science community taxonomy of training effects (Zatsiorsky, 1995) includes:

(a) *acute effect*, referred to changes induced by a single bout of exercise;

(b) *immediate effect*, referred to changes induced by one workout session or training day;

(c) *cumulative effect*, the result of a series of workouts overall certain time frame;

(d) *delayed effect*, referred to changes that occur over a given time interval after a certain goal oriented specific program;

(e) *residual effect*, which operates with the retention of changes induced by systematic workloads after the cessation of training beyond a certain time period.

The *residual* training effect is well-observed but the least studied phenomenon of the athletes' response to the training load. In 1991 Counsilman & Counsilman (1991) introduced and conceptualized residual training effect, however, this effect still is not well-accepted and understood among sport practitioners today. The following paragraph contains a few details regarding this effect with respect to "injury-free" planning of training programs as discussed.

Overall, the phenomenology of the *residual* training effect is closely related with the process of *detraining*, which in the past was defined as saturation of progress or "loss trained functions" when training is stopped and/or interrupted for some reason. In fact, *detraining* in elite athletes from different sports may occur selectively and "targeted" toward only certain abilities (i.e., loss of strength) when it does not receive sufficient strength input. For example, it is well-documented that in elite and highly trained endurance athletes the maximum oxygen uptake decreases when the total weekly volume is reduced below a certain level (Steinacker, 1993; Steinacker et al., 1998). Interestingly, Wilmore & Costill (1993) reported a considerable decrease of swimming-specific strength after four weeks off practice. It was suggested that the risk of *detraining,* in general, and loss of aerobic endurance may occur despite the large volume of highly intensive exercises (Mijika, 1999). To reduce potential *detraining*, the consecutive rather than simultaneous development of sport-specific abilities approach should be utilized in elite athletes (Bondarchuk, 1981; Issurin & Kaverin, 1985). Proper prediction of duration and amount of *residual* effect of previous training should be taken into account in order to define the rational sequencing and timing of different training cycles. From this perspective, it is extremely important to know which factors and in what manner may influence the duration of *training* residuals. Some details in terms of duration and underlying physiological mechanisms associated with residual effect of sport-specific abilities are summarized in Table 3. It should be noted, however, that more search is still needed to justify the advantages of consecutive approach for development of sport-specific abilities.

Table 3. The Duration and Underlying Physiological Mechanisms of the Residual Training Effect for Different Physical (Motor) Abilities (Issurin & Lustig, 2004)

Physical (motor) ability	Residual's duration, days	Physiological background
Aerobic endurance	30 ± 5	Increased amount of aerobic enzymes, mitochondria number, muscle capillaries, hemoglobin capacity, glycogen storage, higher rate of fat metabolism
Maximal strength	30 ± 5	Improvement of neural mechanism, muscle hypertrophy mainly due to the muscle fibers' enlargement
Anaerobic glycolitic endurance	18 ± 4	Increased amount of anaerobic enzymes, buffering capacity and glycogen storage, higher possibility of lactate accumulation
Strength endurance	15 ± 5	Muscle hypertrophy mainly in slow-twitch fibers, improved aerobic/anaerobic enzymes, better local blood circulation and lactic tolerance
Maximal speed (alactic)	5 ± 3	Improved neuromuscular interactions and motor control, increased phosphocreatine storage

The coaches' knowledge about training *residuals* and temporal *detraining* is extremely important for planning transition at some stage of athletes' preparation from simultaneous to consecutive development of the sport-specific fitness components and abilities. The overall rule of thumb is that coaches should remember the necessity of transition from simultaneous to successive/consecutive development of the training program allowing the enhancement of the *residual* effect of exercise and prevent detraining. By doing so, the principle of variability of training programs can be implemented and aimed at achieving injury-free peak performance at proper time with no indication of over-training. The lack of appreciation of training effects may have serious consequences for elite athletes' well-being.

On a final note, the general principle of differential effects of training may also be applicable for rehabilitation programs of injured athletes. It is well-accepted in the clinical practice that several important steps should be utilized during sport injuries rehabilitation programs for athletes suffering from various orthopedic injuries. Specifically, control of inflammation after knee injury should precede specific exercise aimed at increasing range of motion (flexion first and then extension). The next consecutive step should be specialized strength training followed by endurance and implementation of sport specific drills to restore pre-injury functional capacities. A simultaneous approach aimed at recovering multiple functions may be detrimental for athletes' recovery protocol, prolong reacquisition of sport specific functions and, most importantly, may put athletes at high risk for recurrent injuries. Within the conceptual framework of rehabilitation *residuals*, comprehensive research is needed to examine the type of exercise, its duration and physiological mechanisms underlying maximal positive effects for various sport-related injuries. *"How much is not enough and how much is too much"* in terms of the volume, intensity and duration of breaks between rehabilitation sessions should be a primary concern of medical professionals to fully rehabilitate injured athletes. It should be noted that this important principle of training/rehabilitation residuals has not been seriously considered and appreciated in the clinical setting.

4. BLOCK COMPOSITION CONCEPT: OVERVIEW

There are many contradictions and controversies in current practice and sport science regarding the issue of *periodization,* in general, and specifically in differential training effects. In attempts to resolve these existing contradictions and to achieve at least some consent among sports science experts and coaches in charge of elite athletes preparation, the revised approach to training *periodization* has been recently proposed (Issurin, 2004). This approach, so-called the *Block Composition Concept* (BCC), summarizes the general principles of elite athletes preparation and provides guidelines for alternative training *periodization*, with specific focus on short-term planning.

Specific focus of the training workloads is the most important fundamental principle of the BCC. Indeed, empirical evidence and anecdotal facts suggest that only highly-focused training workloads can produce sufficient stimulation for the development of required functional properties and the acquisition of skills/motor abilities in elite athletes. The concept of block-mesocycles implies that at least three differential effects of training load should be considered and carefully controlled by the coaching staff. The first one is *accumulation*, where acquisition of basic fundamental motor skills and abilities should be a main focus of the training session. Specifically, basic jumping skills, proper body alignment and posture, and hand-torso-leg coordination should be acquired and consolidated first before transition for acquisition of more complex skills such as forward and backward somersaults in gymnastics. The lack of aforementioned basic skills and premature transition to more advanced motor modalities may put athletes at higher risk for injury due to improper techniques.

The second effect is *transmutation*, characterized by creating a training environment which affords maximal positive transfer and utilization of previously acquired fundamental skills to sport-specific drills. For example, proper arm swing techniques synchronized with proper upright posture during the initiation of vertical jumps should benefit skillful execution of back single and double somersaults in gymnastic and springboard diving. Another example from the sport of gymnastics is that the proper head position during a vertical jump may significantly help to prevent disorientation and potential serious injuries during multiple somersaults, which is a well-known phenomenon. As discussed in the previous text, the *identical-elements theory* (Thorndike, 1914) and principal of similarities between stimuli (type of exercise) and responses (developed skills) may maximally foster positive transfer of basic to more specific skills. Also, the dissimilarity between acquired basic and sport-specific skills may induce numerous improper movement forms and techniques which are detrimental for athletes' growth and development.

Finally, *realization* effect of the training load assumes that an athlete should utilize his/her acquired fundamental and sport-specific potentials and skills and reach the optimal plan results at the peak of performance targeted on the most important event of the season. Again, this is the so-called "medium-size training cycles" or "block-mesocycles" characterized by the successively/consecutively focused development of fundamental (earlier in the cycle) and more sport specific (later in the cycle) abilities/skills. The training effects with respect to residuals can be illustrated in the following manner (See Figure 2 below).

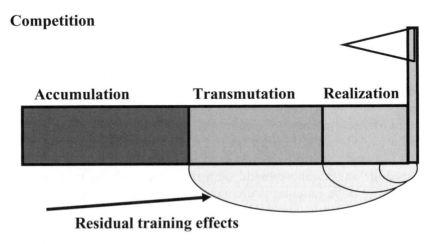

Figure 2. Superposition of the residual training effects induced by the sequenced blocks-mesocycles (adapted from Issurin & Shkliar, 2002 with permission the primary author)

4.1. The Annual Cycle Compilation

Similar to the classic approach, the annual cycle planning starts with the selection of major target-competitions of the season usually determined by international and national sport authorities. The specific item of the revised training approach is the subdivision of the entire annual cycle into a number of training stages. Each stage should contain a consecutive combination of extensive work on fundamental skills/abilities at the beginning of the season and more intensive work on sport-specific abilities with reduced volume as competitive stage approaches. Types of workout within the BCC aimed at achieving the accumulation, transmutation, and realization effects are shown in Table 4.

Table 4. The Main Characteristics of the Three Types of Blocks Meso-cycles

Main characteristics	Accumulation	Mesocycle type Transmutation	Realization
Targeted motor and technical abilities	Basic abilities: aerobic endurance, muscular strength, basic coordination	Sport-specific abilities: special endurance, strength endurance, proper technique	Integrative preparedness: modeled performance, maximal speed, event specific strategy
Volume-intensity	High volume, reduced intensity	Reduced volume, increased intensity	Low-medium volume, high intensity
Fatigue-restoration	Reasonable restoration to provide morphological adaptation	No possibility to provide full restoration, fatigue accumulated	Full restoration, athletes should be well rested
The tests' battery	Monitoring of the level of basic abilities	Monitoring of the level of sport-specific abilities	Monitoring of maximal speed, event-specific strategy, etc.

The rational sequencing of the meso-cycles within the training stage allows the optimal superposition of the residual training effects to be obtained, if properly planned. Figure 1 above shows the principal possibility of obtaining optimal interaction of the training residuals allowing high level of competitive performance of previously acquired both fundamental motor and specific technical abilities. It should be noted from Table 4 that training residuals of fundamental skills and abilities last much longer then residuals of more specific abilities, while the residuals of maximal speed and event-specific readiness are the shortest ones (Table 4). Following this knowledge, the duration of the training stage is determined by the length of the training residuals and should be close to two months. In fact, the training stages can be shorter (near to peak season, for instance), or longer (at the season's beginning or due to specific needs).

Proper control of athletes' responses to the training load is an essential attribute of the science of training, allowing the prediction/prevention of athletic injuries. Accordingly, it is strongly recommended that the test battery should be reproduced in each stage of athletes' preparation for competition. In conjunction with actual results of competition, proper testing of athletes' physical, functional and psychological parameters may provide important monitoring and feedback information that can be used for future training corrections. As suggested, the number of training stages in an annual cycle is a sport-specific decision and depends on the number of external (i.e., number and location of major competitions) and internal (i.e., psychological status, predisposition for injuries, age and gender) factors.

Clearly, the *high volume and intensity of the training load required by modern sports is one of the major causes of injury in elite athletes*. Thus, proper realization of BCC may provide a number of benefits, including the fact that the Block Composition model allows the total mileage and time expenses for training to be reduced, without substantially changing the total number of workouts. The other benefits of BCC include:

- monitoring the detraining effect and focusing on a reduced number of abilities/skills successively acquired at each stage;
- providing appropriate tests, the   so-called "dose-response-effect" analysis to control for maladaptive physical/physiological responses;
- providing psychological climate allowing a focus on the reduced number of targets; consequently the mental concentration and motivation level can be maintained more effectively throughout the entire season;
- controlling nutritional aspects requiring a high protein diet in order to enhance the anabolic effect of strength training; carbohydrate nutrients are particularly important in meso-cycles for special and strength endurance (Wilmore & Costill, 1993).

4.2. Bases of the Short-term Planning

The general propositions of BCC make more sense when considering short-term planning and properly designing the training micro-cycles and several workouts. At least two basic prepositions of the BCC immediately affect both the process and outcome of the short-term planning, namely:

- specific focus and concentration of the training loads on the minimum abilities-targets;
- consecutive step-by-step development and maintenance of abilities-specific targets.

That said, there are three major aspects of the short-term planning that should be considered by coaches when planning individual workouts to control for fatigue and maintain athletes' high level of motivation, including: (a) load-related *differentiation* of workouts; (b) *compatibility* of different training loads in several and adjacent workouts; and (c) basics of the training micro-cycle *compilation*.

<u>Load-related differentiation</u> emphasizes the importance of both physiological adaptation and mental concentration. For practical purposes, it is necessary to point out three general functions of workouts: *development, retention, and restoration*. The appropriate load level should be selected corresponding to these aims.

<u>Compatibility</u> of the specific training modalities within the single workout and within the workout series emphasizes possibility of both negative (interferences) and positive interaction of several immediate training effects. The BCC allows the prevention or at least reduction of negative interactions by the use of a compatible combination of exercise routines within certain training modalities (i.e., strength, flexibility, endurance).

<u>*Basics of the training micro-cycle compilation*</u> contain a number of specific statements including (a) no more than three training modalities (usually one dominant, the second – compatible with the main purpose, the third – modalities of restoration exercises) should be implemented simultaneously; it is postulated that 65-70% of the entire training time within one training session should be allocated to one or two purposed training modalities; (b) high intensity of workout depends on the context and outcome of the previous session focused on key targets. For example, the session with primary focus on strength training should be followed by a high intensity session focused on flexibility training and significant reduction of workout targeting the strength training; (c) minimizing the number of training modalities is particularly important and typical for elite athletes. The daily program for less experienced and particularly for junior athletes may be more diversified in order to maintain a high level of motivation and attractiveness.

CONCLUSION

There is always a trade-off between high-achievement and probability of overtraining as well as high risk of injury among elite athletes. Proper planning, specificity and individualization of the training program are key factors to consider. There are tendencies in modern sports to (a) standardize the training program within certain sports; and (b) modify the exercise content to achieve maximal adaptation and reduce the probability of accommodation.

The first tendency proposes the use of more or less standardized workload combinations within meso- and micro-cycle programs. The positive aspect of this tendency is the possibility of comparing the results and responses obtained in different training cycles with the same (or similar) workload combinations. This provides prerequisites for a current training control and improvement of sport specific training technology. The negative aspect is that the possibility of excessive accommodation when the athletes' response to a continuing stimulus decreases followed by a decrease in the training effect as well. This may force the coaches to reconsider the initial training routine with emphasis on an increased training load and ultimately putting athletes at high risk for overtraining and injury.

The second tendency relates to the effect of novelty when the unaccustomed exercises induce more pronounced adaptive responses. However, there is still a problem as to how to increase the effect of stimulus novelty when an athlete is accustomed to repetitive sport-specific exercises. Indeed, additional research, enhanced coaches' experience and quality observations are necessary to overcome existing controversies in training programs aimed at maximizing performance enhancement without jeopardizing the safety and well-being of athletes.

Acknowledgments

This chapter was prepared based on ideas and writings of Professor Vladimir Zatsiorsky and Dr. Issurin with further elaboration by the author. Personal consultations with these prominent leaders in the field of athletes training are highly appreciated.

REFERENCES

Zatsiorsky, V. M. (1995). *Science and practice of strength training*. Champaign, IL: Human Kinetics.
Thorndike, E. L. (1914). *Educational Psychology: Brief course*. New York: Columbia University Press.

Balui, I. (1995). Planning, periodization, integration and implementation of annual training programs. Presentation to and in proceedings of the Australian Strength and Conditioning Association National Conference, (pp. 40-66). Gold Coast, Australia.

Dawson, B. (1996). Periodization of speed and endurance training. In P. Reaburn & D. Jenkins (Eds.), Training for speed and endurance, (pp.76-96). St. Leonards, Australia: Allen & Unwin.

Issurin, V. (2003). *Aspekten der kurzfristigen Planung im Konzept der Blockstruktur des Training. Leistungsport, 33*, 41-44.

Nadori, L., & Granek, I. (1989). Theoretical and methodological basis of training with special considerations within a microcycle. Lincoln: National Strength and Conditioning Association.

Matveyev, L. (1977). *Fundamentals of sport training.* Moscow: Progress Publishers.

Ozolin, N. G. (1970). *The modern system of sport training.* Moscow: FIS Publishers. (Russian).

Bompa, T. (1984). *Theory and methodology of training. The key to athletic performance.* Boca Raton, FL: Kendall/Hunt.

Bompa, T. (1999) *Periodization: Theory and methodology of training* (4th ed.). Champaign, IL: Human Kinetics.

Dick, F. (1980). *Sport training principles.* London: Lepus Books.

Letzelter, M. (1978). Trainingsgrundlagen. Training. Technik. Taktik. Rowolt Verlag GmbH, Hamburg.

Harre, D. (Ed.). (1982). *Principles of sport training.* Berlin: Sportverlag.

Platonov V.N. (1997). *General theory of athletes' preparation in the Olympic sports.* Kiev: "Olympic Literature". (Russian).

Issurin, V., & Kaverin, V. (1985). *Planning and design of annual preparation cycle in canoeing.* In "Grebnoj Sport" (Rowing, Canoeing, Kayaking), Ph.S.; Moscow, p. 25-29.

Issurin, V., & Lustig G. (2004). *Klassification, Dauer und praktische Komponenten der Resteffekte von Training. Leistungsport. 34*, 55-59.

Issurin, V., & Shkliar, V. (2002). *Zur Konzeption der Blockstuktur im Training von hochklassifizierten Sportlern. Leistungsport, 6*, 42-45.

Counsilman, B. E., & Counsilman, J. (1991). The residual effects of training. *Journal of Swimming Research , 7*, 5-12 .

Steinacker, J. M., Lormes, W., Lehman, M., & Altenburg, D. (1998). Training of rowers before world championships. *Medicine and Science in Sports and Exercice, 30*, 1158-63.

Wilmore, J. H., & Costill, D. L. (1993). *Training for sport and activity. The physiological basis of the conditioning process.* Champaign, IL: Human Kinetics.

Mujika, I. (1999) The influence of training characteristics and tapering on the adaptation in highly trained individuals: a review. *International Journal of Sports Medicine, 19*, 439-446.

Bondarchuk, A. P. (1981). *The physical preparation designing in power disciplines of track and field.* Kiev: Health Publisher (Zdorovie, Russian).

CHAPTER 3

BALANCE AS A RISK FACTOR FOR ATHLETIC INJURIES

1. INTRODUCTION

Human upright posture is one of the fundamental skills necessary for successful acquisition of "super-postural activities" such as grasping, reaching, walking, jumping, catching, etc. Most importantly, postural stability allows proper balance preventing numerous health-related problems including traumatic injuries. Charles Darwin pointed out that one of the most important implications of the attainment of *uprightness* was that it freed the hands from locomotion function, so that they became available for sustained use in other directions, such as implementing instrumental activities. Nevertheless, our bodies are still subject to what Arthur Keith (1923) called the "illness of uprightness". Although the upright posture has been the subject of considerable research, our knowledge about balance and human postural control, in general, and poor balance as predisposing factor for athletic injuries, in particular, is still incomplete. In this chapter, a brief perspective of some key issues in the postural control of humans will be provided. An emphasis will be given to current research on the role of cortical function, specifically the cerebral cortex in balance control. Finally, current perspectives on links between improper balance and sport-related injuries will be discussed.

2. POSTURE AS A FUNDAMENTAL HUMAN ACTIVITY

2.1. Posture Definition

Human upright posture, traditionally defined as "…body segments" configuration at any given time (Thomas, 1940), has been extensively studied, but is still a poorly understood phenomenon. There are several reasons why existing models of posture (i.e., *genetic*, *hierarchical* etc.) are unable to explain the enormous complexity of mechanisms of human postural control. One reason is that there is still a strong notion, similar to classical Cartesian dualism, of considering human posture as the "…genetically determined *reference posture* or, integration of all stretch reflexes of the body…" in order to maintain equilibrium (see Massion 1992; Horak & Macpherson, 1996 for review) on one side, and "…voluntary goal-directed movement, as the activity of higher centers superimposed on the bodily mechanisms" (Magnus, 1925; Sherrington, 1910), on the other side. There is still a debate in the literature whether a single central control process is responsible for both movement and its associated postures or alternately, whether there is a dual coordinated control system for voluntary movement and posture (Massion et al., 2004).

Human posture is a product of an extremely complex dynamical system with many degrees of freedom, which like any other fundamental activity, undergoes dramatic changes throughout the life span (Bernstein, 1967). As such, the classical question as to how the performer constrains the many degrees of freedom to preserve and coordinate various body segments' configuration while performing goal-oriented movement is still unclear. In other words, the nature of postural dynamic is extremely complex, since posture includes not only the *antigravity control* (e.g., under static conditions, the center of gravity should remain within the support surface, Winter, 1990) but also the *interface between perception and action* via fine postural adjustments associated with super postural activity, stability control during locomotion and the ability to assess the body's spatial-temporal orientation in the surrounding environment (Riccio, 1993). Accordingly, the primary objective of upright stance performer is to explore the dynamical properties of perceptual-motor workspace and to avoid the actions that exceed its operational limitations (Newell et al., 1989).

2.2. Sources of Information for Postural Control

Considering posture as a complex dynamical assemblage, it is reasonable to assume that all sensory systems can contribute information relevant to the control of posture. It was hypothesized that processes within the postural control system involve organizing, integrating and acting upon redundant visual, vestibular and surface-somatosensory inputs to provide orientation information to the postural control system (Bechterev, 1882). The contribution of sensory information to providing the maintenance of stable posture has been characterized as a hierarchical process in which the congruence of both support surface somatosensory and visual inputs is compared to an inertial-gravitational reference, established by vestibular reference inputs. Thus, balance is a multidimensional ability in which a range of factors such as eye-motor coordination, kinesthetic response, ampular sensitivity, vertical semicircular canal function, and so forth, are crucial (Klack & Watkins, 1984).

In cases of inter-sensory inconsistencies, inputs not congruent with vestibular reference inputs are suppressed and greater perceptual significance is attached to those inputs in accord with vestibular sources (Shumway-Cook & Woollacott, 1985). Overall, human upright posture is a product of a complex dynamical system that relies on integrating input from multimodal sensory sources, including: (a) visual system; (b) vestibular system; and (c) a great number of sensory receptors embedded in the muscles, tendons, joint capsules and skin that we commonly refer to as proprioceptive systems (see Mergner et al., 2003 for review presenting a multi-sensory postural control model based on experiments with normal subjects and those with vestibular loss).

2.3. Neural Basis for Multi-sensory Sources of Balance

A neural basis for the multi-sensory influences on postural motion and the whole body orientation has been proposed. Specifically, convergence of vision, audition, somatosensation and proprioception at the cellular level in the vestibular nuclei has been shown by Costikova (1939). Also, cells have been identified in the vestibular nuclei where whole discharge rates were influenced by both body motion, as signaled by vestibular end organs, and visual motion (Gurfinkel et al., 1965; DiZio & Lackner, 1987). Recent evidence suggests that the cortex, in some way, plays a role in processing multisensory information and modifying human postural control (Quant et al., 2004a&b; Adkin et al., 2006). Moreover, it has been suggested that vestibular and somatosensory input may be integrated within a distributed cortical network, including the temporal-parietal cortex supplementary motor area and prefrontal cortex, in order to process input related *ego-motion* and counteract a loss of balance (de Waele et al., 2001). A more recent study with young controls has also identified fMRI correlates of self-perceived postural instability (Slobounov et al., 2006). Specifically, significant activation of several brain areas, including the parietal cortex, anterior cingulate and cerebellum was observed when young subjects visually recognized gravitational vertical via computer animated model of stable versus unstable postures. Indeed, several studies investigating the effects of human cerebral lesions on posture suggest that perception of the visual vertical involves the insula (Brandt et al., 1994), and perceived gravitational vertical requires healthy function of the thalamus (Karnath et al., 2000, 2005), superior parietal cortex (Blanke et al., 2000; Johannsen et al., 2006), and insula (Johannsen et al., 2006). In addition, lesions of the temporal-parietal junction (a region of multi-modal sensory integration) lead to poor equilibrium control on an unstable support (Perennou et al., 2000).

3. UPRIGHT POSTURE CONTROL

3.1. Assessment of Postural Stability

There are several ways to define the stability limits during postural stances. One is to estimate the area of the base of support defined by the lateral edge of a subject's feet. This approach is recommended while considering the issue of postural stability with respect to the vertical projection of the center of mass on the base of support. In other words, there is a notion that stable posture is maintained while gravito-inertial forces are in alignment with direction of balance. Accordingly, when the body orientation deviates from the direction of balance, the torque is produced to restore the appropriate body orientation with respect to gravity. The vertical projection of the center of mass to the base of support roughly may be assessed by calculating the center of pressure derived from the force platform according to the following approximations. The center of pressure (CP) motion is calculated from the moment of force about the given axis (Mx) and (My) divided by the vertical ground reaction force (Fz). Instantaneous locations of the center of pressure on the base of support are computed independently for X and for Y coordinates as shown in Fig. 1 below:

- ## CPy=Mx/Fz, CPx=My/Fz
- ## Index of Stability =
  ## [CPy$_{max}$-CPy$_{min}$]/[length of feet]
- ## Subject Facing Y direction

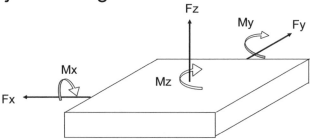

Figure 1. Basic schematics for parameters derived from the force platform and calculation of the center of pressure and stability index.

Most common, the area of the center of pressure (CP-area) in addition to CP displacement and CP velocity are considered in clinical practice to estimate the subject's postural stability.

There is another way to define postural stability boundary during upright stances. In particular, standing space is structured by stability limits that are invariant with respect to a body position. These stability limits vary as a function of the body's morphology and the base of support. Furthermore, stability limits depend on mechanical properties of the body, its environment and on response latency to external perturbations. Thus, the strategy to maintain balance is to reduce the movement space needing the exploration from multiple degrees to one degree of freedom, and varying only one parameter at a time. More specifically, strategy to estimate stability limits during upright stances is to monitor the forward/backward and side-to-side translation of the center of pressure derived from the force plate. Maximal deviation of the center of pressure trajectory for the Y and X coordinates during whole body postural movement is supposed to reflect the subject's limits of postural stability. It should be noted that there are numerous reports, papers and book chapters written and can be considered for details.

3.2. Virtual Reality for Assessment of Postural Stability

Perceiving optic flow (structural pattern of visual motion) contributes to balance control because it specifies changes in body position or postural sway relative to gravity (Gibson, 1966). It is well known since Mach (1985) that a moving visual scene may induce a sensation of ego-motion. Adult normal subjects perceive their direction of heading from optic flow to within 1 deg under a variety of circumstances (Keshner & Kenyon, 2000), and they are similarly sensitive to variations in optic flow in controlling balance (Lee & Lishman, 1975). This phenomenon has been attributed to a conflict between the changing visual input and vestibular and proprioceptive information (Lestienee et al., 1977) and has been extensively studied using the so-called "moving room paradigm" (Lee & Lishman, 1975; Stoffregen & Smart, 1998). Subjects' exposure to optic flow with different visual stimulus patterns consistently induces an increase in postural sway not only in humans (van Asten et al., 1988; Ehrenfried, 2003) but also in animals (Ikeda & Takahashi, 1977). Therefore, concurrent analysis of optic flow variables and subjects' self-motion may help understand the visual control of posture (Beer et al., 2002).

Our recent research clearly demonstrated that the moving room paradigm using 3-D VR technology proved to be effective in inducing subjects' self motion or postural adjustments when exposed to manipulations of visual scenes (Slobounov et al., 2006). This is consistent with other numerous studies reporting that the body tends to tilt and/or rotate in the direction of moving visual field (Lee & Lishman, 1975; Lestienne et al., 1977; Ehrenfried et al., 2003). Postural effects were most pronounced when low frequency (0.2 Hz) visual scene motion was introduced, which supports the notion of "exploratory response" to possible destabilizing visual input implying a perturbation of stance. Also in replication, self-motion and "self-presence" were reported by all subjects who experienced actual whole body postural adjustments as evidenced by the center of pressure and trunk kinematics data. This finding is consistent with previous studies suggesting that an immersive dynamic visual field induces a postural reorganization as reflected in the subjects' head, trunk and ankle responses (Keshner & Kenyon, 2000).

This specific issue of postural control as assessed by Virtual Reality technologies may be found in a number of recently published reports and beyond the scope of this chapter. Overall, the feasibility of Virtual Reality technologies (virtual room experiments) to examine the postural stabilization responses to optical flow is now documented; and behavioral data suggest that healthy subjects adjust their postural movement to direction of the optic flow and thus are capable of preserving the balance in a visually confused virtual environment. Figure 2 below shows the VR experimental set-up used for assessment of postural stability in our laboratory.

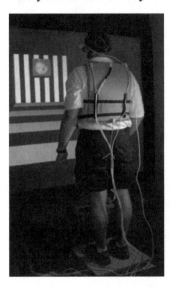

Figure 2. Virtual reality (VR) set-up for assessment of visuo-kinetsthetic integration involved in the control of upright postures. Subjects' postural responses are obtained from force plate and Flock of Birds whole body postural movement while viewing 3D motion of the "virtual room".

3.3. Quiet Stance Domain

During quiet stance the body experiences some very small variations in its position. It oscillates in anterior-posterior (AP) and medial-lateral (ML) direction. This oscillation is quantified by different methods such as image recording where the whole body movement is compared along the time dimension or by the use of the force platform. By using a force platform one can record the time-to-time position of the center of pressure (COP). COP movements are commonly used to assess the body oscillation during stance. Normally in stance, COP migrates approximately 0.4 cm in AP direction and 0.18 cm in ML direction (Winter et al., 1998) while COM displacements are somewhat smaller. The differences in migration of COP and COM in AP direction have been associated with the generation of torques at the ankle joints, while ML displacements have been associated with activity of hip muscles (Winter et al., 1996). Several models have been suggested to explain and describe COP migration in quiet stance. Three of these models are now briefly described.

3.4. Models

Upright human posture has been frequently modeled as an inverted pendulum (Fitzpatrick et al., 1992b; Winter et al., 1993; Winter et al., 1998; Morasso & Schieppati, 1999). Modeling the human body as an inverted pendulum is based on the assumption that joint motion is only at the ankle joints, body sway is very small and that the feet do not move (Zatsiorsky & King, 1998). Since the stability of the system requires that the center of mass (COM) falls within the base of support it is believed that the COM is the controlled variable of the unstable body system. COM is regulated through a continuous movement of the COP and stabilized around a fixed reference point. In a single pendulum model the difference between the COM and center of pressure (COP) will be proportional to the acceleration of the body COM.

A two-process random-walk model was suggested by Collins and De Luca (1993), where the maintenance of vertical posture could be viewed as part of a stochastic process. This model assumed the movement of the COP as a correlated random walk and determined at which time intervals COP displacements are either positively or negatively correlated. It was found that for short time intervals (less than 1 second) COP displacements were positively correlated. That is, displacements of the COP in one particular direction were followed by displacements in the same direction. At longer time scales, negative correlations were found. Overall it was suggested that these COP characteristics correspond to an *open loop control* at short time scales and a *closed loop control* at longer time scales.

A more recent model for the control of posture has been formulated by Zatsiorsky and Duarte (1999). They introduced a method of decomposing COP trajectories into two components, termed *rambling* and *trembling*. The decomposition first identified instant equilibrium points (IEP). An IEP is the position of the COP when the horizontal forces are zero. At these moments, the projection of the COM onto the base of the support coincides with the COP position. The individual IEPs, connected through a spine fitting function, form the rambling trajectory, while the difference between rambling and COP trajectories is called the trembling trajectory. The authors suggest that the rambling trajectory describes the motion of a moving reference point with respect to which the body's equilibrium is instantly maintained, while the trembling trajectory describes body oscillation around the reference point trajectory.

3.5. Preprogrammed Reactions

To date, a vast body of research in human posture has focused on neural organization, or muscular contraction synergy associated with postural control (Diener et al., 1988; see also Horak et al., 1994, for review). Theoretical insights have suggested that a pattern of organization is pre-programmed by executive commands according to the particular motor task in order to constrain the large number of functionally related muscle synergy (Arutyunyan et al., 1969). Following the imposition of brief surface displacement, the automatic postural adjustments were recorded and characterized by relatively fixed temporal and structural patterns of ankle and thigh muscle activity (Forssberg & Nashner, 1982). Preprogrammed reactions consist of a combination of muscle activation patterns specific for a given perturbation and act at a time delay of about 50-100 ms. They are different from simple reflexes in the sense that they depend on the instruction to the subject and that their magnitude can be independent from the length changes in the muscle (Latash, 1993). A vast literature exists on the regulation of posture under external perturbations and the modulation of preprogrammed reactions (see: Nashner, 1976; 1977; Nashner & Woollacott, 1979; Nashner et al., 1989; Horak et al, 1986-1990).

3.6. Anticipatory Posture Adjustments (APA)

It should be noted that Hess (1943) first pointed out that a proactive (i.e., anticipatory) postural stabilization is required for performance of goal-oriented volitional movement producing an internal disturbance of equilibrium. It was shown that the ensemble of the sequential postural adjustments obey certain rules (i.e., flexible postural synergy) that are modifiable in order to optimize the balance. Since that time, the preparatory nature of postural responses is considered to be centrally programmed and independent of any feedback from the moving limb. These preparatory (i.e., anticipatory) postural adjustments associated with volitional movement were intensively studied in relation to dynamic asymmetry and muscle forces generating movement that can perturb postural equilibrium (Bouisset & Zattara, 1987; Ramos & Stark ,1990; Massion, 1992). The disturbing effects of voluntary movement seem to be anticipated by the CNS. This anticipation is clearly demonstrated in studies revealing muscle activation about 50-150 ms prior to the movement initiation when no reflexes can be triggered (i.e. absence of external stimulus). Anticipatory postural adjustments have been described considering different movements performed by arm, leg, trunk and head (Belinkiy et al., 1967; Cordo & Nashner, 1982, Gurfinkel et al., 1988, Vander Fits et al., 1998), including those for forearm loading and unloading tasks (Aruin & Latash, 1995a, 1995b; Aruin & Latash, 1996; Shiratori & Latish, 2000). It has been suggested that APA are generated by the CNS in a feed-forward fashion in an attempt to predict the upcoming postural perturbation and to generate approximate corrections.

4. CORTICAL CONTROL OF HUMAN POSTURE

The role of the cerebral cortex, in general, and contribution of higher cortical functions in human postural control has been less studied and thus, controversial. Historically the neuromuscular control of "automatic" postural responses was thought to arise from brainstem and spinal circuit with limited consideration for the role of cerebral cortex (Magnus, 1926; Sherrington, 1910). The original proposition that "the whole righting apparatus is arranged sub-cortically in the brainstem, and in this way made independent of direct voluntary influences" (Magnus, 1926), has persisted with time. The idea that postural responses are regulated subcortically was based upon the notion that postural responses are triggered automatically, without any voluntary intent, and therefore, are initiated more quickly and with less variability then cued voluntary movement (Diener et al., 1984; Keck et al., 1998). However, the onset of postural responses occurs at longer latencies than those of stretch reflexes (Matthews, 1991), indirectly suggesting that postural responses exhibit greater potential for modulations and modifications by neural centers hierarchically residing higher along the neural axis (Jacob & Horak, 2007). Moreover, unlike stretch reflexes, postural responses involve synergistic activation of muscle throughout the whole body, and these are also more context-specific, flexible and adaptable than spinal proprioceptive reflexes (Horak & Macpherson, 1996).

4.1. Behavioral Studies

There is considerable behavioral evidence suggesting the contribution of the cerebral cortex and higher cognitive functions in the control of upright posture. Indeed, postural response have been shown to be modified by various cognitive-motor processes "represented" in the cerebral cortex, including those involved in anticipatory postural adjustments (Bouisset & Zattara, 1981). Postural responses are modified by: (1) changes in cognitive load and attention when performing dual postural tasks (Brown et al. 1999; Carpenter et al. 2004; McIlroy et al. 1999; Maki et al. 2001; Brauer et al. 2002; Norrie et al., 2002; Quant et al., 2004a), (2) changes in a subject's intentions to respond with a specific strategy (Burleigh et al. 1994; Burleigh

& Horak 1996; McIlroy & Maki 1993; 1995), (3) learning and modification of postural responses with prior experience (Horak & Nashner, 1986; Horak et al., 1989; McIlroy & Maki 1993; Quintern et al., 1985); and (4) with changes in initial conditions (Chong et al., 1999; Henry et al., 2001; Tjernstrom et al., 2002). Considering that attention, mental calculation and memory properties have been attributed to represent high-order cognitive functions controlled by the cerebral cortex (Dehaene et al., 2004; Kaiser & Lutzenberger, 2005; Naghavi & Nyberg 2005), it is reasonable to suggest the direct cortical involvement in control of postural equilibrium (Jacobs & Horak, 2007).

4.2. Brain Imaging Studies

There are several lines of research that have directly demonstrated the involvement of cerebral cortex in human postural control. Variation of slow negative potentials in the primary motor cortex preceding the onset of postural adjustment was observed in Saitou et al. (1996) EEG study. A more recent EEG study by Slobounov et al., (2005) has also documented that the initiation of self-paced postural movement is preceded by slow negative direct current (DC) shift, similar to movement-related cortical potentials (MRCP) accompanying voluntary goal-oriented movement. Also, a brief burst of gamma (40Hz) EEG activity preceded the initiation of compensatory postural movement when balance was in danger (see Fig.1A). The spatial distribution of EEG patterns in postural actions approximated that during previously observed postural perceptual tasks (Slobounov et al., 2000). Similarly, change in cortical excitability, revealed by slow negative DC potentials just prior to the anticipated postural perturbations has been reported by Jacobs and Horak (2007). Similar to DC potentials that observed 1-2 s prior to a voluntary self-initiated postural movement (Saitou et al., 1996; Slobounov et al., 2005), they reported EEG readiness potentials with significant negative variation only in the cue condition. In this study, the cue-related differences in the subjects' readiness potentials were highly correlated with the cue-related improved postural stability. Experimental set-up to record EEG collectively with postural response is shown in Fig. 3B.

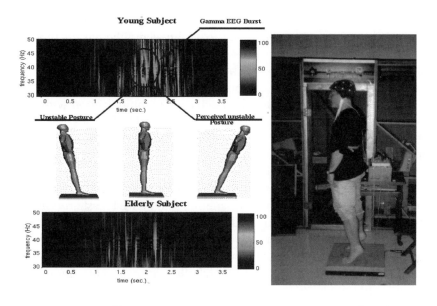

B   **A**

Figure 3. (B) Burst of EEG 40 Hz gamma activity associated with prediction of postural instability; (A) Experimental set-up to record brain electrical activity in conjunctions with whole body postural movement.

Overall, EEG studies support the hypothesis that cerebral cortex may play a role in the optimization of postural responses to external perturbations thorough changes in anticipatory central set. Therefore, the postural movement once considered to be automatic, might be susceptible to voluntary cortical control.

There are several EEG studies examining the role of cerebral cortex on triggered postural responses after a perturbation (Dietz et al., 1985; Ackermann et al., 1986; Dimitrov et al., 1996; Quant et al., 2004; Adkin et al., 2006). These studies have shown that EEG potentials following postural perturbations (known as perturbation-evoke responses, PER) become altered with changes in the central set, such as with changes in the predictability of a perturbation (Adkin et al., 2006) or with a secondary motor task (Quant et al., 2004). The PER are thought to represent cortical processing of sensory input related to the balance disturbance (Dietz et al., 1985) that arises as an error-or conflict-related signal (Adkin et al., 2006). Specifically, the first negative peak between 100 and 150 ms (N1) may represent sensory disturbance, and late PERs may be linked to the sensorimotor processing of balance correction, similar to the error-related negativity responses observed for incorrect responses in decision-making paradigms (Pailing & Segalowitz, 2004; Yasuda et al., 2004). These studies clearly demonstrate that postural set influences the characteristics of the cortical contribution to the control of balance after perturbation.

Evidence suggests that the cerebellar-cortical loop is responsible for adapting postural responses based on prior experience and the basal ganglia-cortical loop is responsible for pre-selecting and optimizing postural responses based on current context. Thus, the cerebral cortex most likely influences longer latency postural responses directly via corticospinal loops and shorter latency postural responses indirectly via communication with the brainstem centers that harbor the synergies for postural responses, thereby providing both speed and flexibility of pre-selecting and modifying environmentally appropriate responses to a loss of balance (Jacobs & Horak, 2007).

Finally, the neural substrates for maintaining standing postures in humans have been investigated using the mobile gantry PET system (Ouchi et al., 1999). Interestingly, compared with the supine posture, upright standing with feet together activated the cerebellar anterior lobe and the right visual cortex, while standing in tandem was accompanied by activation within the visual association cortex and cerebellar vermis. This finding is consistent with previous research suggesting that the human cerebellar vermis may coordinate the timing in keeping the center of gravity within the stability margins, therefore, it may provide a control of postural scaling and central set in stance (Horak & Diener, 1994).

5. ABNORMAL POSTURAL CONTROL

5.1. Impaired Postural Control in Mild Traumatic Brain Injury (MTBI)

Several previous studies have identified a negative effect of MTBI on postural stability (Lishman, 1988; Ingelsoll & Armstrong, 1992; Wober et al., 1993). Generally, balance problems in MTBI subjects have been attributed to disruption of various CNS functions responsible for postural stability (Guskiewicz et al., 2003; Slobounov et al., 2002; 2005; Thompson et al., 2005). A growing body of experimental studies has demonstrated postural stability deficits, as measured by the Balance Error Scoring System (*BESS*, a clinical test that uses modified Romberg stances on different surfaces) on post-injury day 1 (Guskiewicz et al., 2001; 2003). An increased velocity of the center of pressure and an overall weight-shifting speed indicating both static and dynamic instability in concussed subjects has also been shown by Geurts et al (1999). It was suggested that the recovery of balance occurred between day 1 and day 3 post-injury for most of the MTBI subjects (Peterson et al., 2003).

The initial 2 days after MTBI are the most problematic for subjects standing on the foam surfaces, which was attributed to sensory interaction deficits in the use of visual, vestibular and somatosensory systems (Valovich et al., 2003; Guskiewicz, 2003). However, Cavanaugh et al. (2005 a, b; 2006) have shown that the Approximate Entropy (ApEn) method may detect changes in postural control in subjects with "normal" postural stability as determined by conventional balance testing long after a cerebral concussion. This is consistent with other recent studies demonstrating long-lasting residual balance abnormalities in MTBI subjects (Slobounov et al. 2005), which is most evident during dynamic postural tasks.

5.2. Virtual time-to-contact (VTC)

Recently, the notion of virtual time-to-contact (VTC) that specifies the spatial-temporal proximity of the center of pressure to the stability boundary, as an informational property in the regulation of posture was proposed (Slobounov et al. 1997). The original speculation that VTC may be a low-dimensional informational control variable in postural regulation (Carello et al., 1985) was supported in a serious of experiments (Martin, 1990; Riccio, 1993; Haibach et al., 2007). This new approach to understanding the nature of posture regulation places the emphasis on information for control not on departures from a stability point within the equilibrium region of the potential base of support, as in inverted pendulum models of posture (Mergner et al. 2003), but rather on the temporal safety margin, as specified by the virtual time to collision with the stability boundary (Lee, 1976). A significant consequence of this approach is that the control variable for posture is defined over the organism-environment-task interaction, rather than simply a product of the organism (Newell 1986; Riccio, 1993).

Slobounov et al. (1997) labeled the measure they created as virtual time to contact (VTC) which is the instantaneous time to the functional stability boundary defined on the dynamics of each point in the time series. The word *virtual* was used because the individual does not want to make contact with the stability boundary. Thus, VTC is an estimate of the time to the boundary *should* it occur, which would only happen in the case of a loss of stability as in a fall. In this approach a time series of the virtual time to contact can be determined that is based on the dynamics of the time to contact relative to the boundary rather the relative position of the center of pressure to the stability boundary. See Fig. 4 below for details of computation of VTC.

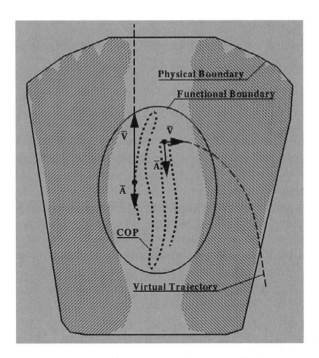

Figure 4. Schematic of VTC computation with respect to two-dimensional physical stability boundary. Clearly, it is temporal-spatial properties of the instantaneous COP values and combination of position, velocity and acceleration vectors rather than just position of the COP within the stability boundary are important fully appreciate VTC as an informational variable in control of posture. As can be seen, the virtual trajectory (VT) has a parabolic shape if the direction of the initial velocity and acceleration vectors is not collinear. However, the VT is linear if the initial velocity and initial acceleration have the same direction or if either of them is equal to zero.

In a series of recent studies, VTC has been shown to be a more sensitive index of postural stability than other traditional COP-based measures in aged and traumatic brain injured subjects (Haibach et al., 2007; Slobounov et al., 2008). Van Wegen and colleagues using a variation of this virtual-time to contact measure have shown similar age-related properties of time-to-contact in the control of postural stability (Van Wegen et al., 2001; 2002). Hertel et al. (2005) have shown the robustness of the virtual time to contact measure in control of uptight posture in human single leg quiet standing. More recently, Hertel and Olmsted-Kramer (2007) revealed alterations of time-to-boundary measures (TTB) in subjects suffering from chronic ankle instability (CAI), and suggested that TTB measures may detect postural control deficits related to CAI that traditional measures (i.e., COP range and velocity etc.) do not. Clearly, further research is needed to examine neurocognitive and physiological basis of VTC in control of human posture.

5.3. Abnormal VTC in Concussed Athletes

VTC methodology has been recently applied to detect long-lasting residual postural abnormalities in concussed athletes. It appears that both ApEn and VTC measures can detect postural abnormalities that cannot be observed using traditional balance testing. The concussed subjects were able to accomplish dynamic postural tasks in the presence of signs of altered measures of postural control and may be explained by enormous brain plasticity to control movement in spite of deficiency. Again, residual postural abnormalities in concussed individuals may be undetected using conventional research methods. Indeed, VTC measures were altered in the absence of signs of postural instability based on traditional measures of postural control. Most importantly, VTC alterations were more prominent when postural stances were challenged by introducing dynamic stability tasks that utilize more of the functional reserve. Both minimal/mean and mode values of VTC increased along with enhanced VTC variability in subjects suffering from concussion. This may imply that following concussion the subjects conserved enlarged "temporal safety margin" (Carello et al., 1985) in the regulation of posture in order to preserve balance in dynamic situations. Collectively, recent research shows that the residual deficits of concussion are most sharply revealed when the postural system is challenged by more demanding tasks.

Previous research has provided support for the proposition that VTC, which specifies the spatiotemporal proximity of the center of pressure to the stability boundary, is an informative variable in the regulation of human posture (Slobounov et al., 1997; Haibach et al., 2007). A more recent study provides additional evidence for this proposition (Slobounov et al., 2007). Specifically, the nominal values of VTC regardless of subjects' injury status were significantly lower during dynamic postural tasks compared to static quite stances. This may indicate a reduced "safety margin" (Corello et al., 1985; Hertel et al., 2006; van Wegen et al., 2002) associated with more complex postural tasks resulting in less time available to initiate compensatory postural adjustments (Riccio, 1993) to preserve balance. This interpretation is in agreement with recent conceptualization of anticipatory postural adjustment (APA) reversals that emphasizes the important role of safety in generation of postural adjustments associated with voluntary movement.

It should be noted that the shape and structure of VTC time series distribution during A-P sway before concussion is similar to those after concussion. VTC consistently increased with enhanced variability as a subject approached the stability boundary (see Fig. 5), meaning that the same strategy was implemented to preserve balance in the dynamic task regardless of injury status. Finally, it should be emphasized that although

VTC values were reduced during dynamic postural tasks with respect to those during quite stance, these were significantly increased after concussion. This suggests that concussed subjects may be more conservative in terms of preserving a larger "safety margin" in order to avoid "collision" with stability boundary and fail to accomplish the task. That the overall shape of VTC distribution was preserved after concussion, though its nominal values and variability were increased, provide further evidence that VTC may serve as a low dimensional controlled variable in regulation of upright posture (Slobounov et al., 1997).

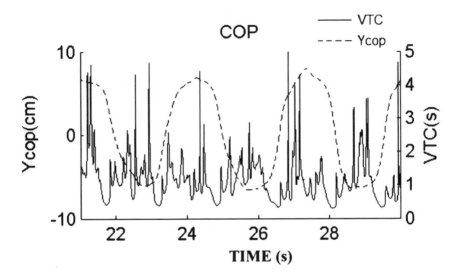

Figure 5. The VTC time-series with respect to COP position.

Collectively, the current findings suggest that the alteration in the regulation of postural movement in concussed individuals may not be detected using conventional assessment tools. Whether this alteration is relatively transient resulting in the acquisition of more conservative compensatory strategies to preserve balance, or a long-term persistent residual postural abnormality is yet to be determined. The clinical implication of our findings is in agreement with Cantu (2006) that the athletes who prematurely return to play based solely on conventional symptom resolution criteria may be highly susceptible to future and possibly more severe brain injuries. Indeed, a combination of various assessment methods and tools should be used by clinicians in order to make an accurate decision in terms of return to play and to identify athletes at risk for re-current concussions.

6. BALANCE-RELATED TRAUMATIC INJURIES IN ATHLETICS

Well preserved balance and proper whole body posture are important fundamental components for all sports and recreational activities. Clearly, the demands imposed by various sports (i.e., execute complex drills on the narrow, slippery and inclined bases of support, such as gymnastics, figure skating, skiing and a number of X-sports) require that athletes should possess highly developed postural control. On the other hand, specificity of training programs in these sports, in general, and specific drills and routines, in particular, may contribute to the acquisition of advanced balance system in those athletes. In fact, in our early work (Slobounov & Newell, 1994), a number of postural stability tasks were administered to athletes (i.e., gymnast; figure skaters, divers) and non-athletes. Interestingly, differences in balance measures were observed only during more challenging tasks (i.e., single-leg stance with closed eyes).

More recently, the results of our early work have been confirmed by Bressel et al. (2007) study. This study focused on comparison of static and dynamic balance among collegiate athletes competing or training in soccer, basketball, and gymnastics. To assess static balance, participants performed 3 stance variations (double leg, single leg, and tandem leg) on 2 surfaces (stiff and compliant) using the Balance Error Scoring System. For assessment of dynamic balance, participants performed multidirectional maximal single-leg reaches from a unilateral base of support, using normalized leg reach distances from the Star Excursion Balance Test. In fact, these clinical tests are commonly used by AT practitioners to assess balance problems after traumatic injuries to low extremities. Interestingly, Balance Error Scoring System error scores for the gymnastics group were 55% lower than for the basketball group. This was also true for the Star Excursion Balance Test scores. This clearly indicates that gymnasts have more advanced postural control, probably due to specificity and training and demands imposed by gymnastics. Another interesting finding from this study is that the Star Excursion Balance Test scores were 7% higher in the soccer group than the basketball group. Overall, gymnasts and soccer players did not differ in terms of static and dynamic balance, which is quite surprising. In contrast, basketball players displayed some signs of less advanced postural stability that may be a predisposing factor for the ankle instability/injury. Clearly, further research is needed to examine differential effect of sport activities on balance and postural control.

Improper balance is most dangerous in extreme X-sports, requiting execution of complex drills on the narrow base of support. A recent pilot study by Major et al. (2007) examined the effect of the skater experience and lower extremities biomechanics on energy absorption and observed balance strategies used during two basic tricks. In these tricks, the skater is required to jump onto an elevated rail and maintain balance while standing in a single position (stall) or sliding along the rail (grind). Lower extremity joint kinematics, impact force characteristics and general movement behavior were examined during landing and balance phases. The vertical impact force was found to decrease with increasing skater experience in stalls and grinds. Similar to drop landing experiments, peak impact force decreased with increasing knee flexion during stalls in experienced skaters. During stalls, skaters demonstrated classic balance maintenance strategies (ankle, hip, or multi-joint) depending on trick length. During grinds, the skater's centre of mass never passed over the rail base of support, suggesting the use of momentum produced from obliquely approaching the rail. Overall, experienced skaters have more advanced techniques to prepare body positioning prior to impact to prevent falls and more importantly to reduce the risk of severe injuries. Interestingly, the less-experienced skaters were more concerned with maintaining balance than refining the technique to minimize impact force. This means that dual-task requirements (i.e., maintenance of proper balance and pure execution of drills) may be a limiting factor of novice skates. Consequently, proper balance training is crucial for injury prevention among X-sport athletes.

6.1. Balance Impairment as a Result of Traumatic Injury

Postural control deficits and abnormal balance after traumatic injuries in athletes have been a focus in a number of recent studies. As can be concluded from the previous discussion, there are multiple contributions to postural stability from visual, somato-sensory and vestibular systems. Any damage to these systems may directly contribute to impaired postural control. For example, in the case of sport-related traumatic brain injury, athletes with TBI may experience short-term deficits in the visual system (i.e., blurred vision). Not surprisingly, abnormal balance is the major symptom of MTBI. In fact, as mentioned in the previous text, there is growing evidence of long-lasting postural abnormalities in athletes who suffered from even mild TBI.

Another example is empirical evidence of impaired balance as a result of reduced muscle strength, muscle sensitivity (e.g., proprioceptive sense) and developed functional joint(s) instability in athletes with traumatic injury to lower extremities. Specifically, a recent study by Docherty et al. (2006) has examined the postural control deficits in athletes suffering from functional ankle instability among Division 1 collegiate athletes by means of Balance Error Scoring System (BESS). In this study, the BESS test battery requires participants to stand unsupported on two different surfaces (firm and foam) in three different stances (double stance, single-one-leg stance, and in tandem). The major findings from this study is that athletes with functional ankle instability had poorer balance on all testing conditions and had lower BESS scores compared to normal controls. In other words, postural control deficits were identified in athletes suffering from functional ankle instability using the BESS test batteries. It was suggested that these deficits could be a contributing factor to the repeated episodes of instability and giving way that often occurs following an inversion ankle sprain in athletes.

Balance deficits in athletes with low extremities injury also are assessed by the Biodex Balance System. Akbari et al. (2006) examined the effect of acute lateral ankle sprain on postural stability using two clinical tests: the Functional Reach Test and the Star-Excursion Balance Test under both vision conditions (eyes open versus eyes closed). More severe balance impairment as a result of acute lateral sprain was observed when vision was not available regardless of clinical tests. This implies that the impaired proprioceptive sense as a result of ankle sprain was a major contributing factor influencing abnormal balance in injured athletes. Therefore, proprioceptive sense training in addition to strength training and conditioning should be implemented as soon as possible to speed up the recovery of balance and to prevent recurrent injuries.

6.2. Abnormal Balance as a Risk Factor for Injury

There is growing evidence that abnormal balance may be a predisposing factor for injury, especially, the lower extremities injuries. It seems reasonable to expect that preseason ankle instability due to a) residual effect from previous injuries, b) reduced range of motion following injury, c) diminished strength of the joint, and d) reduced joint position sense may put athletes at risk for recurrent injuries not only to the low extremities but also to the other parts of the body. An athlete may develop abnormal movement patterns/techniques and "bracing behavior" which is an act of preparing or positioning for impact or danger during execution of athletic drills. For example, in track, a hurdler may land differently on the less "stable" leg out of subconscious fear of re-injury due to impact forces with the track surface. In springboard diving, a diver may initiate a somersault by differentially pressing the board by "stable" versus "unstable" legs producing asymmetric rotational whole body motion. This is not only detrimental in terms of diving technique but most importantly is extremely dangerous putting the diver at high risk to hit the board. All of these examples represent "abnormally acquired movement patents" and possibly long-term "bracing behaviors" which are extremely harmful for athletes' physical and psychological well-being. Psychological factors, including the fear of injury, will be discussed elsewhere in this text.

Another important factor, indirectly related with poor balance as a risk factor for injury, is muscle fatigue. A few recent studies examined the effect of fatigue and balance level in athletes. According to Rose et al (2000), athletes are not differed in balance levels at pre-fatigue levels, but there are differences between fatigued and non fatigued athletes at the post-fatigue level. Balance due to fatigue still does not pose a major problem for athletes, since peripheral muscle fatigue, and accordingly balance, usually restore within 20 minutes post extensive exercise program (Susco, 2004). On the contrary, balance usually becomes a problem for most injured athletes. It is obvious, and as discussed in the previous text, that lower extremity injuries cause balance deficits. But the same negative effect may occur for injuries of the upper body since the athletes' center of mass is controlled and maintained by both the core and the upper/lower extremities. Not surprisingly, even a broken arm could cause deficit in balance (Rose et al, 2000). As athletes recover from their injuries, as evidenced by increased range of motion, strength and endurance, the balance improved as well (Rose, 2000).

Although abnormal balance has been proposed as a risk factor for sport-related injuries, few well-controlled studies have been conducted to examine this relationship. Sporadic evidence suggests that Star Excursion Balance Test (SEBT) may detect the risk of low extremities injury among high school basketball players (Plisky et al., 2006). Logistic regression models used in this study indicated that players with an anterior right/left reach distance difference greater than 4 cm were 2.5 times more likely to sustain a lower extremity injury. Interestingly, female athletes with a composite reach distance less than 94.0% of their limb length were 6.5 times more likely to have a lower extremity injury. In fact, this is consistent with numerous recent studies clearly indicating higher risk for ACL injuries in female athletes. Overall, it was suggested that the SEBT can be incorporated into pre-season physical examinations to identify players at higher risk for injury due to impaired balance.

There were very few prospective studies focusing on identifying the risk factors that predispose an athlete to ankle-ligament trauma. According to Beynnon et al., (2002) review, there is some agreement among researchers regarding the risk factors for ankle-ligament injury. However, considerable controversy remains. Although female athletes are at a significantly greater risk of suffering a serious knee sprain, such as disruption of the anterior-cruciate (ACL) ligament, this does not appear to be the case for ankle-ligament sprains. Specifically, according to According to Vrbanic et al. (2007), Croatian female athletes participating in high-risk sports (volleyball and handball) suffer anterior cruciate ligament (ACL) knee injury at a 4- to 6-fold greater rate than do male athletes. ACL injuries in females result either from contact mechanisms or from certain unexplained non-contact mechanisms occurring during daily practices. The occurrence of non-contact injuries points to the existence of certain factors intrinsic to the knee that can lead to ACL rupture. When knee joint movement overcomes the static and the dynamic constraint systems, non-contact ACL injury may occur. Certain recent results suggest that balance and neuromuscular control play a central role in knee joint stability, protection and prevention of ACL injuries. These authors suggest the Sport KAT 2000 testing system may be used to monitor balance and coordination systems and to estimate risk predictors in athletes who withdraw from sports due to lower sports results or ruptured ACL.

Balance and injury in elite Australian footballers was also a focus of Hrysomallis et al (2007) research. Consistent with other studies, low pre-season balance ability was shown to be associated with an increased risk of ankle ligament injury. Moreover, successful prediction of susceptibility to ankle sprain injury by means of a single leg balance (SLB) test prior to season was reported by Trojian & McKeag (2006). Specifically, high school varsity and intercollegiate athletes with positive SLB test results not taping the ankle imposed an increased risk of sprain over the season. Overall, gender does not appear to be a risk factor for suffering an ankle-ligament sprain. In addition, athletes who have suffered a previous sprain have a decreased risk of re-injury if a brace is worn, and the consensus is that generalized joint laxity and anatomical foot type are not risk factors for ankle sprains. However, the literature is divided with regard to whether or not height, weight, limb dominance, ankle-joint laxity, anatomical alignment, muscle strength, muscle-reaction time and postural sway are risk factors for ankle sprains.

6.3. Balance Training/Retraining Following Injury

Low extremities injuries are common among athletes both in contact and non-contact sports. This results in postural instability, asymmetry and probability of further residuals. Upon the general assumption that the maintenance of proper upright posture is essential not only for competing at a high level but also for prevention of injury in athletics, it is reasonable to suggest that a) coaches should implement balance training programs for prevention of injuries, and b) medical practitioners should seriously consider rehabilitation programs aimed at restoring balance after injury. In fact, several recent studies have clearly demonstrated a beneficial effect of balance training in athletes suffering from various injuries. Specifically, the Star Excursion Balance Training (SEBT) implemented with both eyes open and eyes closed paradigms was shown to be more effective to restore functional stability than conventional therapy after ankle sprain (Chaiwanichsiri et al., 2005). Similarly, a 4 week balance specific training program may benefit the athletes suffering from chronic ankle instability

(Hale et al., 2007). In addition, 6 weeks of balance training using either a mini-trampoline or a dura disc tool may improve postural stability in athletes suffering from lateral ankle sprain (Kidgell et la., 2007).

More recently, balance training has been adopted to try and prevent injuries to the ankle and knee joints during sport activities (Hrysomallis, 2007). As a single intervention, balance training has been shown to significantly reduce the recurrence of ankle ligament injuries in soccer, volleyball and recreational athletes; however, it has not been clearly shown to reduce ankle injuries in athletes without a prior ankle injury. Balance training on its own has also been shown to significantly reduce ACL injuries in male soccer players. Taking into consideration growing concern regarding the high risk of major knee injuries in female soccer players and overuse knee injuries in female volleyball players, it is reasonable to suggest that specialized balance training along with conventional conditioning may benefit female athletes at high risk for low extremity injuries.

A high body mass index and previous ankle sprains have been shown to increase the risk of sustaining non-contact inversion ankle sprains in high school football players (McHugh et al., 2007). Accordingly, it was hypothesized that stability pad balance training (SPBT) may reduce the incidence of these injuries. Players balanced for 5 minutes on each leg, 5 days per week, for 4 weeks in preseason and twice per week during the competitive season. Post-intervention injury incidence was compared with pre-intervention incidence (107 players-seasons) for players with increased risk. Interestingly, injury incidence for players with increased risk was 2.2 injuries per 1000 exposures before the intervention and 0.5 after the intervention. This represents a significant (77%) reduction in injury incidence as a result of implemented balance training for athletes at risk for ankle sprains.

Goal-oriented and task specific training has been shown to improve diminished functions. However, it is often a challenging task to maintain patients' motivation and stick with rehabilitation routines. One way to meet this challenge is to employ the game-based exercise rehabilitation programs. In fact, this was implemented by Betker et al (2007), who used game-based exercises for dynamic short-sitting balance rehabilitation of patients with chronic spinal cord and traumatic brain injuries. The patients in this study exhibited increases in practice volume and attention span during training with the game-based tool. In addition, they demonstrated substantial improvements in dynamic balance control. Similarly, upright balance function can be improved following balance specific training performed in a supine position in the Virtual Reality (VR) environment providing the perception of an upright position with respect to gravity (Oddson et al., 2007). Overall, a supervised goal-oriented balance training program should not only be used after injuries to low extremities but also as a preventive intervention for athletes at high risk for injuries.

CONCLUSION

This chapter presents a brief perspective of previous work with special emphasis on (a) current research regarding the role of cerebral cortex in human postural control, (b) balance assessment tools, (c) abnormal postural control in athletes suffering from various injuries, and (d) benefits of specific balance training. Considerable behavioral evidence clearly suggests the contribution of the cerebral cortex and higher cognitive functions in the control of upright posture. This was documented "indirectly" via modification of postural responses induced by various high-order cognitive functions and/or learning that are thought to be "represented" in the cerebral cortex. There is also a growing body of more recent research "directly" demonstrating the involvement of cerebral cortex in postural movement, similar to many other voluntary movements under study. This was primarily documented via EEG patterns preceding (movement-related cortical potentials, MRCP) self-initiated postural movement and following (perturbation-evoke responses, PER) postural perturbation. It should be noted, however, that this newly evolving understanding of the role of cortical processes in the control of posture in humans is still in its infancy. Future research implementing current advances in cellular recording and brain imaging techniques may be warranted to further explore differential contribution and functional connectivity of various brain structures in control of human posture.

Clearly, abnormal balance is present as a result of traumatic injury to peripheral structures and/or to the brain (MTBI). On the other hand, abnormal posture and improper body orientation may result in impairment of overall movement form and athletic techniques. This may contribute to higher risk of injuries in the athletics. The residual postural abnormalities may not be clearly seen and most often are overlooked when traditional posture assessment tools are implemented. Therefore, advanced methods, including brain imaging studies and Virtual Reality (VR), should be implemented to assess balance in athletes. If postural abnormalities are warrant, appropriate balance training/retraining is highly recommended to prevent high possibilities of traumatic injuries.

Acknowledgments

This chapter was inspired by long-term collaboration with Dr. Karl Newell, who introduced me to the field of human postural control. Some empirical evidence provided in this chapter has been collected and published with Karl in numerous journal and book chapters.

REFERENCES

Keith, A. (1923). Hungarian lectures: Man's posture: its evolution and disorder. *British Medical Journal, 1*, 451-499.

Thomas, A. (1940). Equilibre et Equilibration. Masson, Paris.

Massion, J. (1992). Movement, posture, and equilibrium: interaction and coordination, *Progress Neurobiology, 38*, 35-56.

Horak, F.B., Macpherson, J.M. (1996). Postural orientation and equilibrium. In: Rowell LB, Shepherd JT (eds), *Handbook of Physiology, Sec 12, Exercise: regulation and integration of multiple systems*. Oxford University Press, New York, pp 255-292.

Magnus, R. (1926). Physiology of posture. *Lancet 11*, 531-585.

Sherrington, C.S. (1910). Flexion-reflex of the limb, crossed extension-reflex, and reflex stepping and standing. *Journal of Physiology, 40*, 28-121.

Massion, J., Alexandrov, A., Frolov, A. (2004). Why and how are posture and movement coordinated? *Progress in Brain Research, 143,* 13-27.

Bernstein, N. (1967). *The coordination and regulation of movement*. New York: Pergamon.

Winter, D.A. (1990). *Biomechanics and motor control of human movement*. (2nd ed.). New York: John Wiley & Sons. Inc.

Riccio, G. (1993). Information in movement variability about qualitative dynamics of posture and orientation. In K.M. Newell & D. Corcos (Eds.), *Variability and motor control* (pp. 317-359). Champaign, Ill. Human Kinetics.

Newell, K. M., Kugler, P. N., van Emmerik, R.E.A., & McDonald, P.V. (1989). Search strategies and acquisition of coordination. In S.A. Wallace (Ed*.), Perspectives on the coordination of movement* (pp.85-122). Amsterdam: North Holland.

Bechterev, V.V. (1882). Physiology of balance. *Military-Medical Journal, 7*, 178.

Klack, J., & Watkins, D. (1984). Static balance in young children. *Child Development, 5*, 854-857.

Shumway-Cook, A., & Woollacott, M. (1985). The growth of stability: Postural control from developmental perspectives. *Journal of Motor Behavior, 17*, 131-147.

Mergner, T., Maurer, C., Peterka, R.C. (2003). A multisensory posture control model of human upright stance. *Progress in Brain Research, 142*,189-201.

Costikova, S.I. (1939). Postural stability and geometry of the joint angles. *Biophysica, 12*, 12-24.

Gurfinkel, V., Koz, Y., Shik, M. (1965). *Human posture regulation*. Moscow: Nauka, (in Russian).

DiZio, P. A. , & Lackner, J. R. (1986). Perceived orientation, motion, and configuration of the body during viewing of an off-vertical, rotating surface. *Perception and Psychophysics, 39*, 39-46.

Quant, S., Adkin,A.L., Staines, W.R., Maki, B.E., McIlroy, W.E. (2004a). The effect of a concurrent cognitive task on cortical potentials evoked by unpredictable balance perturbations. *BMC Neuroscience, May 17*, 5-18.

Quant, S., Adkin, A.L., Staines, W.R., McIlroy, W.E. (2004b). Cortical activation following a balance disturbance. *Experimental Brain Research, 155*, 393-400.

Adkin, A.L., Quant, S., Maki, B.E., McIlroy, W.E. (2006). Cortical responses associated with predictable and unpredictable compensatory balance reactions. *Experimental Brain Research, 172*, 85-93.

de Waele, C., Baudonniere, P.M., Lepecq, J.C., Tran Ba Huy, P., Vidal, P.P. (2001). Vestibula projections in the human cortex. *Experimental Brain Research, 141(4),* 541-51.

Slobounov, S., Wu, T., Hallett, M. (2006). Neural basis subserving the detection of postural instability: An fMRI study. *Motor Control, 10 (1),* 69-89.

Brandt, T., Dieterich, M., Danek, A. (1994). Vestibular cortex lesions affect the perception of verticality. *Annals of Neurology, 35*, 403-412.

Karnath, H.O., Ferber, S., Dichgans, J. (2000). Neural representation of postural control in humans. *Proceedings of National Academy of Science, U S A 97*, 13931-13936.

Karnath, H.O., Johannsen, L., Broetz, D., Kuker, W. (2005). Thalamic hemorrhage induces "pusher syndrome". *Neurology 64*, 1014-1019.

Blanke, O., Perrig, S., Thut, G., Landis, T., Seeck, M. (2000). Simple and complex vestibular responses induced by electrical cortical stimulation of the parietal cortex in humans. *Journal of Neurology Neurosurgery & Psychiatry, 69*, 553-556.

Johannsen, L., Broetz, D., Naegele, T., Karnath, H.O. (2006). "Pusher syndrome" following cortical lesions that spare the thalamus. *Journal of Neurology, 253*, 455-463.

Perennou, D.A., Leblond, C., Amblard, B., Micallef, J.P., Rouget, E., Pelissier, J. (2000). The polymodal sensory cortex is crucial for controlling lateral postural stability: evidence from stroke patients. *Brain Research Bulletin, 53*, 359-365.

Gibson, J.J. The senses considered as perceptual systems. Boston: Houghton Mifflin, 1966.

Mach, E. Grundlinien der Lehre von den Bewegungsempfindungen. Leipzig: Enhelman, 1975, p.129

Keshner, E.A., Kenyon, R.V. (2000). The influence of an immersive virtual environment on the segmental organization of postural stabilizing responses. Journal of Vestibular Research, 10, 207-219.

Lee, D.N., Lishman, J.D. (1975). Visual proprioceptive control of stance. *Journal of Human Movement Science, 1*, 87-95.

Lestienne, F., Soeching J., Berthoz A. (1977). Postural readjustments induced by linear motion of visual scenes. *Experimental Brain Research, 28*, 363-384

Stoffregen, T., Smart, J. (1998). Postural instability precedes motion sickness. *Brain Research Bulletin, 47*(5), 437-448.

Van Asten, W.N., Gielen, C.C., Denier, V. (1988). Postural adjustments induced by simulated motion of differently structured environment. *Experimental Brain Research, 73,*371-383.

Ehrenfried, T., Guerraz, M., Thilo K., Yardley, L., Gresty, A. (2003). Posture and mental task performance when viewing a moving visual field. *Cognitive Brain Research, 1*, 34-45.

Ikeda, T., Takahashi, M. (1997). Active posture control experimental motion sickness in guinea-pigs. *Acta Otolaryngologia, 117*(6), 815-818.

Beer, J., Blakemore, C., Previc, F. (2002). Areas of the human brain activated by ambient visual motion, indicating three kinds of self-movement. *Experimental BrainResearch,143*, 78-88.

Slobounov, S., Sebastianelli, W., Tutwiler, R., Slobounov, E. (2006). Alteration of postural responses to visual field motion in mild traumatic brain injury. *Neurosurgery, 59(1)*, 134-139.

Slobounov, S., Hallett, M., Wu, T., Shibasaki, H., Newell, K. (2006). Neural underpinning of postural responses to visual field motion. *Biological Psychology, 72*, 188-197.

Slobounov, S., Newell, K., Slobounov, E. (2006). Application of virtual reality graphics in assessment of concussion. *CyberPsychology & Behavior, 9(2)*, 188-191.

Winter, D.A., Patla, A.E, Prince, F., Ishac M., Gielo-Perczac, K. (1998). Stiffness control of balance in quiet standing. *Journal of Neurophysiology, 80*, 1211-1221.

Winter, D.A., Prince, F., Frank, J.S., Powell, C., Zabjek, K.F. (1996). Unified theory regarding AP and ML balance in quiet stance. *Journal of Neurophysiology, 75*, 2334-2343.

Fitzpatrick, R.C., Taylor, J.L., McCloskey, D.I. (1992b). Ankle stiffness of standing humans in response to imperceptible perturbation: reflex and task-dependent components *Journal of Physiology, 458*, 69-83.

Winter, D.A., MacKinnon, C.D., Ruder, G.K., Wieman, C. (1993). An integrated EMG/biomechanical model of upper body balance and posture during human gait. *Progress in Brain Research, 97*, 359-367.

Morasso, P.G., Schieppati, M. (1999). Can muscle stiffness alone stabilize upright standing? *Journal of Neurophysiology, 82,* 1622-1626.

Zatsiorsky, V.M., King,D.L. (1998). An algorithm for determining gravity line location from posturographic recordings. *Journal of Biomechanics, 31,* 161-164.

Zatsiorsky, V.M, Duarte, M. (1999). Instant equilibrium point and its migration in standing tasks: rambling and tremblig components of the stabilogram. *Motor Control, 3,* 28-38.

Diener, H.C., Horak, F.B., Nashner, L.M. (1988). Influence of stimulus parameters on human postural responses. *Journal of Neurophysiology, 59,* 1888-1905.

Horak, F.B., Diener, H.C. (1994). Cerebellar control of postural scaling and central set in stance. *Journal of Neurophysiology,72,* 479-493.

Arutyunyan, G.H., Gurfinkel, V.C., & Mirski, M.L. (1969). Organization of movement on execution by man of an exact postural tasks. *Biofizica, 14,* 1162-1167 (In Russian).

Forssberg, H., & Nashner, L.M. (1982). Ontogenetic development of postural control in man: Adaptation to altered support and visual conditions during stance. *Journal of Neuroscience, 2,* 545-552.

Latash, M.L. (1993). *Control of human movement.* Human Kinetics, Champaign.

Nashner, L.M. (1976). Adapting reflexes controlling the human posture. *Experimental Brain Research, 26,* 59-72.

Nashner, L.M. (1977). Fixed patterns of rapid postural responses among leg muscles during stance. *Experimental Brain Research, 30,* 13-24.

Nashner, L.M., Woollacott, M. (1979). The organizational of rapid postural adjustments of standing humans: an experimental-conceptual model. In: Talbot, R.E., Humphrey, D.R. (eds). *Posture and Movement.* Raven Press, New York, 243-225.

Nashner, L.M.,Shuppert, C.L., Horak, F.B., Black, F.O. (1989). Organization of postural controls: an analysis of sensory and mechanical constraints. *Progress in Brain Research, 80,* 411-418.

Horak, F.B., Nashner, L.M. (1986). Central programming of postural movements: adaptation to altered support-surface configurations. *Journal of Neurophysiology, 55,* 1369-1381.

Horak, F.B., Diener, H.C., Nashner, L.M. (1989). Influence of central set on human postural responses. *Journal of Neurophysiology, 62,* 841-853.

Horak, F.B, Nashner, L.M., Diener, H.C. (1990). Postural strategies associated with somatosensory and vestibular loss. *Experimental Brain Research, 82,* 167-177.

Hess, W. R. (1943). Teleokinetischer und ereismatisches Kraftesystem in der Biomotori. Helver Physiologisches und Pharmacologisches Acta: C62-C52.

Bouisset, S., Zattara, M. (1987). Biomechanical study of the programming of anticipatory postural adjustments associated with voluntary movement. *Journal of Biomechanics, 20,* 735-742.

Ramos, C.F., Stark, LW. (1990). Postural maintenance during movement: simulations of a two joint model. *Biological Cybernetics, 63,* 363-375.

Belink'ii, V.E., Gurfinkel, V.S., Pal'tsev, E.I. (1967). On the elements of control of voluntary movements. *Biofizika, 12,* 135-141 (in Russian).

Cordo, P.J., Nashner, L.M. (1982). Properties of postural adjustments associated with rapid arm movements. *Journal of Neurophysiology, 47,* 287-302.

Gurfinkel, V.S., Lipshits, M.I., Lestienne, F.G. (1988). Anticipatory neck muscle activity associated with rapid arm movements. *Neuroscience Letters, 94,* 104-108.

Vander Fits, I.B.M., Klip, A.W.J, Van Eykern, L.A., Haddres-Algra, M. (1998). Postural adjustments accompanying fast pointing movements in standing, sitting and lying adults. *Experimental Brain Research, 135,* 81-93.

Aruin, A.S., Latash, M.L. (1995a). Directional specificity of postural muscles in feed-forward postural reactions during fast voluntary arm movements. *Experimental Brain Research, 103,* 323-332.

Aruin, A.S., Latash, M.L. (1995b). The role of motor action in anticipatory postural adjustments studied with self-induced and externally triggered perturbations. *Experimental Brain Research, 106,* 291-300.

Aruin, A.S, Latash, M.L. (1996). Anticipatory postural adjustments during self-initiated perturbations of different magnitude triggered by a standard motor action. *Electroencephalography and Clinical Neurophysiology, 101,* 497-503.

Shiratori, T., Latash, M.L. (2000). The roles of proximal and distal muscles in anticipatory postural adjustments under symmetrical perturbations and during standing on rollerskatters. *Clinical Neurophysiology, 111,* 613-623.

Diener, H.C., Dichgans, J., Bootz, F., Bacher, M. (1984). Early stabilization of human posture after a sudden disturbance: influence of rate and amplitude of displacement. *Experimental Brain Research, 56,* 126-134.

Keck, M.E., Pijnappels, M., Schubert, M., Colombo, G., Curt, A., Dietz, V. (1998). Stumbling reactions in man: influence of corticospinal input. *Electroencephalography and Clinical Neurophysiology, 109,* 215-223.

Matthews, P.B. (1991). The human stretch reflex and the motor cortex. *Trends Neuroscience, 14,* 87-91.

Jacobs, J.V., Horak, F.B., Fujiwara, K., Tomita, H., Furune, N., Kunita, K. (2007). Changes in activity at the cerebral cortex associate with the optimization of responses to external postural perturbations when given prior warning. *Gait Posture Supplement.*

Bouisset, S., & Zattara, M. (1981). A sequence of postural movements preceded voluntary movement. *Neuroscience Letters, 22,* 263-270.

Brown, L.A., Shumway-Cook, A., Woollacott, M.H. (1999). Attentional demands and postural recovery: the effects of aging. *Journal of Gerontology: Biological Sciences & Medical Science, 54,* M165-M171.

Carpenter, M.G., Frank, J.S., Adkin, A.L., Paton, A., & Allum, J.H. (2004). Influence of postural anxiety on postural reactions to multi-directional surface rotations. *Journal of Neurophysiology, 92,* 3255-3265.

McIlroy, W.E., Maki, B.E. (1995). Early activation of arm muscles follows external perturbation of upright stance. *Neuroscience Letters, 184,* 177-180.

Maki, B.E., Zecevic, A., Bateni, H., Kirshenbaum, N., McIlroy, W.E. (2001). Cognitive demands of executing postural reactions: does aging impede attention switching? *Neuroreport, 12,* 3583-3587.

Brauer, S.G., Woollacott, M., Shumway-Cook, A. (2002). The influence of a concurrent cognitive task on the compensatory stepping response to a perturbation in balance-impaired and healthy elders. *Gait Posture 15,* 83-93.

Norrie, R.G., Maki, B.E., Staines, W.R., McIlroy, W.E. (2002). The time course of attention shifts following perturbation of upright stance. *Experimental Brain Research, 146,* 315-321.

Quant, S., Adkin,A.L., Staines, W.R., Maki, B.E., McIlroy, W.E. (2004a). The effect of a concurrent cognitive task on cortical potentials evoked by unpredictable balance perturbations. *BMC Neuroscience, May 17,* 5-18.

Burleigh, A., Horak, F. (1996). Influence of instruction, prediction, and afferent sensory information on the postural organization of step initiation. *Journal of Neurophysiology, 75,* 1619-1628.

Burleigh, A.L., Horak, F.B,. Malouin, F. (1994). Modification of postural responses and step initiation: evidence for goal-directed postural interactions. *Journal of Neurophysiology, 72,* 2892-2902.

McIlroy, W.E., Maki, B.E. (1993). Task constraints on foot movement and the incidence of compensatory stepping following perturbation of upright stance. *Brain Research, 616,* 30-38.

Quintern, J., Berger, W., Dietz, V. (1985). Compensatory reactions to gait perturbations in man: short- and long-term effects of neuronal adaptation. *Neuroscience Letters, 62,* 371-376.

Chong, R.K., Horak, F.B., Woollacott, M.H. (1999). Time-dependent influence of sensorimotor set on automatic responses in perturbed stance. *Experimental Brain Research, 124*, 513-519.

Henry, S.M., Fung, J., Horak, F.B. (2001). Effect of stance width on multidirectional postural responses. *Journal of Neurophysiology, 85*, 559-570.

Tjernstrom, F., Fransson, P.A., Hafstrom, A., Magnusson, M. (2002). Adaptation of postural control to perturbations--a process that initiates long-term motor memory. *Gait Posture 15*, 75-82.

Dehaene, S., Molko, N, Cohen, L., Wilson, A.J. (2004). Arithmetic and the brain. *Current Opinion Neurobiology, 14*, 218-224.

Kaiser, J., Lutzenberger, W. (2005). Cortical oscillatory activity and the dynamics of auditory memory processing. *Review Neuroscience, 16*, 239-254.

Naghavi, H.R., Nyberg, L. (2005). Common fronto-parietal activity in attention, memory, and consciousness: shared demands on integration. *Conscious Cognition, 14*, 390-425.

Saitou, K., Washimi, Y., Koike, Y., Takahashi, A., Kaneoke, Y. (1996). Slow negative cortical potential preceding the onset of postural adjustment. *Electroencephalography and Clinical Neurophysiology, 98*, 449-455.

Slobounov, S., Hallett, M., Stanhope, S., & Shibasaki, H. (2005). Role of cerebral cortex in human postural control: an EEG study. *Clinical Neurophysiology, 116*, 315-323.

Slobounov, S., Tutwiler, R., Slobounova, E., Rearick, M., Ray, W. (2000). Human oscillatory activity within gamma-band (30-50 Hz) induced by visual recognition of non-stable postures. *Cognitive Brain Research, 9*, 177-192.

Dietz, V., Quintern, J., Berger, W., & Schenck, E. (1985). Cerebral potentials and leg muscle e.m.g. responses associated with stance perturbation. *Experimental Brain Research, 57*, 354-384.

Ackermann, H., Diener, H.C., & Dichgans, J. (1986). Mechanically evoked cerebral potentials and long-latency muscle responses in the evaluation of afferent and efferent long-loop pathways in humans. *Neuroscience Letters, 66*, 233-238.

Dimitrov, B., Gavrilenko, T., Gatev, P. (1996). Mechanically evoked cerebral potentials to sudden ankle dorsiflexion in human subjects during standing. *Neuroscience Letters, 208*, 199-202.

Adkin, A.L., Quant, S., Maki, B.E., McIlroy, W.E. (2006). Cortical responses associated with predictable and unpredictable compensatory balance reactions. *Experimental Brain Research, 172*, 85-93.

Painling, P. E., Segalowitx, S. J. (2004). The effects of uncertainty in error monitoring on associated ERPs. *Cognitive Brain Research, 56*, 215-233.

Yasuda, A., Sato, A., Miyawaki, K., Kumano, H., Kuboki, T. (2004). Error-related negativity reflects detection of negative reward prediction error. *Neuroreport, 15*, 2561-2565.

Ouchi, Y., Okada, H., Youshikawa, E., Nobezava, S., Futatsubashi, M. (1999). Brain activation during maintenance of standing postures in human. *Brain, 122(2)*, 329-338.

Horak, F.B., Diener, H.C. (1994). Cerebellar control of postural scaling and central set in stance. *Journal of Neurophysiology, 72*, 479-493.

Slobounov, S., Slovounova, E., Newell, K. (1997). Virtual time-to-collision and human postural control. *Journal of Motor Behavior, 29*, 263-281.

Carello, C., Kugler, P.N., Turvey, M.T. (1985). The information support for the upright stance. *Behavioral Brain Science, 8*, 151-152.

Martin. E.J. (1990). An information based approach in postural control: the role of time-to-contract with stability boundary. Unpublished Master Thesis, University of Illinois at Urbana-Champaign.

Haibach, P.S, Slobounov, S.M, Slobounova, E,S., Newell, K.M. (2007). Virtual time-to-contact of stability boundaries as a function of support surface compliance. *Experimental Brain Research, 177*, 471-482.

Lee, D. (1976). A theory of visual control of braking based on information about time-to-collision. *Perception, 5*, 437-459.

Mergner, T., Maurer, C., Peterka, R.C. (2003). A multisensory posture control model of human upright stance. *Progress in Brain Research, 142*,189-201.

Newell, K. (1986). Constraints on the development of coordination. In: Wade MG, Whiting HTA (eds) *Motor skill acquisition in children: aspects of coordination and control*, Martinies NIJHOS, Amsterdam.

Slobounov, S., Sebastianelli, W., Cheng, C., Newell, K.M. (2008). Residual deficits from concussion as revealed by virtual time-to-contact measures of postural stability. *Clinical Neurophysiology*.

Van Wegen, E.E.H., van Emmerik, R.E.A., Wagenaar, R.C., Ellis, T. (2001). Stability boundaries and lateral postural control in Parkinson's disease. *Motor Control, 5*, 254-69.

Van Wegen, E.E.H., van Emmerik, R.E.A., Riccio, G. (2002). Postural orientation: age-related changes in variability and time-to-boundary. *Human Movement Science, 21*, 61-64.

Hertel, J., Olmsted-Kramer, L.C., Challis, J. (2006). Time-to-boundary measures of postural control during single leg quite standing. *Journal of Applied Biomechanics, 22*, 67-73.

Hertel, J., Olmsted-Kramer, L.C. (2007). Deficits in time-to-boundary measures of postural control with chronic ankle instability. *Gait Posture 40(2)*, 69-70.

Lishman, W.A. (1988). Physiogenesis and psychogenesis in the post-concussional syndrome. *Biological Journal of Psychiatry, 153*, 460-469.

Ingelsoll, C.D., & Armstrong, C.W. (1992). The effect of closed-head injury on postural sway. *Medicine in Science, Sports & Exercise, 24*, 739-743.

Wober, C., Oder, W., Kollegger, H., Prayer, L., Baumgartner, C., & Wober-Bingol, C. (1993). Posturagraphic measurement of body sway in survivors of severe closed-head injury. *Archive of Physical Medical Rehabilitation, 74*, 1151-1156.

Guskiewicz, K.M., McCrea, M., Marshall, S.W., Cantu, R.C., Randolph, C., Barr, W., Onate, J.A., & Kelly, J.P. (2003). Cumulative effects associated with recurrent concussion in collegiate football players: the NCAA Concussion Study. *Journal of the American Medical Association, 290*(19), 2549-2549.

Slobounov, S., Sebastianelli, W., Simon, R. (2002). Neurophysiological and behavioral Concomitants of Mild Brain Injury in College Athletes. *Clinical Neurophysiology, 113*, 185-193.

Slobounov, S., Sebastianelli, W., Moss, R. (2005). Alteration of posture-related cortical potentials in mild traumatic brain injury. *Neuroscience Letters, 383*, 251-255.

Thompson, J., Sebastianelli, W., Slobounov, S. (2005). EEG and postural correlates of mild traumatic brain injury in athletes. *Neuroscience Letters, 377*, 158-163.

Slobounov, S., Sebastianelli, W., Moss, R. (2005). Alteration of posture-related cortical potentials in mild traumatic brain injury. *Neuroscience Letters, 383*, 251-255.

Guskiewicz, K.M., Ross, S.E., & Marshall, S.W. (2001). Postural stability and Postural stability and neuropsychological deficits after concussion in collegiate athletes. *Journal of Athletic Training, 36*(3), 263-273.

Guskiewicz, K. M., Weaver, N. L., Padua, D. A., & Garrett, W. E. (2000). Epidemiology of concussion in college and high school football players. *American Journal of Sports Medicine, 28*, 643-650.

Guskiewicz, K.M., McCrea, M., Marshall, S.W., Cantu, R.C., Randolph, C., Barr, W., Onate, J.A., & Kelly, J.P. (2003). Cumulative effects associated with recurrent concussion in collegiate football players: the NCAA Concussion Study. *Journal of the American Medical Association, 290*(19), 2549-2549.

Geurts, A., Knoop, J., & van Limbeek, J. (1999). Is postural control associated with mental functioning is the persistent postconcussion syndrome? *Archive Physical Rehabilitation, 80*, 144-149.

Peterson, C., Ferrara, M., Mrazik, M., Piland, S., Elliott, R. (2003). Evaluation of neuropsychological domain scores and postural stability following cerebral concussion in sport. *Clinical Journal of Sport Medicine, 13*(4), 230-237.

Valovich, T., Periin, D., Gansneder, B. (2003). Repeat administration elicits a practice effect with the balance error scoring system but not with the standardized assessment of concussion in high school athletes. *Journal of Athletic Training, 38*(10), 51-56.

Cavanaugh, J., Guskiewicz, K., Stergiou, N. (2005a). A nonlinear dynamic approach for evaluating postural control: new directions for the management of sport-related cerebral concussion. *Sport Medicine, 35*(11), 935-950.

Cavanaugh, J., Guskiewicz, K., Giuliani, C., Marshall, S., Mercer, V., Stergiou, N. (2005b). Detecting altered postural control after cerebral concussion in athletes with normal postural stability. *British Journal of Sports Medicine, 39*(11), 805-811.

Cavanaugh, J.T., Guskiewicz, K.M., Giuliani, C., Marshall, S., Merser, V.S., Stergion, N. (2006). Recovery of postural control after cerebral concussion: new insights using approximate entropy. *Journal of Athletic Training, 41*(3), 305-313.

Slobounov, S., Newell, K. M. (1994). Postural dynamics as a function of skill level and task constraints. *Gait and Posture, 2,* 85-93.

Bressel, E., Yonker, J.C, Kras, J., Heath, E.M. (2007). Comparison of static and dynamic balance innm female collegiate soccer, basketball, and gymnastics athletes. *Journal of Athletic Training, 42*(1), 42-46.

Major, M.J, Beaudoin, A..J, Kurath, P., Hsiao-Wecksler, E.T. (2007). Biomechanics of aggressive inline skating: Landing and balancing on a grind rail. *Journal of Sport Sciences, 25*(12), 1411-1422.

Docherty, C.L., Valovich, McLeod, T.C., Shultz, S.J. (2006). Postural control deficits in participants with functional ankle instability as measured by the balance error scoring system. *Clinical Journal of Sport Medicine, 16*(3), 203-208.

Akbari M, Karimi H, Farahini H, Faghihzadeh S. (2006). Balance problems after unilateral lateral ankle sprains. *Journal of Rehabilitation Research Development, 43*(7), 819-824.

Plisky, P.J., Rauh, M.J., Kaminski, T.W., Underwood, F.B. (2006). Abnormal Balance as a Risk Factor for Injury in Athletes. *Journal of Orthopaedics & Physical Therapy, 36*(12), 911-919.

Beynnon, B.D., Murphy, D.F,, Alosa, D.M. (2002). Predictive Factors for Lateral Ankle Sprains: A Literature Review. *Journal of Athletic Training, 37*(4), 376-380.

Vrbanić, T.S., Ravlić-Gulan, J., Gulan, G., Matovinović, D. (2007). Balance index score as a predictive factor for lower sports results or anterior cruciate ligament knee injuries in Croatian female athletes--preliminary study. *College Anthropology, 31*(1), 253-258.

Hrysomallis C, McLaughlin P, Goodman C. (2007). Balance and injury in elite Australian footballers. *International Journal of Sports Medicine, 28*(10), 844-847.

Trojian, T.H., McKeag, D.B. (2006). Single leg balance test to identify risk of ankle sprains. *British Journal of Sports Medicine*, 40(7), 610-613.

Chaiwanichsiri, D., Lorprayoon, E., Noomanoch, L. (2005). Star excursion balance training: effects on ankle functional stability after ankle sprain. *Journal of American Association Thai, 88*(4), S90-S94.

Hale, S.A., Hertel J., Olmsted-Kramer, L.C. (2007). The effect of a 4-week comprehensive rehabilitation program on postural control and lower extremity function in individuals with chronic ankle instability. *Journal of Orthopaedics & Sports Physical Therapy, 37*(6), 303-31.

Kidgell, D.J., Horvath, D.M., Jackson, B.M., Seymour, P.J. (2007). Effect of six weeks of dura disc and mini-trampoline balance training on postural sway in athletes with functional ankle instability. *Journal of Strength and Conditioning Research, 21*(2), 466-469.

Hrysomallis, C. (2007). Relationship between balance ability, training and sports injury risk. *Sports Medicine, 37*(6), 547-556.

McHugh, M.P., Tyler, T.F., Mirabella, M.R., Mullaney, M.J., Nicholas, S.J. (2007). The effectiveness of a balance training intervention in reducing the incidence of noncontact ankle sprains in high school football players. *American Journal of Sports Medicine, 35*(8), 1289-1294.

Betker, A.L., Desai, A., Nett, C., Kapadia, N., Szturm, T. (2007). Game based Exercises for Dynamic Short-Sitting Balance Rehabilitation of People With Chronic Spinal Cord and Traumatic Brain Injuries. *Physical Therapy*, (August, 21).

Oddsson, L.I., Karlsson, R., Konrad, J., Ince, S., Williams, S.R., Zemkova, E. (2007). A rehabilitation tool for functional balance using altered gravity and virtual reality. *Journal of Neuroengineering Rehabilitation, 10*(4), 25.

CHAPTER 4

FATIGUE-RELATED INJURIES IN ATHLETES

1. INTRODUCTION

During a maximum contraction of many mixed muscles, it is common to see a 50% reduction of force over a period of a few seconds. This is known as *fatigue* (Rothwell, 1994). The term *fatigue* though can refer to both physical and mental exhaustion due to prolonged stimulation or exertion. As such, it is a phenomenon that is of interest to many scientific disciplines, including the science of coaching, as is used in a variety of contexts. Of particular interest is localized progressive muscle fatigue which has been defined as an inability to maintain required force level after prolonged use of muscle (Gandevia et al. 1998). While reduction in force production is obviously detrimental in many circumstances, progressive muscle fatigue has also been shown to impair postural stability (Johnston et al., 1998), muscle coordination (Carpenter et al., 1998) and control of limb velocity and acceleration (Jaric et al., 1997).

Fatigue has traditionally been attributed to the occurrence of a "metabolic endpoint", where muscle glycogen concentrations are depleted, plasma glucose concentrations are reduced and plasma free fatty acid levels are elevated (cf: Meeusen et al., 2006). However, one of the major complications that arise in studying muscle fatigue is that both peripheral and central mechanisms contribute to the manifestations of muscle fatigue (Enoka & Stuart, 1992). These mechanisms are highly interactive in nature, and should both be acknowledged as a complex phenomenon. Unfortunately, both their independent and collective contributions to progressive muscle fatigue are still poorly understood.

One of the problems with conceptualization of fatigue is research methodology. In a clinical practice most often fatigue is assessed based on subjects' self-reports (SR)/responses to questionnaires. Some of the traditional instruments to measure fatigue include: Rhoten Fatigue Scale (RFS, one-item 11 point scale), Profile of Mood States (POMS, having vigor, 8-item and fatigue, 7-item subscale with 5 point scales), Multidimensional Fatigue Inventory (MFI, 20-item, 7-point scale, 6 dimensions), Fatigue Symptom Inventory (FSI, 13-item, 0-10 scale, 2 dimensions), Piper Fatigue Scale Revised (PFS, 22-item, 0-10 scale, 4 dimensions), Multidimensional Fatigue Symptom Inventory (MFSI, 83-item, 5 point scale, 5 dimensions) and Brief Fatigue Inventory (BFI, 9-item, 0-10 scale, single dimension). One of the common shortcomings of these scales is the lack of test-re-test validity; they are lengthy and often confusing to the patients. It should be noted that there is NO single standard accepted as a clinical practice instrument today. Moreover, no single SR instrument has been shown to correlate with other biological/physiological and/or neural markers of fatigue. For example, Dimeo et al. (1997) evaluated a correlation between fatigue assessed by POMS and the Symptoms Check List (SCL-90-R) and physical performance of 78 patients. They observed a week correlation (r= -0.30) between the SP rating and physical performance (stress test on treadmill). However, there was stronger correlation between SR fatigue rating and depression (r=0.68) and/or anxiety (r=0.63). Similarly, low correlation between SR fatigue rating and physical performance (walking) was also reported by Simmonds (2002). Clearly, biological markers in conjunction with commonly accepted clinical practice indices of fatigue should be considered for pure assessment of fatigue.

Numerous anecdotal facts and observations in sport environment indicate a negative effect of fatigue on performance of athletic drills. Clearly, fatigue induces performance deterioration due to reduction force production, lack of accuracy and reduced speed of motor responses. Possible degradation of the aiming precision of a whole-body pointing task as evidenced by impaired movement coordination due to selective muscle fatigue has been shown by Schmid et al., (2006). Another recent study of the instep football kick have shown significantly slower ball velocity observed in the fatigue condition due to both reduced lower leg swing speed and poorer ball contact (Apriantono et al., 2006). This well-controlled study that reduced leg swing speed, represented by a slower toe linear velocity immediately before ball impact and slower peak lower leg angular velocity, was most likely due to a significantly reduced resultant joint moment and motion-dependent interactive moment during kicking. These results suggest that the specific muscle fatigue induced in the present study not only diminished the ability to generate force, but also disturbed the effective action of the interactive moment leading to poorer inter-segmental coordination during kicking. Moreover, fatigue obscured the eccentric action of the knee flexors immediately before ball impact. Interestingly, this might increase the susceptibility to injury.

However, initial sign of fatigue may be overlooked due to enormous compensatory strategies that athletes may employ to achieve the goal despite the fatigue. For example, a swimmer may switch from predominant arms to predominant legs work intensity to be able to maintain required speed of propulsion. Gymnasts often switch practices from "swinging" (i.e. high bar) to supporting (i.e. parallel bar) apparatuses to floor exercise in order to distribute work load and use different muscle groups to continue the workout despite fatigue. In other words, "re-allocation of resources" is one of the strategies to maintain the motor task productivity despite progressive muscle fatigue.

In fact, this "resources re-allocation" concept was supported by empirical research on progressive muscle fatigue (Johnston et al. 2001). Specifically, it was shown that the initial stage of muscle fatigue associated with non-significant reduced force along with increased the amount of muscle activation allowing production of the required force level. Interestingly, progressive muscle fatigue was also accompanied with significant enhancement of motor cortical activation (as revealed by the brain electrical activity, EEG) enabling the subject to accomplish the force production accuracy task despite fatigue. Considering this fact, it is reasonable to suggest that compensatory adjustment occurred at the central nervous system. On the one hand, it allowed the athletes to "keep going" to meet the sports/coaches requirements. On the other hand, this may put athletes at risk for injury due to exhaustion of both physical (i.e. central and peripheral) and psychological resources. In the following sections, first (a) current conceptualization of central and peripheral mechanism of progressive muscle fatigue; and second, (b) fatigue-injury relationship in athletic environment will be discussed.

2. MECHANISMS OF MUSCLE FATIGUE

2.1. Peripheral Fatigue

There at least three major reasons why muscle fatigue might occur following a sustained muscle contraction. Namely: (a) *failure in transmission* at the nerve-muscle junction, on in conduction at the fine terminal branches of motor axons, (b) *failure of the contractile machinery* in the muscle, and (c) *reduction in the central drive* to motoneurones below that necessary to sustain maximal muscle activity. The relative roles of these three phenomena depend on the type of contraction, the muscle group under study and some psychological factors (these will be considered in the appropriate section of this chapter, see also Rothwell, 1994).

Factors thought to be important in the development of peripheral fatigue during sustained muscle contractions include the depletion of muscle glycogen, which results in limiting the rate of adenosine diphosphate rephosphorylation, and the progressive loss of body fluids. This, in turn, results in increased cardiovascular, metabolic and thermoregulatory strain (Bergstrom et al., 1967). However, failure of neuromuscular transmission was believed to be a major factor in the development of fatigue for many years. In humans, such failure was demonstrated by supramaximal electrical stimulation of a motor nerve inducing a mass muscle action potential (the so-called M-wave) due to synchronous activation of all the units in the muscle (Merton, 1954). Repeated electrical stimulation after 20 seconds produced a significant decline in the amplitude of M-wave, indicating that action potentials are not longer generated in all units of the muscle under study. It should be noted however, that the amplitude of M-waves during supramaximal electrical stimulation of the muscle may not be equal to those during maximal voluntary contraction of the same muscles. This indicates that neuromuscular transmission may be retained during progressive muscle fatigue. But, fatigue is most probably caused by the failure in the production of contractile force (Rothwell, 1994).

In fact, the anaerobic metabolisms can cause the decline in contractile function of muscles during intensive exercises. For a long time, the accumulation of lactic acid was believed to be an important factor of muscle fatigue. However, recent discoveries challenged the role of lactic acid as an important cause of muscle fatigue. Reduced pH may have little effect on contraction of mammalian animal muscles. Inorganic phosphate rather than lactic acid appears to be a major direct cause in the decline of contractile functions of muscles. The increased pH may affect muscle fatigue in several ways. It may act directly on the myofibrils and decrease cross-bridge force production and myofibrillar $Ca2+$ sensitivity. The pH may also act directly on the SR $Ca2++$ release channels, increase their open probability and increase the tetanic $Ca2++$ in the early stage of fatigue, and inhibit the ATP-driven SR $Ca++$ uptake and reduce tetanic calcium in late fatigue by entering the SR, precipitating with $Ca2+$, and thereby decreasing the $Ca2+$ available for release. A more detailed discussion of the muscle contractile function may be found in other texts and far beyond the scope of the chapter.Peripheral muscle fatigue can be assessed by an experimental procedure using electrical stimulation-evoked twitch force (TF) elicited before (fresh state) and immediately after (fatigue state). The first classical study using TF procedure was performed by Merton in 1954 and later used by other researchers (e.g. Marsden et al., 1983) to study muscle fatigue. The stimulation is usually applied to the biceps brachii muscle, a major elbow flexor (of course, depending on the research protocol). The stimulus parameters (i.e. amplitude and frequency of stimulation) are kept the same before and after sustained motor task (SMT) and all twitch force values are normalized with respect to maximal voluntary contraction obtained from a subject before SMT. Since muscle fatigue is reliably indicated as a decline of its ability to generate force (see Gandevia, 2001 for review), the amount of twitch force reduction immediately after the SMT is considered as physiological index of peripheral muscle fatigue. Another line of research within the scope of peripheral fatigue is related with properties of, (a) group-III afferents that are sensitive to changes in both the mechanical state and the metabolic environment of the muscle, and (b) group-IV afferents that are most responsive to the chemical milieu in the muscle (Enoka & Stuart, 1992; Gandevia, 1998).

The common research methodology used in the studies of the influence of fatigue on the feedback delivered by group III-IV afferents is to compare the recovery of function, when blood flow is normal and when it is impeded. During ischemia, the metabolites that accumulate in the fatigued muscle will continue to stimulate the group III-IV afferents. Moreover, the depression of discharge rate for motor units in the biceps brachii after a sustained maximal voluntary contraction (MVC) did not recover during the 3 minutes that blood flow was occluded but did recover to control values within 3 minutes after blood flow was restored. This result may suggest that a peripheral reflex mediated by group III-IV afferents from the fatigued muscle contributed to the decrease in the discharge rate. It should be noted however that the central connections of group III-IV afferents can evoke diverse responses. Todd et al., (2007) investigated the contribution of feedback by group III-IV afferents to the fatigue experience during a sustained 2 minutes MVC with the elbow flexor and extensor muscles as the end-effectors. The research protocols involved comparing the amplitude of potentials evoked in muscle with TMS of the corticospinal tract during the MVCs and during recovery, when blood flow to the muscle was occluded and when it was not. When the fatiguing contraction was performed with the triceps brachii muscle, the amplitude of the evoked response decreased during the MVC and remained depressed during ischemia, but recovered within 15 seconds after the removal of ischemia. The amplitude of the evoked potentials in triceps brachii also decreased after the fatiguing contraction was performed with the elbow flexor muscles. In contrast, the amplitude of the evoked potentials in biceps brachii increased after a fatiguing contraction with triceps brachii. These results indicate that group III-IV afferents depressed the excitability of the motor neurons of triceps brachii, but facilitated those that innervate biceps brachii. Overall, the contribution of feedback from group III-IV afferents to the decline in motor unit activity during a fatiguing contraction probably differs for flexor and extensor muscle. Additional research is needed to further explore differential contribution of feedback from group III-IV afferents to progressive muscle fatigue.

2.2. Central Fatigue

The *central* fatigue is a form of fatigue that is associated with specific alterations of the CNS functioning that may influence, (a) the *central neural drive* to the muscles, (b) *mood* and *sensation* of effort, and (c) *tolerate pain* and *discomfort*. The notion that the central neural system (CNS) contributes in the development of muscle is not new. Pioneering work by Alessandro Mosso (1904) clearly demonstrated a reduced capacity to perform sustained muscle contractions following a mental activity/effort, resulting in the development of the term "mental fatigue" (cf: Meeusen et al., 2006).

The currently accepted *"central fatigue hypothesis"* is based on the assumption that during prolonged exercise the synthesis and metabolism of central monoamines, specifically serotonin, dopamine and noradrenalin (norepinephrine) are influenced. It was initially proposed by Newsholme et al. in 1987 that during prolonged exercise, increased brain serotonergic activity may augment lethargy and loss of drive, resulting in a reduction in motor unit recruitment at the peripheral level. This, in turn, may influence both physical and mental efficiency of the person performing the exercise, referred to as a central fatigue. While there have been many other neurobiological mechanisms proposed to explain reduced neural drive referred to as "central" fatigue, the neurotransmitter hypothesis has received the greatest recognition to date. In recent years, a number of studies have attempted to alter the central serotonin levels via various dietary supplementations with specific nutrients. This issue will be discussed elsewhere in this book. It is also possible that the interaction between brain serotonin and dopamine during sustained exercise could play a regulative role in the onset of fatigue (Davis & Bailey, 1997). Since dopamine was known to influence CNS functioning in terms of alteration of mood, memory, attention, and motivation, it is reasonable to accept the revised central fatigue hypothesis that an increase in the brain ratio of serotonin to dopamine is associated with feelings of tiredness and lethargy, accelerating the onset of fatigue. In contrast, a low ratio of serotonin to dopamine may favor improved performance via the maintenance of high levels of motivation and anxiety (Davis & Bailey).

While substantial support exists for metabolic and biomechanical changes in motor units recruitment and firing rates, changes in reflex mechanisms occur with muscle fatigue. Evidence has been provided as changes in cortical excitability (e.g. central fatigue). For example, to assess the excitability of the motor cortex, transcranial magnetic stimulation (TMS) has been applied to the brain and changes in both the motor evoked potential (MEP) in the muscle and the ensuing silent period (SP) seen in the EMG after an MRP has been examined (Taylor et al., 1996). More recently, the contribution of central and peripheral fatigue during sustained low intensity elbow flexion in healthy subjects using TMS has been reported (Sogaard, 2006). The sustained contraction induced both muscle and progressive central fatigue. This was reflected in subjects' inability to maximally contract the muscles. Surprisingly, the central fatigue recovered much faster (10min versus 25 min) compared to the peripheral fatigue as evidenced by recovery of resting twitch force. Despite apparent contradictions, the rather broad conclusion drawn from numerous TMS studies are that changes occurring in the MEP amplitude and SP duration associated with muscle fatigue are due to complex interactions of both excitatory and inhibitory processes in the motor cortex.

There are several lines of research aimed at investigating the origin of central fatigue, in general, and chronic fatigue syndrome (CFS), in particular (see Bailey, 2007 for review). Reduced post-exercise motor cortical excitability along with observed perfusion defects in the frontal and temporal lobes was documented. Evidence for central fatigue was also provided based on observed impairment of cerebral blood flow abnormalities in white matter signal intensity associated with slower reaction time. Central fatigue may be also studied by Electroencephalography, using both subdural electrodes and by means of surface-recording EEG signals (Johnston et al., 2001). For example, researchers used implanted electrodes in the motor cortex of monkeys and observed a direct link between motor cortical cells and motor unit activity as EEG post spike effects in the target muscle. This finding provided direct evidence regarding the involvement of motor cortical cells in motor unit recruitment during fatigue.

The changes in electro-cortical activity both in time and frequency domains associated with global and local muscle fatigue have also been observed in humans (Ivanova, 1990). Specifically, reduction of EEG alpha power (8-12 Hz) in fronto-central areas and an increase in amplitude of movement-related potentials was documented at early stages of progressive muscle fatigue. These EEG findings are consistent with other research on the effect of muscle fatigue on movement-related potentials (Freude etla., 1987) and more recently replicated by Johnston et al. (2001). A significant increase in motor-related cortical potentials was accompanying progressive muscle fatigue as evidenced by reduction of force production and reduced EMG responses. More recently, these EEG results were supported by fMRI findings indicating the contribution of central mechanisms to muscle fatigue (Liu et al., 2003).

Considering the fact that movement-related cortical potentials reflect cortical output neurons (Kristeva, 1990) that are recruited to strengthen the descending commands, it is reasonable to suggest that modulation of brain activation patterns reflect compensatory central mechanisms to overcome the muscle fatigue. However, it should be mentioned that a contradictory report by Shibata et al. (1997) claimed no increase of electro-cortical activation preceding muscle contraction. Shibata et al. used an arterial occlusion technique to induce both metabolic changes in the muscle and force deficits in task performance similar to that seen in muscle fatigue. The team reported a significant increase in brain activation during the maintenance phase of the isometric contraction associated with increased EMG activity under condition of arterial occlusion. It was suggested that changes in brain activation patterns under arterial occlusion may reflect the recruitment of additional motor units to compensate for the reduction of force production. Other researchers have also reported NO alteration of EEG signals associated with planning phase of force production, but reduction of alpha (8-12 Hz) and beta (14-18 Hz) during the sustained phase of force production (Lui et al., 2005a). In other words, these researchers suggested that maximal force production induced fatigue has differential effects on cortical signals during the motor task execution compared to its maintenance. Overall, regardless of directionality of changes in patterns of brain activation (increase versus decrease) there is convincing evidence in the literature that progressive muscle fatigue has central sources and mechanisms.

Additional information regarding the alteration of cortical activities during muscle fatigue was received from the fMRI studies by Liu et al. (2005b). These researchers observed that during a sustained (2-minute) maximal-effort handgrip, handgrip force and EMG signals declined in parallel during the course of muscle fatigue, the fMRI-measured BOLD signals first substantially increased and then decreased. The brain areas involved in such activities include primary sensorimotor areas and the secondary and association cortices (supplementary motor, prefrontal, and cingulate areas). This fMRI study shows the nonlinear changes of brain activation patterns that may reflect an early adjustment to strengthen the descending command for force-loss compensation and subsequent inhibition by sensory feedback as fatigue became more severe. It could explain at least to some extent why in TMS studies the excitability and inhibition in motor cortex both increase and the voluntary drives from the motor cortex continuously decreased during fatiguing exercise and did not recover after exercise if the blood flow was occluded.

There are limited reports in the literature on the cortical modulation of muscle fatigue and/or fatigue related cortical dysfunctions in clinical populations. For example, abnormal EEG findings (both event-related desynchronization "ERD" and event-related synchronization "ERS") have been reported by Zamarian et al. (2006) studying Parkinson patients. Similar findings have been reported by Babiloni et al. (2000) studying Alzheimer patients. Assuming that these clinical populations usually suffer from chronic fatigue syndrome (CFS), it is reasonable to suggest that EEG abnormalities observed in these patients may somehow reflect alteration of central drive to control muscle contractions. Abnormal cortical activation during fatiguing voluntary movement in Multiple Sclerosis (MS) was also reported by Leocani et al. (2001). Indeed, prevalent fatigue is the most common clinical symptom of patients suffering from MS. Normal controls and MS patients with and without fatigue symptoms performed self-initiated extensions of the thumb for prolong period of time. Post-movement ERS within high beta frequency band (18-22 Hz) was found to be significantly lower in the MS fatigue group compared to the MS non-fatigue patients. Moreover, ERD at the same frequency band (18-22 Hz) was more widespread in the frontal area of the brain for the MS fatigue group. Overall, it was suggested that the alteration of ERS/ERD patterns may be a result of inhibitory circuits acting on the motor cortex after termination of movement and/or decrease connectivity between cortical neurons causing reduced capability for producing synchronized oscillation of brain signals.

Depressive symptoms are the strongest predictor of fatigue in clinical population. However, fatigue cannot solely be explained by depression (Bower, 2005). Clearly, chronic fatigue is an important problem in cancer survivors. However, the extent to which the incidence and prevalence of fatigue among these patients differ from similar adults without cancer history remains an important question. A recent epidemiological study has reported that chronic fatigue, sometimes even in severe forms of exhaustion, along with persistent symptoms of depression are the most common consequences of cancer (Reyes-Gibby et al., 2006). In a prospective study of breast cancer patients undergoing radiation, pretreatment fatigue, depressed mood and anxiety scores were solid predictors of fatigue 2.5 years after treatment (Geinitz, et al., 2004). Similarly, women with higher levels of depressive symptoms in the five years after diagnosis were at greater risk for chronic fatigue, even after being able to control the initial fatigue (Bower et al., 2006). Overall, this link between depression and fatigue found in clinical populations should be of particular concern in athletes suffering from traumatic injuries and those who had experienced various forms of psychological trauma. Specifically, the stress athletes induced by traumatic injury may be associated with greater immune dysregulation, provoking the maladaptive effects of stress and depression on inflammation (there are current trends to propose that inflammation may be one mechanism underlying chronic fatigue in clinical populations). It should be noted that a number of health-related behaviors are associated with both fatigue and higher levels of proinflammatory cytokines, including pain, physical activity, body mass, sleep, diet, and comorbid health conditions (Bower, 2005).

3. FATIGUE-INJURY RELATIONSHIP

3.1. Injuries due to Central Fatigue

There are numerous well-documented reports and anecdotal facts clearly indicating both direct and indirect links between fatigue and various forms of injury both in athletes and non-athletes. For example, it is very common and well-known that a driver's fatigue is a major risk for road accidents that can often result in injury and death. However, considerable debate still exists concerning factors associated with driver fatigue. Recently, a study investigated both physiological and psychological determinants of the drivers' fatigue (Wijesuriya, et al., 2007). Three fatigue outcome measures were used, including a physiological, psychological, and a combined physiological and psychological measure. Significant factors associated with physiological fatigue included higher levels of baseline EEG delta activity (i.e. an index of generalized slowness of brain functions and sleepiness) and an extraverted personality.   Factors related to the psychological fatigue outcome measure included sleepiness, low healthy lifestyle status, an extraverted personality, tension-prone personality, and negative mood states.   The combined fatigue outcome measure was associated with factors such as a tension-prone and extraverted personality, low systolic blood pressure and negative mood states.

Another relevant recent study examined the prevalence of fatigue in an outpatient spinal cord injury population, as well as the additional clinical variables contributing to that fatigue (Fawkes-Kirby et al., 2007).   In this study, data was collected from 76 individuals admitted to the GF Strong Outpatient SCI Program (Vancouver, British Columbia) using the Fatigue Severity Scale (FSS).   A majority of these patients were admitted for medical reasons and had pain, spasticity, incomplete injuries and/or were on more that one medication with a known side effect of fatigue.   Fatigue among individuals with spinal cord injury who are seeking outpatient rehabilitation is very common.   The authors concluded that the severity of fatigue was greater for individuals with incomplete lesions.   Moreover, pain was also a potentially important covariate of fatigue.   The effect of fatigue on the functional recovery of patients with spinal cord injury is unknown at this time.

Signs of chronic fatigue are also prevalent in patients with traumatic brain injuries.   In a recent study, Bushnic et al. (2007) have evaluated the association between neuroendocrine findings (testing including thyroid, adrenal, gonadal axes and growth hormone (GH) after glucagon stimulation) and fatigue after traumatic brain injury using Fatigue Severity Scale (FSS) and Global Fatigue Index (GFI).   Higher GH levels were significantly associated with higher FSS scores.   There was a noted trend between lower basal cortisol and higher scores on both the FSS and GFI.   It was concluded that an observed relationship between higher GH levels and greater fatigue contradicted the prevailing hypothesis that post-acute TBI fatigue is associated with GH deficiency.   However, the relationship between lower basal cortisol and greater fatigue was in the expected direction.   Overall, the fatigue derived from neuroendocrine abnormalities alone may be masked by fatigue induced by other factors commonly experienced following TBI. Again, the exact contribution of GH deficiency to diminished quality of life post-TBI remains unclear.

Fatigue injury relationships can also be mediated by impaired cognitive functions associated with prolonged exercise leading to mental exhaustion. For example, the athletes' diminished ability to focus on key elements of drills, make fast and timely decisions, predict and anticipate forthcoming events, particularly at the end of practice/competition, may result in impairment of movement forms and technique, putting these athletes at high risk for injury. In fact, this was partly documented in a number of recent studies. For example, combined effect of fatigue and decision making on female athletes risk for the anterior cruciate ligament (ASL) injury due to improper landing has been shown by Borotikar et al (2007). Specifically, fatigue caused significant increases in initial contact hip extension and internal rotation, and in peak stance knee abduction and internal rotation and ankle supination angles. Fatigue-induced increases in initial contact hip rotations and in peak knee abduction angle were also significantly more pronounced during unanticipated compared to anticipated landings. Overall, the integrative effects of fatigue and decision making may represent a worst case scenario in terms of ASL injury risk during dynamic single leg landings, by perpetuating substantial degradation and overload of central control mechanisms.

Injury may be caused by the lack of anticipation and prediction of possible collision and/or impact with objects or players due to progressive central and peripheral fatigue. For example, warning prior to lifting helps to properly prepare appropriate muscle groups to prevent injury. It was shown that warning did not alter the level of trunk muscle activity prior to sudden loading when subject experienced progressive fatigue (Mawston et al., 2007). These findings indicate that warning prior to sudden loading may enhance postural responses, reduce ranges of joint motion and increase stability. However, the benefits of prior warning for reducing ranges of joint motion may not be present when a person is fatigued. In other words, fatigue may induce the athlete's ability to predict and anticipate the amount of impact and exact timing of landing (for example landing in gymnastics after multiple somersaults).

In fact, numerous ankle injuries in gymnasts are caused by improper landing. The author's numerous personal interviews with injured gymnasts clearly indicate that most of these injuries happened at the end of the practice sessions and were linked to fatigue and an unprepared landing on hard surface. The lack of prediction of the impact forces (i.e. cognitive appraisal and appropriate readiness to initiate proper preparation) during lifting the weight is another example of fatigue/cognition/injury link in athletes. In addition, fatigue generally increases initial and peak knee adduction, abduction, and peak knee internal rotation, with the later being more pronounced in female athletes (McLean et al., 2007). Specifically, female athletes usually landed with more initial ankle plantar flexion and peak-stance ankle supination, knee abduction and knee internal rotation compared to male athletes. Females also demonstrate greater quadriceps-hamstrings co-activation ratios than males, regardless of the fatigue condition. There are the muscle mass (or strength) and associated level of vascular occlusion, substrate utilization, muscle composition as well as neuro-muscular activation hypotheses to explain gender differences in muscle fatigability (see also Lariviere at al., 2006 for details).

Only females showed increased knee flexion at initial contact after fatigue during hopping/landing (Padua et al., 2006). Fatigue induced alteration of low-limb control during landing may be a contributing factor to alteration of knee stabilization resulting in higher risk of ASL injuries in female athletes. Several other EMG studies differentiated landing strategies in female athletes with low and high risk for ASL injuries (see Nyland et al., 1997 for review). Fatigued recreational athletes (especially females) have also demonstrated the altered motor control strategies, which may increase anterior tibial shear force, and cause strain on the anterior cruciate ligament (Chappell et al., 2005). Clearly, regardless of gender, constant monitoring for any signs of mental fatigue, especially at the end of the practice sessions may benefit athletes in terms of predicting/preventing injuries.

Lack of prediction of timing during drop landing due to fatigue may also alter the agonist/antagonist muscle co-activation patterns, which play an important role for stabilizing the knee joint (Kellis & Kouvelioti, 2007). Localized muscle fatigue (including contribution of central mechanisms) may alter the tibial response parameters, including peak tibial acceleration, time to peak tibial acceleration and the acceleration slope, measured at the knee during unshod heel impacts (Holmes & Andrews, 2006). These facts should be also considered while plyometric drop-jump training which is currently used in numerous sports. A recent study examining effect of fatigue on tibial impact properties and knee kinematics in drop jump have clearly shown that neuro-muscular fatigue induced by treadmill running protocol caused a significant increase in tibial impact acceleration and peak angular velocity in drop jump (Moran & Marshall, 2006). This, in turn, may considerably increase the risk for chronic and overuse injuries of low extremities.

The reduced leg swing speed, represented by a slower toe linear velocity immediately before ball impact and slower peak lower leg angular velocity, was most likely due to a significantly reduced resultant joint moment and motion-dependent interactive moment during kicking (Apriantono et al., 2006). These results suggest that the specific muscle fatigue induced in the present study not only diminished the ability to generate force, but also disturbed the effective action of the interactive moment leading to poorer inter-segmental coordination during kicking. Moreover, fatigue obscured the eccentric action of the knee flexors immediately before ball impact. This might ultimately increase the athletes' susceptibility to injury.

Finally, a sudden unexpected loading (i.e. lack of preparatory activities) and fatigue arising from manual handling practices in the workplace have been identified as contributing factors to the risk of low back injury (Mawston et al., 2007). Fatigue-induced reduction in active muscle stiffness necessitated increased antagonistic co-contraction to maintain stability resulting in increased spinal compression with fatigue. Also, fatigue-induced reduction in force-generating capacity limited the feasible set of muscle recruitment patterns, thereby restricting the estimated stability of the spine (Granata et al., 2004). In this study, Electromyographic (EMG) and trunk kinematics from healthy participants were recorded during sudden-load trials in fatigued and un-fatigued states. Empirical data supported the model predictions, demonstrating increased antagonistic co-contraction during fatigued exertions. Overall, warning prior to sudden loading may enhance preparation for postural responses, reduce ranges of joint motion and increase whole-body postural stability resulting in better control for fatigue-related injuries.

3.2. Abnormal Postural Control and Injury due to Fatigue

In the previous text we discussed abnormal postural control and balance deficiency as both, (a) *predictor for high risk* of injury, and (b) significant *factor of residual deficit* due to injury. In the following discussion it is an intention to provide additional evidence that progressive muscle fatigue may contribute to postural control abnormalities, ultimately putting athletes at high risk for injury. It should be noted that balance and stability is not restricted to whole body postural stability but may involve any joint(s) contributing to execution of movement.

Clearly, control for whole body and selective body segments for stabilization during movement is affected by intact proprioception and joint position sensitivity. Therefore it is feasible to suggest that sustained and prolong physical activity may reduce proprio-sensation along with joint position sensitivity. As a result, the lack of sensitivity due to fatigue may be a contributing factor for joint(s) instability and associated injuries. There are several recent studies in support of this assumption. Specifically, the contribution of impaired stretch reflex responses in conjunction with fatigue-related ankle injuries was warranted by Jackson et al. (2007) recent study. This finding clearly indicates that reducing dynamic stability of the ankle joint is due to alteration of reflexive response. Thus, the lack of sensitivity of ankle joint stability during landing (e.g. floor exercise in gymnastics) and/or impact with moving objects (catching the springboard in diving) may cause frequently observed ankle sprain injury in these sports. In another relevant study, the effect of muscular fatigue on knee joint proprioception was examined (Lattanzio et al., 1997). In this study, subjects were instructed to perform a two-legged squat to specific knee flexion angles. There was significant decline in proprioceptive function after the fatiguing exercise (e.g., 80% VO2max until maximal exhaustion). Interestingly, female athletes experienced more trouble to reproduce required range of knee motion under fatigue condition.

Similarly, the lack of sensori-motor control between upper extremity stability and mobility due to fatigue not only deteriorated overhead throwing movement in athletes, but also may be considered as a contributing factor to overused types of shoulder injuries (Tripp et al., 2007). These researchers reported that fatigue occurred after an average of 62 +/- 28 throws and increased 3-dimensional variable error scores (i.e. decreased acuity) of the entire upper extremity and all joints in both positions. Fatigue increased errors (ranging from 0.6 degrees to 2.3 degrees) at arm cock for scapulothoracic internal-external rotation, upward rotation, and posterior tilt, glenohumeral internal-external rotation and flexion-extension, elbow flexion-extension, wrist ulnar-radial deviation and at ball release for scapulothoracic internal-external rotation and upward rotation, glenohumeral horizontal abduction-adduction, elbow pronation-supination, and wrist ulnar-radial deviation and flexion-extension. Also, deceleration time was shown to be significantly increased by the fatiguing intervention in the Bowman et al., (2006) study. It was suggested that the decreased ability to decelerate may be an adaptive response by the subjects to dissipate a lower percentage of force per second. Although, this adaptive response may cause impaired overhead motion and ultimately lead to injury. Overall, this suggests that functional fatigue affects the acuity of the entire upper extremity, each individual joint and multiple joint motions in overhead throwers. Clinicians and coaches should definitely consider the deleterious effects of upper extremities fatigue when designing the injury prevention training and rehabilitation programs.

There are few direct reports suggesting the negative effect of fatigue on whole body postural control ultimately linking to injury in athletes. Specifically, baseline and post-fatigue postural stability scores were assessed using the isokinetic fatiguing contractions protocol (Harkins et al., 2005). Postural sway velocity, as an index of postural instability, was significantly higher when the 30% fatigue protocol (70% of decrease in strength) was introduced compare to 50% (50% of decrease in strength) fatigue protocol. In another relevant study, deficits in static postural control as a result of fatigue in conjunction with chronic ankle instability was examined (Gribble et al., 2004a). The Star Excursion Balance Tests clearly indicated disrupted dynamic postural control, most notably by altering control of saggital-plane angles proximal to the ankle as a function of progressive fatigue. These overall findings were replicated by Gribble & Hertel (2004b) using a single-leg stance research protocol. Specifically, fatigue about the hip and knee had great adverse effect on postural control properties. Moreover, it was shown that fatigue and deficits in postural control may predispose musculoskeletal injury, with greater effect of localized fatigue of the frontal plane movers of the hip compared to the ankle on maintenance of postural control in a single-leg stance (Gribble & Hertel 2004c). Clearly, more well-controlled research is needed to fully elaborate on moderating effect of fatigue on injury in athletes due to abnormal postural control.

CONCLUSION

Despite significant advances in understanding fatigue associated with sustained exercise, there are still controversies regarding the contribution of peripheral (pure muscular) and central (super-spinal factors) to progressive muscle fatigue. The rate of progress in the field of muscle fatigue has not matched expectations expressed in the early 1990s, since the paper published by Enoka & Stuart (1992) and the international symposium in Miami (Gangevia et al., 1995). Clearly, peripheral mechanisms including muscle glycogen concentration, transmission at the nerve-muscle junction and contractile machinery in muscles have been well-documented. However, significant progress associated with advanced brain imaging technologies, such as fMRI and TMS, clearly demonstrated differential contribution of various brain regions to this phenomenon. The question whether the mechanisms predominantly responsible for fatigue is located in the exercising muscle or somewhere else in the CNS remains to be answered. As was stressed by Nybo & Secher (2004 sf: Barry & Enoka, 2006), due to the mutual interaction of central and peripheral mechanisms, the dichotomy is not particularly useful and should be avoided.

Another point to be made regarding muscle fatigue is the "research methodology". As was mentioned in the previous text, subjective scales and numerical categories used to assess the contribution of central mechanisms are most often lacking ecological validity, they are confusing to patients and rarely match other biological and neural markers of fatigue. These numerous confounding factors, such as task dependency, subjects' perception of exhaustion, individual differences in terms of tolerance to pain, current mental status and nutritional factors may bias the results of "pure" experimentation of muscle fatigue.

That said, impairment of motor skill production, multiple compensations in terms of techniques used to achieve the desired goal due to progressive muscle fatigue, is a serious predisposing factor for injuries in athletics. There are more questions than answers, such as, (a) *how to reduce the risk factor for injury due to fatigue*, (b) *what are pure markers of threshold for progressive muscle fatigue* when it is time to stop the loading to prevent injury, (c) *how to reduce fatigability* of the muscle to meet the demands of current sports, and (c) *how to speed-up recuperation* of athletes between practice sessions, just to name a few. Indeed, multidisciplinary effort of exercise physiologists, psychologists, coaches and medical practitioners is needed to properly address these and many other important questions related to reducing the risk of injury in athletes due to fatigue.

On the final note, according to Dr. O'Brien, dealing with elite athletes, coaches and medical practitioners need to constantly monitor the fatigue level, including both muscle and mental fatigue to determine the training load, intensity and volume of proposed exercise programs. The lack of recuperation control may lead to the cumulative effect of fatigue which is a direct cause of injury in elite athletes.

REFERENCES

Rothwell, J. *Control of human voluntary movement*. Second edition. Chapman & Hill, 1994.

Gandevia, S.C., Enoka, R.M., McComas, A.J., Stuart, D.G., Thams, C.K. (Eds.), *Fatigue: neural and muscular mechanisms*. New York: Plenum Press, 1998.

Johnston, J., Rearick, M. Slobounov, S. (2001). Movement-related cortical potentials associated with progressive muscle fatigue in a grasping task. *Clinical Neurophysiology, 112*, 68-77.

Carpenter, J.E., Blasser, R.B., Pelizzon, G.G. (1998). The effects of muscle fatigue on shoulder joint position sense. *American Journal of Sports Medicine, 26*, 62-65.

Jaric, S., Radovanovic, S., Milanovic, S et al. (1997). A comparison of the effects of agonist and antagonist muscle fatigue on performance of rapid movement. *European Journal of Applied Physiology, 76*, 41-47.

Meeusen, R., Watson, P., Hasegawa, H., Roelands, B., Piacentini, M. (2006). Central fatigue: The serotonin hypothesis and beyond. *Sports Medicine, 36*)10), 881-909.

Enoka, R.M, Stuart, D.G. (1992). Neurobiology of muscle fatigue. Journal of Applied Physiology, 72, 1631-48.

Dimeo, F., Stieglitz, R.D., Novelli-Fischer, U., Fetscher, S., Mertelsmann, R., Keul, J. (1997). Correlation between physical performance and fatigue in cancer patients. *Annals of Oncology, 8*, 1251-1255.

Simmonds, M.J. (2002). Physical function in patients with cancer: Psychometric characteristics and clinical usefulness of a physical performance test battery. Journal of Pain Symptom Management, 24, 404-414.

Schmid, M., Schieppati M., & Pozzo, T. (2006). Effect of fatigue on the precision of a whole-body pointing task. *Neuroscience, 139* (3), 909-20.

Apriantono, T., Nunome, H., Ikegami, Y., & Sano, S. (2006). The effect of muscle fatigue on instep kicking kinetics and kinematics in association football. *Journal of Sports Science, 24*(9), 951-60.

Bergstrom, J., Hermansen, L., Hultman, E. et al. (1967). Diet, muscle glycogen and physical performance. *Acta Physiologica Scandinavia, 71*(2), 140-150.

Merton, P.A. (1954). Voluntary strength and fatigue. *Journal of Physiology, 123*, 553-564.

Marsden, C.D., Meadows, J.C., & Merton, P.A. (1983). 'Muscular wisdom' that minimizes fatigue during prolonged effort in man: peak rates of motoneuron discharge and slowing of discharge during fatigue. *Advanced Neurology, 39*, 169-211.

Gandevia, S.C. (2001). Spinal and supraspinal factors in human muscle 309-14 fatigue. *Physiological Review, 81* (4), 1725-89.

Todd, G., Taylor, J.T., Butler, J.E., Martin, P.G., Gorman, R.B., Gandevia, S.C. (2007). Use of motor cortex stimulation to measure simultaneously the changes in dynamic muscle properties and voluntary activation in human muscle. *Journal of Applied Physiology, 102*(5), 1756-1766.

Mosso, A. *Fatigue*. London. Swan Sonnenschein, 1904.

Newsholme, E.A., Acworth, I., Bloomstrand, E. (1987). Amino acids, brain neurotransmitters and a function link between muscle and brain that is important in sustained exercise. In: Benzi G (Ed). *Advances in myochemistry*. pp. 127-133. London: John Libbey Eurotext.

Davis, J.M. & Bailey, S.P. (1997). Possible mechanisms of central nervous system fatigue during exercise. *Medicine in Science Sports Exercise, 29*(1), 45-57.

Taylor. J.L., Butler, J.E., Allen, G.M., Gandevia, S.C. (1996). Changes in motor cortical excitability during human muscle fatigue. *Journal of Physiology, 490*, 519-529.

Sogaard, K., Gandevia, S.C., Todd, G., Petersen, N.T., Taylor, J.L. (2006). The effect of sustained low-intensity contractions on supraspinal fatigue in human elbow flexor muscles. *Journal of Physiology, 573*(Pt 2), 511-523.

Bailey, A. (2007). Chronic fatigue syndrome: a tiresome illness. British Journal of Nursing, 16(16), 967.

Ivanova, M. *Cortical control of voluntary movement.* Moscow: Academy of Science, 1990. (Russian).

Freude, G., Ullsperger, P. (1987). Changes in bereitschaftpotential during fatiguing and non-fatiguing hand movement. European Journal of Physiology, 56, 105-108.

Liu, J.Z., Zhang, L.D., Shan, Z.Y., Sahgal, V., Brown, R.W., Yue, G.H. (2003). Brain activation during sustained and intermittent submaximal fatigue muscle contractions. *Journal of Neurophysiology, 90*, 300-312.

Kristeva, R., Cheyne, D., Lang, W., Lindenger, G., Deecke, L. (1990). Movement-related potentials accompanying unilateral and bilateral finger movements with different inertial load. *EEG and Clinical Neurophysiology, 75*, 410-418.

Shibata, M., Oda, S., Moritani, T. (1997). The relationship between movement-related cortical potentials and motor unit activity during motor contraction. *Journal of Electromyography and Kinesiology, 7*(2), 79-85.

Liu, J.Z., Yao, B., Siemionow, V., Sahgal, V., Wang, X.F., Sun, J., Yue, G.H. (2005a). Minimal changes in cortical command but substantial declines in muscular output during severe muscle fatigue. *Brain Research, 1057*, 113-126.

Liu, J.Z., Zhang, L.D., Yao, B., Sahgal, V., Yue, G.H. (2005b). Fatigue induced by intermittent maximal voluntary contractions is associated with significant losses in muscle output but limited reductions in functional MRI-measured brain activation level. *Brain Research* 1040, 44-54.

Zamarian, L., Visani, P., Delazer, M., Seppi, K., et al., (2006). Parkinson's disease and arithmetics: the role of executive functions. *Journal of Neurological Sciences, 25*(248), 124-130.

Babiloni, F., Babiloni, C., Carducci, F. et al. (2000). Movement-related electroencephalo-graphic reactivity in Alzheimer disease. *Neuroimage, 12*(2), 139-146.

Leocani, L., Colombo, B., Magnani, G. et al. (2001). Fatigue in multiple sclerosis is associated with abnormal cortical activation to voluntary movement: EEG evidence. *Neuroimage, 13*(6Pt 1), 1186-1192.

Bower, J.E. (2005). Prevalence and causes of fatigue after cancer treatment: the next generation of research. *Journal of Clinical Oncology, 23*(33), 8280-2.

Reyes-Gibby, C.C., Aday, L.A., Anderson, K.O., Mendoza, T.R., Cleeland, C.S. (2006). Pain, depression, and fatigue in community-dwelling adults with and without history of cancer. *Journal of Pain and Symptom Management, 32*(2), 118-128.

Geinitz, H., Zimmermann, F.B., Thammm, R., Keller, M. Busch, R., Molls, M. (2004). Fatigue in patients with adjuvant radiation therapy for breast cancer: long-term follow-up. Journal of Cancer Research and Clinical Oncology, 130(6), 327-333.

Bower, J.E., Ganz, P.A., Desmond, K.A. et al. (2006). Fatigue in long-term breast carcinoma survivors: A longitudinal investigation. Cancer, 106(4), 751-8.

Wijesuriya, N., Tran, Y., Craig, A. (2007). The psychophysiological determinants of fatigue. *International Journal of Psychophysiology, 63* (1), 77-86.

Fawkes-Kirby, T.M., Wheeler, M.A., Anton, H.A., Miller, W.C., Townson, A.F., Weeks, C.A. (2007). Clinical correlates of fatigue in spinal cord injury. *Spinal Cord*

Bushnik, T., Englander, J., Katznelson, L. (2007). Fatigue after TBI: association with neuroendocrine abnormalities. *Brain Injury, 21* (6), 559-66.

Borotikar, B.S., Newcomer, R., Koppes, R., McLean S.G. (2007). Combined effects of fatigue and decision making on female lower limb landing postures: Central and peripheral contributions to ACL injury risk. *Clinical Biomechanics*. Sep 21.

Mawston, G.A., McNair P.J., Boocock, M.G. (2007). The effects of prior warning and lifting-induced fatigue on trunk muscle and postural responses to sudden loading during manual handling. *Ergonomics*, 1-14.

McLean, S.G., Felin, R.E., Suedekum, N., Calabrese, G., Passerallo, A., Joy, S. (2007). Impact of fatigue on gender-based high-risk landing strategies. *Medicine & Science in Sports & Exercise*, *39* (3), 502-14.

Larivière, C., Gravel, D., Gagnon, D., Gardiner, P., Bertrand Arsenault, A., Gaudreault, N. (2006). Gender influence on fatigability of back muscles during intermittent isometric contractions: a study of neuromuscular activation patterns. *Clinical Biomechanics, 21* (9), 893-904.

Padua, D.A., Arnold, B.L., Perrin, D.H., Gansneder, B.M., Carcia, C.R., Granata, K.P. (2006). Fatigue, vertical leg stiffness, and stiffness control strategies in males and females. *Journal of Athletic Training, 41* (3), 294-304.

Nyland, J.A., Caborn, D.N., Shapiro, R., Johnson, D.L. (1997). Fatigue after eccentric quadriceps femoris work produces earlier gastrocnemius and delayed quadriceps femoris activation during crossover cutting among normal athletic women. *Knee Surgery Sports Traumatology, Arthroscopy, 5* (3), 162-7.

Chappell, J.D., Herman, D.C., Knight, B.S., Kirkendall, D.T., Garrett, W.E., Yu, B. (2005). Effect of fatigue on knee kinetics and kinematics in stop-jump tasks. *American Journal of Sports Medicine, 33* (7), 1022-9.

Kellis, E., Kouvelioti, V. (2007). Agonist versus antagonist muscle fatigue effects on thigh muscle activity and vertical ground reaction during drop landing. *Journal of Electromyography and Kinesiology*.

Holmes, A.M., Andrews, D.M. (2006). The effect of leg muscle activation state and localized muscle fatigue on tibial response during impact. *Journal of Applied Biomechanics, 22*(4), 275-84.

Moran, K.A., Marshall, B.M. (2006). Effect of fatigue on tibial impact accelerations and knee kinematics in drop jumps. *Medicine & Science in Sports & Exercise, 38* (10), 1836-42.

Granata, K.P., Slota, G.P., Wilson, S.E. (2004). Influence of fatigue in neuromuscular control of spinal stability. *Human Factors, 46* (1), 81-91.

Jackson, N.D., Gutierrez, G.M., Kaminski, T. (2007). The effect of fatigue and habituation on the stretch reflex of the ankle musculature. *Journal of Electromyography and Kinesiology*.

Lattanzio, P.J., Petrella, R.J., Sproule, J.R., Fowler, P.J. (1997). Effects of fatigue on knee proprioception. *Clinical Journal of Sports Medicine, 7* (1), 22-7.

Tripp, B.L., Yochem, E.M., Uhl, & T.L. (2007). Functional fatigue and upper extremity sensorimotor system acuity in baseball athletes. *Journal of Athletic Training, 42* (1), 90-8.

Bowman, T.G., Hart, J.M., McGuire, B.A., Palmieri, R.M., Ingersoll, C.D. (2006). A functional fatiguing protocol and deceleration time of the shoulder from an internal rotation perturbation. *Journal of Athletic Training, 41* (3), 275-9.

Harkins, K.M., Mattacola, C.G., Uhl, T.L., Malone, T.R., McCrory, J.L. (2005). Effects of 2 ankle fatigue models on the duration of postural stability dysfunction. *Journal of Athletic Training, 40* (3), 191-4.

Gribble PA, Hertel J. (2004a). Effect of lower-extremity muscle fatigue on postural control. *Archives of Physical Medicine and Rehabilitation, 85*(4), 589-92.

Gribble, P.A., Hertel, J. (2004b). Effect of hip and ankle muscle fatigue on unipedal postural control. *Journal of Electromyography and Kinesiology, 14*(6), 641-6.

Gribble PA, Hertel J, Denegar CR, Buckley WE. (2004c). The Effects of Fatigue and Chronic Ankle Instability on Dynamic Postural Control. *Journal of Athletic Training, 39*(4), 321-329.

Gandevia, S.C., Allen, G.M., & McKenzie, D.K. (1995). Central fatigue: critical issues, quantification and practical applications. *Advanced Experimental Medical Biology, 384,* 281-294.

Barry, B., Enoka, R. (2006). The neurobiology of muscle fatigue: 15 years later. *Integrative and Comparative Biology, 47,* 465-473.

CHAPTER 5

NUTRITION AS A RISK FACTOR FOR INJURY IN ELITE ATHLETES

1. INTRODUCTION

During the past two decades there have been scientific breakthroughs in understanding the role human metabolism plays in exercise, physical performance and athletic injuries. Studies have shown that specific forms of dietary behaviors may potentially be linked to health benefits or problems and their association to athletic performance. The field of sports nutrition has indicated that athletes have greater demands for macro and micronutrients than inactive humans. These findings have dictated the dietary recommendations of individuals participating in sports. This innovative and intergraded science has shifted from practical studies investigating the effects of dietary restrictions and supplementation, to the direct investigation of the biochemical basis of specific nutritional demands for elite performance and injury mechanics. This review chapter is based on sports nutrition and its association with sports injuries. Various topics of the nutritional demands of exercise are reviewed in their biochemical and metabolic relationships to athletic performance.

As sport induced injuries are on the rise, sports medicine specialists and sports nutritionists have been trying to determine how nutrition is related to injury. On the contrary, many exercise physiologists believe "fuel is fuel" and it doesn't matter what comes in. As human performance becomes more advanced and elite athletes are becoming more dependent on their team nutritionist, it is becoming evident that proper nutrition is essential for proper performance during practice and competition. Specific nutrients are critically important for enhancing the quality of performance, conditioning, practice time, recovery from fatigue, and avoiding sports induced injuries. For an athlete, improving biomechanical performance and avoiding the disturbance of homeostasis by strenuous demands by their specific sport is crucial. Since athletes require more nutrients than the recommended daily allowances (RDAs), it is important that they not only eat a well-balanced diet consisting of carbohydrates, protein, fat, vitamins, and minerals, but meet the nutritional demands and supplementation required before and after rigorous exercise.

2. NUTRITIONAL SUPPLEMENTATION AND INJURY

2.1. Carbohydrates

Carbohydrate metabolism is one of the most energy contributing biochemical pathways. Although the exchanging of carbohydrates and lipids for fuel utilization plays a role in endurance exercise, regulating the consumption of these two substrates can be a determinant in exhaustion and fatigue times. Glycogen, liver and skeletal muscle glycogen is an obligatory substrate for sprinting, jumping, lifting and other high power outputs demands. Skeletal muscle glycogen becomes a limiting factor in strenuous exercise due to its limited storage amounts and rapid utilization. Fatigue occurs when glycogen levels are significantly depleted in active muscles. Strenuous exercise may also exhaust liver glycogen, leading to a low blood glucose concentration and impaired endurance performance. Therefore, an effective way to improve endurance while prolonging the effects of fatigue is to increase glycogen stores in liver and glycogen muscle prior to initiating prolonged exercises, such as a marathon and/or long distance swimming.

This may be achieved by ingesting carbohydrates, preferably monosaccharides or oligosaccharides, due to their rapid absorption and transport to peripheral tissues, either before or during exercise. For instance, it has been reported that glycogen stores can be increased by eating a low-carbohydrate diet for 3 to 6 days prior to competition. It was reported that following by a high-carbohydrate diet during the final 3 days results in the storage of 1.5 times more glycogen than normal (Evans, 1985).

2.2. Protein

L-carnitine in the form of a nutritional supplement is believed to have potentially positive effects on muscle injury. The lysine-derivative is synthesized in the body and is obtained from red meat. Its primary role is to transfer fatty acids across the mitochondria matrix for beta-oxidation of fatty acid. It also plays a role in synthesis of adenosine triphosphate (ATP). Volek et al. (2002) conducted a study on trained males who were supplemented with 2 grams of L-carnitine or a placebo prior and during a 4 day long training session. Muscle-damage markers were measured to evaluate the effects of L-carnitine. Results showed that the L-carnitine subjects experienced less DOMS, lower blood CK concentrations and less muscle damage according to MRI imaging.

2.3. Carbohydrates and Protein

With an estimated 80,000 Anterior Cruciate Ligament (ACL) tears in the United State and 56,000 of them occurring during sport, athletes are prone and unfortunately expected to experience just an injury. ACL damage is the major and prominent injury in most contact sports, such as football, basketball and ice hockey. Foremost, a rehabilitation specialist's top priority is to restore muscle mass, strength, and flexibility, along with a full range of motion. In conjunction with physical therapeutic interventions, oral supplementations have been recognized as rehabilitation sessions.

Muscle mass and strength is enhanced by increasing the protein synthesis to breakdown ratio in muscle. This is accomplished through resistance training, protein intake, as well as carbohydrate intake. If no nutrients are endogenously ingested, then the net protein balance will remain negative. A positive net balance is required for significant protein syntheses. Holm et al. (2005) conducted a twelve-week rehabilitation session on patients with ACL injuries who had experienced quadriceps and hamstring atrophy and weakening. The patients were supplemented with a mixture of protein and carbohydrate immediately after rehab sessions. Results conclude that there where greater hypertrophic responses in the distal region of the quadriceps muscle. This site of the large muscle is associated with knee-joint stability and maximal function.

2.4. Antioxidants

Mild traumatic brain injuries (MTBI), commonly referred to as concussions, are considered the least serious type of traumatic brain injury. Yet, repetitive MTBI can lead to neurodegenerative diseases by means of axonal injury and even death. Traumatic head injury studies have incorporated the use of animals to determine the effects of traumatic injuries. This is due to the importance of neuronal function and degeneration as factor for head injuries in humans.

One theory of damage is based on the effects of reactive oxygen species (ROS), which are a major cause of secondary neuronal tissue damage following physical head trauma. Research has indicated that ROS are capable of damaging or altering the expression of numerous genes, proteins and macromolecules (Halliwell & Gutteridge, 1999). ROS directly affect the brain following injury through oxidation of DNA and proteins, and inducing peroxidation of membrane lipids. ROS also activates pathways which lead to different cascades of cellular messengers and transcription factors that promote axonal inflammation and apoptosis. In addition, they are believed to be involved in the aging process causing neurodegenerative diseases such as Alzheimer's and Parkinson's.

A nutritional intervention currently being used to speed the recovery process, increase the oxygen concentration in the brain and prevent further damage after a traumatic head injury is the supplementation of antioxidants. Clinicians are now supplementing patients and rats in animal models with Vitamins A, C, E, all B vitamins, and minerals such as selenium and zinc to confirm their positive effects against free- radicals. Studies performed to prove the free-radical theory have determined that antioxidant enzymes, such as superoxide dismutase (SOD), catalase and glutathione peroxidase (Gpx) are altered in head-injured patients. The evidence suggests that polyunsaturated fats are a major constituent and substrate for lipid peroxidation, which results in free radicals. High levels of iron in the substantia nigra are shown to promote radical formation. Furthermore, the human brain contains significantly low concentrations of catalase, glutathione, glutathione peroxidase and vitamin E. The oxidative metabolism of dopamine also has the potential to generate radicals.

Neuronal tissue has cellular and extracellular defense mechanisms against ROS. These mechanisms including enzymes such as superoxide dismutase and glutathione peroxidase, catalase to reduce the overall load of oxidative stress (Halliwell & Gutteridge, 1999). Supplementation of large doses of Vitamin E, Vitamin C, Selenium (as part of glutathione peroxidase) and cysteine (as part of glutathione) have shown to protect axonal tissue from the further damage caused by reactive species.

2.5. Skeletal Muscle Injury

Most muscular injuries are caused by high counteracting forces and/or high resistance bouts. As the availability of athletic events increase at the amateur, high-school, collegiate and professional level, individuals are experiencing more injuries resulting in everyday muscular pain and impaired physical performance. In a society where the norm seems to be an "easy fix", many Americans are turning to nutritional supplementation to eliminate the negative feeling experienced due to muscle injury. Although there are hundreds of beneficial claims reported by consumers, many of these claims are due to placebo effects. The recommendations and scientific evidence of these supplements are being manipulated by the fitness industry and athletics media, causing false beliefs.

To date, there have been numerous studies on the effects of vitamins (antioxidants) on exercise-induced skeletal muscle injury. Damaged skeletal muscle can be measured by several direct and indirect markers. These markers include magnetic resonance imaging (MRI), a decreased range of motion (ROM), delayed-onset muscle soreness (DOMS), increased 3-methylhistidine in urine (3-MH), increased creatine kinase (CK) and lactate dehydrogenase (LDH) levels in the blood and increased inflammatory biomarkers such as C-reactive protein and interleukin-6 (Bloomer, 2004).

According to scientific experiments, the water-soluble vitamin C (ascorbic acid) and the fat-soluble vitamin E (alpha-tocopherol) have shown to effectively reduce certain signs and symptoms or degree of muscle injury, not eliminate it. Over a decade ago, Kaminsky and Boal (1992) studied the effects of ascorbic acid on delayed muscle soreness (DOMS). In this double-blind, cross-over study, subjects were supplemented with vitamin C or placebos for 3 days before and after their prescribed exercise. DOMS was measured by a visual scale. Results showed that DOMS was lower in the vitamin C individuals than the placebo group. Other experiments on vitamin E alone had negative results, meaning that the vitamin did not have significant effects on muscle damage and pain. However, after failing to conclude positive effects of vitamins C and E individually, researchers began to combine the two antioxidants to treat muscle injury. At the molecular level, vitamin C helps vitamin E maintain its reduced and most biologically active form. Bloomer et al. (2004) performed a study on untrained women who were supplemented a antioxidant mixture composed of 1 gram of vitamin C, 400 IU of vitamin E, and 90 micrograms of selenium or a placebo before and after their prescribed exercise. Results showed that both CK and DOMS levels were lower in the antioxidant group.

2.6. Creatine

Creatine supplementation in athletes has had a negative stigma. Creatine monohydrate has been associated with dehydration, bloating, water retention and kidney strain. Yet, no studies have concluded such results. Some studies demonstrate that creatine supplementation increases sprinting performance, fat-free mass and the rate of muscle contractions. Greenwood et al. (2003) examined the effects of creatine supplementation on cramping, dehydration and various musculoskeletal injuries in NCAA Division 1 football players during the 1999 season. Results indicated that athletes who utilized creatine experienced less cramping and injury than the non-creatine users. Creatine users also improved their work capacity. In another creatine-based study, Franaux et al. (2001) measured maximum power output of knee extensions in males and females. The results were also beneficial in creatine users. Mean power output in knee extensions was increased and time to fatigue was also improved in both sexes. However, females were significantly favored.

3. VITAMIN SUPPLEMENTATION
AND PERFORMANCE

Since vitamins do not directly provide immediate energy as carbohydrate and fats do, it would be controversial and highly debatable to ask the question "does vitamin supplementation improve performance?" The true question should be "do vitamin deficiencies impair athletic performance?" Exercise alone can increase an individual's resting metabolic rate (RMR). In contrary, diet alone decreases an individual's RMR. By increasing one's RMR, the demands for energy expenditure and protein synthesis are increased. Therefore, metabolic pathways that utilize B vitamins such as thiamine (B1), riboflavin (B2) and Pyridoxine (B6) are stressed and in turn, elevate vitamin demands of an athlete.

Studies displayed in table 1 indicate that active individuals tend to have inadequate vitamin statuses. This might be due to high energy demands and low nutrient dense foods. Nutritional researchers have performed studies on clinically vitamin depleted individuals to observe if vitamins deficiencies affect the ability to physically perform at maximal capacity. van der Beek et al. (1994) depleted men of thiamine, riboflavin and pyridoxine. Results showed a 12% decrease in maximal work capacity (VO2 max) and a 7% increase in blood lactate accumulation. Suboticance et al. (1990) assessed teenage boys with poor vitamin B6 and riboflavin levels. Their physical performance on a bicycle ergometer improved after being supplemented with vitamin B6 and riboflavin. With existing studies, it is safe to say that vitamin deficiencies negatively affect maximal work capacity and athletic performance.

4. ORIENTAL SUPPLEMENTS AND PEROFRMANCE

For many years now, Americans in the athletic world have had a sense of wonder about Asian dietary habits. Both how they eat and what they eat. Countries like China and Japan seem to lack the epidemics, such as cancer and obesity, occurring in the United States. Yet their elite athletes tend to follow that path in reference to fatiguing and recovery. Chen et al. (2002) studied the effects of Huangqi Jianzhong Tang, a medication used to treat peptic ulcers and convulsions in the ancient oriental regions, on fatigue time in athletes. The athletes were supplemented with the herb in powdered form and a placebo. Results confirmed that Huangqi Jianzhong Tang effectively reduced muscular fatigue in the herb group. Because of this, muscular strength also increased. The scientists believe Huangqi Jianzhong Tang reduces fatigue by increasing the oxygen uptake in aerobic systems.

Ma-huang, commonly called ephedra or ephedrine, has been used by the Chinese for centuries. Purposes have been for respiratory medications and weight loss through its thermogenic effects. Its alkaloid form stimulates the central nervous system. Ginseng is another commonly used herb. It is known for its adaptogenic effects on the sympathetic nervous system. It is believed to reduce stress, increases immunity, decrease the sensation of fatigue, and improve reaction time.

Incidence of Athletes Worldwide With Low Nutritional Intakes

Table 1. Studies conducted on athletes worldwide to determine if dietary recommendations were met.

Authors	Gender/Sport	Lower than Required Recommendation	Country
Clark et al.	Females/Soccer	Vitamin E, Folate, Copper, Magnesium	United States
Rico-Sanz et al.	Males/Soccer	Calcium	Puerto Rico
Rankinen et al.	Males/Skiing	Vitamin D and E, Zinc and Magnesium	Finland
Zieglar et al.	Females/Figure Skating	Folate, Iron, and calcium	Unites States
Papadopou lou et al.	Females/Volleyball	Calcium, Iron, folate, magnesium, Zinc, and Vitamin A and E	Greece
Beals et al.	Volleyball	Folate, B vitamins, vitamin C, Iron, Calcium, Magnesium, and Zinc	USA
Kim et al.	Females/Judo	Calcium and Iron	Korea

5. NUTRITION AND FATIGUE

Rothwell (1994) defined the term *fatigue* as both physical and mental exhaustion due to prolonged stimulation or exertion. Fatigue resulting from exercise and physical activity has been related to central and peripheral factors. These factors have been indentified to be influenced by the intensity and duration of the activity, the athlete's physiological status and nutrition intake. Most text books and studies concentrate on peripheral factors and known biochemical pathways that lead to fatigue, such as the depletion of muscle and liver glycogen, the inhibition of Calcium ions in muscle physiology by hydrogen proton from the accumulation of lactic acid, water loss and impaired electrical impulses in muscle during performance. Ovulation in females has also been associated with exercise fatigue (Drake, 2004). However, neurobiological factors that directly and/or indirectly influence central fatigue are less known and lack scientific proof (Nybo,

2004). Amino acids and 5-hydroxytryptamine (5-HT), known as serotonin, have been closely identified to be influential factors in causing central fatigue.

5.1. Serotonin (5-HT)

Fluctuations in serotonin levels in the brain have been involved in the control of sleeping patterns, arousal and clinical depression. 5-HT has been suggested to play a role in mental fatigue during physical activity. The first study to demonstrate that serotonin is influenced by exercise was published in the early 1960's by Barchas and Freedman (1963). After exhausting laboratory rats with exercise protocols, they determined an increased concentration of 5-HT in the rats' brains. Several other studies have indicated that 5-HT is released in the hippocampus and frontal cortex during and after intense exercise (Frusztajer et el., 1990; Kim et al., 2002; Kirby et al., 1995). The controversy about serotonin being involved in exercise is currently being debated.

However, the relationship between serotonin release and central fatigue during exercise has been supported by animal studies where animal brain 5-HT levels have been pharmacologically manipulated. Running performance in rats was improved by the administration of a peripheral 5-HT antagonist (Bailey, 2003). Experiments with human subjects are controversial. Some indicate the involvement of 5-HT in fatigue and some do not. Differences in pharmaceutical agents and their time of administering, differences in frequency, intensity and duration of exercise, or individual variation in neuroendocrine responses, might explain opposing results (Blomstrand, 2006).

5.2. Amino Acids

During the past decade, the ingestion of branched chain amino acids (BCAAs) and their influence on the transportation of tryptophan and 5-HT metabolism has been investigated. Several studies on BCAAs have focused on their plasma concentrations and increasing effects of free tryptophan. In theory, the supplementation of BCAAs, decrease the transport of tryptophan into the brain, 5-HT synthesis and the perception of fatigue (Bromstrand, 2006). The intake or administration of tryptophan is predicted to elevate free tryptophan levels and accelerate the perception of mental fatigue. Supporting the involvement of tryptophan was the administration of tryptophan to rats and horses by Farris et al. (1998) and Soares et al. (2002). The increased free and total blood tryptophan concentration in the animals was associated with a reduction in performance, thus supporting the involvement of tryptophan and 5-HT in fatigue. Blomstrand et al. (1997) conducted a study where human subjects were supplied with a mixed dose of BCAAs. Results concluded that perceptions of exhaustion and mental fatigue were decreased. In another study conducted by Mittleman et al. (1998), physical performance in a warm environment was evaluated as time to exhaustion. Human exhaustion was improved from 137 to 153 minutes. However, the debate between whether BCAAs should be supplemented with carbohydrates is still under research. BCAAs supplementation has also shown to be DOMS in females (Shimomura et al. 2006).

6. NUTRITION AND FEMALE ATHLETES

Since Title IX of the Education Amendments of 1972 (20 United States Code section 1681) was enacted in 1972 by the Office of Civil rights (NCAA), high school females have increased their participation in sports by 904% and college females have increased their participation by 456% (WSF, 2007). Although these statistics show advancement in our fight for gender equality in athletes, it also demonstrates a growing field of possible gender specific injuries and trends. A common diagnostic tool and unfortunate issue that female athletes are being regularly recognized as having is the famous female athlete triad. It is a syndrome composed of anorexic behaviors, amenorrhea and osteopenia, early stages of osteoporosis. The signs and symptoms include decreased appetite, irregular periods, stress fractures, re-injuries, chronic fatigue and impaired endurance.

The female athlete triad is commonly recognized in females who frequently are involved in disordered eating by restricting calories and protein (meat) in their diet, and often lack rest. Specific sports have physical requirements, such as obtaining a smaller body frame and less body fat. Gymnastics, track and field and competitive cheerleading fall under that category. Dieting is common in competitive dancers and results in amenorrhea (Warren, 1986). Frusztajer et al. (1990) conducted a study on ballet dancers with similar endocrine profiles to determine if stress fractures differed with respect to nutrition. Results indicated that dancers with stress fractures were more likely to consume diets meeting less than 85% of the RDA. This trend in dieting was to avoid high fatty foods while consuming more low-calorie substitutes.

Several studies have also shown that female athletes with higher incidences of stress fractures have been associated with lower intakes of calcium and less frequent usage of oral contraceptives (Kiningham, 1996). The importance of calcium is observed in bone metabolism. While the recommended dietary allowance of calcium is 1,200 to 1,500 milligrams per day for females within the age range of 11 to 24 years of age, national surveys of teenage females have indicated that average daily intakes of calcium if less than 900 milligrams per day. Benefitting calcium metabolism is vitamin D. An additional daily supplementation of 400 to 800 international units (IU) is believed to facilitate the absorption of calcium (National Institute of Health). Vitamin K supplementation has also shown to improve bone metabolism in amenorrhoeic female athletes (Craucuin, 1988). When analyzing the female triad, one must strongly consider the culture and background of the athletes. Okano et al. (2005) concluded there is a lower prevalence of disordered eating in Chinese female athletes than Japanese. Still, both nations have significantly lower indices of eating disorder than that of the United States. This is partly explained by the lack of socio-culturally desire to be thin and low frequency of dieting during their lifetime.

CONCLUSION

There is no doubt whether nutrition plays a crucial role in athletic performance. Research has proven that macromolecule and vitamin metabolism is intertwined with elite performance. If such scientific fact was debatable, Olympic and professional teams would not be competing with highly trained and educated nutritionists on their sidelines. Manipulating what metabolic fuels to utilize, which cellular components to improve and which recovery substances to ingest has allowed physically active individuals to enhance their quality of athletic performance and better prevention of injury at the physical and psychological levels. In general, the purpose is to provide an athlete with easily digestible food to provide adequate energy and carbohydrates for glycogen replacement and ensure proper recovery. Introducing proper nutrition and its contribution to training and optimal performance should be a top priority for all coaching staff. Especially in sports requiring low body weight and a lean body composition such as gymnastics, cross-country, diving and swimming and competitive cheerleading and dance. Vegetarian athletes must comprehend the reality of being at risk due to their low protein intakes and low-density foods consumption. In regards to the female triad, some of its components are often undetected and misdiagnosed. Prevention is crucial and recognizing the syndrome's risk factors is a key. Clinicians should be cautious with symptoms such as anemia, electrolytes imbalances, fatigue and signs of depression. One must also understand that young athletes, especially female athletes, are prone to adapt maladaptive eating behaviors when body image and composition are emphasized by their coaches or instructors. Therefore, diagnosing such problems may play a role in managing this condition. For optimal treatment of the female triad, physicians and other professionals with subspecialties in nutritional counseling should not be the only individuals running the rehab plan. Coaches, trainers and family members closest to the athlete should also partake in the process. Although human performance can be benefitted through nutrient supplementation, an athlete's heredity is still the greatest factor in determining elite, and most importantly, injury-free performance.

Acknowledgments

This chapter was written by Julio Cezar Gomes, Penn State undergraduate student with double major in Kinesiology & Nutritional Science under the author's direct supervision and guidance.

REFERENCES

Evans, W.J. & Hughes, V.A. (1985). Dietary carbohydrates and endurance exercise. *American Journal of Clinical Nutrition, 41*, 1146–1154.

Volek, J.S., Kraemer, W.J., Ruben, M.R., et al. (20020). L-carnitine L-tartrate supplementation favorably affects markers of recovery from exercise stress. *American Journal of Physiology-Endocrinology and Metabolism, 282*, E474-82.

Holm, L., Esmarck B., Mizuno, H., Hansen H., Suetta, C., Holmich, P., Krogsgaard M., Kjaer M. (2005). The Effect of Protein and Carbohydrate Supplementation on Strength Training Outcome of Rehabilitation.

Halliwell, B., Gutteridge, J.M.C. *Free Radicals in Biology and Medicine*, third ed. Oxford University Press, Oxford, 1999.

Bloomer, R.J., Goldfarb, A.H., McKenzie, M.J., et al. (2004). Effects of antioxidant therapy in females exposed to eccentric exercise. *International Journal of Sport Nutrition and Exercise Metabolism, 14*, 377-88.

Kaminsky, M., Boa, L. R. (1992). An effect of ascorbic acid on delayed-onset muscle soreness. *Pain, 50*, 317-21.

Greenwood, M., Kreider, R.B., Greenwood, L., Byars, A. (2003). Cramping and Injury Incidence in Collegiate Football Players Are Reduced by Creatine Supplementation. *Journal of Athletic Training, 38*(3), 216-219.

Franaux., M., Louis, M., Sturbois, X., & Poortmans, J. R. (2001). Effects of creatine supplementation in males and females. *Medicine and Science in Sports and Exercise, 33*, 5.

van der Beek, E.J., van Dokkum, W., Wedel, M., Schrijver, J., van den Berg, H. (1994). Thiamin, riboflavin and vitamin B6: impact of restricted intake on physical performance in man. *Journal of the American College of Nutrition., 13*(6), 629-640.

Suboticanec, K., Stavljenić, A., Schalch, W., Buzina, R. (1990). Effects of pyridoxine and riboflavin supplementation on physical fitness in young adolescents. International Journal for Vitamins and Nutrition Research., 60(1), 81-88.

Chen, K.T., Su, C.H., Hsin, L., Su, Y. (2002). Reducing fatigue of athletes following oral administration of Huangqi Jianzhong Tang. *Acta Pharmacologia, 23*(8), 757-761.

Clark, M., Reed, D.B., Crouse, S.F., et al. (2003). Pre- and post-season dietary intake, body composition,and performance indices of NCAA division I female soccer players. *International Journal of Sport Nutrition and Exercise Metabolism, 13*(3), 303–319.

Rico-Sanz, J., Frontera, W.R., Mole, P.A., et al. (1998). Dietary and performance assessment of elite soccer players during a period of intense training. *International Journal of Sport Nutrition, 8*(3), 230–240.

Rankinen, T., Lyytikainen, S., Vanninen, E., et al. (1998). Nutritional status of the Finnish elite ski jumpers. *Medicine & Science in Sports & Exercise, 30*(11), 1592–1597.

Ziegler, P.J., Jonnalagadda, S.S., Nelson, J.A., et al. (2992). Contribution of meals and snacks to nutrient intake of male and female elite figure skaters during peak competitive season. *Journal of the American College of Nutrition, 21*(2), 114–119.

Papadopoulou, S.K., Papadopoulou, S.D., Gallos, G.K. (2002). Macro- and micro-nutrient intake of adolescent Greek female volleyball players. *International Journal of Sport Nutrition and Exercise Metabolism, 12*(1), 73–80.

Beals, K.A.. (2002). Eating behaviors, nutritional status, and menstrual function in elite female adolescent volleyball players. *Journal of the American Dietary Association, 102*(9), 1293–1296.

Kim, S.H., Kim, H.Y., Kim, W.K., et al. (2002). Nutritional status, iron-deficiency-related indices, and immunity of female athletes *Journal of the American Dietary Association* 18(1), 86–90.

Rothwell, J. (1994). *Control of human voluntary movement.* Second edition. Chapman & Hill.

Drake, S.M., Evetovich, T.K., Eschbach C., Webster, M., Whitehead, T. (2004). The effect of menstrual cycle on electromyography and mechanomyography during fatigue. *Medicine and Science in Sports and Exercise, 36,* 5.

Nybo, Z. & Secher, N.H. (2004). Cerebral perturbations provoked by prolonged exercise. *Progress in Neurobiology, 72,* 223-261.

Barchas, J.D., Freedman, D.X. (1963). Brain amines: response to physiological stress. *Biochemical Pharmacology, 12,* 1232-1235.

Frusztajer, N.T., Dhuper, S., Warren, M.P., Brooks-Gunn, J. & Fox, R.P. (1990). Nutrition and the incidence of stress fractures in ballet dancers. *American Journal of Clinical Nutrition, 5*(1), 779-783.

Kirby, L.G., Allen, A.R., Lucki, I. (1995). Regional differences in the effects of forced swimming on extracellular levels of 5-hydroxytryptamine and 5-hydroxyindole-acetic acid. *Brain Research, 682,* 189-196.

Bailey, S.P., Davis, J.M., Ahlborn, E.N. (1993). Serotonergic agonists and antagonists affect endurance performance in the rat. *Internaitonal Journal of Sports Medicine, 14,* 330-333.

Blomstrand, E. (2006). A Role for Branched-Chain Amino Acids in Reducing Central Fatigue1-3. *The Journal of Nutrition, 136*(2), 544-548.

Farris, J.W., Hinchcliffe, K.W., McKeever, K.H., Lamb, D.R., Thompson, D.L. (1998). Effect of tryptophan and of glucose on exercise capacity of horses. *Journal of Applied Physiology, 85,* 807-816.

Soares, D.D., Lima, N.R., Coimbra, C.C., Marubayash,i U. (2002). Evidence that tryptophan reduces mechanical efficiency and running performance in rats. *Pharmacological Biochemistry Behavior, 74,* 357-362.

Blomstrand, E., Hassmén, P., Ek, S., Ekblom, B., Newsholme, E.A. (1997). Influence of ingesting a solution of branched-chain amino acids on perceived exertion during exercise. *Acta Physiologica Scandinavica, 159,* 41-49.

Mittleman, K.D., Ricci, M.R., Bailey, S.P. (1998). Branched-chain amino acids prolong exercise during heat stress in men and women. *Medicine in Science, Sports & Exercise, 30,* 83-91.

Shimomura, Y., Yamamoto, Y., Bajotto, G., et al. (2006). Nutraceutical effects of branched-chain amino acids on skeletal muscle. *Journal of Nutrition, 136*(2), 529S-32S.

www.womenssportsfoundation.org

www.ncaa.org

Warren, M.P., Brooks-Ounn, J., Hamilton, L.H., Warren, L.F., Hamilton, W.O. (1986). Scoliosis and fractures in young ballet dancers: relation to delayed menarche and amenorrhoea. *New England Journal of Medicine, 314,* 1348-1353.

Kiningham, R.B., Apgar, B.S., Schwenk, T.L. (1996). Evaluation of amenorrhea. *American Family Physician, 53*, 1185-1194.

National Institiute of Heatlth Consensus conference. (1994). Optimal calcium intake. NIH Consensus Development Panel on Optimal Calcium Intake. *JAMA, 1272*, 1942-1948.

Cracuin, A.M., Wolf, J., Knapen, M.H., Brouns, F., Vermeer, C. (1988). Improved bone metabolism in female elite athletes after vitamin K supplementation. *International Journal of Sports Medicine, 19*, 479-484.

Okano, G., Holmes, R.A., Mu, Z., Yang, P., Lin, Z., Nakai, Y. (2005). Disordered Eating in Japanese and Chinese Female Runners, Rhythmic Gymnasts and Gymnasts *International Journal of Sports Medicine, 26*, 486-491.

PART II: COACHES AND ATHLETES' PERSPECTIVES OF INJURY

CHAPTER 6

INJURY IN ATHLETICS: COACHES' POINT OF VIEW

1. INTRODUCTION

There are numerous causes and a variety of physical, behavioral and psychological consequences of athletic injuries. Coaching errors are commonly cited as one of the major causes of athletic injuries. Generally speaking, there are two types of coaching styles that make a tremendous impact on the physical and psychological atmosphere in the training/competition environments. As identified by Greg Louganis, three time Olympic Champion, the first one is judgmental and critical, which is characterized by the situation when coaches are trying hard (maybe with good intention) to identify as much error in performance as possible, then present these errors in a critical manner. "You are not listening..., how many times have I told you to keep your eyes on the ball..., you are still not doing that..., you'll never get this, why don't you try to play golf instead...," to name just a few examples of this coaching style. It could be expected that tremendous tension can be anticipated in the coach-athlete relationship over time. Inevitably, the breaking point will be reached and the relationship will end due to a deficient coaching style. (Personal communication with Greg Louganis, 2007 US Diving National Training Camp, Indianapolis, Ind.)

The second style of coaching is characterized by the atmosphere where the coaches observe and assess an athlete's performance, with the intention to identify both progress and still existing errors in performance. "Good effort..., right direction to think..., you should feel better than yesterday..., keep trying.., still losing contact with the ball...," to name just a few comments within this style of coaching. Not surprisingly, this second coaching style creates an extremely positive learning/training environment benefiting both the physical and mental well-being of the athletes.

In addition to issues with the overall coaching styles identified above, there are several fundamental coaching problems that directly and/or indirectly cause high risk for injury in athletic environments. Specifically, inadequate assessment of athletes' physical skills, misunderstanding of psychological coping resources, rushing with acquisition of new techniques, overtraining and overloading causing accumulated muscle and mental fatigue, and early return to sport participation after injury are just a few examples of coaching errors that increase the risk of injury/re-injury in athletics. The following text contains some common coaching mistakes identified by clinical sport practitioners and psychologists, including:

Overreacting when the athlete makes an error or does not perform according to the coach's instructions and expectations. It should be noted that there are at least two types of errors of overly observed performance: (a) error of execution of skill, when a clear picture of an ideal performance is intact but physical capacities are not adequate and/or sufficient enough to meet high demands of performance; and (b) error of planning, when a clear image of an ideal performance is lacking due to inadequate cognitive assessment of the situation and demands, but the athlete's capacity in terms of strength, flexibility and endurance is intact. Accordingly, clear identification of overall performance is critical to avoid coaches *overreacting*, which in fact an indication of inefficient judgment and/or too critical a style of coaching.

Demanding too much time or commitment from athletes so that they are continually injured. Clearly to be competitive at the collegiate level to say nothing of Olympic level, it is necessary to allocate enormous amounts of time and effort to practice and work outs. For example, currently elite Chinese Olympic divers practice on average eight hours a day, six days per week. My personal observation of the Chinese divers' training session during the 2007 Diving World Series revealed that they performed on average 120 competitive dives per day in addition to post-practice conditioning training for almost two hours. Suffice it to say that this observation took place during the high peak competitive season; one may wonder how many dives the Chinese divers perform and/or for how many hours they practice during the pre-competitive preparatory phase of training program. Thus, in order to be competitive in the international sporting arena, coaches must demand that athletes commit most of their time to sport. One thing to consider is all the competing, confounding factors (i.e., financial compensation, professional versus collegiate status, cultural/societal differences, etc) before demanding excessive time and/or commitment to sport to meet its growing demands.

Relentlessly putting a high amount of pressure on the athletes, causing every practice to become a life-or-death situation or requiring that athletes are constantly at boundary level of being over-trained and/or burn-out. There is common "goal-oriented philosophy" that sport psychologists call "outcome orientation". Rarely do coaches encourage the amount of effort an athlete puts into action to accommodate his/her coaches' demands. Accordingly, so-called "mastery orientation," which emphasizes the importance of the process of skill acquisition and step-by-step improvement, is a rare practice in a coaching environment. As Greg Louganis believes, "... *it is not necessary to be perfect every time performing the dives both at practice and even at competitions. Everybody makes mistakes, nobody is perfect. There is always room for improvement and this is an endless process. Thus, coaches should focus on progress, even non-significant, rather than on ideal performance routines. This coaching style may reduce a lot of pressure even if the demands of the sport are exhausting*." (Personal communication, August 2007, US Diving Training Camp, Indianapolis).

*Specifically for collegiate athletics, **NOT respecting** that student-athletes need to have **balance in their life** – time for school, work, family, friends and rest. Despite the regulation imposed by NCCA which restricts the time for practices/competitions, and promotes the balanced life and well-being of student-athletes, external demands/expectations as well as the self-imposed high achievement attitude of collegiate athletes may put these athletes at risk for burn-out and injuries. Sporadic evidence and numerous observations suggest that the transition from high school to college frequently induces a higher risk for injury due to a disrupted life-style. This factor should be taken into consideration, especially when dealing with freshman collegiate athletes who just left their home, parental support, friends, etc. As Penn State gymnastic coach Randy Jepson stressed, "*... it is really difficult to work with incoming freshman athletes. They do not have realistic expectations and do not have enough confidence in us as coaches. It takes time to develop a relationship and demonstrate that our major concern is their safety and well-being, not what they can or cannot do in the gym.*"

*Overemphasizing **body weight**, especially in complex coordination sports with female athletes can lead to self-image problems and even serious eating disorders. This is a tremendous concern not only for females but also for male athletes. My personal view is as follows: it is not body weight that should be used as a possible limiting factor for performance enhancement but rather an athlete's ability to control his/her body and possess enough strength, flexibility and endurance to properly perform all necessary drills and routines. It is more important to emphasize healthy nutrition and eating habits than to be preoccupied with body weight. Coaches must focus on fitness and body shape rather than body weight. That stated, it is important to note that if any signs of eating disorders are present this should be considered as a serious clinical problem requiring treatment by specialists. It is essential that coaches acquire the knowledge to be able to detect signs of eating disorders in order to prevent the possibility of catastrophic consequences.

***Mistreating** the athletes for being lazy, not trying hard enough or not placing high enough and dragging the whole team down the slope is another detrimental style. Athletes are human and should be treated accordingly. As stressed by coach Jepson, "*People should be treated as people, not like machines and robots having a goal to be best in their sport. I treat gymnasts on my team as people and sport is just a part of their life.*" Truly, this coaching philosophy may be an instrumental part of the success of the Penn State Men's gymnastic program, which won three National Championships over the last ten years under the leadership of coach Randy Jepson.

Losing perspective on the whole purpose of sports and being completely preoccupied with winning at any cost and putting athletes under tremendous pressure and stress. As stressed by the coach Ron O'Brien, who has been coaching multiple Olympic Champions for more than 40 years, "… *there are too many unfortunate circumstances that are far beyond our control influencing a winning versus losing situation. So, why bother? Wouldn't it be better to spend our time more productively focusing on technical details and ways to improve an athlete's skill and experience?*" This so called "mastery orientation approach" is rewarding and a mutually beneficial approach from both athletes' and coaches' perspectives. This motive is a driving force of numerous great coaches who think of the well-being of the athletes. Similarly, as clinicians, we are not treating generalized anxiety disorder, or ACL per se, but rather we are taking care of people with physical and/or mental problems.

Another important problem to consider in terms of the coaches' role in prevention of injury in athletes is the "**knowledge-attitude relationship issue**". Clearly, there is a differentiation of function in modern sports. Coaches are coaching, fitness experts are taking care of athletes' conditioning and doctors are involved if injuries happened. Not surprisingly, the primary concern of coaches is to stay current in the areas directly related to performance enhancement. There are different sources to improve professional qualification of coaches. Not surprisingly, huge problems appear when *coaches' ignorance meets with arrogance*. That said, the content of the coaches' knowledge should be spread to some extent to understand the causes and consequences of injuries in athletes. As was described in the previous chapters of this book, there are common overuse injuries in certain sports due to their nature that could be predicted or maybe prevented by using appropriate training programs. For example, the knowledge of the causes of chronic ankle dislocations in gymnastics (i.e., improper landing after somersaults) may develop the proper attitude in the coaches, who would then employ specific exercise routines to prevent this problem.

There is some relevant research examining the relationship between knowledge and attitude in other domains. A few previous studies (Simonds, 2005; Sefton, 2003; Livingston & Ingersoll, 2005; Kaut et al., 2004) have examined the knowledge of and/or attitudes about concussion in college athletes, and only one study has examined the prevalence of the underreporting of concussion in athletes (McCrea et al., 2004). Although the specifics of these studies can be found in greater detail elsewhere in the relevant literature, the findings generally show that athletes' knowledge and attitudes about concussion vary. There are some domains that athletes are proficient in and others that they appear to be deficient in (Simonds, 2005; Sefton, 2003; Livingston & Ingersoll, 2005). Additionally, there is a subset of athletes whose attitudes about concussion appear to be unsafe (Simonds,

2005; Sefton, 2003). Finally, the behavior of a significant portion of athletes in reaction to concussion appears to be congruent with this unsafe attitude, as about half of the high school athletes surveyed by McCrea et al. (2004) failed to report a concussion after it occurred. Thus, although there has been some data from various studies about concussion knowledge, attitudes, and risky behaviors (i.e., playing while experiencing the symptoms of concussion), there have been no studies conducted that have examined the relationships between concussion knowledge, attitudes, and risky behaviors occurring after concussion.

Research focusing on "knowledge-attitude" relationships within the scope of prevention may be sufficiently similar to known literature pertaining to sexual behavior in adolescents. For example, Kirby (2002) reviewed the literature pertaining to outcomes of educational interventions intended to decrease unprotected sex and pregnancy rates. The researcher reviewed seventy-three studies conducted in the United States and Canada that were conducted using adolescents ages 12-18. Each study evaluated the effectiveness of a sex and/or HIV education program. These programs were intended to increase knowledge, alter attitudes and influence behavior. A number of indicators of program effectiveness were examined in the review: initiation of intercourse, frequency of sexual intercourse, number of sexual partners and contraceptive/condom use.

Participants in the studies were exposed to a particular intervention and were followed up for a period of time after the end of the intervention. The results of the interventions were promising in that about one-third of the programs measuring the initiation of intercourse contributed to a delay in the initiation of intercourse. No changes in initiation of intercourse were seen in most of the other studies. About 25% of programs resulted in reduced frequency of sex, and most of the other studies showed no change in the frequency of sexual intercourse. In 30% of the studies examining the number of sexual partners, the number of partners decreased. As with other outcome indicators, no changes were seen in the other studies in the frequency of sexual intercourse. Participants in about two-thirds of the studies that measured changes in condom use after the end of the educational program reported significant increases in condom use, and most of the other studies reported no change in condom use. It should also be noted that the vast majority of these seventy-three studies either reported reductions in risky sexual behavior or no change in the frequency of the sexual practices. The lack of change in sexual practices is considered to be an indicator of an effective educational program because it is believed that the knowledge gained by students likely contributes to their decision to maintain their current sexual practices, rather than increasing the likelihood of pregnancy or the transmission of HIV due to an increase in number of sexual partners or an increase in unprotected sex (Kirby, 2002).

This review begins to establish a link between the knowledge gained from the educational programs and the influence that this information may have on the diminution or stabilization of potentially unsafe and, therefore, risky sexual behaviors.   However, because the study lacked direct information about the relationships between knowledge, attitudes and behaviors, the nature of these relationships is unclear.

An additional study was conducted to address this relationship and further establish a connection between sex and HIV education programs and changes in risky sexual behavior (Kirby, Laris, & Rolleri, 2007).   The researchers reviewed eighty-three studies, many of which were conducted in the United States.   Moreover, several studies were conducted in other developed and underdeveloped countries. Thus, this study included a more diverse and, potentially, representative sample of adolescents and young adults. The age range of participants involved in the studies included in the literature review was between nine and twenty-four years of age.

The findings pertaining to risky sexual behavior were roughly similar to those found in the Kirby (2002) review. More importantly, almost all of the studies examining knowledge about sexual intercourse and safe sex practices reported increases in knowledge and about two-thirds of the studies examining attitudes about sex reported safer attitudes towards sex. About 40% of the studies examining student perceptions of sexual behavior in peers (an influential factor in the decision making of adolescents; Millstein & Halpern-Felsher, 2002) reported that students tended to believe that their peers were engaging in safer sexual behaviors. The authors concluded that the educational programs tended to increase knowledge and alter attitudes about sex and change perceptions of peer sexual behavior. These cognitive and emotional changes then influenced changes in risky behavior (Kirby et al., 2007).

These literature reviews suggest that the likelihood of engagement in a risky behavior is associated with the knowledge that an individual possesses about that behavior. Furthermore, the evidence points to a link between an individual's attitudes about risky behavior and engagement in the behavior. Although it is unclear as to how knowledge and attitudes about sport-related injuries affect the likelihood of reporting an injury upon recognition of the symptoms or how these factors influence adherence with medical treatment recommendations after injury, it seems possible that either knowledge or attitudes, or both, may lead to higher injury reporting rates and greater adherence among injured athletes. Overall, it is the responsibility of clinical practitioners to expand the athletes' and coaches' knowledge of the causes of athletic injury.

2. INTERVIEW WITH ELITE COACHES

Personal interviews with several world class Olympic coaches as well as with collegiate coaches were conducted, and aimed to explore their perspectives on causes and consequences of injuries in elite athletes. Upon request, and due to the sensitivity of the issues, some of the last names of the coaches are omitted.

Q1. Injury is a common risk and unfortunately an unavoidable part of athletics. Most athletes, regardless of sport, experience some type of injury during their athletic careers ranging from mild to severe. Despite technological advances and improved sports equipment, advanced coaching expertise and knowledge about particular sports, understanding nutritional and psychological factors contributing to athletes' progress and well-being, the number of injuries continues to rise. Could you please elaborate on why injury is still an unavoidable part of athletics today? What elements do you feel are most essential in coaching elite athletes to prevent risk of injury?

Coach Ron O'Brien (USA Diving): Dr. O'Brien has coached nineteen different Olympians to twelve Olympic medals (five gold, three silver, four bronze) and produced US National Champions for twenty-four straight years. In Olympic, World Championship, World Cup, Pan Am, University, USA Diving and NCAA, his divers won 196 gold, 113 silver and 106 bronze medals. He coached Greg Louganis to four Olympic gold medals. Responding to the questions, Dr. O'Brian thinks coaches need to continuously monitor the fatigue level, including both muscle and mental fatigue, and determine the training load accordingly. The two most common contributors to injury are overtraining and inadequate use of rest which would allow the athletes to recover from previous training loads. In other words, the lack of recuperation control may lead to the cumulative effect of fatigue which is a direct cause of injury in elite athletes.

Coach Yembo (China): In China, we have the luxury to select, not recruit, children at an early age for particular sports. We believe that prevention of injury starts from the selection of children with proper physical and psychological properties, allowing us to work hard and create an injury free environment. First, the most essential component of our training program is that every single kid selected for elite programs would go to a two-to-three year gymnastic program in order to acquire fundamental motor skills, such as body awareness, coordination, flexibility and strength. Second, we are fully aware that if an athlete suffers an injury requiring surgery, there is no way for this athlete to be fully functional, even after recovery, despite advanced medicine. For this reason we do everything possible, including proper nutrition, to prevent injury. Finally, over the years of training hard, Chinese coaches have acquired the knowledge and experience of how to control for overtraining and proper recuperation. Of course, injury may happen as bad luck, but not due to coaching errors.

Coach Georgeo (Italy): I have been coaching elite athletes for more than twenty years and believe that injury can and should be prevented. My injury prevention strategy is "trusting" my athletes' feelings. I encourage them to "listen and feel their body," and as early as excessive muscle soreness happens, they should let me know. Then we immediately switch to other types of activities, without reduction of volume or intensity of the pre-planned training session. Another aspect of injury prevention is to "overload" the body in the preparation phase so when major competitions start, there won't be any pressure due to fatigue. At this time, we basically flatten the load and intensity of the training program and constantly control for possible drop of conditioning. Whenever it happens, we stop specific training and return to basic skills. Finally, emotional happiness of the athlete is critical to prevent over-training despite current demands in elite sports.

Coach Michail (Russia): There is one way to reduce injury in elite sports to date: Reduce training volume to simplify exercise routines and to reduce the longevity of athletes' careers. By doing so, however, there won't be elite sports anymore. In other words, every single athlete who achieved world class level is at high risk for severe injuries or multiple injuries, including catastrophic injuries. They should accept it or quit; there is no way around it. That said, we do everything possible to make sure to maintain and control high levels of fitness the whole year around, to reach the flexibility level up to two times higher than necessary and to perform competitive routines five times more than required at the competition. The other important factor to consider in terms of injury prevention is proper warm up prior to the main part of the training session. In fact, boys are needed for a longer time and require more effort than girls because they are bigger and stronger. Again, injury is an unavoidable part of elite athletes. They should acquire experience to compete with injuries, because injury can happen at any time, even during the Olympic Games and there is no way to miss the most important event in an athlete's life.

Coach Julio (Cuba): Due to excessive pressure from international elite athletes and tremendous pressure from our society to win at all cost, we have to practice harder and harder. Unfortunately, there are limits of human resources, then these limits are ignored and the injury happens. It should be stressed that we often know there is a limit of athletes' capacities but voluntarily take a risk to explore it. If we are lucky, we are at the top of the world, if not, that is our destiny.

Coach Josean (Spain): Injury happens often in elite athletes mostly because coaches are not patient. They force the athletes to higher standards before they are ready to meet new challenges, both physical and most importantly psychological. I believe that 90% of injuries in elite athletes happen because of an abnormal mental status of the athletes, when they get tired at the end of season. As coaches, we know how to "load" athletes and make them physically strong. Unfortunately, we do not know how to "load" athletes mentally to make them psychologically strong. Unless we get this knowledge, injury will be a part of elite athletics.

Coach Ganter (PSU, Football): I think the injuries are inevitable because in contact sports things have to happen. There has to be some give and something is going to give when you are dealing with a contact sport. Obviously, the knowledge of the game is an important contributor to injury. It is important to make sure that athletes are in proper position. The strength training is the most essential element in coaching college athletes in order to prevent the risk of injury. I think conditioning and strength training probably outweigh the other two, because you could be out of position or you could be in an awkward position and still your strength training should carry you through any serious injury, at least as far as prevention goes. So, I think strength training, including its proper gain and control, is the key element in terms of prevention of injury in football.

Coach Jepson (PSU, Gymnastics): I would like to address the questions regarding the injury as an unavoidable part of athletics in gymnastics, and what we can do in order to prevent traumatic injuries in our sport. It is my strong belief that physical preparation of gymnasts is the most essential factor of injury and injury prevention. Functional abilities, general strength and conditioning and flexibility and specific skills are most important. The lack of these athletic properties is a predisposing factor for injury. Especially in my sport, you have to have general strength and specific flexibility in order to be injury free. So, you as a coach are going to set up the situation when injury should be under control. The second thing, looking from an injury prevention perspective, is the selection of elements that a given gymnast can learn, consistently perform and be comfortable with in his competitive routine. Athletes and coaches should have realistic expectations of demands and personal capacities to meet these demands. You need to prepare athletes both physically and emotionally so that expectations should be reasonably adequate and acceptable both from coaches' and athletes' perspectives. Do not expect good performance from athletes who are well prepared physically, but not ready emotionally for upcoming events. Even the amount of work load would be differentially accepted and perceived by athletes with different levels of "emotional" readiness. I would say that you can avoid injury if you train properly. As far as risk is concerned, there is no question that gymnastics is a risky sport by mother's nature, however, proper training of gymnasts is a key factor of injury-free environment in gymnastics.

Coach Shephard (PSU, Women Gymnastics): Every time the body is in motion, it needs to overcome the inertia, which is difficult to control. Especially in my sport of gymnastics, as soon as the gymnast left the base of support, it is really difficult to change the movement trajectory. So, inappropriate take-off, slight errors in movement initiation may lead to an unavoidable risk of injury. In addition, you must stop the movement fast, the hard landing, deceleration and all other mechanical properties of required skills are difficult to control, which is another major reason of injury, including traumatic brain injury. In other words, objective demands for the sport of gymnastics, high risk associated with skill performance and the necessity to maintain focus on what is supposed to be done – all these are predisposing factors for injury in modern gymnastics. You subject your body to abnormal forces all the time which lead to overuse injuries as well to traumatic injuries. You land incorrectly, you land hard on one leg, you fall, and this will be always present in gymnastics. Therefore, I agree that risk of injury is unavoidable and an inherent part of athletics in general, and in gymnastics in particular. There is no way totally to eliminate it. However, if you train smart and correctly, you can definitely reduce the number of injuries, especially overuse injury, by appropriate planning and controlling the training process. The planning of training programs, the assessment of physical skill and psychological status of an athlete are important, but the most important issue, at least in our sport of gymnastics is body composition. Overall physical fitness and conditioning and the lack of these properties must be the most important priority of gymnastic coaches. If you take care of it, you can dramatically reduce the number of injuries. Second, systematic approach, periodization, knowing how to smartly push athletes to achieve their potentials are also crucial. Overall, we can in some way control the injury, but in terms of avoiding the injury, this is a big question for me.

Coach Battista (PSU, Ice Hockey): Size and strength, F=MA. They are simply pushing the limits. Equipment is lighter but not necessarily more protective. I also feel seasons are too long and both mentally and physically it is taking its toll on these kids. Coaches need to work closer than ever with strength coaches, nutritionists, sport psychologists and trainers. Hydration issues, recovery issues, relaxation techniques, flexibility training, curbing over training, time management, stress related issues – there are so many more stressors today.

Coach Ross (PSU, Women's Volleyball): I have two thoughts on why injuries occur, and the first is that players come in to college unprepared for the physical demands of practice, and the frequency and intensity of preparing at the next level. The second is that some injuries occur because the sport is demanding. For example, hand injuries occur from blocking (vs. bigger, stronger, more experienced hitters), and ankle or foot injuries may occur from the repetitions, which increase the chance of trauma. A goal would be for the players to come in to college healthy and prepared for the demands both physically and mentally.

Coach Kaidanov (PSU, Fencing): As coaches, we should face the reality that injury is indeed a real problem in athletics. The demands of the sport are so high that it is beyond the abilities of athletes to meet these demands. We have to train harder and harder to be competitive and athletes' effort is so high that this at one time or another may lead to traumatic injury. I should stress that both physical effort, related with volume and intensity of the training program as demand of the sport is so high, as well as psychological perceived effort are contributing factors to injury. In other words, physical demands of the sport are much higher than athletes' capacities and capabilities to meet these demands. Therefore, I should say that unfortunately, traumatic injuries in sports are still an unavoidable part of athletics today. That said, I should stress that coaching strategies aim to at least reduce the risk of injury, including appropriate all-season strength training, flexibility and endurance. Athletes should not only be physically ready at the beginning of the competitive season, but most importantly, should maintain this level of physical conditioning throughout the entire season. This is a direct responsibility of coaches which is definitely a crucial factor for the prevention of injury. In other words, the stronger the athlete is physically, the less probability of injury we should expect.

Q2. According to a survey of 482 athletic trainers, almost 50% responded that they believed every single injured athlete suffered psychological trauma (Larson et al., 1996). The survey also indicated that 24% of trainers referred an athlete to counseling for situations related to their injury. Recent studies demonstrated that the probability of psychological problems dramatically increases in athletes suffering from three or more minor injuries. Do you agree that every injury may cause psychological trauma and therefore athletes should seek psychological counseling shortly after injury?

Coach Ganter: My overall answer to this question is NO. This is coming from my personal experience and based on what I have seen when kids were coming back from serious injury. The majority of cases in football players are able to find ways out from injury by their own. For the most part, I have been surprised by how reckless athletes are when they come back. I remember one of our players throwing his knee brace over the fence which he had to wear till the end of the season after injuring his ACL. He wore it for about ten minutes in practice then took it off and threw it over the fence and would never put it on again. So, I look at a guy who was coming back eleven months after an ACL surgery and notice that there is no psychological problem there. No fear, no intimidation at all. Another player was recovering from a knee injury. On the first day when he was allowed to participate, he went out on the field and yelled across to Coach Joe Paterno. He then did zigzag exercises for about fifty yards, planted his knee down and then shrugged his shoulders to show that it was nothing. There are far more guys that lack concern as to whether they are going to experience injury again. It is interesting that the trainers feel the other way; it surprises me.

Coach Jepson: I disagree that every single injury induces psychological trauma; it depends of the type of injury. For example, in gymnastics we have a lot of minor injuries, inducing bruises, scratches, etc., that are not traumatic, but induce more discomfort, unlike serious traumatic injuries requiring both medical and psychological attention. If an athlete has only discomfort or muscle soreness, it should not be recommended to see a psychologist. However, if an athlete has traumatic injury, psychological services could definitely help. My approach is to recommend injured athletes for psychological evaluation and possible treatment if the injury is classified as moderate and severe, otherwise, the athlete should be able to fully recover as the symptoms of physical trauma are resolved.

Coach Shephard: Most athletes are mature enough to deal with injury and fully recover through physical rehabilitation programs without extra psychological attention. Injury is a part of the sport. However, serious, season ending injuries can be very psychologically demanding and damaging. This situation is often associated with a severe psychological impact requiring the involvement of psychological personnel to deal with this issue. It definitely depends on the severity of the injury; athletes may or may not be referred for psychological evaluation, counseling and even treatment. So, referring athletes with mild injury to a psychologist may create even more problems, since athletes may develop the symptoms of pre-occupation with injury.

Coach Battista: Each athlete will react differently, but the question should be approached. Athletes who are afraid of losing their spot on the depth chart due to injury, spending the extra time needed to rehabilitate and experiencing loneliness due to an extended rehabilitation all play a role. I would caution against the "self-fulfilling prophecy mentality" of some athletes looking for an excuse to get out of practice and/or games which may mask bigger issues. My generation was taught to "suck it up" at any cost and sacrifice everything for the team. The pendulum may have swung too far in the other direction and now we are being overly cautious.

Coach Ross: Career-ending injuries need to be evaluated with the assistance of counselors. Players understand the expectations of their participation and are accountable for their physical and mental health. The occasions that we did seek external help was for eating disorders. I seek tough players with endurance and weed out the players who are dependent on psychological stroking.

Coach Kaidanov: It depends on the classification of injury. Severe injuries should definitely be treated differently from a psychological perspective, than mild or moderate injuries. The overemphasis or preoccupation with injury and the overestimation of the impact of injury may cause even more negative mental consequences. Fencing athletes are extremely mentally strong and can handle the emotional consequences of injury. They would be frustrated, but strong enough to overcome the temporary emotional problems. Overall, different approaches should be used on different athletes for the psychological recovery from injuries. As coaches, we should do our best to prevent multiple injuries or at least to predict athletes at risk for multiple injuries rather than consider how an athlete, can cope psychologically from injuries.

Q3. Injured athletes usually return to sport participation based upon clinical symptoms resolution and upon recommendation of medical staff. However, there is a notion among medical practitioners that clinical symptoms resolution may NOT be the injury resolution. Incomplete rehabilitation following injury may lead to the development of so-called bracing (self-protecting) behavior. This is a dangerous situation that may lead to more severe injuries. Through your coaching experience could you describe the signs and symptoms of bracing behavior among your athletes? What would be your coaching strategies to prevent, and if observed, to eliminate the symptoms of bracing behavior?

Coach Ganter: Some players are self-protective. As a coach, you can tell when an athlete is not ready to return; I have told doctors and trainers that an athlete is not ready even after they considered the athlete ready to return. If an athlete is not 100%, then I do not allow his return. It is also important to believe the athlete; if he tells me he is hurt, then I believe him. Some guys need to be pushed to get back onto the field, but if an athlete tells me he is hurt, or cannot play at 100%, then I respect that. It is hard for me to believe that a kid at this level would not want to be out there, so when they say they are hurt I usually believe they are hurt.

Coach Jepson: This type of behavior in gymnastics occurs a lot. It happens most often when there are still residual physical symptoms that occupy the gymnast's attention. So the question is when does the injury resolve completely and when is the athlete ready for 100% participation? The presence of bracing behavior is an indication that we have to be careful to prevent re-injury. Due to bracing and protecting, it is damaging to movement dynamics and the abnormal techniques that ultimately develop and causes chronic, long-term problems. Gymnastics is a very conservative process and the coaches must develop the relationship of trust. It is a fundamental element in my philosophy of coaching, and I will not allow an athlete to perform if not at 100%. My relationship with the medical staff is also important, and sharing a mutual understanding to eliminate confusion is a goal that everyone shares.

Coach Shephard: Physical symptoms resolution does not necessarily mean that an athlete is ready to compete and is fully recovered from injury. In gymnastics, we re-train injured athletes. After a leg injury, an athlete would have the tendency to land on the non-injured leg, putting more pressure on it during a landing. This creates danger of the overuse syndrome and leads to further injury. This situation can also develop bad habits and lead to a serious predisposition for further injuries. So, we do progressive skills of

gradually landing gradually increasing the height of the blocks, or changing
the surface of support from soft to hard, etc. So, they know that the injured
leg can "take it", so we build the confidence that no injury would happen if
done properly. This is just one coaching strategy to deal with negative effect
of "bracing behavior".

Coach Battista: A perfect example would be a former standout athlete in
hockey, four-time first team all-American Josh Mandel, broke his foot
blocking a shot and was casted in early December. It was the second year in
a row for the same foot at the same time of the year. Although cleared to
practice and play in early January at 100%, he skated gingerly on his foot for
the first two days of practice and was ineffective. We had to decide whether
to take him on a trip to Arizona or to take someone else. We met with Josh,
asked him to do a few non-contact drills and to not think about the foot and
just go all out. He slowly gained confidence and by the end of practice was
essentially back to his normal level of play. The psychological barrier had to
be overcome. Basically we want the athlete to be honest and give us
feedback. If we can do testing to reassure them that they are indeed okay,
then we should. Eventually they have to "get back in the pool" and give it a
try. I have found that over the years most athletes tend to come back too
soon, but usually don't do further damage. When it takes them longer to
recover they face the inevitable questioning by teammates with regard to
their commitment and toughness especially in the more physical sports. It's
all about communication with the coaching and medical staffs and
developing that level of trust that the athlete feels they are not going to be
put into a harmful situation.

Coach Ross: The question is how we determine if a player is 100% ready to
return to full participation without getting them into the competitive arena.
The only way to tell is to test them at game speed and this may result in a re-
injury, or players risking a new injury because they are afraid to go hard.
We try to have the players increase their effort and push the injured body
part in small group settings before we return them to full group participation.
A sign of bracing in volleyball would be a player returning from a leg injury,
and either hurting the other leg, or a different lower body joint by
compensating. We have had players try to change their mechanics because
of their rehab regimen, and they not only lose power, but confidence in their
ability to succeed.

Coach Kaidanov: Frankly, I underestimated this particular aspect of
consequences of injury, until I start recently thinking about it. First of all we
should be certainly sure that no physical signs of injury present before we
allow our athletes to practice again. Though, it is important to note that
upon return to practices it is reasonable to suggest that non-injured body
parts should be "activated" first, to gradually regain athletes' confidence that
they are fully ready for new challenges. This is very important issue, and I
saw in my practice that a lot of athletes "brace" or protect their used-to-be

injured leg, leading to enormous technical problems, new skills learning, and possibly to injury. Overall, full rehabilitation is a key to prevent possibility of development of bracing behavior. And as soon as the injured athlete returns to participation, we need to start again from fundamentals and gradually re-learn all pre-injury skills. The other interesting thing, the use and/or abuse of actual braces, cast for example also should be considered within the scope of this question. Actual braces are necessary to protect injured joint from overuse. This may also enhance athletes' psychological confidence. Although, these braces should be removed as soon as athletes are fully physically recovered from injury. Otherwise, athletes could develop abnormal dependence on these braces, which may create numerous problems.

Q4. Holistically, sport medicine specialists as well as most coaches have been concerned primarily with physical aspects of injury and injury rehabilitation. Thus, athletes who attained a pre-injury physical level are assumed to be fully prepared for safe return to practices and competitions. Do you think that athletes' psychological adaptation to injury may play a role in the rehabilitation process? Do you think that medical symptom-free, post-injury athletes are fully ready for 100% sport participation? Please elaborate.

Coach Ganter: There are two parts to this question. The last part, are they usually ready, I would say yes. That is just based on my coaching experience. Yes, even just looking at one of our offensive linemen, in the training room just coming off of ACL surgery, I do not think there is any question that psychology will play a role in how he recovers and how he approaches his rehabilitation and whether he is ever going to play again. I think that is psychological and that would help either make him rehab at a higher rate versus taking his time and maybe not doing it at all. I do not think there is any question about that.

Coach Jepson: I really believe that psychological adaptation to injury plays a very important role in athletes' rehabilitation post-injury. It is important in order to prepare athletes for the work load as well as for demands of sport requirement both physical readiness, confidence, in believe in their own abilities to meet high demands of the sport. Without this adaptation, athletes will be frustrated, coaches will be frustrated with lack of achievement and accomplishment. It has happened a lot in gymnastics. We use in this case a lot of spotting techniques, set-back and return to fundamental skills, re-learn every single element of their used-to-be automatic skills. Basically, I understand psychological adaptation as the return to basics, and gradually regain the confidence in ability to perform the pre-injury routines.

Coach Shephard: Psychological adaptation is an important component in injury rehabilitation in sport. For example, you can be physically ready from medical doctor's perspective, but athletes may be afraid of performing the certain skills that cause previous injury. So, regardless of physical symptoms resolution, athletes should acquire psychological status as well to be fully ready to compete. I can give you example, two years ago, one of my gymnasts dislocated her elbow during "Tkachev" vault. After completed treatment and rehabilitation, she was capable of doing this vault again, though, she was so afraid of doing that, we were forced to change her vault to less a complex vault, so she would be comfortable with it. Again, the psychological part of it is huge, so, you go back to fundamentals to regain the confidence of doing it consistently, and sometimes, as in this case, you should change the routine.

Coach Battista: I firmly believe that athletes must feel they are ready to go mentally as well as physically. Whether it's the concern of reinjuring or further injuring they must feel safe and feel they have the support of the staff and team. However I think there simply needs to be that extra communication which establishes the level of trust between the staff and the athlete. They must be psychologically ready to engage the competition.

Coach Ross: I think the severity of the injury and the athlete's previous exposure to injury has a significant impact on how they perceive their readiness to return to full participation. I think it is important for the medical professionals to be familiar with the demands of the sport in question. It is possible that a player is cleared to play because she is functional, but that may not be adequate for the player to fully compete at a high level.

Coach Kaidanov: This decision definitely should be made by experienced professional, that is a coach. However, medical professionals should be also involved in this process, they should be familiar with demands of our sport and help the coach with final call in terms of level of recovery from the injury. I should say that full physical recovery may not be indication of athletes' 100% readiness. Other aspects of preparation, including athletes' responses to training demands, attitude, motivation and emotions are important factors to consider as well. I believe that sports psychologist may play important role in estimation of athletes' psychological readiness. Also, sports psychologist can help athlete to regain his status as full-time fully recovered individual.

Q5. Who should be responsible for the final decision in terms of an athlete's return to full sport participation: coach or medical doctor? Do you think that there should be different criteria in terms of athletes' readiness for returning to practices versus returning to competitions?

Coach Ganter: I would say, first the player, and second the doctor, in no way the coach. As long as there is good judgment used in practice, I can see a green cross routine (where a green cross indicates a partial level of activity). I think that is fine. I think getting a kid ready for participation means all out 100%. I had used the old adage about it: "Do what you can and go till you cannot go any more, until it hurts or whatever; if it starts to bother you, then get out".

Coach Jepson: As I said before, I have great relationships with physicians and trainers. This is mutual decision, and I trust our medical staff, because they are well aware of our sport of gymnastics. They attend our practices and competitions and they know some specifics of the sport that allow them to make accurate decision when athletes are ready to return to sport participation. There is no way I push athletes to do something they are not ready to do. Now, it is my responsibility to communicate with medical staff, so athletes have good representation from both sides. It should be noted that sometimes medical staff do not fully understand our sport. So, they should be educated as well. They have to know the actual mechanisms involved in movement, so the decision in terms of return to play will be made according to this knowledge. So, this is a responsibility of coaches to educate medical personnel about specificity of sport. If physician knows the medicine, but does not know the sport, it could be difficult to make appropriate decision. On the other, I am not a doctor, and do not know a lot of medical aspects of injury. Thus we have to work together for the safety of our athletes. In fact, as I mentioned before, our medical staff are in the gym with us, and this creates a lot of trust, which is critical in terms of athletes' rehabilitation from injury in general, and return to participation in particular.

Coach Shephard: I think that the coach and the doctor should collaborate together and make a decision of return to participation based upon physical status of the athlete, first, and other athletic characteristics, second. Some doctors, especially in gymnastics, do not have enough knowledge what athletes can do after injury, and injury rehabilitation. For example, if a gymnast has ankle injury, she still can do full bar routine, except landing. She also can do a lot of conditioning exercise for upper body. That said, she was not allowed to participate in practices by medical personnel. They say "no participation", and eliminate a lot of things that athletes could be doing for faster return to pre-injury status. This situation for sure may create conflict between medical doctors and coaches. So, doctors may be very knowledgeable in terms of physical aspects of injury, but at the same time, they could be ignorant in terms of understanding the sport and gymnastics and possibilities of compensatory training programs for injured athletes without aggravating the injury. So, the final call from a legal standpoint should be from a doctor, but from the practical standpoint, the coach should be responsible for final decision regarding the return to participation.

Coach Battista: I really believe it needs to be a collaborative decision. But that the coach with the athlete and the doctor's input should be given the opportunity to make the case for an athlete's return. But, in this litigious day, I would have to say it is up to the team doctors, especially in cases involving serious injuries which could prove life threatening (hydration, weight issues, heart and lung, concussions, etc.). I believe the athletes should have some means by which we can "test" their mental and physical readiness for returning to action, whether practice or competitions.

Coach Ross: I think that the final decision on whether a player is ready to play rests with the coach, however, the medical staff has to be comfortable that the player is capable of competing injury free. The physician should not release a player to a coach unless they are sure the player is ready. Too many times I see a player cleared to play, but nowhere near ready to truly compete. There is a difference between contributing and excelling, and in some situations a coach may want the contribution and they can limit the demands placed on the player until they are fully able to compete. Both parties share the same goal, and that is a healthy athlete who can offer their best effort.

Coach Kaidanov: It should be definitely mutual decision. The medical doctor should clear the athlete for participation based on their clinical signs and regulation. However, coach should clear the athletes for specific types of activities without compromising the possibility of re-injury. Having said that, the medical doctor overall is responsible for final decision based upon clinical symptoms resolution and his or her knowledge and experience dealing with specific sport activities. The medical doctor should predict possibilities of re-injury and make their call accordingly.

Q6. There are a number of interventions recommended by sports psychology practitioners -- including negative thought stoppage, cognitive restructuring, healing imagery, muscle relaxation, goal setting, etc. -- to speed up rehabilitation of injured athletes. What kind of coaching strategies would you recommend to enhance athletes' readiness for returning to a full range of sports participation?

Coach Ganter: I am an old school guy, and I do not have any knowledge or experience in any type of psychological rehabilitation. I think the greatest motivator is playing time and if they want to get back on the field and play, they are going to hurry up and get better. I think, if injured athlete is worried about regaining a position or playing time, sometimes that will speed the process up too. I have no experience with people giving psychological coaching or anything like that in rehabilitation of my kids.

Coach Jepson: I do a lot of visualization, like I said, specifically focused on physical skills that were associated with injury. I truly believe that major cause of injury in gymnastics is improper techniques and errors in performance of complex skill. So, gymnasts should be clear minded in terms of understanding the fundamental mechanics of skills they perform. Also, skill progression, especially in case of injury, is critical to return to pre-injury status. We teach gymnasts to focus on the positive, rather than to think about possibility of re-injury.

Coach Shephard: I think that properly framed, gymnastics-oriented and injury recovery-focused visualization is a tremendous coaching resource to speed up the rehabilitation of injured athletes. Visualization should include not only visual imagery per se but also skill imitation, feeling, sensing the recovery, feeling the pressure and tension in the injured joint producing required skills. We use this often in our program, not only for performance enhancement but also as a part of psychological intervention program for injured athletes. For example, we have a gymnast with Achilles tendon injury, so I required that she should do visualization every day at least ten minutes per session, with the accent on acquisition of feeling that her tendon get stronger and stronger every day, becoming more flexible. Positive thought process and positive thinking about progress of recovery is a great contributing factor speeding up the whole process. It is important to stress again, that it should be specific and recovery goal-oriented visualization. This should be trained similar to physical skill training.

Coach Battista: First, I remind them that they are athletes and are in most cases in much better shape mentally and physically than the average person. Most doctors are going to err on the side of the conservative diagnosis. I try to keep them active in team meetings and activities so they look forward to getting back as soon as possible. I firmly believe in the importance of communication with all parties to develop both a written and verbal game plan that helps the athlete remain focused. We are big into goal-setting, imagery and relaxation exercises, and use our sports psychologist whenever the athlete is willing to participate.

Coach Ross: I think that there is a full array of interventions that can assist with the development of an athlete. The use of these items may work with some athletes, and I would encourage their use. The player may have to deal with the fear of returning to full participation and I think anything that can reassure them that they are ready to go is valuable. My communication with my players is that they need to test for themselves before they can get full clearance from me. They need to feel comfortable and capable of reentering the sport.

Coach Kaidanov: As a coach, I am in charge of modification of practices, considering the level of recovery from injury. I also change and modify the goals that athletes should set for themselves. They should be realistic, but challenging enough, so athletes would return to full participation sooner. We could also contract a certain routine of injury-free exercise that focuses primarily on involvement of non-injured parts of the body. For example, if an athlete is recovering from hamstring muscle pull, we could recommend a series of exercise programs related with abdominal conditioning. So athletes are fully involved in the training program but accent of the training is modified and shifted to the upper body conditioning.

Q7. As ascertained in various studies, it is clear that gender differences in athletes are highly influential in shaping the psychological and emotional experience of injury. For example, females report higher levels of fear related to injury due to movement than males. Are there different coaching strategies for dealing with female athletes as opposed to male, particularly in regard to recovery from injury?

Coach Ganter: I do not know, I do not have any experience with female athletes or how they are coached. I just know a little bit since I have been around the female coaches we have here at PSU. I know for sure, they are probably tougher than we are. You said that they have a little more fear of injury than the males. Maybe the good female coaches have to be tougher.

Coach Jepson: It is clearly not my area of expertise, though I think that there are fundamental differences in coaching males versus female athletes. I guess female athletes are more emotional and sensitive, therefore coaching strategies in female sports should be oriented on creating extremely positive learning environment.

Coach Shephard: Gender differences are absolutely essential issues to consider when coaching female athletes. They learn differently, they feel differently, they are more sensitive to critique and coaching styles. You have to be very sensitive towards the mentality of the female athletes. I think it is essential that they should have daily team meetings to discuss various aspects of their life, not only athletic life. They should be happy and psychologically well to respond to enormous pressure to be student-athletes. There are delicate issues such as body weight, body image, self-esteem, that extremely important for athletes, especially for female gymnasts. Thus, my primary responsibility is to maintain psychological well-being of my gymnasts in any way I can. Unfortunately, not much research is presently available for coaches on how to deal with female athletes, therefore we mostly orient on our personal experience and experience of my female assistant coach.

Coach Battista: Certainly not my area of expertise, but definitely a factor since the culture of women's sports is inherently different (cultural influences, relatively new and few female coaches who can relate). My gut feeling would be a higher need for communication and reassurance.

Coach Ross: I am aware of some research claiming that female and male athletes display similar levels of confidence, psychological maturity and toughness when tasks are appropriate for females, when females and males have similar experiences and physical abilities, and when clear evaluation criteria and feedback are present. I fully agree and believe in enormous potential of female athletes in terms of dealing with training load, athletic demand, discipline and commitment to sport. This is at least the philosophy in our team at Penn State. It should be noted that concern about body image definitely affects all women including student-athletes. Athletes just as any other women are extremely sensitive to the general societal pressure towards unhealthy thinness. We as coaches should be also sensitive to how we communicate with female athletes about this issue. I suggest that we should follow nutritional guidelines and focus on healthy eating behavior rather than on weight issues. I also should say that the most important aspect of coaching is to treat the athletes with respect and dignity regardless of gender, race and social preferences.

Coach Kaidanov: I should say that I agree with the notion that gender influences all aspects of coaching including practices, competition, and coaches' interaction with athletes. Gender influences coaches' interpretation of athletes' responses to work load, their expectations and effects of psychological pressure. It is important to note that gender is an important individual characteristic. Accordingly, if you would like to follow the principle of individualization, you should directly tie this to gender. To my knowledge, there is not much research and recommendations how to deal with gender issues in coaching practice.

Q8. Among athletes it is common to hide fear in order to avoid appearing weak. However, it is known that in previously injured athletes, fear of subsequent injuries may induce erratic emotional responses, avoidance reactions, and bracing behaviors. In your opinion, and in terms of psychological recovery, do you think that athletes recover better, or faster, from an injury if the injury is given more attention, or less attention? Also, do you feel that there is a difference in response to attention paid to the injury in male vs. female athletes?

Coach Ganter: I would have to say more attention is needed when you coach injured athletes. I can not picture an injury getting less attention, there is no motivation there, there is no "we got to get you back." That is a tough one, but I would have to say more attention would help promote the quicker recovery. I think the better rehabilitation and more attention a guy would get and more encouragement would get them back quicker. I have no opinion regarding the gender differences.

Coach Jepson: I think that fear is important protective mechanism and plays an important role in athletics. As I said, gymnastics is extremely risky sport and, not surprisingly, fear is present every time a gymnast is preparing for or performing a routine. Most importantly, however, to dissociate fear per se, from ability to control fear. Successful gymnasts, regardless of gender can control fear, trust their body and their coaches. I also believe that fear comes from uncertainty due to lack of consistency in performing the routines. Thus, the more consistent and reliable the skills, the less fear might be at place. Unlike other less risky sports, if there are less risky sports, we should think of fear in terms of fear of being hurt, versus fear of being embarrassed, or fear of failure.

Coach Shephard: I agree that fear is necessary component of sports environment. Usually female athletes are more open in terms of expressing fear. They are honest and expressive if they are afraid to learn or to perform new skills. So, my responsibility as a coach to take into consideration the fear factor, and progressively reduce it though consistency of performance of risky routines and acquisition of confidence. Also important, is that athletes should know that we are good "spotters" and are able to protect the athletes in case of falling. In essence, fear can be controlled, if properly trained.

Coach Battista: All players are different. Most players would respond better to a coach or trainer who gave positive feedback: "Looking better already," "almost there," "can't wait to have you back." Some however need to have it downplayed while others need to be told straight up that it's not that bad; get over it! I definitely feel that females respond differently than males (some are tougher and more stubborn!!!). But the key is still communication. Generally, I would think the females like more than less information.

Coach Ross: Fear of injury is not a common emotional reaction in volleyball, unlike fear of failure performing certain skills. My coaching approach to deal with fear of failure is to stress that I reward learning progress, commitment to sport and to team rather than winning or losing, issues that unfortunately dominate modern sport, including collegiate sport. On the other hand fear of re-injury as a result of premature return to play is an important factor to consider. Therefore, we have to explore the root of the problem rather than to treat the consequences of our erroneous assumptions about the injury and its impact on athletes' well-being, both physical and psychological. In other words, we should be crystal clear about the severity of injury, its impact on athletes' status and most importantly about the current emotional status of a previously injured athlete. An athlete should be not only physically injury free at the time to return to play, but also should not experience any signs of irrational thoughts, anticipated pain due to movement, etc. These observable signs of premature return to play should be used as red flags for coaches requiring termination of situations when these signs are present and maybe additional physical rehabilitation and/or psychological counseling.

Coach Kaidanov: There should not be any behavioral signs of fear of injury, otherwise it could potentially lead to injury. Neither should there be any irrational thoughts and expectation of injury. This extremely negative emotional reaction distracts athletes from major focus of control, technical skills, competitive strategies and decision making processes, especially during competition. I guess, if an athlete may develop this emotional distraction, he or she should be referred to professionals dealing with this issue.

Q9. Sport-related concussion has received significant attention in recent years. Despite some advances of studying concussions, important questions are still to be answered including:

*Which concussion grading scale and return to sport participation guidelines are sufficient to prevent more severe secondary and multiple concussions?

*After how many concussions should an athletic career be terminated?

*Are there long-term cognitive and behavioral deficits after single and especially multiple concussions?

Collegiate athletes are at high risk for sport-related brain injuries. The likelihood of brain injury is a function of head impact (or sudden acceleration/deceleration) within the context of sport participation. The concussion may occur in any activity, regardless of the nature of this activity, and when the brain injury occurs, it has potential for a lasting effect on the athlete. What do you think the collegiate coach should know about concussions and what should be done from a coach's point of view in order to prevent concussions?

Coach Ganter: What they should know is the dangers of and what is a real concussion. I know even personally, I can think of times when I must have had a concussion and remained in the activity, went back out with severe headaches, you know either into the practice or even the next day. I could remember that. I think now, when our kids have severe headaches and maybe they got a blow on the head, it is worth to keep them out for a day or two. I am saying, jeez, I am sure we could have a lot of problems having concussed kids back in the field. Anyway, I think knowledge, what is a concussion, what are the dangers if he continues to participate with any symptoms of a concussion, is critical for coaches. What we can do to prevent concussion, is to make sure that we are doing safe drills when players do not have a helmet and just being smart and taking precautions. So that we do not have any unnecessary concussion because of lack of protection. We do too many things without a helmet. Our summer football camp is without a helmet drills. There are just so many things that worry me about that. Really, the more I think about it, the more I worry about it. I have become more knowledgeable about concussions, and what can happen down the road is really worries me.

Coach Jepson: A couple of cases we had in our gymnastic team. I would like to stress that I am not a neurologist and, practically, have little knowledge about concussion. Therefore, I think that medical professionals should treat athletes with concussion, regardless the level of injury. I know that there are gradations in terms of mild, moderate or severe concussions. I think this is very serious injury and every single case of brain injury should be considered from our coaches' perspectives as severe injury, requiring immediate medical attention and treatment. I am aware of possible consequences of concussion including learning problems in student-athletes suffering from single and multiple brain injuries. I think that we as collegiate coaches should be more educated about signs and symptoms of concussion, especially about long-lasting residual abnormalities.

Coach Shephard: This is definitely a confusing injury not only for athletes but also for us as coaches, because unlike other injuries, you often do not observe obvious physical evidences of injury, such as broken arm, cast, etc. Unfortunately coaches do not have enough knowledge about this serious type of injury, the brain injury. My understanding was that this injury is temporary, at least in the mild form. Therefore, I thought that athletes suffering from concussion should be ready to start practices within one week post-injury. However, my recent experience with one of my gymnasts, who suffered mild concussion five weeks ago and still experiencing problems, has convinced me that this is more serious injury than I have ever thought. Therefore, more education is needed for coaches to fully realize the danger of brain injuries. I was not aware of procedures, scales, assessments etc., and still do not know the details about this injury. It is important for us to understand long-term consequences of concussion to realistically expect the injured athletes to be back for full participation.

Coach Battista: We should be educated on all the most up to date information on concussions and recovery from concussions. As someone who has dealt with this both as a youth and adult athlete, it is a primary concern of mine. Any and all data should be collected and analyzed to help determine the short and long term effects of concussions as well as the appropriate time needed to recover. Until such a time that affordable, portable "EEG" machines that are capable of quickly giving feedback on brain patterns are available, we need to develop the best alternatives possible. Baseline testing prior to tryouts, "litmus" tests that give some sense of the magnitude of the concussion administered by trainers/doctors should be on hand. In general I favor a conservative approach. I do believe we have to be careful how the test is administered (e.g. the first question asked shouldn't be "do you have a headache" or "do you remember what happened," it should be generic like "how do you feel?"). I worry that sometimes we make suggestive comments that the athlete simply reacts to in an affirmative way. We had an athlete who answered, "I guess I have a headache," and jokingly said, "I don't know what hit me," and it turned out he was fine.

Coach Ross: The concussion issue is one that I feel needs to remain in the medical community. I don't think the coaches are trained to evaluate this condition. Although it is rare for a volleyball player to suffer from this injury, I have had a few players that have, and they were monitored and regulated by the physicians and athletic trainers. Not much in my sport can be done to prevent the occurrence of this injury, but certainly adequate instruction in the floor skills area can reduce the exposure to hitting one's head on the playing surface.

Coach Kaidanov: I should say that concussion is a very unusual and seldom traumatic injury in fencing. But in case if concussion would happen in my sport, I would definitely refer concussed athletes to professionals. I truly believe that this is a serious trauma, regardless the grade, symptoms and/or symptoms resolution. I also suggest that coaches should be educated in advance about this type of injury, so, if it would happen, appropriate actions should be implemented. This is particularly important since we are dealing with student-athletes who should go to school, study, acquire a lot of intellectual knowledge that requires memory, concentration, and other mental abilities. Thus, overlooking the symptoms of concussion may cause dramatic consequences in terms of student-athletes' ability to successfully graduate. Again, highly professional medical people should be involved in case of concussion in athletes.

Q10. Many athletes who have had single concussions recover quickly and are able to return to play. However, athletes who have had a history of concussion may exhibit certain symptoms such as an episode of concussion, including headache, dizziness, nausea, emotional liability, disorientation in space, impaired balance and postural control, altered sensation, photophobia, lack of coordination and slowed motor responses. As a coach, do you think it is possible to discern these symptoms as irregular or abnormal in an affected athlete, and if so, how would you adjust your coaching methodology?

Coach Ganter: I feel that at least personally, I am better educated in what a concussion is, what causes one and the symptoms of concussions, mostly because a previous player of mine had to give up football because of headaches. Because of the emphasis put on by the medical personnel, I think that most coaches are better educated about concussion, what causes one, and what are some of the symptoms. I do not know how you do it manually, but just visually you know the stories and interviews I have had with kids about headaches and inability to sleep, and the inability to concentrate when they study, having to get up in the middle of the night to take a shower just to get some heat on their head because their head was killing them, the headaches. I heard guys talk about blurred vision and I mean we have enough of them around here that I think by observation. If you notice any change in the way the practice and their performance and then you talk to them, I think most coaches would be able to discern that there is something wrong here. This kid maybe got dinged yesterday and better have a doctor look at him.

Coach Jepson: Again, my expertise and experience dealing with concussions is limited. Therefore, I would follow directions from medical professional how to treat brain injured gymnasts. One thing I know for sure, I would be very careful coaching gymnasts with concussion, because of nature of our sport requiring abrupt change of direction of head motion, hard landing and possible falls. This may cause the situation when previous brain injured gymnasts could suffer from another and more severe concussion. Having said that, it should be noted, if an athlete with previous brain injury would be cleared for participation, I do not think that I would treat this athlete differently. I would keep eye on it, but would not overemphasize my concern. I would consider this as typical injury. I should be sure that their mind is clear, they know what they are doing, are in control of their body and mind, can focus and concentrate on skill performance etc. For example, I had gymnast who suffered from mild concussion and weeks later he could not remember what he did, and had long lasting memory problems. Of course, he was not ready to come back and we did not allow this from happening. So, I watched him very closely. Actually, I watch every single athlete very closely, regardless of whether s/he is injured or not. So, I know if something wrong with them, I just do not allow them to take the risk. This is my common procedure, concussion included.

Coach Shephard: First of all, I would like to stress again that my expertise and experience are limited, therefore I would follow the recommendations of medical people regarding the treatment and coaching brain injured athletes. For sure, I would monitor these athletes very closely and will be watching for any signs of lack of concentration, attention, fatigue, reduced motivation. If this happens I would terminate their practices and refer these athletes to medical people for evaluation. I would not push these athletes further without proper assessment of impact of the injury. Again, most coaches have no idea about this type of injury. Therefore, coaches' education is a critical factor in terms of prevention of multiple concussions in athletics.

Coach Battista: Tricky area. I think it depends on your own background as both an athlete and how your coaches dealt with it. In the old days we simply said shake it off or you just got your bell rung you'll be fine so there is a macho thing here. Someone bruises an arm or a knee we put ice on it and everyone feels sympathy for the athlete. Someone complains of a headache and they are either consciously or subconsciously considered a wimp. I do believe coaches who really know their players can spot differences in behaviors and motor skills but with a large squad it is not always possible to detect the smaller changes. Again communication is a priority. Working with the training staffs and having assistant coaches on the same page with regard to creating a safe and caring environment. Helping the athletes to feel comfortable about speaking up if they have a concern without fear of losing their spot in the line-up or having the confidential information become public. Educating the players about the potential long term effects without scaring the hell out of them and taking away the aggressive mind set.

Coach Ross: As I mentioned in my previous responses, concussion is rare injury in our sport. I personally do not have enough experience and expertise to deal with concussed players. I truly believe that in general, this injury should be carefully treated, evaluated and re-evaluated in order to prevent residual long-term debilitating effects. It is known that symptoms of concussion may reappear long time after the accident, so close monitoring of these symptoms is important.

Coach Kaidanov: As I mentioned in my responses to previous question regarding the concussion, I would be very conservative in terms of dealing with post-concussed athletes. I would closely monitor for any signs of abnormal movement patterns, such as occasional loss of attention, progressive fatigue, unexpected mood swing, inability to concentrate. If these symptoms are present, I would immediately terminate practice and send this symptomatic athlete for professional evaluation and possible treatment.

Q11. Currently, it is still being debated whether concussions result in permanent neurological damage or in transient behavioral and psychological malfunctions. This controversy is due in part to the lack of assessment of the development of fear of re-injury, bracing reactions and overall avoidance behaviors. Is it obvious during practice or in competition, to you as a coach, if a previously injured athlete has developed bracing behaviors? Do you have particular strategies for dealing with athletes who exhibit this type of behavior in response to injury?

Coach Ganter: In terms of concussion, I would say no. I have never noticed that from the guy coming back from a concussion. I definitely have seen guys who are ginger for a knee or maybe an ankle and who maybe self-brace. Remember how I said that I think an athlete knows better when it is time to go all out. That is the way I feel, so I have noticed it from that standpoint, but not from the standpoint of the concussion. For the most part the kids almost, I can not think of anybody except for one player, really who said "I need a couple of days because of my head." I can think of dozens that have said I need another week or I am not ready yet or this is not good for other parts of their bodies. I have had no experience with a guy saying my head still hurts or I am not all there or I am not ready.

Coach Jepson: In my experience with concussed gymnasts, I observed some cases of "self-protecting" behaviors, similar to those following other traumatic injuries. I do not have enough experience in order to elaborate specific coaching strategies for concussed gymnasts, therefore I would treat these athletes similarly that I am treating gymnast with serious traumatic injuries.

Coach Shephard: I guess I should study more about concussion and educate myself about this injury. One thing I know for sure, that from now on I would consider even mild concussion as severe injury, due to accumulated knowledge of long-term disabilities resulting from brain injury.

Coach Battista: See my responses to Q10. There is a point where as a coach I think we need to simply put the kids' long term health ahead of short term gains. I have experienced multiple concussions and have dealt with the aftermath. I am not sure of the long term effects of concussions that are spread out over time, but I have no doubt that short term effects can be a hindrance to performance as well as daily functions. Multiple concussions in a short term period of time are of even greater concern to me. Since I have experienced this firsthand I am more sensitive to the issue than others may be, thus my dealings with this issue are certainly biased. Some athletes are better than others at hiding their real feelings about things. Again fear of being removed from the line-up, being singled out as a wimp, factor in. Some athletes have a much higher threshold for pain so I really think this is more of an art than a science to some degree. Coaches get to know which athletes tend to cry wolf and which ones try to be tough guys. When I do come across the kids who are using bracing reactions we try to make them feel as confident as possible that they will be okay and that the idea of holding back or slowing down may actually put them in a more vulnerable position.

Coach Ross: I do not know whether concussion induces permanent neurological damage, or transient functional abnormalities in brain and behavior. But I know for sure that improperly treated concussion, premature return to practices after concussion is not permissible, as in case of any other traumatic injuries in sport. Clear and accurate assessments performed by qualified medical personnel are essential to prevent risk of brain injury in athletics.

Coach Kaidanov: Again, I do not have enough experience dealing with concussion in my sport of fencing. Therefore, I cannot further elaborate on this issue. Although, I guess, there certain types of protective patterns that athletes may develop to prevent the second or multiple concussions.

Q12. What advice would you give to upcoming coaches today regarding how to identify athletes at risk for injury and ultimately to prevent injuries among student-athletes?

Coach Ganter: I think the number one thing is the strength training. If you put a kid in the position where he is overmatched from a strength standpoint and any type of physical overmatch, you are certainly risking an injury. The second thing is position, especially for a young coach if he is coaching young people. You need to teach them proper hitting position, body position, how to fall -- just how to protect themselves. So, I think the strength training is first, conditioning is probably second, because if they get tired they probably are more vulnerable to an injury, and then position.

Coach Jepson: The biggest thing is there is no short cut. Physical preparation, you have to learn groundwork. Important thing, if athlete is psychologically not ready to do certain things, do not over push. Again, you have to build good foundation. Athletes should understand what is proper way of their preparation, and this what we do as the coaches. If athletes understand this, it means they are coachable, and if so, they are learnable and can reach their potentials whatever it is. Holistic understanding, physical, mental, emotional, understanding that some injury may happen and if so, they should find some "advantages of it," of being tougher and more knowledgeable as athletes and most importantly as individuals. At this point, I think athletes should know that they can trust me, because my primary responsibility is not to make national champions, but to develop quality people. I have their best interest in mind, and they know it. And if they believe that, I am accomplishing my mission as a coach properly. People should be treated as people, not like machines and robots having a goal to be best in their sport. I treat gymnasts in my team as people and sport is just a part of their life.

Coach Shephard: Avoid overtraining and emphasize conditioning, especially pre-season when most of the athletes are not in good physical shape. You cannot do just gymnastics to be in good shape for gymnastics. You should do a variety of conditioning programs before you do gymnastics. You have first to prepare your body to absorb the impact during landing, you have to prepare your abdominal muscles to perform the bar routine. And most importantly you have to plan properly given the time you have for preparation. Physical readiness is not the only component of a successful season. Athletes should be ready psychologically as well. Proper motivation, psychological skill training, individual goal setting, stress management skill are just a few attributes of injury free training environment. There is large preparation prior to competition. You have to build proper confidence with proper progression of physical skills and general fitness. Slowly, brick by brick build various aspects of athletes' progression, with no rush. You must build fundamentals and certain discipline and commitment.

Coach Battista: First, knock off the old school macho stuff and be more concerned about safety, factoring in water breaks, taking into account environmental conditions (heat and humidity, lightning, air quality). Educate them about the value of mental training and help them buy into relaxation and feedback as a skill no different than skating or shooting. To be resourceful by utilizing school supplied or community volunteer experts that can help with nutrition, psychology, strength and condition (making sure you do background checks and follow-ups on suspicious behavior, e.g. a volunteer strength trainer recommending supplements without your knowledge). Teaching the kids the value of proper warm-up and flexibility (not just stretching!) is an essential issue. Urging the kids to play within the rules and to not "cheat" by hitting away from the play or pushing the rules to extremes which may incite retaliations (in the more physical sports and the stick yielding sports especially) is another fundamental rule. Educating parents about the new research and findings is crucial also. Numerous times we have had parents argue with us when we held their son out of competition due to concussions and in their minds the kids need to just "suck it up and tough it out." Parents should understand the coaches' and the team's policies in advance.

Coach Ross: I think my advice would be to monitor the amount of jumping used in training and instruction because I think many injuries occur because of lack of sufficient strength to handle the training level. When athletes tire, they become more susceptible to injury, and the coach needs to pull back their demands as opposed to pushing the players harder. There is no question that proper instruction in the performance of the necessary skills correlates with a safer environment, and coaches are responsible for making sure the training facility is safe and properly maintained. In closing, it is critical for the coaches to work with the health professionals in assuring that their athletes are ready to restart their participation and not listen only to the athlete.

Coach Kaidanov: I would say that the most important thing in coaching is to be patient. Encourage challenge when athletes are ready to meet this challenge. Do not expect quick success rather than build fundamental skill, conditioning and character.

Coach O'Brien: Through personal interaction at the beginning and end of each training day, coaches should determine the athlete's fatigue level: extreme – high – moderate – low, and adjust the training plan for that day or the next day accordingly. If the coach sees a significant deterioration in the athlete's performance level, use rest to allow recovery before injury occurs. We do not have a chance to risk, and every coaching mistake may lead to severe consequences for athletes' well-being. My other advice to young coaches is that knowledge about your sport, care about your athletes and commitment to your coaching duties are among most important contributors to injury free environment.

3. COMMONALITIES AND DIFFERENCES

Injuries are commonly blamed on the coaches. Inadequate assessment of athletes' physical skills, misunderstanding of psychological coping resources, rushing with acquisition of new techniques, overtraining and overloading causing accumulated muscle and mental fatigue, early return to sport participation after injury are just a few examples of coaching errors that increase the risk of injury/re-injury in athletics.

If a coach is being too critical of an athlete, this could result in a negative relationship and possibly even injury. These overly judgmental and critical coaches create an atmosphere that is detrimental for the athlete. They try too hard and present their errors in a critical manner. The second coaching style, which creates a positive training environment benefiting both physical and mental well-being of the athletes, is the observing and assessing one. This technique is more effective for collegiate athletes.

When different Penn State athletic coaches were questioned about injuries, their responses were similar in some cases, yet different in others. This journal shows how there are common causes of injuries yet different opinions about them.

Regarding the first question, many of the coaches had similar answers when it comes to the most essential aspects in coaching in a manner that prevents injury. Most of them stated that their athletes can avoid injuries by becoming physically ready for the demands of their sport. In order to become physically ready, they have to do training before, during and even after their season. One of the coach's comments that I found interesting was when the women's volleyball coach admitted that when her first year players start off in the program, they tend to get injured because they are not as physically and mentally prepared as the upperclassmen.

When asked whether the coaches believed their injured athletes need to see a sports psychologist after each injury, the opinions were similar yet different. When Coach Ganter mentioned one of the players who threw his knee brace over the fence after ten minutes of practice, it reinforced my opinion of how mentally tough the football players are. The football coach was not hesitant to say "NO" to this question right away. The women's volleyball coach stated something similar, "We weed out the players who need psychological stroking and seek the tough ones." The general consensus of the coaches was that a sports psychologist is needed if it is a moderate to severe injury, but not for any of the minor ones.

In response to the third question, regarding the bracing behavior once an athlete returns to his/her sport, there were some interesting responses. The women's gymnastics coach explained the strategy of "progressive skills of landing," by gradually increasing the height of the blocks, or changing the surface of support from soft to hard. This seems like an effective method and more coaches should use that type of approach if they do not do so already. The example from Coach Battista was interesting too, where the hockey player had to "break the psychological barrier" and ease back into the practice, starting with non-contact drills.

Overall, the coaches at Penn State have similar responses to these questions. They all seem to have these answers because this is a Division One intercollegiate school, and therefore athletics are taken seriously. Some of the variations observed are due to the demands of the different athletic teams.

If I were to give advice to a new coach for a collegiate athletic team, I would tell them that it is better to be safe than sorry when handling an injured athlete on their team. Even though an injury such as a concussion is not blatantly obvious, the athlete should be given enough recovery time and the proper treatment to prevent further problems in the future. It is known now that athletes can get even epileptic seizures for up to ten years after receiving many head injuries from their sport. If a coach were to make an injured athlete compete again too soon after an injury, they could be affecting their health in later years without realizing it and therefore it is better to be safe than sorry.

On high school and professional athletic teams, the athlete can typically decide whether or note he or she is able to compete again after an injury. On collegiate teams, that is not always the case. When the coaches were asked when an injured athlete should return to play and competition, there were varied opinions on when the time was appropriate. Gender differences are another aspect that needs to be taken into account when handling injuries. For example, facial injuries are considered more traumatic for female athletes. Coaches need to be sensitive to the gender differences since there are many of them. In reading the Penn State coaches' opinions, they seem to be sensitive to this subject.

International coaches' opinions regarding injury differed depending on what country they were from. As Coach Yembo from China stated, injury may happen as bad luck, but not due to coaching errors. This differs to what some experts in the US believe. They have the ability to select athletes when they are very young in order to train them for their sport, rather than waiting to recruit them at a later age. The most essential component of their training program is that every single kid selected for their program would do two-to-three years of gymnastics. This is very interesting, and they do this so the young athletes acquire fundamental motor skills, such as body awareness, coordination, flexibility and strength. Also, they believe that if an athlete suffers an injury requiring surgery, there is no way the athlete will be fully functional after recovery.

Coach Georgeo emphasizes the importance of the athlete's happiness to prevent over-training. He overloads the body in the preparation phase so that when major competitions start, there will not be any pressure due to fatigue. Georgeo also encourages them to "listen and feel their body", and as soon as even excessive muscle soreness happens they should let him know. His techniques are different than those of the coach in China, and seem to work well for him considering he has worked with elite athletes for over 20 years.

Coach Michail of Russia stated they do everything possible to make sure to maintain and control a high level of fitness the whole year around, to reach a flexibility level up to two times higher than necessary, and to perform a competitive routine five times more than is required at the competition. He mentioned how injury in elite athletics is unavoidable and being prepared to perform with an injury may be necessary, especially at the Olympic level. Another method for injury prevention Michail mentioned is a proper warm up prior to the main part of the training session.

Coach Julio from Cuba stated how hard work is what contributes to injury prevention in his coaching. Knowing the limits of human capacity is important so that you do not push an athlete too hard. Coach Josean of Spain made a bold statement saying, "I believe that 90% of injuries in elite athletes happened because of abnormal mental status of athletes, when they get tired at the end of season." Mental exhaustion could definitely have an impact on an athlete's performance. He also mentioned that when coaches are impatient, athletes tend to get injured. Josean also blamed the inability to "load" athletes mentally to make then psychologically strong, for athletic injuries. All of these coaches described their personal techniques for preventing injuries for their athletes. They all mentioned different aspects and methods, which is very interesting. Some prevent injuries by gymnastics, happiness, hard work, overloading, and fitness. Where these coaches are influences their techniques and effectiveness on their athletes.

Overall, I think that elite coaches' opinions and perspectives on injury are important to consider when developing individualized training programs for athletes.

CONCLUSION

Injury in athletics is a growing concern. Despite technological innovations, coaches' advanced knowledge about their loved sports, progresses in scientific research on various aspects of athletes' preparation, the number of injuries with long lasting consequences has a tendency to progressively increase. Premature return to sport participation based upon resolution of only physical injury symptoms considerably enhances the risk for re-injury. Moreover, multiple traumatic injuries induce psychological trauma that is often overlooked in making decisions regarding the return to participation. This psychological trauma is evidenced by cognitive impairment, sensory-motor disabilities, and overall behavioral properties that may lead to development of so-called "bracing reactions" or "self-protective responses." This is a dangerous symptom that ultimately may contribute to high risk of re-injury.

Our current research and the interviews with elite coaches clearly support our previously formulated notion that resolution of only physical injury symptoms is not indicative of injury resolution (Slobounov & Sebastianelli, 2006) allowing the athletes to fully participate in their loved sports. Incomplete recovery of either physical/functional (i.e., strength, range of motion, endurance) or psychological (emotional status, irrational thoughts, preoccupation with possible injuries, motivational attributes, inadequate goals) functions are definite warning signs for possibility of re-injury. The most important message from the coaches' responses is that education and knowledge about injuries is currently lacking. The most serious concern is lack of knowledge about brain injuries in athletics. Most coaches rely on professional opinions regarding the impact of brain injury and the time frame for return to play. Taking into account that symptoms of traumatic brain injury my reappear months after the incident, meaning that there are could be long-terms functional disabilities even after mild brain injuries, it is essential that coaches should be properly trained and educated in terms of the devastating effects of concussion. Overall, coaches' advanced knowledge about injury is an important factor that may influence their athletes' attitude toward safety and ultimately could reduce the risk of injury in athletics.

REFERENCES

Simonds, C. B. (2005). Development of a questionnaire about concussion and return to assess knowledge and attitudes about concussion and return to play criteria in college athletes. (Doctoral dissertation, LaSalle University, 2004) *Dissertation Abstracts International: Section B: The Sciences and Engineering, 65*(7-B), 3724.

Sefton, J. M. (2003). An examination of factors that influence knowledge of and reporting of head injuries in college football. Unpublished master's thesis, Central Connecticut State University, New Britain, Connecticut.

Livingston, S. C., & Ingersoll, C. D. (2004). An investigation of collegiate athletes' knowledge of concussions. *Journal of Athletic Training, 39*(Suppl. 2), S-17-S18.

Kaut, K. P., DePompei, R., Kerr, J., & Congeni, J. (2003). Reports of head injury and symptom knowledge among college athletes: Implications for assessment and educational intervention. *Clinical Journal of Sports Medicine, 13*(4), 213-221.

McCrea, M., Hammeke, T., Olsen, G., Leo, P., & Guskiewicz, K. (2004). Unreported concussion in high school football players: Implications for prevention. *Clinical Journal of Sports Medicine, 14*(1), 13-17.

Kirby, D. (2002). Effective approaches to reducing adolescent unprotected sex, pregnancy, and childrearing. *Journal of Sex Research, 39*(1), 51-57.

Kirby, D. B., Laris, B. A., & Rolleri, L. A. (2007). Sex and HIV education programs: Their impact on young people throughout the world. *Journal of Adolescent Health, 40*, 206-217.

Millstein, S. G. & Halpern-Felsher, B. L. (2002). Perceptions of risk and vulnerability. *Journal of Adolescent Health, 31*(Suppl.) 10-27.

Slobounov, S. & Sebastianelli W. (2006). *Foundations of Sport-Related Brain Injuries*. Springer Publishing Company.

CHAPTER 7

INJURY FROM ATHLETES' PERSPECTIVES

1. INTRODUCTION: ATHLETES' RESPONSES TO INJURY

There are numerous predisposing factors for athletic injuries, both intrinsic (i.e., physical/biological/psychological status, including fitness level, personality type, availability of coping resources, history of stressors) and extrinsic (i.e., type of sport, coaching errors, psycho-social environment). Indeed, from a practical perspective, it is impossible to control these mutually dependent factors since there is no solid theoretical foundation for predicting or preventing sports-related injuries. There was an attempt to separate physical/biological and psychological factors related to injury and to develop a multi-component theoretical model of stress and injury (Andersen & Williams, 1988). This initial model implies a direct link between stress induced by sports participation and/or injury resulting from stress responses. Due to numerous critiques of the initial "stress-injury" model, these authors proposed a revised version of the model which emphasizes "bidirectional links" between the athlete's personality and his/her coping resources (Williams, 2001, see also Figure 1).

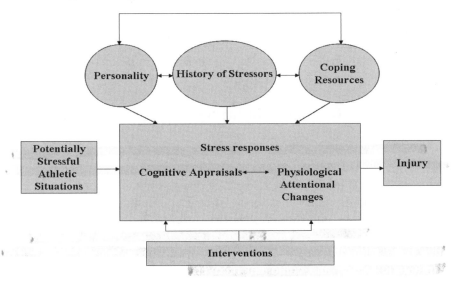

Figure 1. Revised version of "stress-injury" relationship model (adopted form Williams, 2001, with permission from John Wiley & Sons, Inc.)

The central core element of this revised model is a bidirectional relationship between the athlete's cognitive appraisal of the situation and stress responses as evidenced by physiological changes. For example, self-perception of sports-related demands is challenging and realistically may consolidate an athlete's resources both physically and psychologically. This can help him or her stay focused and creates an injury-free situation. When cognitive appraisal is inaccurate and distorted by irrational thoughts and there is the belief that resources are inadequate to meet the demands, the injury risk is increased due to "bad distress" (see: Williams, 2001 for details of this model). Stress, related to attention deficits in this model, is attributed to increased generalized tension, narrowing of the visual field, increased distractibility, reduced sustained attention and the reduced capability to extract meaningful information from background noise. Not surprisingly, sustained muscle tension in conjunction with reduced central resource mechanisms may result in fatigue. As discussed in the previous chapter, fatigue may be a serious risk factor for injury.

Since controversy and confusion surround the classification/definition of injury, it is difficult to properly assess the clinical value of the "stress-injury" model. In fact, it was suggested that the "stress-injury" model is probably most appropriate for acute injuries. For injuries such as overuse injuries, the causes and the mechanisms are largely unknown (see: Williams, 2001 for details of this model). Unfortunately, multiple predisposing factors are never operating in isolation. The ways and manners in which these multiple factors are interacting and collectively influencing athletes' responses are unknown for acute injuries as well. Similarly, it is almost impossible to make any reasonable comparisons between studies examining "stress-injury" relationships due to different research methodologies, types of injury under study, differential severity and the number of injuries experienced over time, age and gender.

1.1. Cognitive Responses to Injury

Cognitive appraisal of the injury in general and its impact on athletes' short and long-term responses to injury specifically, is highly individual. The same impact and amount of damage due to trauma for one athlete could be a career ending injury, but for another it could simply be an annoyance. Numerous times elite athletes have suffered a bone fracture during competition, finished the game and then gone to seek medical attention. At other times elite athletes have developed psychological trauma, even without a traumatic injury, and could not recover, forcing the termination of their athletic careers. It should be stressed that emotions play a critical role in athletes' perception of injury. Despite individual differences, there is commonality in athletes' cognitive responses to injury, as was previously identified by Beck & Emery (1985) and summarized in Heil's (1993) text:

Catastrophizing – exaggerating the severity of injury. A couple of common examples of irrational thoughts are: At the time of injury: "*I'll never be able to play again.*" And then after a pain flare-up: "*I'll never get over this pain.*"

Overgeneralization – incorrectly extending the impact of injury to aspects of playing ability and/or daily activities that are not likely to be affected by injury. A few examples are: "*Because of this shoulder injury I'll probably never be swimming at full speed again,*" or, "*With this injury, my girlfriend will leave me.*"

Personalization – experiencing undue personal responsibility for injury or exaggerating the meaning in relation to other teammates or coaches. For example, injured athletes may be preoccupied with the idea, "*Why me and not others,*" or, "*I was working harder than anyone in the team, but I am the one who always gets injured*".

Confusion, or, "mental block," defined as an injured athlete's acute response, characterized by the inability to cognitively comprehend the injury. "*I was in the greatest shape of my entire career. I cannot believe this happened.*" "*Why did this injury happen when I was ready for the best?*" Unfortunately, injuries always happen at unfortunate times.

Selective abstraction – focusing on specific aspects of an injury that have little meaning in the overall context of the injury. For example, an irrational thought such as: "*My team mate had an ACL and it was a career ending injury for him; it would be for me too.*" Another example is: "*If I were allowed to warm up longer this would not happen.*"

Absolutistic/dichotomous thinking – unreasonable and complex thoughts related to injuries of all categories. This is typically a reflection of injured athletes' lack of appreciation for painful experiences. An athlete is often confused about real or neurotic types of pain and experiences, and has thoughts such as: "*My painful injury is neither physical nor mental.*"

In should be noted that in addition to individual differences in cognitive responses to injury, the type and severity of injury, athletes' positive versus negative experiences with previous injuries and recovery, gender and many other variables should be considered when examining the evolution of athletes' responses to injury. For example, a successful rehabilitation and 100% return to sports participation after an ACL (e.g., severe and often season ending injury) injury is a significant factor in an athlete's appraisal of acute ankle sprain.

1.2. Affective Responses to Injury

The cognitive appraisal of injury never occurs in isolation from its emotional content. Direction of thoughts, beliefs of prompt recovery and an athlete's overall mental trends are hardly influenced by emotions. The feeling of guilt, uncertainty about the course of recovery and the fear of pain are all emotional responses which influence the self-perception of injury. That said, there was a proposal to consider the personal, emotional reaction to traumatic injury in isolation as a grieving process. (Hardy & Crace, 1990). This is similar to a series of stages experienced by terminally ill people (Kubler-Ross (1969). This initial seminal thinking was outlined in *On Death and Dying* and proposed that patients typically experience disbelief, denial and isolation, anger, bargaining, depression, acceptance and resignation. Similarly, injured athletes may follow a five stage grief response, including:

1. *Denial*, as a sense of disbelief as well as varying degrees of failure to accept the seriousness of the situation and severity of injury.
2. *Anger*, as mental and/or physical aggression/irritation due to various attributions.
3. *Bargaining*, as a sense of hope that injury is not as serious as it looks.
4. *Depression*, as a result of anxiety, fear of uncertainty and anticipation of trouble, often manifested in rapid mood swing.
5. *Acceptance* and *reorganization*, as a sense that reality is threatening life goals and personal values.

Among other cumulative reactions to injury, as described by Petitpas & Danish (1995) may be the following:

Feeling of identity loss and/or loss of an important social role such as not being recognized as an athlete any more and being disengaged from sport roles. This may have serious consequences affecting the athlete's self-concept.

Fear and obsessive thoughts of not being able to recover to the pre-injury physical status and performance level. Since the injured athlete cannot practice or compete, there is plenty of time for worry and experience anxiety.

Drop of *confidence level*, due to temporary and/or a prolonged restriction to practice. This in turn may result in decreased motivation for rapid recovery and return to sports participation.

Performance decrement, especially evident upon initial return to sports participation, may be because of lowered confidence and missed practices.

1.3. Signs and Symptoms of Poor Adjustment to an Injury

There are possibilities that the long-term cumulative effect of previous injuries that may lead to a problematic adjustment to the current injury and recuperation process. Some common signs of poor adjustment to injury are identified by Petitpas & Danish (1995) and summarized as follows:

- Evidence of situational anger, depression as evidenced by alteration of mood, confusion and/or apathy.
- Obsessive thoughts of not being able to recover and obsessed with asking when s/he can return.
- Denial, reflected in irrational remarks and thoughts, such as, 'injury is just an annoying delay and not a big deal.' It is a reflection of an athlete's effort to convince him/herself or other people that the injury does not really matter.
- An injured athlete's desire to return to sports participation prematurely based on upon an irrational perception of the injury, and driven by fear and denial.
- Exaggerated storytelling and/or bragging about past athletic accomplishments. Obsessive thoughts of fast, unrealistic recovery and ignorance of the severity of the injury.
- Dwelling on minor somatic complaints, not mentioning restricted ability to move and accomplish basic skills.
- Guilt, shown by constant remarks about not contributing to the team effort.
- Becoming dependent on the therapist or on the therapy process; displaying a lack of involvement in the treatment process and being a passive recipient of the treatment protocol.
- Isolation and withdrawal from teammates and coaches. Making excuses for skipping therapy and medical appointments.
- Acting helpless and hopeless; asking "Why should I even try?"

Both emotional and cognitive athletes' responses and problematic adjustment to injury may be assessed by interviews, observations of athletes in various sports and their daily living environment, and by providing psychological tests. Only interviews performed by experienced professionals may have an implication for the clinical practice in dealing with injured athletes. Accordingly, in the following text, definitions, common strategies of effective interviewing techniques and basic procedures of initial interviews will be discussed.

2. METHOD OF INITIAL INTERVIEW

Clearly, the most direct approach to assessing the individual experiences of injured athletes is to interview these athletes about their feelings, behavior, attitude and the likes. However, in order to obtain valid information about athletes' experiences with injury and their personal views on injury, <u>robust</u> interviewing techniques must be used. And most importantly, the manner and atmosphere while interviewing injured athletes must be preserved. It should be noted that injured athletes might have personal reasons to interpret certain questions in a variety of ways. She may want to please or displease the clinician for unknown reasons, and participate in the interviewing process accordingly. Therefore, biases and interviewing errors are important factors to consider before making any assumptions and conclusions regarding the effect of injury on both athletes' physical status and their psychological well-being.

First, it is important to provide a definition of "interview" and dissociate it from "conversation". In general, interview can be defined as "...face-to-face verbal encounters or exchange of ideas and opinions...but the interview interaction is designed to achieve a consciously specific purpose" (Wiens, 1983). The initial interview is more purposeful and organized than just conversation; however, it is less formalized than a pure psychological test which is aimed at revealing covert or hidden features that cannot be assessed by the interview or behavioral observations. In other words, a unique characteristic of the interview method is the wider opportunity it provides for an individualized approach that will be effective in eliciting data from a particular person (Pharez, 1988). That said, the relative flexibility represents both the strength and the limitations of many interviewing techniques. Not surprisingly, the interview method is regarded as a combination of art and science. Despite the flexibility of an interview as a method to elicit information about the patient, there are three common principles that dissociate it from conversation. The interview must be: (a) *goal-orien*ted; (b) *carefully planned*; and (c) *skillfully executed*.

In terms of *goal-orientation* of the initial interview, two things are important to consider. First, there are different types of interview that will be discussed in some details in the following text. Second, prior to the interview, an initial hypothesis should be formulated regarding the nature of the problem under study. For example, in dealing with injured athletes it may be hypothesized that:

"Every single and moderate to severe acute traumatic injury results in the development of psychological trauma"

"Only severe traumatic injury results in the development of psychological trauma"

"Multiple even minor traumatic injuries result in the development of fear of re-injury"
"Female athletes experience more severe psychological trauma than male athletes suffering from the same type of injury"
"There are multiple symptoms of injury, which should be resolved prior to return to sport participation. The presence of some of the symptoms may result in more severe recurrent injuries."

Accordingly, while conducting the interview it is an intention of the clinical practiner to rule out and/or confirm the initially formulated hypotheses. It is anticipated that at the end of interviewing session, the steps for solving the problem will be clearly formulated and intervention strategies are proposed.

In terms of the necessity to *carefully plan* the interviewing session, it should be considered that:

(a) It is inappropriate to conduct an interview without scheduling the time and place beforehand. Accordingly, depending on the place (i.e., office and/or training room) an appropriate surrounding atmosphere should be prepared.

(b) It is the clinician's responsibility to get knowledge about the sport the injured athlete is engaged in. Accordingly, not only general knowledge about this sport but also common terminology should be learned prior to conducting the initial interview with the injured athlete.

(c) History of previous injuries should be requested from medical practitioners, upon receipt of the athlete's consent, to consider the possibility of chronic cumulative abnormalities.

(d) Background information about the actual incident of injury and an athlete's initial response to injuries should be requested from the athletic trainer and the medical doctor.

(e) If possible, team mates may be interviewed as well to get more information about the injured athlete.

2.1. Types of Initial Interview

There are different types of initial interview commonly accepted in clinical practice. They differ less in techniques and methods than in goal and purposes, namely to elicit valuable information about the case prior to administering intervention strategies. As mentioned before, a quality interview is an essential part of the treatment protocol. As was noted by Berg (1954), "… it is no exaggeration to assert that a bungled intake interview can prolong treatment while an effective one can shorten it." (cf: Pharez, 1988, p. 149)

The intake-admission interview, aimed at: (a) determining why an injured athlete has come to the clinician; and (b) judging whether the clinician with his/her competence may meet the needs and expectations of the injured athlete;

Case-history interview, aimed at acquiring concrete facts, dates, events, sites, type of competition/practice. Overall, as much information as possible should be obtained in the first initial session with the injured athlete;

Diagnostic interview, aimed at provoking and assessing an injured athlete's mental status. It is important to note both verbal and non-verbal (facial) expression, alteration of postural responses, present mood and emotional status, abnormal mental trends (preoccupation with the injury, hopelessness/helplessness). Some examples cited from Phares (1988) are as follows:

(a) *Expression and Posture* (e.g., mask-like expression, silly smile, grimacing, etc.)

(b) *Behavior during interview* (e.g., hostility, lack of insight, cooperation, etc.)

(c) *Mood and emotions* (e.g., flat affect, euphoria, excitement, etc.)

Pre-test, post-test interview, aimed at exploring the preference and possibility of administering psychological inventories, if necessary, to assess the psychological profile of injured athletes in more details. For example, verbal or visual tests may be more appropriate to further assess the "covert" signs and symptoms of psychological trauma as a result of injury;

Research interview, aimed at elicitation of information about this specific injury and/or athlete. The results of this interview may be presented as a case study describing the athlete's unpredicted responses to injury;

Pre-therapy interview, aimed at detecting the preferable psychological therapeutic techniques and/or intervention. For example, depending on the athlete's direction of irrational thoughts, behavioral/physiological/psychological symptoms of trauma discovered during the initial interview, a program consisting of one or more of the following should be recommended: behavioral modification, healing imagery, negative thought stoppage, or stress inoculation.

Crisis interview, aimed at detecting the potential for disaster and preventing crisis. Most often this type of interview is conducted during severe, disabling and career ending injuries in athletes. Proper and immediate referral to a qualified clinical practitioner is necessary following the crisis interview. Obviously, the main goal of the crisis interview is to meet the problems as they occur and to provide an immediate resource.

2.2. Interviewing Skills: Types of Questions

It should be noted that regardless of the types of interview identified in the previous text, a quality interview depends on the types of questions, which may become more structured as the interview proceeds. The type of question should be a priori designed and carefully formulated to facilitate clinician-injured athlete communication during the interview. In the following text types of questions aimed at assessing the status of the injured athlete, as initially proposed by Maloney & Ward (1976) for general clinical practice, are provided.

Open-ended questions. Regardless of the type of initial interview, the interviewing session should start with a series of *open-ended* questions, basically to prevent "yes-no" answers, indicating that the interview is over before it gets started. These types of questions should provide a rich foundation for constructing a hypothesis and allowing the clinicians to observe the injured athlete's cognitive and emotional status. A few examples of this type of question include: *Why are you here? Tell me about your sport. Tell me about your experience with your sport-related injuries.* Again, *open-ended* questions are the essential part of the whole interviewing process and should be planned accordingly.

Facilitative comments and questions. These are also known as "*supportive comments*" used in the interview. For example, in order to encourage the athlete to elaborate on some details that may be critical to assess the impact of the injury, the following questions may be asked: *Do you remember what happened prior to the collision with another athlete when you had concussion? I suspect that it is important...Could you expand on your feelings right after the injury? Did you experience different symptoms after the second concussion, than after the first one?* It is important to note that facilitative comments and questions should be addressed in such manner that direct "**yes** or **no**" answers are prevented.

Confronting questions. It is common for injured athletes to consciously or subconsciously provide inconsistent or even contradictory statements. In this case, to direct an interview in the right direction and to eliminate confusion, confronting questions may be asked, such as: "*...Well, I must admit that I may misunderstand your previous statement, but you earlier said that ...*". That said, it should be noted that a contradictory statement may contain a lot of meaning and make a lot of sense and may suggest a desire to mislead the clinician. Thus, confronting questions should be used with great caution to maintain the proper psychological climate during the interview.

Direct questions implying direct "**yes** or **no**" answers should be used primarily at the end of interview, or when the clinician has achieved a certain level of trust and sense of understanding of the situation and intends to terminate the interview. Here are some direct questions to consider: *Did you see a neurologist after the concussion? Did you pass neuropsychological testing? Have you lost consciousness at the site of the injury? Does it still hurt?* It is important to note that overuse of direct questions may tend to shift the entire responsibility for the course of the interview onto the clinician. This could result in an interview that evolves as a series of short questions and short answers of no use.

2.3. Rapport and Non-Verbal Communication

2.3.1. En rapport

Apparently, the most important factor of a successful interview is the clinician's ability to establish a mutually beneficial and supportive relationship with the injured athlete. Everything matters when it comes to injured individuals, because it is not the injury per se but the injured athlete as a whole that needs proper treatment and care. The special term used to characterize the clinician-patient relationship is *rapport*. Berg (1954) defined *en rapport* as the comfortable atmosphere surrounding the interview which is permissive, reasonably harmonious and characterized by mutual interest. It is not a requirement for the sake of rapport that every patient would become a friend, but it's required that proper attitudes, understanding, sincerity, acceptance and empathy from both parties are achieved. When an injured athlete realizes that the major goal of the clinician is to understand the problem and provide help, then a solid background for rapport is established. In other words, *en rapport* assumes a relationship founded upon respect, trust and confidence that the problem will resolved with mutual effort from both clinician and injured athlete. It is important to realize that the injured athlete should be an <u>active participant</u> of the treatment protocol, rather then a <u>passive recipient</u> of the clinician's instructions.

There are several predispositions and ingredients that make *en rapport* possible. Clearly, the clinician's values, background and experience are among the most important factors in good rapport and an overall effective interview. Among other factors of *en rapport* are the following:

Maximally effective **communication**, both verbal and non-verbal aimed at bidirectional delivery and processing of the information and to achieve the goal of the interview;

An **Initiation** of the interviewing session aimed at establishing a comfortable atmosphere and in which the plan of action is proposed. It is recommended to start the interview with some irrelevant issues to assess the injured athlete's current mental and physical state and reduce anxiety and tension.

The use of **proper language**, considering the injured athlete's background, education and sport, is critical for a good rapport. As mentioned in the previous sections, it is the responsibility of the clinician to educate him/herself about the sport in which the injured athlete was involved and to get accustomed to sport-specific terminology. It is not recommended to use scientific jargon to demonstrate the clinician's "competence" with the intention of getting respect from the patient.

Silence may be indicative of the injured subject's resistance, but also may have a lot of meaning. This may indicate the patient's confusion and inability to think rationally. That said, prolonged silence may ruin the rapport and if this happens, it is recommended that clinician should introduce new line of inquiry.

Effective communication is not a monolog but rather a dialog aimed at understanding, assessment and acceptance. Therefore, **listening skills** and evaluation of both the cognitive and emotional content of the injured athlete's behavioral responses is one of the key ingredients of en rapport. In fact, the main reason for a "broken link" with the patient is the failure to listen due to multiple reasons, including: (a) distraction by irrelevant stimuli/information/memories; (b) preoccupation with irrelevant thoughts; (c) concern with appearance and the patient's impression of the clinician. These are just a few elements that can cause the clinician's inability to obtain and process the information valuable to the assessment of the situation. Therefore, it is important to develop a habit of being an "active" listener.

2.3.2. Non-verbal communication

It was long time ago when Leather (1976) declared that non-verbal communication has great functional significance in our society, for several major reasons:

1. Non-verbal rather than verbal information is the major determinant of meaning and attitudes. Specifically, tone of voice, facial expression and whole body posture may provide more clues than an athlete's expression "this hurts a lot". Pain experienced as a result of injury can be better expressed by the facial expression of grimacing.
2. In addition, non-verbal means more accurately convey the intensity of feeling and emotions than do verbal expressions of being sad and/or frustrated by the injury.
3. It was well-know for centuries than non-verbal clues usually transmit meaning and intentions that are relatively free from deception, because these are less under the conscious control of the patient. Conceptualization and thinking about possible answers and their consequences distort and modulate the patients' responses and create confusion about the situation.
4. Non-verbal clues may provide additional information, if properly assessed, and serve to clarify the patient's intention, direction of thoughts and meaning of the words.

5. There are <u>less redundancy</u>, repetitions and inconsistencies in non-verbal clues

6. Finally, non-verbal clues are highly <u>suitable for suggestion</u>. That said, it should be noted that non-verbal clues should be used with caution due to the element of subjectivity and biases. Overall, non-verbal communication is an important attribute in a successful interview, but the interpretation of the clues may vary from case to case.

Clearly, since Dr. Leather's declaration, the clinical value of non-verbal communication and skills to extract meaning have become even more appreciated. Indeed, the skilled clinician is one who has learned to extract meaning from signs and gestures, from facial expressions and grimacing, from whole body posture and tone of voice. It is a "marriage" of science and art that makes the clinician a valuable asset to an injured athlete's treatment.

CONCLUSION

Unfortunately, there is a lack of well-controlled research on athletes' responses to injury. Numerous confounding factors well out of the control of the researchers basically make it impossible for clinical professionals to rely on speculations obtained by the researchers. For example, several signs of "poor adjustment to injury" were identified by sport psychology researchers. A list of symptoms as warning sings of poor adjustment to injury was proposed. It included:

(a) Feelings of anger and confusion;
(b) Obsession with the question of when one can return to play;
(c) Denial, as the direction of thoughts that injury is not a big deal;
(d) Guilt about letting the entire team suffe;
(f) Rapid mood swings, from hope to depression;
(g) Hopelessness that 100% recovery may happen.

Would it be feasible to propose that these aforementioned symptoms are not an indication of "poor adjustment to injury" but rather the common patterns of elite athletes' responses to acute injury? Are there any "good adjustments" to injury? Over twenty years of coaching and dealing with injured athletes, I've never seen the symptoms of "good adjustment" to injury, especially if the injury requires prolonged treatment and rehabilitation. That said, with proper assessment of the impact of the injury, personality predispositions, and available coping resources, well-grounded interventions may be successfully administered.

REFERENCES

Andersen, M.B. & Williams, J.M. (1988). A model of stress and athletic injury: Prediction and prevention. *Journal of Exercise & Sport Psychology, 10,* 294-306.

Beck, A.T., & Emery, G. (1985). *Anxiety Disorders and Phobias: A Cognitive Perspective.* New York: Basic Books.

Heil, J. (1993). *Psychology of Sport Injury.* Champaign, Ill: Human Kinetics.

Hardy, C. J., & Crace, R. K. (1990). Dealing with injury. *Sport Psychology Training Bulletin, 1*(6), 1-8.

Kubler-Ross, E. (1969). *On death and dying.* New York: Macmillan

Petitpas, A., & Danish, S. (1995). Caring for injured athletes. In S. Murphy (Ed.), *Sport Psychology Interventions* (pp.255-281). Champaign, Ill: Human Kinetics.

Wiens, A.N. (1983). The assessment interview. In I.B. Weiner (ed.), *Clinical Methods in Psychology.* (2nd ed.). New York: Wiley-Interscience.

Pharez, J. (1988). *Clinical Psychology. Concept, Methods, & Profession.* Third Edition. Chicago, Illinois: The Dorsay Press.

Berg, I. A. (1954). The clinical interview and the case record. In L.A. Pennington & Berg (eds.), *An Introduction to Clinical Psychology.* New York: Roland Press.

Maloney, M.P. & Ward, M.P. (1976). *Psychological Assessment: A Conceptual Approach.* New York: Oxford University Press.

Leathers, D.G. (1976). *Nonverbal Communication System.* Boston: Allyn & Bacon.

CHAPTER 8

INTERVIEWS WITH INJURED ATHLETES

Interview 1: Baseball

Tom T. a junior at Penn State University, has been playing baseball for more than fifteen years. Tom majors in Recreation and Tourism Management. He was the starting pitcher at his high school, Council Rock. During his career in baseball, he has suffered many injuries. Tom has broken his right wrist twice, left wrist once, and his right elbow twice. Altogether in his life he has incurred thirteen broken bones (note that some of the breaks were not sports-related). Unfortunately, during his last season of baseball in high school, Tom broke his elbow for the second time, eliminating him from being able to play for college. He told the story of how this occurred. He was running and slid into home plate, and the catcher fell on top of him and broke his elbow, ending his career. Even though Tom did go through rehabilitation, he felt that he could not pitch as fast and hard as he used to and removed himself from the sport. Still today, he cannot pitch in certain ways due to the injury that happened in his senior year. In this interview, as you will read, I focused on the aspects of what he endured during his season, obtain the view through his eyes as a coach and what he has seen from his coaches, and how still to this day he is dealing with not having the same ability he had once had.

Q1: As you know, injury is a common risk and is an unavoidable part of sport. Can you elaborate on why injury is still an unavoidable part of athletics today? What elements do you feel are more essential in your training as a coach on how to prevent the risk of injury? Putting yourself in your coaching shoes, what do you feel are the more crucial elements in reducing this risk?

Tom T: I strongly feel that injury in sport is inevitable. Most of injuries that do endure are mainly freak accidents. Take for example on how my last injury happened and ended with me not pursuing in baseball anymore. All I was doing was sliding into home plate and the catcher fell on me. He did not do it on purpose, but just by accident. We, as athletes, can only do so much to be cautious, but no one is perfect and can avoid the risk of injury. Participating in sports there is always that risk. The coaches can only teach and do so much to prevent this from happening. The elements that I feel are essential would be position, getting the technique down right and also, a training program. Having a program, such as strength training really helps build up the core areas of your body. Also, having the right state of mind during practice and competition, this can help eliminate the risk. Some injuries do happen, because athletes are not focusing on technique, or what others are doing around them. If I were to put myself in the coach's shoes, I would strongly encourage my players to get into a great strength training program, teach them the basics with proper form and positioning, and get my players to have trust in me. Having trust is a major component, because if you did not have trust I strongly believe that will happen more frequently . As a coach, they are the ones who are aware of things, had different experiences, and have dealt with these situations.

Q2: Psychological trauma has become a very important issue in sport today. Have you ever dealt with or seen another one of your team mates have sport-related psychological trauma? Do you feel that every single injury may cause some type of psychological trauma and that athletes need to seek counseling after injury?

Tom T: In certain aspects, I felt like I did go through psychological trauma, but not to the extent that I needed to seek any counseling for it. After seeing physically, I could not throw a pitch as good as I once had really got to me. It did have the affect that had led me to end my baseball career, but still I don't feel that I needed help for it. Everything happens for a reason. Yes, mine was due to injury, but as a look back on it I probably would have gotten burnt out from the sport and just quit later. I have seen some teammates go through psychological trauma, but it was not sport-related, such as depression, bipolar and other disorders. To answer the second part of the question, I do feel that certain injuries due cause that psychological trauma. By having the top star in whatever sport have their talent taken from them in a matter of minutes and have nothing left, then I feel they need that counseling. The fact that something is taken away from you in the matter of moments does take a serious toll on the athlete, especially if they have nothing to fall back on in life. So, basically it depends on how much the sport weighs on the athlete that they would have to endure some type of counseling.

Q3: Injured athletes who return to sports sometimes feel that they may not be psychologically ready to come back. They start to display a symptom called "bracing behavior." Coming back from rehabilitation have you ever experienced yourself displaying this sort of behavior? What signs and symptoms could you identify in your behavior (if any) or in other athletes on your team? How would you prevent this behavior and how would the coaches deal with this problem?

Tom T: I honestly have to say that I have never done or thought I had displayed that type of behavior. After all my injuries in baseball and any sport I did, I tried to come back early. By me coming back as early as I had done, they thought I had re-injured my arm. The main reason that I wanted to have a fast recovery is that I wanted to play. I always had the motivation to get right back into the swing of things. I loved baseball it just kept me wanting to get back in right away. Some "bracing behaviors" that I saw from my teammates would be just complaining that they were still hurt, therefore the coach would sit them for practice and not play them in the game. With the coach not allowing them to play in the game, they would complain of why they were not getting in at all. Since I don't think I displayed the behavior, I would not know how to prevent. Seeing that my coach had to deal with some players like that, he would just go back to the basics, you practice before you play. Honestly if they did not practice after an injury, I feel that they are more prone to re-injury again. It all comes back to having that coach-athlete trust. You do not want to have the risk of re-injury and neither does your coach. The coach has to do what he feels is the best for you and the team.

Q4: Do you think that athletes' psychological adaptation to injury may play a role in the rehabilitation process? Do you feel that if you are 100% symptom-free you are ready for sport participation?

Tom T: Psychological adaptation does play a huge role in the rehabilitation process. If you are not mentally ready then physically you would not be either. These two comes hand-in-hand. If you are only physically ready, then you are probably more likely to milk the injury (or display that bracing behavior that we just talked about earlier). The more mentality you have the stronger you are going to be as a person and in the sport. You mental ability keeps you motivated. Always get you going in whatever you do. Now if you are 100% symptom-free in both physical and mental aspect, then yes you are absolutely ready to come back. The only thing that would hold you back if you lack one of those elements, but other than that I would think and feel that you are ready to get back into the game.

Q5: Who should be responsible for the final decision in terms of an athlete's return to full sport participation: coach or medical doctor? Do you think that there should be different criteria in terms of an athlete's readiness for retuning to practice versus returning to competition?

Tom T: I feel that the decision should be made by the athlete, medical staff, and then the coach. In making my comeback in sport I made the decision to come back early, even though it may not have been the best decision made by me, but it's solely up to the athlete. I mean we are the ones who know our own body and what it is capable of doing. The medical staff and coach can only see how you're doing though what you tell them or what they see in rehabilitation or practice. For the criteria of coming back with the athlete's readiness you have to perform well and let the coaches see and know that you are ready and are able to compete at the level you did before the injury had occurred. If you are not performing top notch then the coach would not let you start like you were before. You have to start from the basics and move your way up. My coach for baseball let me practice and in the beginning of making my way back to competition, he would only put me in for one or two games. The coach solely believed if I was in there too long, then the injury would emerge again.

Q6: Sports psychology has come up with a number of interventions to help speed up the rehabilitation process of injuries. Have your coaches and medical staff ever used an intervention to help you return to full sport participation? Which technique did you feel worked the best for you?

Tom T: Once my injury was healed, the rehabilitation staff sets certain goals and had me learn the basics again. You can not just jump right into the sport again. Everything needs to take time, especially dealing with an injury that has set you back for awhile. After the medical staff said I was good to go, it was then up to my coach. Coach still took it slow with me, because I was still getting used to things again. He had me go over the basics and use the imagery technique. Once I started to see it in my head, and start to show results on the field, he then started to make me set some goals during practice, which got me ready for competition. Having these interventions helped speed up the repairing process, but only to a certain extent, because having an injury can be either mental or physical. No one has the magic touch to be like; okay your arm is not broken or that you are not afraid to play again. We just need to take our time in the recovery process and not speed things up that could lead to problems later on in life.

Q7: Studies have shown that there are clear gender differences among athletes dealing with the psychological and emotional aspects of an injury. Even though you have played in male sports all your life, do you feel that there are different strategies that coaches take for female athletes verses male athletes? Do you feel that female athletes express different emotions than male athletes do?

Tom T: I have no prior experience in this area, since I have been only playing with all male athletes. Maybe there can be different strategies that coaches give to female athletes, because they do tend to be more moody and emotional than males. I mean females do go through more psychological barriers, such as eating disorders, image, self-esteem issues, whereas male athletes do not usually see this type of barrier besides trying to perform their best. For male athletes, we hardly ever show emotion or affection, because if we did, we would be made fun of or be labeled as the weak one on the team. I feel that coached towards female athletes might be a little harder to eliminate these biases against them. If you really think about it, in the end it might make the female athlete stronger and come over these drawbacks.

Q8: In your opinion, and in terms of psychological recovery, do you as an injured athlete think you would recover better or faster if given more attention or less attention? Also, do you feel that there is a difference in response to attention paid to the injury in male verses female athletes?

Tom T: In dealing with my injury it helped that the coach and medical staff gave me the attention that was needed. It helped me pull through the process better, but I wouldn't say faster. The reason for that is, because I think I got myself motivated come back faster since I wanted to bad so badly. Obviously, getting attention helps even more, because it gives you more motivation to do better and hopefully comeback fully recovered. When coaches, medical staff, or even parents show attention to the injured it creates a sense of trust. Trust is a major issue that comes along with sport. If you don't have trust than what do you have? Makes you feel that you don't have a stable person to fall back on. I don't feel that there is a difference response to attention paid females verses males. There are certain coaches respond to situations differently. Dealing with female athletes, you just have to watch what you say and not treat them like babies.

Q9: Another issue that concerns many athletes and coaches is brain-injuries, mostly concussions. Have you ever dealt with a head trauma? Did your coaches ever talk to you about concussions and were they well informed? Do you feel that coaches can try and prevent concussions?

Tom T: I am lucky enough to say I have never dealt with any concussions or severe head traumas. The coaches did talk to us about certain injuries, but did not discuss specifically about concussions. He made it very general, and instructing to be safe during practice and competition. They talked about the risk and dangers involved, but that was summing up all physical injury that could transpire. When the coach was talking about the risks involved I would assume that they are well-informed in some situations, but they are not part medical staff so how much should they know? I mean you hear of injuries all the time when you're growing up, so some just assume that you are aware of the risk. Coaches can try and prevent concussions, but can only do so much. Make sure they wear proper equipment for protection, and make sure they are paying attention and focusing. Other than what I said, I do not think there is much more that can be done.

Q10: Some athletes exhibit certain symptoms after having a history of concussions, including headache, dizziness, nausea, emotional liability, disorientation in space, impaired balance and postural control, altered sensation, photophobia, and lack of coordination and slowed motor responses. Do you feel that your coach labeled these symptoms as irregular or abnormal in the affected athlete?

Tom T: I do not personally feel that I have that ability or knowledge about these symptoms, because I'm not a medical doctor and have not experienced a concussion before in my life. If I recall, I do not think that the coach would label it some of those symptoms as irregular or abnormal, because half those symptoms are common in any sport atmosphere. If you run too much, you may feel nausea or dizzy. Not hydrating yourself can lead you with some of these symptoms. So basically, I feel the trust issue comes back into play. Honestly if the athlete feels that they are not acting like themselves or talking about how their habits have changed, then I would think that the coach would see it as abnormal.

Q11: Since you haven't endured a concussion, do you feel that it is obvious during practice or competition that a previously injured athlete has developed "bracing behaviors?" Did your coach have particular strategies for dealing with athletes who exhibit this type of behavior in response to injury?

Tom T: I have never seen one of my teammates milk a concussion, but I'm sure some do. They probably feel that there could be some physiological damage that could happen. If the athlete feels like they are not ready to come back, because they had a previous concussion maybe they need to take more caution in their sport. If I was showing that behavior, I probably would go and talk to my coach and see what he has to say to me. He would probably send me to the medical staff, which I guess they would do some tests to see if I was alright, but I can't really give you an answer since I have no prior history of having a concussion. The strategies coaches might exhibit to the athlete might include some psychological interventions or just have the athlete taken over to the medical staff, so that they can discuss the fear the athlete has about this issue.

Q12: What advice would you give athletes to identify risk of injury or prevent injuries to happen?

Tom T: Listen to what your coaches tell you. Also, drink lots of milk, and have a balanced diet. You need a well-balanced diet to get all the vitamins and minerals to help with injury and the healing process. Stay positive and always keep your head up even in the worst situations, because you have the ability to recover and give it all you have. I would emphasize having a good training program, so that you are physically able to meet the demands from your coach, because as you move up in life and start to play at a higher level the demands become more complex and enduring than when you were little.

Q13: Do you regret your choice of ending your career in baseball?

Tom T: I do not regret the choice I've made, solely because I can still play for fun. I played baseball for fifteen years and I feel if I would of when further that the fun and thrill would be taken out of the sport for me. So, I have no regrets, but just maybe coming back into the season too soon that my range of motion is not back to normal. Every athlete just needs to realize that it takes time and you just need to wait and be patient, because you will be able to play the sport again whether it's at a collegiate level or just for fun.

Interview 2: Lacrosse

No matter what sport the athletes play, they are more than likely to suffer some sort of injury during their career. The recovery process can be more demanding, painful and psychologically difficult than the actual injury. This is particularly true for traumatic injuries. To examine some of the effects a serious injury has on an individual with a history of severe injury, a

Penn State student and club lacrosse team member Jeremy E. (junior-Chemical Engineering) was interviewed.

Q1: *What was your injury and how did it happen? Could you see it coming?*

Jeremy E: During the fall of my sophomore year (2001) I tore my left ACL and MCL playing football. I played running back and was getting tackled. I don't remember any of the details except I was running and I got hit from behind. Next thing I know is that I am lying on the ground with extreme pain and then my knee went numb. I was helped off the field and the trainers looked at it. I ended up going to the hospital and having an MRI which is when I found out the extent of the injury.

Q2: *When did you have your surgery and were there any complications during rehab? How long did you have to wait before you could rehab?*

Jeremy E: I had to wait until the swelling went down to have surgery. I hurt my knee in September and had surgery in October. It sucked though, because I began rehab for about a month after the surgery and then I had a complication since my MCL did not heal right the first time. So I had to go in and have that fixed and rehab for another 5 ½ months. I was able to get back in a little under 6 months and play in some of my lacrosse games.

Q3: *Wow, that seems fast to be fully recovered from an ACL. How did your rehab go and what did you have to do?*

Jeremy E: Nah, it seemed like it was forever. I've always been an athletic and competitive person and couldn't wait to play again. I played football and lacrosse in high school and I still play lacrosse for the PSU club team. I just couldn't stand not playing. In rehab I just had to get my strength and flexibility back after the surgery. It was very painful and sucked a lot. The physical therapist and medical staff was very nice though. I had exercises with weights and a woman sat on my leg to bend it to regain motion and strength. Probably was one of the most painful experiences in my life. I hope I don't have to do it again. My knee still hurt some when I was cleared, but I just figured I had to play through it to get back in shape.

Q4: *While rehabbing did you have a specific goal in mind?*

Jeremy E: I just wanted to get back as fast as I could to full speed 100% and I think I've done that.

Q5: *Did you have any problems getting back into sports physically or psychologically? Did you have to "brace" yourself or compensate for the injury? If so, how?*

Jeremy E: Physically it hurt a little bit when I started running again. I was hesitant to plant and cut like I used to for a bit. I just wore a knee brace for a year after the injury and then I realized I didn't need it anymore and it just got in the way. If I ever got hit the right way it stung some, but now it is basically back to what it was before I got hurt. It does hurt now though sometimes when I run on pavement a lot. I'll always have sort of permanent tendonitis for the rest of my life.

Q6: But no major psychological difficulty?

Jeremy E: I just tried to forget about it for a bit when I was cleared to play and tried to play normal. However, that's much easier said than done. It probably took me a year to get fully comfortable again. My stats suffered some because I was not as aggressive.

I decided I was never going to play football again. I just couldn't do it. With lacrosse I figured I had less of a chance since there is less contact. I have to admit though, I was a little bit worried I could get hurt again, but I was like 'it happens.' If it's gonna happen, it will... I can't do anything about it. I struggled a bit with my conditioning for a while when I got back, but not from any major psychological fear or anything. I never went to see a psychologist or anything. I don't even think about my knee now.

Q7: How did your personality change during this traumatic time and how did your family react?

Jeremy E: My family was real supportive and helped me out a lot and I thank them. I was probably real annoying to my mom. I complained a lot and was real stubborn since I really couldn't do anything other than sit around. I pushed myself too hard at some points I think. Luckily since it was my left leg I was able to still drive. Otherwise, I would have been a pain in the ass for her having to chauffeur me around all the time. I just learned to appreciate the limited opportunities I had in sports more than before.

Q8: How did your coaches help you through your knee injury? Were they well informed about your injury?

Jeremy E: Well it happened during football season and those coaches just told me to take as much time as possible and were understanding. They knew about knee injuries from coaching clinics, but they let the trainer and doctors take care of it. They just offered encouragement mainly. I still went to practice during the season and they let me take part in film sessions. After a while I was able to lift upper body with the team. However, I never played football again after the injury so after the season I didn't have much interaction with them. But my lacrosse coach was very nice. He called me once a week at home to see how I was doing and always offered help if I needed it. I was captain of the lacrosse team and luckily I was able to get back for some of the season. I never felt rushed by any of them. I think I put more pressure on myself than they did.

Q9: Do you think this injury will affect you in your future plans?

Jeremy E: Eh, maybe, but not a lot. Like I said before, I am doing club lacrosse and that is fine. I am also doing Officer Candidate School for the Marines in the summer. I was able to pass all their tests and have my knee cleared to participate. Hopefully, it stays that way. I'll worry about that if it happens.

Q10: If you had to rehab again from this injury, would you treat it the same way?

Jeremy E: Absolutely, I had no problems and I am fine now. I just wish there was a way to get back faster.

Q11: *Do you have any suggestions for athletes, or for coaches, friends and family members of athletes who may suffer a significant traumatic injury in the future?*

Jeremy E: I would just have to say to keep positive. Everything will work out in the end if you put enough effort into the rehab and don't get discouraged if there are any setbacks. As for the coaches, family and friends of an athlete I'd just say to be supportive and include the patient in activities.

Q12: *Many athletes suffer from concussions in sports and they can have severe side effects if not properly diagnosed, how do you think they should be handled by coaches and doctors?*

Jeremy E: I think concussions should be treated safely and with extreme caution. Severe memory loss or psychological effects could occur in the future. Coaches should be informed of concussion symptoms and err on the side of caution rather than risk an athlete's health even if they still can physically compete.

After interviewing Jeremy I found that he reacted to his ACL and MCL tear like I would expect any athlete to. He was in the prime of his career and wanted to recover as quickly as possible to resume playing lacrosse. I got the feeling he is kind of cocky, stubborn and does not like to show any weaknesses. However, he experienced a few setbacks that required additional surgery on his knee. He took them in stride and dealt with them.

When he was cleared to return to the playing field, he was hesitant to go 100%. At first, he was unsure of his abilities, but he was able to quickly regain confidence to be an effective contributor to his team. Eventually, he was able to run, cut and plant with full strength. Jeremy did not have any major or unique psychological fear that required any extra attention or help. He has no noticeable psychological effects from the injury now.

Jeremy's responses were very candid and truthful. His personality has not changed much since the injury. However, he seems thankful that he was able to recover fully.

Interview 3: Cheerleading

The subject interviewed was Ashley F. Ashley participates on the Penn State Club Cheerleading team and has done so for the past three years. Her position on the team is a base, which is the support for stunting and certain poses. Ashley currently has a torn ligament in her left ankle which happened during stunting.   She has had previous cheerleading-related injuries, however not this type of injury. Ashley was asked a number of questions similar to the coaching questions, but from her point of view as an athlete.

Q1: *Could you elaborate on why you believe injury is unavoidable in athletics*?

Ashley F: Injury is unavoidable because when practicing to your full ability, one is not thinking about injury or being cautious. In my sport, girls are being tossed up into the air and caught; without proper techniques and perfect timing injuries can result.

Q2: *Do you agree that every injury could cause psychological trauma and that injured athletes should receive psychological counseling*?

Ashley F: replied that she does not think athletes should receive counseling and that not every injury could result in a psychological trauma. She thinks that athletes should realize that there is a chance of injury and there is a low probability that it will happen again.

Q3: *Do you have any sign of bracing behavior yourself as an injured athlete that you are aware of*?

Ashley F: agreed that she does have bracing behaviors for her injured ankle. She said, "I favor my ankle and try to land more to the right side to take weight off my left injured ankle."

Q4: *Do you think if you are medical symptom-free post-injury that you are fully ready for 100% sport participation*?

Ashley F: did not agree with this statement, instead she replied, "I am not fully ready to go back to competition. I need to rebuild my strength and get back into shape before competing again."

Q5: *Who should be responsible for the final decision in terms of your return to full sport participation: coach or medical doctor*?

Ashley F: responded, "In my situation it should be the doctor's final decision because he knows more about my injury than my coach and the specifics of re-injury to my ankle."

Q6: *What kind of strategies do you do to enhance your readiness to return to sport participation*?

Ashley F: believes that her readiness was enhanced by knowing her position on the team was in jeopardy because there are two teams for the club cheerleading team and the coach could easily pull another base from the other team to fill her position.

Q7: *Do you think there are differences in female athletes versus male athletes coming back from injury*?

Ashley F: believes that females are stereotyped of being more emotional and coaches believe that they will not have the mentality to return to competition. She believes men have to it easier when returning to sports without the hassle from coaches.

Q8: *Do you think you would recover faster or slower with more or less attention given to your injury*?

Ashley F: responded, "I think less attention is better because then you are not constantly thinking about injury while practicing or competing. I also think if more attention was paid to the injury it could increase the risk for re-injury."

Q9: Do you think your coach knows enough about concussions?

"I do not think he knows a lot," **Ashley** replied. "I believe he knows about symptoms of concussions and what to look for if an athlete is concussed since concussions are likely if a flyer falls from a stunt."

Q10: Can you see bracing behaviors within your own teammates?

Ashley F: replied, "I can definitely see teammates favoring one side over another side. Athletes also wear a lot of braces such as an ankle brace or wrist brace to prevent injury or re-injury."

Q11: Do you have any advice for other collegiate athletes in preventing injury or injury in general?

Ashley F: believes that student-athletes should not let coaches push them beyond their limits when they are injured. She also believes that athletes should not jump back into full participation if they are not ready.

Interview 4: Football

There are many factors linked to this increase in sport related injury, but to get a better opinion of why these injuries still continue to the haunt the lives of many athletes, a formerly injured collegiate athlete was interviewed and asked to elaborate on the question asked by many, "How can one prevent injury in sport?"

Christopher B: is a talented freshman Penn State varsity football player. He was a highly recruited wide receiver coming out of the state of Virginia. He has played the game of football for a number of years and is definitely no stranger to football related injuries. Christopher has had a number of sprained ankles, several broken bones and a concussion. So he is definitely one who has his share of injuries and has worked with different medical doctors, trainers, and coaches. With his varied personal experience, Christopher has developed a multitude of opinions on proper techniques for preventing and treating sports related injuries.

Q1: Could you please elaborate with your opinion on why injury is still an unavoidable part of athletics today? What elements do you feel a coach needs to hold most essential when coaching collegiate athletes to reduce the risk of injury?

Christopher B: There are a lot of factors that go into why sport related injuries continue to increase, like fatigue, flexibility and experience. But one of the most important factors I believe is the physical condition of a player. The player may not be in a good physical condition and just because you have very expensive and technologically advanced equipment on does not mean that you are exempt from any possibility of injury. Another factor that most people do not realize is very important is the ability of a player to remain as balanced as possible in all situations in competition. For example, the more balanced you are as a wide receiver, the less likely that you will hit the ground awkwardly and cause an injury. Because I find these two factors to be the most important in preventing physical injury, I think that it would be in the best interest of many coaches to enforce physical health and balance in practice and competition.

Q2: Do you agree that every single injury may cause psychological trauma and therefore athletes should seek psychological counseling shortly after injury?

Christopher B: No, I do not believe that every single injury causes psychological trauma, but I do believe more severe injuries may cause it. If an athlete shows signs of psychological trauma after a severe injury, like broken bones or ligaments, or concussions then I do believe that the athlete should seek psychological counseling shortly after injury.

Q3: Through your experience as an athlete could you describe the signs and symptoms of bracing behavior among yourself or other athletes? What would you think a coach would need to do to prevent or eliminate these symptoms of bracing behavior?

Christopher B: Signs of bracing behavior could range from reduced speed to athletes showing less effort in completing plays that may have involved the injured body part. Signs of fear are common predictors of bracing behavior. As an athlete I think that if a coach observes any bracing behavior in an athlete, he or she should personally talk to the athlete and explain to him or her that the bracing behavior is negatively affecting their performance and if he or she does not let go of it their performance will affect their play and cause re-injury or even a loss of their position on the team.

Q4: Do you think that athletes' psychological adaptation to injury may play a role in the rehabilitation process? Do you think that medical symptom-free post-injury athletes are fully ready for 100% sport participation?

Christopher B: I believe that an athlete's psychological adaptation to injury can play a role in the rehabilitation process. It all depends. Some athletes have the ability to speed up the rehabilitation process by mentally motivating themselves and others do not. I also so not believe that medical symptom-free post-injury athletes are always ready for 100% sport participation. Athletics for the most part need to be both mentally and physically ready before they can return to 100% sport participation. An athlete can be physically ready 100% but only 50% mentally ready and only 75% ready for sport participation.

Q5: Who should be responsible for the final decision in terms of an athlete's return to full sport participation: coach or medical doctor? Do you think that there should be different criteria in terms of athletes' readiness for returning to practices versus returning to competitions?

Christopher B: I believe that neither the coach nor the medical doctor should have the final decision in terms of an athlete's return to full sport participation. I believe that the athlete should have the final decision, because the athlete is the one who knows his or her body the most. The athlete is the one who is going to be returning to the game, not the coach or the medical doctor. But after the athlete's decision, the next important decision is the medical doctor's and then the coach's. I do think that there should be different criteria in terms of athletes' readiness for returning to practices versus returning to competitions. This is a strategy that is already implementing on the Penn State football team.

Q6: What kind of coaching strategies would you recommend to enhance athletes' readiness for returning to a full range of sport participation?

Christopher B: I believe that the coaches should not do anything because it is not their job to come up with strategies that enhance the athletes' readiness for returning to a full range of sport participation. I believe that this job is for the physical therapists and trainers. I think that the coaches should stay out of it because they would only complicate things, especially if they are telling the athlete something that is not parallel to what the physical therapist and trainer is saying.

Q7: In your opinion, and in terms of psychological recovery, do you think that athletes recover better, or faster, from an injury if the injury is given more attention, or less attention?

Christopher B: I think on a broader scale the more attention you pay to an athlete and their injury the more likely that the athlete is going to recover faster than an athlete who has the same injury and received less attention. But on a much narrower scale it actually depends on the athlete. Sometimes the recovery of an athlete has no significant correlation to the amount of attention given to the player.

Q9: What do you think the collegiate coach should know about concussions and what should be done from a coach's point of view in order to prevent concussions?

Christopher B: I am really not an expert when it comes to concussions and therefore have no idea of how any concussion grading scales and return to sport participation guidelines can be used to determine how to prevent more severe secondary or multiple concussions, but I do believe that after three mild concussions or one very severe concussion an athlete should consider ending his or her athletic career. I also really do not have enough information about concussions to know if there are any long-term cognitive and behavioral deficits after single concussions, but I do believe that there is a far greater chance of these effects happening in people who have multiple concussions. I also believe that collegiate coaches should know everything there is to know about concussions, because they happen so often and can cause both short term and long term harm. I do not know what coaches can do to prevent concussions from happening because I really do not know how much research has been done on what causes concussions to happen in the first place.

Q10: Is it obvious during practice or in competition, to you as an athlete, if a previously injured athlete with concussion has developed bracing behaviors?

Christopher B: I believe that it can be obvious at times to coaches and athletes at practice when a member of a team has developed bracing behaviors, because the athlete will change up his style of play. For example, an athlete who is normally runs fast at every practice begins to run noticeably slower after recovering from a concussion and a leg injury.

Q12: What advice would you give to uprising coaches today regarding how to identify athletes at risk for injury and ultimately to prevent injuries among student-athletes?

Christopher B: The only advice that I can give uprising coaches is that they really cannot prevent injuries completely but once they notice that a player is injured he or she needs to provide all necessary physical and psychological treatments to that player to ensure a 100% full mental and physical recovery.

Interview 5: Track and Field: Running

The following interview was conducted with **Meghan N.**, a current Penn State athlete. Meghan is a sophomore who runs for the cross country team and competes in long distance events for the track team. She also is a former high school field hockey player. Meghan suffered an injury to her hamstring muscle during her freshman year, keeping her from competition for a few weeks. She was forced to wear a brace as well. The following reflect her views on injury in collegiate sports.

Q1: Could you please elaborate with your opinion on why injury is still an unavoidable part of athletics today? What elements taught to collegiate athletes do you feel are most essential to prevent risk of injury?

Meghan N: Injury is still unavoidable because even with all the nutritional understanding and coaching expertise, athletes push their bodies to the limit each time they practice/compete. I feel that when athletes are taught to approach their coaches with any minor pains and to see their trainer for preventative physical therapy, they can prevent minor injuries from becoming major. I don't think college athletes are really "taught" any specific elements to prevent risk of injury other than to not let something minor become major.

Q2: Do you agree that every single injury may cause psychological trauma and therefore athletes should seek psychological counseling shortly after injury?

Meghan N: I do not agree that every single injury may cause psychological "trauma". I feel that some injuries, even minor, may get an athlete down, however, I do not feel that every injury causes trauma.

Q3: Through your athletic experience could you describe the signs and symptoms of bracing behavior among your fellow athletes? What strategies could your coach use to prevent, and if observed, to eliminate these symptoms of bracing behavior?

Meghan N: I have seen this "bracing" activity a number of times. Basically, when I see it in runners, if they come back from an injury too early typically run different because they are afraid to run on whatever part of the body was injured. I feel if a coach realizes an athlete is not quite ready to fully participate in workouts on the track they can have them run tempo runs outdoors where no one else is watching the athlete and the athlete can become comfortable with running again without the stress of having to run a certain repeat in a certain time.

Q4: Do you think that athletes' psychological adaptation to injury may play a role in the rehabilitation process? Do you think that medical symptom-free post-injury athletes are fully ready for 100% sport participation?

Meghan N: Yes, I feel that medical symptom free post-injury athletes are ready for participation. However, with any injury regardless of physiological or psychological conditions, the athlete should never go into 100% participation right away. I do feel that an athlete's psychological adaptation to the injury matters in rehab. For example, if they are more positive and able to take things day by day that will probably progress better. But still, regardless of their psychological adaptation, 100% is never good for an athlete right out of injury.

Q5: Who should be responsible for the final decision in terms of an athlete's return to full sport participation: coach or medical doctor? Do you think that there should be different criteria in terms of athletes' readiness for returning to practices versus returning to competitions?

Meghan N: I think the doctor has the ultimate say. They know the athlete's condition and what is best. Yes, as stated above, I think there is a big difference between practicing and 100% competition such as full workouts and competitions. Like anything, after an injury an athlete needs to gain their fitness again. It's like telling someone to go run a marathon without ever training for it. Yeah, I know I could run that marathon and probably finish it, but without training for it, I know I wouldn't be able to walk for many days afterward.

Q6: What kind of strategies do you think coaches should utilize to enhance athletes' readiness for returning to a full range of sport participation?

Meghan N: I really don't think the above activities would work. I think it is more about readiness in training to return to full sport participation.

Mentally, as an athlete, or at least in my personal case, when you get injured, you already set the goal of getting back. From your doctor and trainer, you get an idea of how long it will take to get better. Thus, when you are finally able to train again, you are mentally ready because you have been doing the active steps in physical therapy along the way to get back. Sometimes it is more important for coaches to set guidelines to hold athletes back though. Because I feel like mentally athletes think that they can start right where they left off, and it could be mentally defeating when they realize they can't. As a coach, I feel that the only strategy to maintain an athlete's readiness is to make sure that they have them do workouts and activities in which they can succeed, and to not throw them directly into the fire of competition.

Q7: Do you think coaches use different strategies for dealing with female athletes as opposed to male, particularly in regard to recovery from injury?

Meghan N: Maybe studies have shown this, but I don't personally see women being more fearful then men, at least on my team, so I don't feel there should be gender distinctions.

Q8: In your opinion, and in terms of psychological recovery, do you think that athletes recover better, or faster, from an injury if the injury is given more attention, or less attention? Also, do you feel that there is a difference in response to attention paid to the injury in male versus female athletes?

Meghan N: I think that it is the right medium of "more and less attention" it's the "right attention." If you look at an injury and don't over-obsess or brush it off, and see it for what it really is, I feel psychologically, recovery will be best. NO, I don't feel there is a difference for male versus female.

Q9: What do you think the collegiate coach should know about concussions and what should coaches do in order to prevent concussions?

Meghan N: In my sport, this doesn't really apply.

Q10: Is it obvious during practice or in competition, if a previously injured athlete has developed bracing behaviors?

Meghan N: Like I said, in track I can only see it if someone is running different. I was a field hockey player in high school, and I personally know after tearing my ACL that I played more tentatively. I think it is more obvious to see in team sports.

Q11: What advice would you give to uprising coaches today regarding how to identify athletes at risk for injury and ultimately to prevent injuries among student-athletes?

Meghan N: I would just tell coaches to make sure their athletes take care of their bodies and know when they should start rehabbing minor injuries. There are some things such as bruises that you just don't do anything for, but at the same time, you shouldn't let a nagging injury turn into something much more serious.

Interview 6: Soccer

Interview with **Jeff C.**, Penn State soccer player.

History of Injury:
2 concussions
Turf Toe
Hematoma in both quads
Needs surgery on femur
Knee Cap does not track properly causing problem with cartilage in knee joint
Steps to recovery:
Ultra sound on knee and injected medicine
Rehab
Stretching
Ice
Surgery on femur
Visits with trainer

Q1: Could you please elaborate with your opinion on why injury is still an unavoidable part of athletics today? What elements do you feel are most essential in coaching collegiate athletes to prevent risk of injury?

Jeff C: Athletics, despite certain advancements in equipment, which allow athletes to push their bodies to optimal levels, have remained the same throughout time. Speaking of soccer, equipment and rules have been the same since the turn of the century. Shin guards are actually smaller and the cleats are light weight and the ball plays better and is not as heavy. As we make advancements to be able to perform better and push the bodies to the limit, injuries are unavoidable when we push our bodies to this optimal performance level.

The best thing a coach can do is not push the athlete past a certain point and realize when enough training is enough training. A coach that allows players to recover will have healthier athletes.

Q2: Do you agree that every single injury may cause psychological trauma and therefore athletes should seek psychological counseling shortly after injury?

Jeff C: No, not every single injury causes psychological damage. However, a string of injuries or an injury that will keep you away from the sport for a while definitely has a psychological element. I somewhat disagree that athletes should have to seek counseling after injury; possibly if the athlete will be out for a long period of time.

Q3: Through your coaching experience could you describe the signs and symptoms of bracing behavior among your athletes? What would be your coaching strategies to prevent, and if observed, to eliminate these symptoms of bracing behavior?

Jeff C: Bracing occurs often when a player is returning from injury. Oftentimes it is intentional to protect the athlete until he is positive he feels 100% fit. With knee injuries a player will have a no contact regulation put on him during practice. These players may shy away from tackles or situations that may cause contact. If this causes further bracing behavior you can sit the player for a longer period and allow him to practice outside team practices until he or she feels they are strong enough for contact.

Q4: Do you think that athletes' psychological adaptation to injury may play a role in the rehabilitation process? Do you think that medical symptom-free post-injury athletes are fully ready for 100% sport participation?

Jeff C: No not right away because an athlete may not have his timing down or his confidence level may not be 100%. If you allow a player a longer recovery time and give him more practices before competition these things may return. Timing, I feel, is a large aspect of coming back from injury. Timing, meaning being able to flow with the speed of play, not late on tackles etc. If a player's timing is off he or she could get re-injured easily.

Q5: Who should be responsible for the final decision in terms of an athlete's return to full sport participation: coach or medical doctor? Do you think that there should be different criteria in terms of athletes' readiness for returning to practices versus returning to competitions?

Jeff C: Medical doctor. A medical doctor can look out for the athlete's interests without his own interests involved. I have seen coaches push players to return because he needed a stronger squad even though that player was not fully fit. A coach may not always have the player's interests at heart. There should be different criteria for return to practice and games.

Q6: What kind of coaching strategies would you recommend to enhance athletes' readiness for returning to full range of sport participation?

Jeff C: As a coach, I feel that I would leave rehab up to the athletic trainer but advise the player to follow their instruction closely and push themselves hard in rehab sessions. Also I would advise the player to use his best judgment whether he is 100%. Trainers can sometimes baby the injury and not be aggressive enough during rehabilitation.

Q7: Are there different coaching strategies for dealing with female athletes as opposed to male, particularly in regard to recovery from injury?

Jeff C: I feel that injury and strategies are down to the individual's attitude. I've seen women who can be pushed harder than some men and vice versa. In general however, I feel that you can be more aggressive with rehab for males, and females might take a little longer to recover or rehab both mentally and physically.

Q8: *In your opinion, and in terms of psychological recovery, do you think that athletes recover better, or faster, from an injury if the injury is given more attention, or less attention? Also, do you feel that there is a difference in response to attention paid to the injury in male versus female athletes?*

Jeff C: I think that if you give too much attention to the injury that a player may respond differently when returning to competition. It is cemented into his mind when he is given excessive attention to the injury. If rehab is done and emphasis is put on getting healthy rather than the injury itself a player's attitude changes from not getting hurt again to wanting to get back on the field as soon as possible. This attitude could change the confidence level of the athlete returning to the field. And, I feel, there is a difference between female and male athletes in regards to attention given. I feel women are given more attention.

Q9: *What do you think the collegiate coach should know about concussions and what should be done from a coach's' point of view in order to prevent concussions?*

Jeff C: Coaches should understand the severity that comes with brain injuries. Because you cannot see the injury i.e., swelling, bruising, a coach may not understand the medical implications of the injury. Also a player may appear fine and ready to return but he may be vulnerable to re-injury. Personally, I suffered two concussions on successive game weekends and the coach did not understand why I had to sit out as long as I did.

Q10: *Do you think it is possible to discern these symptoms as irregular or abnormal in an affected athlete, and if so, how should coach adjust his/her coaching methodology?*

Jeff C: No it is really hard to recognize these symptoms and a player may not always be truthful for fear of not playing, losing their starting position, or even being subjected to concussion testing (they are not pleasant and boring).

Q11: *What advice would you give to uprising coaches today regarding how to identify athletes at risk for injury and ultimately to prevent injuries among student-athletes?*

Jeff C: First of all, uprising coaches should learn that winning and losing a game is not what collegiate sport is about. They should understand the life of student-athletes and treat us accordingly.

Interview 7: Track and Field: Javelin

Jesse H. is a member of the Penn State Track and Field team. He has thrown the javelin for eight years and is ranked sixth in the nation and third in Pennsylvania. He has suffered multiple injuries throughout his time as a javelin thrower. Some minor ones included shin splints, damaged back alignment, and turf toe. Some major injuries that added up and eventually ended Hershey's career included orthoscopic cartilage trimming in 2002 due to pain in the shoulder when lifting and throwing. During his freshman year he had Tommy-John surgery on his elbow because he tore his UCL during a throw. Six months after that surgery he had Bankart lesion and capsule repair on his shoulder. Finally his last surgery was in November 2006, when his subscapular tendon burst and part of the Achilles tendon had to be attached because 30% of that tendon was missing. This last surgery was career ending. He is now in physical therapy trying to just gain back normal range of motion and activity in his shoulder. He has taken a coaching position for javelin on the Penn State Track and Field team.

Q1: Could you please elaborate with your opinion on why injury is still an unavoidable part of athletics today? What elements do you feel are most essential to prevent risk of injury?

Jesse H: Injury is an unavoidable part of sports because athletes or coaches do not have enough knowledge of the sport they are involved in. Some elements that are essential to prevent injury are to educate athletes and coaches about the sport and there should be people who know the technique or skills of the sport that can observe the athlete and correct any errors that could lead to injury. I know I was trained with speed before proper technique with the javelin and that is a huge component to why I am injured.

Q2: Do you agree that every single injury may cause psychological trauma and therefore athletes should seek psychological counseling shortly after injury?

Jesse H: Not every injury causes psychological trauma; it depends on its severity. My injuries were severe enough, so I actually went through psychological counseling because I lost confidence in myself and I was unsure of my abilities. I also did not know how to approach the situation I was in.

Q3: Through your experience as an athlete could you describe the signs and symptoms of bracing behavior among athletes? What would be your strategies as an athlete to prevent, and if observed, to eliminate these symptoms of bracing behavior?

Jesse H: After my first injury of my arm I noticed I was bracing when I would throw the javelin because I was not actually fully recovered yet, so I was changing my normal form to reduce chances of hurting myself again. However, that just led to more injury and another surgery. I believe the best strategy to prevent and eliminate bracing is to make sure the athlete is totally recovered and if bracing still occurs after clearance to play, the athlete needs to get psychological counseling or talk to the coach about their fear.

Q4: Do you think that athletes' psychological adaptation to injury may play a role in the rehabilitation process? Do you think that medical symptom-free post-injury athletes are fully ready for 100% sport participation?

Jesse H: Yes, an athlete's psychological adaptation to an injury definitely plays a role in rehabilitation. Some athletes adapt quicker to an injury and want to get back to playing as soon as they're recovered and others may take a little more motivating if they cannot adapt to the situation well. I think that if the athlete has kept up with some kind of training throughout the injury then they should gradually get back into the sport. However, I would not say even though they are medical symptom-free that they should be back in a competitive situation right away.

Q5: *Who should be responsible for the final decision in terms of an athlete's return to full sport participation: coach or medical doctor? Do you think that there should be different criteria in terms of athletes' readiness for returning to practices versus returning to competitions?*

Jesse H: I think that medical staff should have the first opinion on athletes' return to full sport participation. Different criteria are necessary for return to practice versus return to competition.

Q6:   *What kind of strategies would you recommend to enhance the athlete's readiness for returning to full range of sport participation?*

Jesse H: To enhance readiness to return to full play I think you need to practice as soon as possible so you do not lose technique or regress too much. Gradually increase practice skills until recovery to a normal level of play.

Q7:   *Are there different coaching strategies for dealing with female athletes as opposed to male, particularly in regard to recovery from injury?*

Jesse H: Gender does not matter. Every athlete has individual reactions to injury. Women tend to get injured less in javelin, but I do not know the reason behind that because they receive the same training aside from the distance they have to throw it.

Q8: *In your opinion, and in terms of psychological recovery, do you think that athletes recover better, or faster, from an injury if the injury is given more attention, or less attention? Also, do you feel that there is a difference in response to attention paid to the injury in male versus female athletes?*

Jesse H: It is better to have more attention, so the athlete can stay motivated to recover quicker and with more confidence in playing again. I do not think there is difference about the response to attention paid to an injury in male or female athletes.

Q9: *What do you think the collegiate coach should know about concussions and what should be done from a coach's' point of view in order to prevent concussions?*

Jesse H: I do not know a lot about concussions but I would say it depends on the severity of the concussion and how many times it has occurred to terminate an athlete's career. When I played football in high school, coaches gave their players bubble helmets after concussions and I felt that was a good way to prevent future concussions. I think coaches definitely need to know the severity of concussions and their symptoms.

Q10: As an athlete, do you think it is possible to discern psychological symptoms as irregular or abnormal in an affected athlete?

Jesse H: Experiencing a concussion before, I would say those symptoms are normal right when you have the concussion initially but if they persist it would be abnormal. I would refer an athlete to a doctor if I noticed these symptoms much later after the concussion.

Q11: Is it obvious during practice or in competition, to you as a coach, if a previously injured athlete has developed bracing behaviors? Do you have particular strategies for dealing with athletes who exhibit this type of behavior in response to injury?

Jesse H: Since I am currently coaching javelin because I cannot participate, I would say that as a coach, if I were dealing with an athlete who suffered a concussion and they continue to brace themselves I would have a talk with them about confidence in playing. If they need more time to recover I would opt for that, but if they just cannot reduce their fear I might just tell them that maybe playing right now is not the best decision.

Q12: What advice would you give to uprising coaches today regarding how to identify athletes at risk for injury and ultimately to prevent injuries among student-athletes?

Jesse H: Coaches definitely need more knowledge about the sports they are coaching. I know my coaches definitely did not know everything about javelin to show me what was being done wrong. I feel that I have suffered from their lack of knowledge.

Interview 8: Football

Name: **Ross M.**
Sport: Penn State Football, Center
Year: Sophomore Eligibility, Junior Academically

Like many collegiate athletes, Ross M.'s love for the game of football evolved at a very early age, and by age nine he had begun his journey and love affair with the sport that brought him here to Pennsylvania State, University Park. When asked why he loves the game so much he responded, "It's the only remaining team sport besides basketball that you can really individualize." Throughout high school he faced minor thumb injuries and three lumbar fractures in his junior year, all injuries he considered trivial. However, during his first year on the team he tore his right MCL and sprained his left MCL only one year later. Both injuries he considers himself to be fully recovered from, and besides a couple of cases of plantar fasciitis (in the right and left feet), his body sees to be in working condition for next season.

Q1: Could you please elaborate with your opinion on why injury is still an unavoidable part of athletics today? What elements do you feel are most essential in coaching collegiate athletes to prevent risk of injury? As a member of a collegiate team with a big reputation, do you feel that there is additional pressure that leads to injury?

Ross M: Injury is an unavoidable part of athletics because in an ideal world technique would be perfect every play, but unfortunately we don't live in a perfect world. As a lineman, you never know what you'll be working with. Different teams with different movements will always be present to challenge you to adjust. You can practice all you want, but you have to work with what is thrown your way the best way you can. Also, I believe that when players start to fatigue, technique is compromised and injuries occur. At points in the game, you will become tired. It's as simple as that.

As far as prevention goes, coaches are accountable for their team's cardiovascular fitness and should harp on technique as much as possible. Our team has great flexibility, strength, and core stability programs to keep us as injury-free as possible. I believe that flexibility is essential for prevention of injury as well as very useful during rehabilitation for an injury. I think the biggest thing as far as injury goes on this type of team is the amount of risk taken when technique is compromised in order to make big plays. Everyone wants to be that person who made that great play, and that's where a lot of injuries occur. Plus, once a player is injured there is so much pressure to come back as quickly as possible.

Q2: In your opinion, what types of injuries qualify athletes for psychological counseling? Do you think that psychological counseling is the best way to facilitate a recovery and if not what do you think is?

Ross M: Not necessarily. Slight injuries cause apprehension, but not every single injury requires psychological counseling. Injuries that are pretty major, including every degree of a concussion, should require some psychological counseling. While counseling can be very important to some injury cases, physical rehabilitation is the best way to come back from an injury, hands down. Regardless of full recovery or restoration to the exact skill level you were at pre-injury, rehabilitation is essential for you to carry on other areas of your life. You may not be able to play your sport as well as before you were hurt, but you will still want to be able to do other activities without complications.

Q3: Injured athletes usually return to sport participation based upon clinical symptoms resolution and upon recommendation of medical staff. However, there is a notion among medical practitioners that clinical symptoms resolution may NOT be the injury resolution. Incomplete rehabilitation following injury may lead to development of so-called bracing (self-protecting) behavior. This is a dangerous situation that may lead to more severe injuries. Through your experience could you describe the signs and symptoms of bracing behavior among athletes in your sport? What would be your strategies to prevent and eliminate symptoms of bracing behavior? How do your coaches respond to signs of bracing behavior? Why do you think some athlete's develop this behavior while others do not? Do you have any first hand experience with bracing behaviors? If yes, why do you think they occurred?

Ross M: Everyone is a competitor at this level so as soon as an athlete is physically cleared, most are mentally cleared as well. If you are told you are physically ready, I believe that something in your head just clicks and lets you know you can do this. If bracing does occur, it is usually not noticeable during plays especially because of the pace of plays. However, you may see a teammate who has come back from an injury favoring the uninjured side of the body after a play has occurred. As far as my injury, I thought a lot about my knee on the field and in training making sure if was safe at all times. Even when I tried not to, I couldn't help it. I always worried if it was feeling looser than usual and whether or not it would make it more vulnerable during training or in competition.

Coaches definitely can pick up on obvious bracing behaviors. They usually respond by pulling the athlete aside and privately asking them what's going on. I believe that whether or not an athlete develops these bracing behaviors depends specifically on the athlete and the severity of their injury especially if the injury has negative implications for their lives and playing time. Full rehabilitation is the key to avoiding these behaviors.

Q4: Do you think that athletes' psychological adaptation to injury may play a role in the rehabilitation process? Do you think that medical symptom-free post-injury athletes are fully ready for 100% sport participation? What do you think completely qualifies an individual to return to sport after an injury? How do your coaches approach this concept?

Ross M: It definitely takes the right mindset to successfully complete your physical rehabilitation. You need to want to work really hard to get through your injury because without rehabilitation you have to know that you won't 100% recover. With my injury, they made me do heal slides that hurt like hell. I worked my MCL out with so much pain but forced my way through it because I knew I had to. If an athlete is symptom-free and physically cleared, the medical professionals have done their part and the athlete should be 100% ready for sport participation. Only the individual can tell if they are mentally ready for a return so if they don't speak up that is their own fault. It will only hurt them and create the possibility of re-injury.

An athlete must be physically and mentally back to where they were before they were injured to return to play. They need to be especially mentally prepared for the hard work. As far as a coach's role in an athlete returning to play, I believe they leave all of the physical clearance to the professionals. Mentally speaking, coaches ask the players if they are feeling okay and if the answer is yes then they consider that player good to go.

Q5: Who should be responsible for the final decision in terms of an athlete's return to full sport participation: coach or medical doctor? Do you think that there should be different criteria in terms of an athlete's readiness for returning to practice versus returning to competition? How much of a role should the athlete play in deciding when to return play? Is the coach or the medical doctor always right in their assumption that an athlete is ready to return? When do you mostly see flaws in this decision?

Ross M: One hundred percent the medical doctor. The coaches want their best players on the field, and if the doctors say they are ready, the coaches listen. The coach is always right in their assumption that an athlete is always ready to return to play because they are at the doctor's mercy, and any poor decisions would be in the hands of the doctor. In my particular sport, the coaches really trust their medical staff. As far as a difference in criteria in terms of an athlete's readiness for a return to practice versus competition, I would have to say there is none. Our team goes by the motto, "You practice like you play." If you can practice at the high levels, you are definitely ready for competition based on the conditions. An athlete's role in this decision requires that they work hand in hand with the doctor. Together with feedback from the doctor they should decide when returning is appropriate.

Q6: What strategies do you recommend to enhance an athlete's readiness for returning to full range of sport participation? What strategies have you found work the best for you? Are you open minded about different types of intervention suggested to you and if not, why the apprehension?

Ross M: Goal setting is number one. Set little goals such as coming back 10% or 20%, doing a few more squats next time, or adding the treadmill to your workout. Your goals can't be general like "I want to play next week." I also think that a positive attitude is everything. The more optimistic you are about your rehab, the better and faster the end result. The strategies that have worked best for me have been goal setting and focusing on clearing my mind during rehabilitation. I used the mentality that nothing else mattered when I was at rehab accept for getting better so I forgot about everything else for the time being.

I am very open-minded about trying different types of intervention. With returning to the field as a main goal, anything that might help get me there faster is something I will consider. Whether or not it actually works is up for debate.

Q7: Do you think there are different strategies for dealing with female athletes as opposed to male, particularly in regard to recovery from injury? Why do you think this is so? Do you think some of the different coping methods hold athletes back severely?

Ross M: I don't think there is necessarily a difference between females and males, but the strategies for dealing with athletes in general must be very specific to the individual as well as their mental capacity. It is silly to have a cookie cutter rehab program for men and women, especially because each injury is very different and must be cared for in different ways. Certain injuries will affect people more than others, but overall I think any type of negative thoughts will hold any athlete back from rehabbing correctly. The athlete needs to be accountable for a positive attitude and dedication to recovering from their injury.

Q8: Among athletes it is common to hide fear in order to avoid appearing weak. However, it is known that in previously injured athletes, fear of subsequent injuries may induce erratic emotional responses, avoidance reactions, and bracing behaviors. In your opinion, and in terms of psychological recovery, do you think that athletes recover better or faster from an injury if the injury is given more attention or less attention? Also, do you feel that there is a difference in response to attention paid to the injury in male versus female athletes? How do your coaches respond to athletes in the "recovery phase"? Do you agree or disagree with how your coaches respond to injured athletes on your team, and if not what would you do differently?

Ross M: Recovery is very individual specific. However, the more attention given to an athlete, I believe, the better the rehab will go. If coaches aren't paying as much attention to a player during the rehab process, athletes are limited in recovery and are not getting all that they need. I think that if an athlete feels that they are getting all of the attention of a coach they will believe the coach is dedicated to getting them back on the field and will recover quicker. During the recovery phase, coaches are supportive but not as connected to you as they are to their players that are out their making the plays. Football is a business, and they care a lot more about their competing athletes, as much as I hate to say it. I agree with this method only to an extent because at this level I understand that is how things need to be handled. It would be nicer if coaches did actually care more about the psychological factors of the team. From personal experience, during my freshman year when I was out with my first MCL injury, I wished the coaches had more time to pay attention to my injury, especially because I was so worried about coming back.

Q9: Sport-related concussion has received significant attention in recent years. Despite some advances in studying concussions, important questions are still to be answered including:

Which concussion grading scale and return to sport participation guidelines are sufficient to prevent more severe secondary and multiple concussions?

After how many concussions should an athletic career be terminated?

Are there long-term cognitive and behavioral deficits after single and especially multiple concussions?

Collegiate athletes are at high risk for sport-related brain injuries. The likelihood of brain injury is a function of head impact (or sudden acceleration/deceleration) within the context of sport participation. The concussion may occur in any activity regardless of the nature of this activity, and when the brain injury occurs, it has potential for a lasting effect on the athlete. What do you think the collegiate coach should know about concussions and what should be done from a coach's point of view in order to prevent concussions?

How should team members respond to their fellow concussed players?

If there were a seminar held for PSU collegiate coaches to obtain extensive knowledge to help deal with concussed players, do you think your coaches would attend?

Ross M: The general rule of thumb requires an athlete to be symptom free for a week starting from the time that all symptoms were gone, although this was not followed in Anthony Morelli's case. I guess there is a lot of politics involved. Concussions are very serious injuries and should be monitored carefully. Our team undergoes a standard concussion test once a year. This test and the close monitoring of concussed athletes is essential because reports have said that depression, fear of getting hit, negative thoughts, and even thoughts of suicide can result from this injury. Collegiate coaches should understand exactly what a concussion is, what its ramifications are. They need to know how serious this injury is and use this knowledge so as not to push their concussed athletes in any way. Athletes need to be symptom-free and coaches should be aware of the severity of each and every concussion they encounter (based on the medical report). Our brains are incredibly sensitive.

When a concussed player returns to the field, the rest of the team is very supportive. Those who have had them can relate on another level. Extra caution is taken in practice and concussed athletes are not allowed to get hit. They take it easy at first. If there was a seminar held for PSU coaches, I do not think any of our coaches would attend. This is a business and coaches don't want to hear it. They believe that doctors and trainers have the responsibility of knowing every aspect of the concussion.

Q10: Do you think it is possible, as an athletic coach, to discern concussion symptoms as irregular or abnormal in an affected athlete and if so, how do you think a coach should adjust methodology? What are your coach's general guidelines with concussed players?

Ross M: The coach really would not adjust his methodology because without the clearance of a medical doctor, the athlete should not be back on the field. Coaches generally acknowledge options/symptoms that are voiced. If a doctor or player does not speak up, the coach usually will not bring it up. Guidelines for dealing with concussed players involve the one week rule that I talked about before. Basically coaches feel that you should not come out on the field until you are ready because when you do come out you are expected to be able to give 100%.

Q11: Is it obvious during practice or in competition to a coach or even fellow players if a previously injured athlete has developed bracing behaviors? Are there particular strategies that coaches have when dealing with athletes who exhibit this type of behavior in response to injury? In your opinion what are the biggest consequences an athlete could face with an injury of this severity?

Ross M*:* Yes! It is usually pretty obvious if the athlete is showing clear signs. If there is an issue, coaches pull the athlete aside so as not to make a big deal out of it in front of the rest of the team. Coaches will oftentimes talk about strategies to deal with the issue and make sure that they athlete is doing okay. Also, if a coach is concerned he may go to the medical doctor or trainer and ask what is going on with the athlete privately. I feel that concussions are very serious injuries with BIG consequences that can include loss of memory, depression, and anger, with the biggest consequence being suicide. It is so frightening because in one day your life can be completely changed from the sport you love.

Q12: What advice would you give to uprising coaches today regarding how to identify athletes at risk for injury and ultimately to prevent injuries among student-athletes? If you were the coach of your team, how would you handle injury prevention during practice, competition, preseason, and postseason?

Ross M: The best advice a coach can receive is to hire a good training and medical staff for the technical stuff and take the responsibility to harp on technique, technique, technique. Coaches need to talk to their players about the risks of this sport. If I were the coach of my team I would make sure I have the best medical staff possible and spend lots of time on technique. To handle injury prevention I would make sure my team was in good shape year round (which we generally are). In season is the time to keep yourself in shape to prevent injury but off season is key for making improvements in flexibility and strength.

Interview 9: Cheerleading

Coaches and athletes can sometimes overlook the importance of understanding the psychology behind injuries and this can lead to more serious issues that are clearly seen in the following interview with two Penn State cheerleaders. Both Alison B. and Devon C. are sophomores here at Penn State, University Park. They both have been involved in various sports throughout their athletic careers, which made them appropriate to interview.

Alison and Devon have both witnessed and dealt with traumatic athletic injuries.

Q1: Could you please give me a brief background of your athletic career up to and including being the Penn State Cheerleaders.

Alison: I started going to gymnastics when I was three years old and competed from the time I was five years old until tenth grade. When I stopped gymnastics I was at level nine, which is highly competitive. Our gymnastics team required us to take dance classes from preschool until tenth grade. I began cheerleading in third grade when I joined a "midget" squad. I started competitive all star cheerleading when I stopped gymnastics in tenth grade. In high school, I was on the track team sophomore through senior year. In April of my senior year in high school, I tried out for the Penn State cheerleading team. Freshman year of college, I was on Small Co-ed and this year I made Large Co-ed. I compete for Penn State cheerleading currently. We went to the UCA collegiate nationals in Orlando, Florida over winter break but did not place in the top five teams in our division. Our squad was content with just being accepted to participate in this national championship which some of the best teams in the nation go to.

Devon: I took dance class for one year when I was three. I began gymnastics when I was four years old and began competing when I was six. I was on the diving team my freshman year in high school. When I stopped competitive gymnastics in tenth grade, I became a cheerleader at my high school for eleventh and twelfth grade. In April of my senior year of high school, I tried out for Penn State Cheerleading. I made Small Co-ed both years and competed with Small Co-ed last year. I always seemed to like individual sports more throughout my athletic career.

Q2: As a Penn State student athlete, I am sure that you have seen different sport-related injuries in college, as well as on teams when you were younger. Do you feel that coaches you have had in the past, as well as in the present, have been knowledgeable about the psychology behind the sports injuries that you and your team mates had? Are there any specific examples or instances that you can remember?

Alison: From what I remember, in gymnastics when I was younger, my coaches had a better psychological understanding and approach than my current coach. Our coach would rarely make our injuries seem extremely serious when they would occur, and if they actually were he would explain it to us after the initial shock. Typically when we would get injured, we would just get it iced/taped, take some Ibuprofen and just continue on our way. Now our Penn State cheerleading coach has the "better safe than sorry" mentality. He probably has to act this way because if someone were to sue Penn State, there would be much more liability and complications involved. I understand this point of view most of the time, but disagree when he somewhat "babies" injuries that aren't as serious.

Devon: When I was a gymnast, the coaches definitely understood the psychology behind our injuries because when someone would get injured, they would downplay it and make it seem less serious than it actually was. They would do this in order to avoid mentally "freaking us out" and would later on explain how serious the injury was. Now, our coach does not handle injuries in the same manner. He does not attempt to explain the seriousness of the injury in most of the cases and a lot of times will blatantly ignore the injured athlete and just tell us see the trainer right away.

Q3: *Most athletes that participate in high level sports experience some type of injury during their athletic careers. What sports injuries have you had throughout your athletic career as a gymnast and as a cheerleader?*

Alison: I have sprained both of my ankles multiple times. I broke my left arm when I freaked out in the middle of a back hand spring. I broke my right hand and dislocated my toe.

-Did you find that you had more injuries when you had one particular coach or set of particular coaches?

Not necessarily, I had different coaches all throughout my gymnastics career and I was not more likely to get injured with one of them opposed to another. Since I have been a cheerleader at Penn State, I have gotten one serious concussion. This happened when I fell backwards off of one of my team mates' shoulders and hit my head extremely hard on our competition mat. I also sprained my ankle one time since being on the cheerleading team here.

Devon: I sprained my right ankle three times, got a concussion from gymnastics in seventh grade, and popped my bursa sac in my knee in eighth grade. Last year on the Penn State cheer team, I got two concussions. The first concussion was less severe and resulted from falling out of a stunt that came down wrong and hitting one of my other male team mates in the head. The second, and much more severe, concussion I got resulted from falling out of a back tuck basket toss, straight onto the ground. I have never broken a bone and hope that I never will.

Q4: *A number of previous studies have examined athletes' emotional responses to injury (McDonald & Hardy, 1990; Smith et al., 1990), painting an intricate picture of an injured athlete's personal status. Do you feel that sport injuries affect student athletes emotionally and/or psychologically? Can you think of a specific injury you or a team mate had when you were able to observe this?*

Alison: Definitely psychologically, especially when it is a recurring injury. When I was doing an "overshooting" on the parallel bars for gymnastics, I hit my toe on the low bar and dislocated it, I flew out of control, and landed on the mat. This accident hurt very badly, but the initial shock of it was traumatic. Since I have a strong emotional memory of this incident, I feel that it did affect me greatly.

Devon: After an injury, student-athletes can definitely be psychologically affected. Diving was the most psychologically-oriented sport I tried. I was doing fine learning how to dive when I first started and the transition from gymnastics made it easier, but one day that all changed. I dove and hit my legs on the board, and as soon as it happened I thought to myself, "I am never doing that again." Something just clicked and my mentality towards diving changed. This was weird to me because in gymnastics similar incidences happened many times and I would keep on doing it. I am not sure if I thought different because I was older [ninth grade as opposed to five years old] when it happened, but it definitely mentally freaked me out. "Wiping out" messes with your head at the time of the injury and afterwards. It also is emotionally frustrating when you get hurt because when you are sitting out, you have to watch your team mates progress as you can not.

Q5: Do you feel that every injury an athlete gets may cause psychological trauma and therefore require seeing a sports psychologist?

Alison: If an athlete were to see a sports psychologist for every single injury it would get ridiculous. I feel that it is necessary if it is a career-ending injury, but not for less serious injuries. If an athlete is having a particularly painful time with rehab, then it also may be necessary to talk to a sports psychologist. Seeing one would help them discuss their frustrations that come along with the injury.

Devon: No. Being an athlete requires certain toughness in your character. You have to understand that getting injured is a risk when you are a collegiate athlete. I feel that a person can determine for themselves whether it would be appropriate to seek help or not after getting injured.

Q6: Do you think that the psychological mindset of an athlete going through rehabilitation after an injury has an impact on the recovery time?

Alison: Yes. Definitely, the more determined the person is to getting better soon, the faster their recovery would be. I feel that it is hard to sit and watch while being injured, especially when it is a team sport. I feel like it is my fault when I am sitting out and not participating.

Devon: I think that with a positive attitude towards recovery, an athlete can speed up their time in rehabilitation. When children are younger, they feel that it is "cool" to sit out at a practice for a week or two and enjoy the attention. At this level, it is definitely not the same. The faster the athlete wants to and tries to return, the speedier the recovery will be.

Q7: Premature return to sport participation based upon just physical injury symptoms resolution considerably enhances the risk for re-injury. Do you think that the coach or medical doctor should decide when an athlete can return to full sport participation? Is your answer the same when it comes to practices and competitions?

Alison: I think that it is a decision the medical doctor should make, not the coach. My opinion doesn't change when it comes to practices and competitions.

Devon: I feel that it should be up to the medical doctor. When it comes to practices and competitions, the decision should go for both, not one or the other.

Q8: *From your personal experiences in the past, what would you say would be the most common injury in gymnastics? What are some of the more common injuries when it comes to cheerleading? What precautions could be taken to make these injuries avoidable?*

Alison: When it comes to gymnastics and cheerleading, ankle sprains are probably the most common type of injury. Sprains also occur in cheerleading, but not as often as concussions. At the collegiate level, I see many more concussion injuries. I personally have been dropped on my head multiple times. This may result in the fact that we now have males on our team who have never been a part of a cheerleading team of any sort. We also do a lot of basket tosses and stunts in which we are very far from the ground and increasing our chances of getting seriously injured. There are not many precautions that can be made since as long as you are going to have sport you are going to have injuries.

Devon: In gymnastics, ankle sprains were very common among me and my teammates. Now that I am a collegiate cheerleader, it seems that sprains as well as concussions seem to be the most common. In order to prevent serious injuries, the guys on our team could go through mandatory training at which they are taught how to properly throw and catch us when stunting. This would prevent girls being dropped when stunting, which is a main cause of injuries seen on our team.

Q9: *There are a number of interventions recommended by sport psychology practitioners including: negative thought stoppage, cognitive restructuring, healing imagery, muscle relaxation, goal setting, etc., to speed up rehabilitation of injured athletes. How could a coach make returning to practices and competitions after an injury a smooth transition?*

Alison: It would be helpful for the coach to be aware that there may be some difficulty for them at first and to be understanding. It would not be beneficial for a coach to be extremely mean or hard on the athlete right away, especially if it is a psychologically impacting injury.

Devon: If a coach were to avoid placing the blame of the injury on the athlete it would ease the process. They shouldn't baby the injured athlete too much but understand that they did not try to get hurt on purpose.

Q10: *Does the amount and type of attention an injured athlete receives from a coach impact the amount of time that athlete will need to recover?*

Alison: If an injury is treated like it is a big deal, the athlete will begin to believe that as well. Mind over matter theory – the way the coach interacts with the injured athlete will have an impact on their mentality when it comes to getting better.

Devon: If a coach acts like their injured athlete is "dying" that athlete will believe they are "dying." Inversely, if a coach acts like the injury is not a big deal, the athlete may believe that as well. This could be detrimental if the athlete returns into practice or competition too soon and gets injured again. The coach should be neutral when handling their injured athlete in order to get the best recovery time.

Q11: As you stated earlier, concussions are common among collegiate cheerleaders. "The problem with concussion is that with the exception of the unconscious athletes or someone who is severely dazed, it is often very difficult to identify who has sustained a concussion and who has not" (Cantu, 2006). Do you think that it is easy to tell if someone has suffered from a concussion?

Alison: Yes, most of the time. If the concussion is mild it may be more difficult to tell. When I had my serious one earlier this season, I was told how out of it I seemed. When we arrived to the hospital I became very easy going and emotional and even began to cry. I went to the HUB the next day, left and walked to the wrong class.

Devon: It depends on the severity of the concussion when it comes to telling if someone has one. When I got mine for example, I didn't feel that bad until I went home and dry heaved into the toilet.

Q12: What do you think collegiate cheerleading coaches should know about concussions and what can be done to prevent them?

Alison: A coach should know that a concussion is more psychological than physical in most cases. You have to trust your team mates enough to have them throw you in the air and catch you, and that may have to be regained after a traumatic injury. You want to make sure they will be there next time, and not have to wonder. When I got my serious concussion, my coach happened to be the only person standing behind the stunt. I therefore think that the coach should know and practice the ability to catch/spot us while stunting.

Devon: In order to prevent athletes from getting concussed, they should make sure our male team mates know what they are doing. Spotters should be ready when necessary, and be able to do their job correctly.

Overall, Alison and Devon's responses were very similar for many of the questions. This was not very surprising, considering they both have a background of gymnastics and cheerleading, but was interesting to observe. Their responses gave an insight on the opinions and knowledge cheerleading team members have concerning injuries. One of the important messages from these team mates' responses was that education about traumatic brain injury is currently lacking.

CONCLUSIONS

According to athletes with a history of sport-related injuries, most of these injuries can be avoidable if coaches try to increase their knowledge about the skills and techniques of the sports they coach. They know for a fact if they were not trained to be powerful instead of having correct form with javelin throwing, initiating stunts, somersaulting etc, they may still be participating in the sport. Collegiate coaches get so caught up in their team or athlete ranking that they lose sight of what is more important than just winning or losing – which is the athletes' health. They push the athletes too hard or move them through recoveries at such a rapid pace that they end up doing more harm than good for the team and their bodies. Proper training about their sport for coaches, who can then show their athletes correct techniques, is the most essential way to reduce injuries in sports. It is an unfortunate fact that coaching errors, including inability to properly assess the level of skill and/or fatigue and change the loading level, ended up with career ended injuries of their athletes.

Clearly, there was also agreement among injured athletes that bracing can occur after injuries and athletes need to either be completely healthy to return or very confident in returning to avoid further injury. Psychological counseling can be very helpful when an injury is severe enough because it definitely can help athletes get over insecurities and fears. However, athletes are tough enough to overcome psychological problems and emotions when injuries are mild.

As for concussions, most of the athletes did not have a great deal of experience but find it essential for coaches to know the extent and symptoms. That said, a kind of interesting statement was proposed as "… If there was a seminar held for PSU coaches, I do not think any of our coaches would attend. This is a business and coaches don't want to hear about it." Coaches' knowledge about bodily injuries goes hand-in-hand with their attitude towards brain injuries.

In closing, through athletes' eyes we can gather the importance of both technique, flexibility, and strength programs and psychological attributes to build a better prepared and injury free athlete. With insight into coaching views, strategies, and guidelines, we are able to have a better understanding of how both coaches and players manage injury and injury prevention. Whether injury is just a part of athletics or a controllable sport-related phenomenon still requires further discussion.

Acknowledgments

This chapter wouldn't be possible without the contribution of my Penn State undergraduate students, who learned about dealing with injured athletes and conducted these interviews via my class, KINES 497, "Psychology of Injury," in the Spring, 2007.

CHAPTER 9

OVERUSE INJURIES: STUDENTS' POINTS OF VIEW

1. INTRODUCTION

Injuries occur in every sport, at all levels of the game, in various venues across the world. Whether it is an ACL tear in a soccer player in Real Madrid's stadium in Spain or an eleven-year-old Pennsylvanian boy breaking his collar bone on a beat-up field behind his school during a football game, unexpected and often devastating injuries occur. Overuse injuries, or those injuries that occur due to overuse of key body parts necessary to perform skills associated with specific sports, are extremely common among today's athletes. Current notion is that repetitive stress disorder, repetition strain injury, and cumulative trauma disorder are synonyms that are used for an overuse injury. It is important to note that overuse injuries are not caused by a specific injury or accident, but rather by repeated stresses on the body (Difiori, 1999). Due to the prevalence of overuse injuries both in professional sports and recreation activities it is important to increase students' awareness about this issue. Accordingly, the author proposed an assignment for Penn State students majoring in Kinesiology to explore common mechanisms and elaborate on causes and psychological consequences of overuse injuries in sports and recreational activities. This chapter was elaborated by Penn State University KINES 497 "Psychology of Injury" students aimed at discussing overuse injuries in various sports/recreational activities, including tennis, soccer, baseball, football, lacrosse, running, swimming, water polo, and skiing. In addition, some relevant issues related to overuse injuries, such as gender differences and psychological aspects of the recovery process will be also the topic of discussion in this chapter.

2. TYPES OF OVERUSE INJURIES

2.1. Tennis Elbow

The first sport discussed is tennis, a popular sport played across the world, including in France, Australia, and the United States. Tennis originated in the nineteenth century and can be played on grass, clay, or a hard court. Injuries in tennis involve the use of muscles of the arm and forearm and small tears of the tendons. This can often lead to *tennis elbow*, which includes varying degrees of pain or point tenderness at the origin of the wrist extensor muscles near the lateral epicondyle of the humerus (Fedorczyk, 2007). Other factors that contribute to *tennis elbow* include lack of strength, poor technique, increased time or intensity of play. Symptoms of *tennis elbow* include pain on the outside of the elbow, usually during or after intense use. Lifting or grasping can become difficult and pain can sometimes radiate down the arm.

Treatment for this injury would initially be rest, since there have been signs of overuse of the wrist extensors, which are muscles that pull the hand up and down. Activities that cause pain should be discontinued and the R.I.C.E. method (rest, ice, compression and elevation) is helpful to reduce pain and swelling. Icing the elbow for 10-15 minutes and wrapping the forearm near the elbow may help protect the injured muscles as they are healing. Another treatment includes changing stroke mechanics and racquet type and/or size. From injury prevention perspectives it is important that the racquet is sized properly, including grip size. Also observe whether the player is hitting the ball in the center of the racquet and make sure they do not lead the racquet with a flexed elbow, which is a common technical error of recreational tennis players.

Anti-inflammatory medications are an easy solution to control pain and inflammation. Cortisone injections are the next option if the anti-inflammatory medications are insufficient. If more than two injections are taken and there is no relief, additional injections will not benefit the player. According to a study, patients who received steroid injections were statistically significantly better for all outcome measures at follow up. That said, it is a proper technique and prevention, rather than control for pain and medication after the injury has developed, should be recreational players' primary concern. The researchers advocate a steroid injection as the first line of treatment for athletes with tennis elbow demanding quick return to daily activities. Surgery is the final and unfortunate treatment of *tennis elbow*. This is rarely necessary because about 95% of patients with tennis elbow can be treated without surgery. Simple exercises, in general, can be performed just to control symptoms of tennis elbow, and light to moderate muscle strength routines, in particular. Because recurrence of this condition is common, return to activity should not occur too quickly, and preventive exercises that stretch and strengthen the muscles should be done consistently. Some examples of exercise routines are shown in a Figure 1 below. It is important to stress that the volume and intensity of exercise should be prescribed by medical professionals to avoid negative effects and worsening the symptoms.

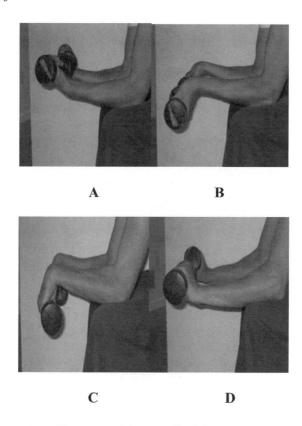

A   B

C   D

Figure 1. Wrist exercise with some weight prescribed for treatment athletes with tennis elbow.

2.2. Overuse Injury in European "Futbol" (Soccer)

Defined as "futbol" in Europe and South America and "soccer" in the USA, *soccer* is known as the sport that unites fans across the globe in the World Cup. This is used to be the most popular male sport and recreation activity in Europe and South America, and has now become an extremely popular female sport as well. To date, boys and girls play soccer across the world. Unfortunately, there are a lot of lower body injuries that can occur in soccer, such as ankle sprains, contusions, muscle strains/pulls, and back pain, to name just a few. A more recent concern with soccer is of undetected **multiple concussions** due to heading the ball. There is current debate in the relevant literature as to whether heading the ball is safe or may cause traumatic brain injury.

Clearly, there are particular injuries that relate to overuse of the muscles and joints. Two injuries occurring in young athletes, due to periods of rapid growth, include *Osgood-Schlatter* disease and *Sever's* disease. Osgood-Schlatter's (see Figure 2 below) is characterized by chronic pain at the top of the shinbone. Ultimately a severe form of tendonitis which comes about from excessive forces on the patellar tendon from jumping and running, Osgood-Schlatter's can lead to separation of the tendon from the bone. Addressed with appropriate rest and modification of training, it usually will resolve (Levengood, 2007).

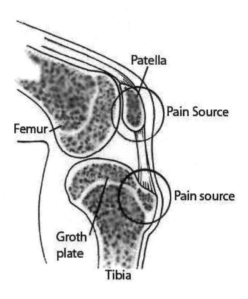

Figure 2. Sources of pain associated with Osgood-Schaletter disease.

Sever's disease (see picture below) is characterized by a chronic pain at the heel where the Achilles tendon attaches. It is an inflammation of the growth plate in the heel bone. X-rays will show a bony change at the site of the Achilles insertion. Appropriate training, shoe modification, and activity modification will be necessary to prevent and treat this common injury in soccer.

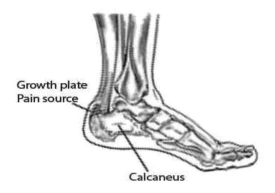

Growth plate
Pain source

Calcaneus

Figure 3. Source of pain associated with Sever's disease.

Older soccer players usually deal with the following common chronic or overuse injuries: patellar tendonitis, stress fractures and shin splints. Patellar tendonitis is a chronic inflammatory response in the patellar tendon secondary to overuse. The athlete will complain of pain just below the kneecap and it is usually very tender to touch. Most often a biomechanical disadvantage due to inflexibility and/or weaknesses is to blame. A good biomechanical evaluation may find what is responsible for the tendonitis, and a specific program set up by a physical therapist can eliminate the pain. Performing an ice massage regularly will also be helpful.

Most stress fractures in athletes occur in the lower limbs and they usually have a slow onset of about two to three weeks, starting with pain only during activity, but then progressing to resting pain. The fractures result from a failure within the bone to adapt to the repetitiveness of certain activities and the torque of the muscles acting across the bone. Many times the fracture cannot be picked up by X-ray, and a bone scan or MRI (a better choice for children) is used for diagnosis. Rest of the area integrated with alternative exercising is essential for the proper healing of this injury.

Pain on the middle, inside, or outside of the lower leg with activity can indicate shin splints. Often shin splints are caused by a rapid increase in running mileage, improper footwear, running on hard surfaces, or poor flexibility.

There are many ways to reduce the susceptibility to these chronic injuries. One way is avoiding the over training syndrome. Many athletes mistakenly think more exercise is better and they fail to get adequate rest. Not returning to soccer too soon is a way to give the injury time to heal. Working on poor techniques or biomechanics is always important because usually these injuries are related to something not working correctly with the muscles or joints. Incorrect footwear is also a huge deal in soccer. Players can play up to ninety minutes a game, so shoes definitely have to be comfortable, supportive, and fit right in order to perform without injury.

There are simple treatments that can eliminate most of these chronic injuries. An inexpensive method is to ice massage the area before and after exercise, stretching the calf muscles. Players may also try inserting cushions in the shoe and having a biomechanical evaluation by a physical therapist who can recommend appropriate shoe inserts and flexibility exercises (The Stone Clinic). Following Hughston Sports Medicine Foundation's idea of P.R.I.C.E., similar to the R.I.C.E. method, involves prevention, rest, ice, compression, and elevation. Prevention is the most important part of the equation. Rest involves giving the injured tissue adequate time to repair itself. Ice is used to decrease inflammation and should be applied before and after practice or games over the injured body part. Compression involves applying an elastic wrap over the injured part to help reduce swelling. Elevation helps to decrease swelling by using gravity to assist in the process. If this method does not relieve the pain players can try nonsteroidal anti-inflammatory drugs (NSAIDs), such as Ibuprofen and Aspirin, to reduce swelling. During periods of acute pain, athletes definitely should consider a stop in play and allow time for the injury to heal.

2.3. Overuse Injuries in Baseball

Popular in North and South America, baseball is known as "America's pasttime" and is played and watched by the young and old alike. In baseball, injuries involving the **shoulder** are highly prevalent and oftentimes potentially career ending. To understand the mechanisms behind such injuries it is important to look at the actions involved in throwing a baseball. Throwing can be broken down into four specific phases. These include: the wind up, cocking, acceleration and deceleration. These actions performed on a routine basis stress the shoulder and soft tissue stabilizers in the joint. Tears or dislocation are common, especially in pitchers. In the figure 4 below the actions involved in throwing a baseball can be viewed:

Wind up   **Cocking**   **Acceleration**   **Deceleration**

Figure 4. Actions involved in the throwing the ball inducing shoulder injury.

The reason that shoulder injuries are so prevalent in baseball is because the shoulder ball fits loosely in the arm socket and is largely unrestricted in movement. While this is beneficial for allowing a wide range of motion, it can lead to significant injury and strain for the player.

Treating this type of injury is often complicated, since it is hard to determine if damage has occurred until a debilitating injury takes place. A case study example for a career ending shoulder injury can be considered by examining Robb Nen. Nen is one of the most famous relief pitchers in baseball, but in 2002 he had surgery to "clean up loose particles in his shoulder" after experiencing pain. However, during surgery a torn labrum, part of the soft tissue in the shoulder, was found. After discovery of this injury, Nen underwent three other surgeries and sat out for a total of eighteen months in order to fix this injury. The reason that this type of shoulder injury is so serious is that doctors have no way to completely fix a torn labrum, and no way to detect it visually without opening up the shoulder because the labrum is located between two bones, making it difficult to X-ray. Below, an image of the labrum of the shoulder can be viewed.

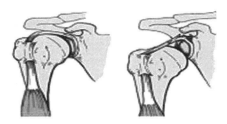

Figure 5. Intact labrum (left image) and damaged labrum (right image)

The best way that athletes can avoid this type of injury is for them to be conscious of a sore shoulder, a loss of velocity, stamina, and poor form when throwing. Coaches should also be aware of these signs and ask players if they are experiencing any pain during play. Shoulder injuries are serious and often career ending for these athletes and should be avoided at all costs. Early signs of shoulder injury must be treated immediately so as to benefit the player.

2.4. Overuse Injuries in American Football

Football, a high impact, highly demanding and physical sport is worth discussing due to its extreme vulnerability to injury in general and overuse injuries in particular. Each year, hundreds of football players on the college and professional levels are injured. In fact, the sport is infamous for generating injury. Sufficient is to stress that the NCCA was created back in 1917 due to the growing concern of injuries in collegiate football. While many injuries are common to football players, including those to the shoulder, leg, and head, injuries to the knee, specifically damage to the **ACL**, are the most debilitating.

The ACL, or anterior cruciate ligament, is the ligament in the knee used to stabilize the leg and help a player change direction quickly. When damage to the ACL occurs a popping sound can usually be heard, and following it, immense pain. One study which examined the frequency of this type of injury in college students found that of those playing Division One football during their collegiate career, approximately 16% have a chance of experiencing this type of injury. The risk of this injury in players is one hundred times higher than in the general population. Below, a figure of the ACL can be viewed.

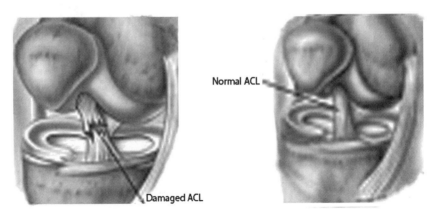

Figure 6. Damaged (left image) and normal (right image) ACL

Increase in age and activity level seem to be the main determinants of an increased risk for this type of injury. These factors also determine the methods for recovery from such an injury. Usually athletes who experience damage to the ACL require rest and physical rehabilitation. The injury usually takes at least six to nine months to heal, which makes it even more devastating to the athlete. Often, surgery is needed to reconstruct the ACL, resulting in a longer rehabilitation and healing process.

Because of the prevalence of ACL injuries in football players, many studies have been conducted to further investigate the consequences of ACL damage. Most of these studies revealed that ACL injury may affect both the physical and psychological well-being of injured athletes. Because of the severity of this injury, significant emotional distress is common. Feelings of fear and depression are also evident in most athletes experiencing this type of injury. Due to the severity of the impact, behavioral treatment such as physical therapy should be accompanied by psycho-therapy interventions to speed up both physical and psychological recovery. Psychological counseling sessions are also recommended, including positive self-talk, imagery, relaxation training and goal setting. It is important that injured athletes should be "actively" involved in the treatment process, for example by constantly monitoring and assessing their psychological status and emotional responses to the rehabilitation process.

An example of an ACL injury on the professional level can be viewed with Chad Jackson, who played for the Patriots. This injury ended this athlete's 2006 season and (as of the writing of this book) no set return is scheduled for this player. Another example can be seen with Javon Walker of the Denver Broncos, who was kept out of the NFL for twelve months due to an ACL injury. A list of elite athletes suffering from an ACL injury might be endless, as ACL injuries continue to occur each year on the professional and collegiate levels. More research should be done on this type of injury and different methods for rehabilitation, including psychological recuperation should be explored. Overall, it is suggested that psychological counseling and emotional support beyond the standard physical therapy and rehabilitation are critical for athletes. In other works, it is highly recommended that every single case of ACL injury should be referred to a qualified sports and/or clinical psychologist for further evaluation and treatment.

Interestingly, one student used a personal experience in his life to explain a different overuse injury that occurred to him several years ago in football. "When I was around fifteen years old about seven years ago I suffered from shin splints. I was a star linebacker for my school and part of my training was running two to three miles a day about three to four days a week. I was a fast, hard hitting line backer who had no problem with putting my shoulder into someone. Then one day at practice while the team and I were running drills I started complaining to the coaches about pain in my shins. The coaches sent me home and I told my parents about the pain. They took me to the doctor and the doctor told me to make an appointment to get X-rays. So I did and the X-rays showed that I had shin splits. I remember asking the doctor how this happened and he told me that my injury was due to stress on my tibia. In other words all the running I was doing led to an overuse injury. The doctor then told me I had to sit out and keep from any physical activity for at least three weeks; I winded up having to stay off the field for only the minimum three weeks. I recovered fully and retuned to play and never lost a step. That was from that injury, but I had many more.

I am now twenty-two and am still suffering from a long-term overuse injury. From all the sports I played and the fact that I was a hard hitting line-backer who had no problem putting my shoulder into a person to make a tackle, I now have bursitis [...the inflammation of a bursa" (Castinel & Prat, 2007)] in both my shoulders. I tell you this because most people think of overuse injuries as a quick fix; where you sit out for a two to six week period and after that you suffer from no more symptoms. When in fact there are many overuse injuries that can stay with you through out the rest of your life. Some of these injuries are bursitis, tendonitis, and forms of arthritis. I can tell you that bursitis is a very painful and uncomfortable injury, and because of it I have been in and out of the doctor for this problem since I was seventeen years old.

1.5. Face and Head Traumas in Lacrosse

One of the more recent sports to gain national attention, lacrosse is a popular sport in the United States. Injuries are common in this sport, most often to **the face and head**. Because of this, protective gear is required to guard the head and upper body from damage. The sport itself is classified "a collision sport" but the prevalence of injuries is estimated at only 4.7 for every 1000 practice sessions and games. According to the NCAA this sport is ranked seventh for number of injuries per game. Several studies examine the impact of head and face injuries in lacrosse players (see: McCulloch & Bach, 2007, for review).

A study conducted by the University of Virginia surveyed athletes on the number of injuries per player. It was found that 85% of lacrosse players experienced some type of injury during the season. Though helmets and facemasks are required during play, concussions are very common in the sport. Diamond and Gale studied the impact of head injuries in lacrosse players and found that collisions with other players are the cause for most of these injuries. In the past, lacrosse players were allowed to play even after a minor concussion, which is dangerous and increases the risk for further damage to the athlete and the risk for re-injury.

A study conducted by the American Alliance for Health in 2000 looked at injury rates in female lacrosse players during a two-year period. Injury reports were given to athletes, trainers, and coaches. Injuries were examined by determining the athlete's position and the amount of protective gear worn during games and practice. This study concluded that the number of head injuries could be reduced by approximately 16.5% if protective head gear were worn (Sherman, 2000).

To reduce the prevalence of these injuries, it is recommended that protective gear be worn at all times for lacrosse players. It also may be beneficial to research the type of gear worn and determine if alternative material might decrease the severity and prevalence of head and face injuries in lacrosse players. Though the number of injuries in this sport may be lower than in others, with the increase in popularity of lacrosse an increase in injuries may be seen in the future.

1.6. Overuse Injury in Running

Tibial Stress Fractures are common among all types of runners. This injury can be very serious in nature and has the potential to be career threatening. A recent study conducted by Milner (2007) titled: "Are knee mechanics during early stance related to tibial stress fractures in runners?" had as its focus common mechanisms and ways of tackling this injury. The authors examined twenty-three runners who had a history of tibial stress fractures. They were compared to a control group that consisted of twenty-three individuals who matched them for age and mileage with no previous bone related injuries. All the subjects were to run 3.7 miles and the data were collected at the foot-strike to the impact peak of the vertical ground reaction force. Subjects were excluded if they had any current injuries or had not yet returned to 50% of their pre-injury mileage, had cardiovascular pathology, had abnormal menses, or were pregnant or suspected they might be pregnant.

The results of this study showed that sagittal plane knee stiffness was significantly greater in individuals who reported having tibial stress fractures. Stiffness was also positively correlated with shock. Knee excursion, knee angle at foot-strike and shank angle at foot-strike were not different between the control group and the group of stress fracture individuals. Knee flexion stiffness was greater in the group who had previous stress fractures than in the control group. The positive relationship between stiffness and tibial shock was more pronounced in the stress fracture group. This difference in knee stiffness, and the lack of between-group differences in the other variables tested, may indicate a role in tibial stress fracture risk.

The findings of this study indicate that there is a positive relationship between mechanics during initial loading and tibial stress fractures in runners. This relationship is important because if you were to retrain the runner in the mechanics of his or her gait, the potential for more stress fractures may decrease. In general, a stress fracture occurs when the forces being exerted are lower than normal, but over time the repetition of these forces causes fractures on the shin, or tibia. Stress fractures can occur in any bone but are usually seen in the foot and shin and occur with athletes that run and jump on hard surfaces, such as runners, ballet dancers, and basketball players. Knowing the history of injury of an athlete plays a key role in diagnosing a stress fracture. X-rays don't usually show a stress fracture; however they will show bony growths around where a stress fracture may have occurred. This would show a healing process. Below are figures of some common exercises to "treat" the runner's overuse injuries: (a) quadriceps strengthening: isometrics; (b) quadriceps strengthening: straight leg lift; (c) hamstring stretch; (d) calf stretch; to name just a few. (Adopted from *American Family Physician*.) Again, the volume, intensity and duration of these exercises should be defined on an individual basis upon consultation with physicians.

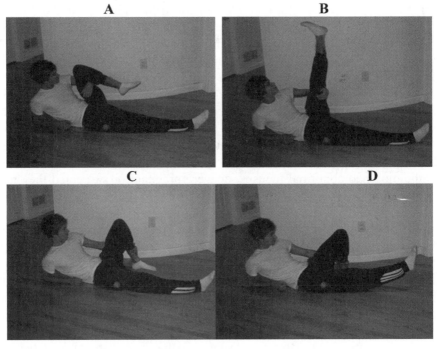

E

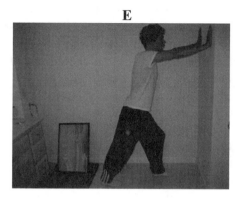

Figure 7. Set of exercise for treatment the runner's overuse injuries.

1.7. Swimmer Shoulder

Another common type of overuse injury is called "swimmer's shoulder." As the name states it is commonly seen in swimmers but also affects springboard divers. Swimmer's shoulder is a condition caused by an impingement of the soft tissues against the coracoacromial arch. This injury, like the others that have been discussed, is most often caused by excessive repetition. The repetitive overhead arm motion of the freestyle stroke is an example of this. Multiple entries into the water in diving are another cause of this injury. Most commonly, this happens at the end of practices when athletes are suffering from progressive muscle fatigue. Often in swimming and diving sports, shoulder injuries occur due to lack of proper warm-up, especially when practices and/or competitions are held in outdoor pools.

There are two types of impingement that may result in swimmer's shoulder. The first occurs in the pull-through phase. At the beginning of the pull-through phase, when the hand enters the water, if the swimmer's hand is across the midline of the body this places the shoulder in a position of horizontal adduction which mechanically impinges the long head of the biceps against the anterior part of the coracoacromial arch. The second type of impingement occurs during the recovery phase of the stroke. As a swimmer fatigues it gets hard to lift the arm out of the water and the muscles of the rotator cuff, which work to externally rotate and depress the head of the humerus against the glenoid, become less efficient. When this happens the supraspinatus becomes impinged between the greater tuberosity of the humerus and the middle and posterior portions of the coracoacromial arch.

Numerous recent studies have been focused on the causes and symptoms of swimmer's shoulder as well as on treatment protocols. Some treatment protocols for pain management were offered including: (a) avoiding painful activities; (b) two weeks of non-steroidal anti-inflammatory medications and ice; (c) decreased anterior capsule stretching and increased posterior capsule stretching; (d) increased rotator cuff exercise with an emphasis on external rotators; (e) scapular-positioning muscle exercises and increasing body-roll (see Weldon & Richardson, 2001 for details).

Increased range of motion is an important factor in minimizing injury in swimmer's shoulders. By allowing the arm more forward elevation, a shoulder with increased range of motion allows the arm and the body to achieve a 180 degree angle. This angle permits the body to be parallel to the surface, minimizing the forward axial surface area and reducing dragging (Weldon & Richardson, 2001).

1.8. Overuse Injuries in Water Polo

Water Polo is a sport that combines swimming and stamina with strength and accuracy. It demands an athlete who possesses both quickness and stamina. To stay above the water, the lower body and core must be very strong and have great stamina. To battle the opposition in order to win the ball or win a shooting lane, the upper body must be equally strong. Finally, and most importantly, the upper extremities must be able to shoot a water polo ball with both speed and accuracy. The ball weighs 400-450 grams (14-16 ounces), and can be either 0.7 or 0.65 meters in circumference, for men or women (Wikipedia, May 4, 2007). Rarely is a shot in water polo truly on balance, like a baseball throw. This is a contributing factor to injuries of the throwing arm, since an off-balance throw will place undue stress on the shoulder and elbow. These injuries might occur because of either overuse or acute trauma (Colville, 1999), depending on the situation. Overuse trauma could occur strictly because of the shooting motion, whereas acute trauma could occur because of the contact element of the game. An arm stopped suddenly in mid-throw could overstress a ligament suddenly and cause acute trauma. The stress on the muscles attempting to internally rotate the shoulder will increase greatly at contact with a resistant force, and overload these muscles even more than they were originally. This mirrors how many injuries are suffered in football, where contact is the source of the injury, not the actual throwing motion.

1.9. Overuse Injuries in Volleyball

Volleyball is a sport that, while it uses overhead motion, is not a throwing sport. Its overhead motion, or "spike" is one in which the "spiker" does not have possession of the ball at any time. The only contact made is at the end of the acceleration stage, when the volleyball is struck with a closed fist. Also, this contact is generally made with the body suspended in the air.

The contact is much more over the top than other sports, which generally demand the throw to be made with the shoulder abducted at 90 to 120 degrees. The best spike in volleyball is made at the highest point, so the contact is made with the shoulder abducted more than 120 degrees. With this level of abduction, not only is the humerus in an unstable position with the supraspinatus pressed against the acromion, but the scapula has moved laterally to accommodate the movement of the shoulder.

This places the scapula in an unstable position also, possibly creating a greater stress on the shoulder. In a study of volleyball players of the Danish Elite Division during the 1993-1994 seasons, 15% of injuries were overuse injuries to the shoulder. Most of the injuries were acute, such as blows to a finger when blocking (Aagaard & Jorgensen, 1996). Volleyball players are susceptible to overuse because of the constant spiking, but there are also other overuse injuries. In a study conducted of world-class beach volleyball players, there were other overuse injuries noted: the most common being back pain and knee pain, followed by shoulder problems. The study concluded that overuse injuries represented a significant source of disability (Bahr & Reeser, 2003).

1.10. Overuse Injury in Skiing

A sport many people may not think of as having an overuse injury associated with it is skiing. Skiing accidents are the most common causes of damage to the ligament, among others, that cause the injury known as **skier's thumb**. When a skier falls on an outstretched hand with a ski pole in the palm it creates the force necessary to stress the thumb and stretch or tear the ligament. Some symptoms of skier's thumb include swelling, inability to grasp, and thumb pain, tenderness, or discoloration. A study was done observing skiers. Within twenty-eight skiing injuries, seventeen were ulnar collateral ligament injuries. The most common cause of this was an uncontrolled fall, associated with failure of the ski pole to be separated from skier's hand forcing the pole to be driven into the web space, causing thumb abduction and hyperextension. To relieve pain the thumb should be iced and movement should be avoided as much as possible (Srittmatter, 2005). If there is a partial rupture of the ulnar collateral ligament it can be treated with four weeks of immobilization. If there is a total rupture, surgery is used to repair the ligament, but there is potential for complications and only partial recovery afterwards (Strittmatter, 2005).

The summary of common overuse injuries in sport and recreational activities is shown below in the Table 1.

Types of Overuse Injuries	Symptoms of Overuse Injuries	Causes of Overuse Injuries
Jumper's Knee	Soreness below the knee or above the shin	The patellar tendon in the knee is repeatedly pulled
Little Leaguer's Elbow or Shoulder	Pain in the shoulder or elbow area	Repeated over head throwing
Osteochondritis Dissecans	Pain and swelling to the knee	It can run in families or can be caused by a metabolic problem
Sever's Disease	Pain to the heel of the foot	Jumping or running activities
Shin Splints	Pain and soreness to the shin	Running on a hard surface
Sinding-Larsen-Johansson Disease	Knee pain	Due to a fracture of the knee cap caused by over extension of the patellar tendon
Spondylolisthesis	Pain to the back	Caused by over flexion and extension of the lower back
Spondylolysis	Pain to the back	Caused by over flexion and extension of the lower back

The chart below shows some risk factors that contribute to overuse injuries, adopted from: www.physsportsmed.com/issues/1997/05may/oconnor.html.

Instrinsic	Extrinsic
Malalignment	Training Errors
Muscle Imbalance	Equipment
Inflexibility	Environment
Muscle Weakness	Technique
Instability	Sports-acquired deficiencies

Finally, below is the Nirschl Pain Phase Scale of Athletic Overuse Injuries adopted from: www.ultrunr.com/painphase.html

Phase 1. Stiffness or mild soreness after activity. Pain is usually gone within 24 hours.

Phase 2. Stiffness or mild soreness before activity that is relieved by warm-up. Symptoms are not present during activity, but return afterward, lasting up to 48 hours.

Phase 3. Stiffness or mild soreness before a specific sport or occupational activity. Pain is partially relieved by warm-up. It is minimally present during activity, but does not cause the athlete to alter activity.

Phase 4. Similar to phase 3 pain but more intense, causing the athlete to alter performance of the activity. Mild pain occurs with activities of daily living, but does not cause a major change in them.

Phase 5. Significant (moderate or greater) pain before, during, and after activity, causing alteration of activity. Pain occurs with activities of daily living, but does not cause a major change in them.

Phase 6. Phase 5 pain that persists even with complete rest. Pain disrupts simple activities of daily living and prohibits doing household chores.

Phase 7. Phase 6 pain that also disrupts sleep consistently. Pain is aching in nature and intensifies with activity.

3. PSYCHOLOGICAL RESPONSES TO OVERUSE INJURIES

The psychological consequences of an overuse injury can be devastating. Most athletes have the ability to fight through pain, while their mind tells them they can but their body says differently. The mindset of an athlete usually follows the saying, "no pain, no gain". For an athlete injury is not allowed, pain is ignored and unacceptable, and there should be no complaints. If an athlete is experiencing pain and it decreases during training, the athlete believes they have worked through the pain; they may even assume that their initial pain was imaginary. Also athletes who are trained to endure pain have difficulty distinguishing normal aches and pains from pain signaling the onset of a possible injury.

A study was done among 280 NCAA athletes using an impact of event scale (IES) to measure the distress of these athletes. Out of all the athletes sampled, 48% were classified as injured. Most of these injured athletes were still participating in their sport (Shuer, 1997). This is really frustrating evidence documenting that motivated athletes, due to numerous reasons, pretend that they are fully functional and capable of making a contribution to team success despite the presence of injury symptoms. This fact raises another question as to how define return-to-sport participation based on athletes' self-perception of injury or the presence of physical evidence of residual damage.

Athletes dread injuries; it places them in a world with no guarantees or predictable outcomes in terms of return-to-play/competition. After an injury, athletes will typically respond with a range of emotions such as denial, shock, anger, and depression, to mention just a few. Athletes are generally committed to their sport; they strive to improve and get stronger. Once an injury sets in some may stay committed to overcoming the injury. In this case they will work hard at every rehab session and maintain a positive attitude. This type of athlete does not view the injury as a crisis but rather a challenge that must be conquered. It is good for an athlete to set goals and develop strategies to reach them. This gives them something to focus on, encouraging motivation and determination. For example, Tim Willoughby, 48, a right-handed tennis player who suffered from a severe case of tennis elbow, decided to learn to hold his racket in his left hand. He taught himself to play in reverse where his backhand became his forehand and vice versa. Tom suffered from a bad case of tennis elbow from playing every day and instead of giving up he re-taught himself the fundamentals of tennis on his left side. Once his arm healed he went back to playing with his right arm but when his elbow begins to flare up he rests it and switches the racket over to the other side (Bateman, 2005).

Some athletes are over confident and have a tendency to try to speed up recovery by going above and beyond, but this ultimately leads to too much too soon for the injured body part. Some athletes may isolate themselves from their teammates and friends, feeling left out and helpless. Others will continue to be an active member of the team by attending practices and competitions.

Most endurance athletes are going to experience an overuse injury. Athletes do not want to be told to rest. So instead the best way to recover from an injury is to continue training without using the injured part. Athletes can cross train to maintain fitness and to heal the injury. If a runner who usually runs eight miles a day is now suffering from runner's knee and is unable to run, they should instead replace the usual running with swimming. This will help to maintain endurance and strength. Athletes are exposed to the potential for injury every day and this is beyond anyone's control. For new swimmers the danger is a rotator cuff or other shoulder injury. Though all these injuries are different they do have one thing in common: they're caused by overuse. If the athlete is unable to cope with the injury they should seek psychiatric help to reconsider their short and long-term goals and to adjust to the trauma.

Another factor that has been researched regarding overuse injuries is the variation between races and genders. Although race is not a determining factor for who will or will not have an injury related to overuse, gender does play a small role in distinguishing overuse injuries. This can be caused by hormonal differences, anatomical differences, and differences in the activities done by men and women. Carrying angles, Q-angles, and body-mass are several other implications of sex differences on overuse injuries (Laker, 2006).

CONCLUSION

Chronic injuries are hard to avoid and prevent; however, most are treatable. Coaches can help to prevent and avoid chronic injuries by being aware of which chronic injuries are specific to their sport. They can also adjust practice duration and frequency to lower the risk of chronic injuries. With chronic injuries, as with any injury, there is a risk of developing a psychological trauma. It is also the coach's job to try and prevent or reduce the risk of a psychological trauma with the help of a sports psychologist. After observing multiple sports and participating in multiple sports over the years, it is clear to me that chronic injuries are an on-going problem. More research is needed to better prevent and treat these injuries.

Acknowledgements

This chapter summarizing some current research and characterization of overuse injuries in sport and recreational activities wouldn't be possible without the significant effort and commitment of my numerous students. I would like to acknowledge the special contribution of Mary-Kate Courtney to this chapter.

REFERENCES

Difiori, J. (1999). Overuse Injuries in Children and Adolescents. *The Physician and Sports Medicine* www.med.umich.ed/1libr/sma/sma_overuse_sma.htm.

Fedorczyk, J. (2007). Tennis Elbow: Blending Basic Science with Clinical Practice. *Journal of Hand Therapy, 19* (2), 146-152.

Levengood, G. (2007). Goal! To Recognize and Prevent Overuse Injuries in Soccer. Hughston Health Alert. Hughston *Sports Medicine Foundation.* 2, http://www.hughston.com/hha/a.soccer.htm.

Castinel, A., Prat, A. (2007). Stress Fracture in the Lumbar Spine in a Professional Rugby Player. *British Journal of Sports Medicine*, May pubmed.com: www.orthopedics.about.com/cs/sportmedicine/a/blbursitis.htm.

McCulloch, P. & Bach, B. (2007). Injuries in Men's Lacrosse. *Orthopedics, 1* (30), 29-35.

Sherman, N. (2000). Head Injuries and protective eyewear in women's lacrosse. *Journal of Education, Recreation and Dance, 71,* 2-9.

Milner, T. (2007). Are Knee Mechanics During Early Stance Related to Tibial Stress Fracture in Runners*? Clinical Biomechanics, 6,* 17-26.

Weldon, E.J., & Allen B. Richardson, A. (2001). Upper extremity overuse injuries in swimming: A discussion of swimmer's shoulder. *Clinics in Sports Medicine 20* (6), 119-129.

Wikipedia; "Water Polo Ball." 04 May 2007. http://en.wikipedia.org/wiki/Water_polo_ball.

Colville, J.M. & Markman, B. S. (1999). Competitive Water Polo. Upper Extremity Injuries. *Clinics in Sports Medicine, 18* (2) 305-312.

Aagaard, H., & Jorgensen, U. (1996). Injuries in elite volleyball. *Scandanavian Journal of Medicine & Science in Sports, 6* (4), 228-232.

Bahr, R, &Reeser, J.C. (2003). Injuries among World-Class professional beach volleyball players. *The American Journal of Sports Medicine 31* (1), 119-125.

Strittmatter, J. (2005). Skier's thumb. Emedicine Health, Denver Health Medical Center.http://www.emedicinehealth.com/skiers_thumb/article_em.htm.

Shuer, M. (1997). Psychological effects of chronic injury in elite athletes. *WJM, 166* (2), 997-2002.

Bateman, N. (2005). Tennis elbow. Medical Encyclopedia. Healthwise Inc. http://health.msn.com/encyclopedia/healthtopics/articlepage.aspx?cp-documentid=100068004.

Laker, S.R. (2006). Overuse Injury. http://www.emedicine.com/pmr/topic97.htm.

CHAPTER 10

FITNESS ASSESSMENT IN ATHLETES

1. INTRODUCTION

One of the major coaching errors in modern sports is the lack of appreciation for proper assessment of the physical fitness of elite athletes. Clearly, with all respect to coaches' knowledge regarding "how to load" athletes to achieve peak performance at the right time, there is a common unsolved problem around and lack of knowledge about "how much is enough, and how much is too much," which is directly related to high risk of injury. One recent well-controlled study among high school student-athletes conducted by Contos et al. (2006) clearly indicated that lack of physical fitness prior to the season is directly correlated with high risk for various traumatic sport-related injuries. There are numerous other studies in support of this observation. A high level of fitness achieved during pre-season must be maintained throughout the whole competitive season. Therefore it should be properly and systematically assessed, allowing for modification of the training program in the case of under-recovery and/or overtraining. It should be noted that fitness assessment among athletes must be sport-specific, since different sports (complex coordination, cyclic, endurance, games etc.) require sport-specific properties. Therefore, in the following text, various models and programs will be discussed within different categories of sport activities. Specific focus in this discussion is the situation with fitness assessment at Penn State Collegiate Athletics.

2. FITNESS ASSESSMENT IN COMPLEX COORDINATION SPORTS

Fitness, a measure of the combination of muscular strength, endurance, power, agility, speed, balance, and flexibility, is sport specific. Strength and power tests vary among different sporting contexts. Different aspects of fitness can also very from one sport to another. For example, a football player may have the capability to lift 150 lbs in a gym, but at the same time be incapable of lifting a 115 lb. girl with one hand as seen in cheerleading. The amount of time devoted strength training and power testing differs from sport to sport. Complex coordination sports such as gymnastics may focus a large portion of their time on flexibility, whereas sports like long distance swimming and running spend a majority of their effort on improving endurance.

Due to sport specific differences in fitness, different sport specific physical assessment tests are useful in order to determine when an athlete has reached a necessary fitness level. These tests can also be helpful in determining when an athlete has fully recovered from an injury and when it is appropriate for him/her to safely return to play. In the next few paragraphs, different complex coordination sports such as gymnastics, figure skating, competitive dancing, and cheerleading will be discussed to varying degrees on topics such as injury, fitness, strength and endurance assessments.

2.1. Gymnastics

The GFMT, which stands for gymnastics functional measurement tool, was developed by physical therapist Mark D. Sleeper to be used in assessing the overall fitness level of gymnasts. The GFMT can also be used to measure the fitness progress of a gymnast and is highly sport specific. In order to test a gymnast's level of muscular strength, endurance, and flexibility, Sleeper designed a series of specific tests that are each awarded a certain number of points depending on how well the athlete performs the task.

Sleeper developed a series of challenges to assess the muscular strength of gymnasts. He measured their strength by using rope climb challenges, hanging pikes, over-grip pull-ups, a push-up test, and a handstand hold. The rope climb tests the level of ability with which the athlete performs the task and also factors in time. For example, if a gymnast was able to climb the rope hand over hand with his/her hips at ninety degrees of flexion within the time frame of 0-10 seconds, he/she would be awarded ten points, the maximum score. However, if the gymnast was only capable of climbing the rope hand to hand, as opposed to hand over hand, and also enlisted the help of his/her legs and completed the climb in fourteen seconds, he/she would only earn five points. Two points for method and three for time. Another muscular strength challenge developed by Sleeper is the hanging pikes test. The hanging pikes test measures abdominal and hip flexor strength. This test begins with the gymnast hanging from a horizontal bar with his/her body in completely straight alignment. The athlete is then tested on the number of times he/she can touch his/her toes to the bar while keeping the legs straight. In order to prevent a momentum advantage, between each leg lift the gymnast must pause in a straight hanging position.

2.1.1. Endurance

In order to test muscular endurance, Sleeper designed a series of six tests. The agility sprint is the first test of endurance. In addition to testing muscular strength, the rope climb, hanging pikes, over-grip pull-ups, the push-up test, and the handstand hold all test muscular endurance as well. The agility sprint is based on time. The gymnast is asked to sprint the diagonal of a regulation competition floor. Two cones are placed on opposite corners of the floor. The athlete must sprint to the cone, touch the cone, run back and touch the opposite cone five times before the clock is stopped. The athlete's level of endurance is based on the speed in which he/she completes this task. The handstand hold tests strength and balance and is another endurance test. The gymnast is given two chances to hold a handstand on the beam for as long as possible. A handstand is defined as the gymnast's total body weight being supported entirely by the hands. If any body part besides the hands touches the floor or beam or if the gymnast moves his/her hands from the original position, the time is stopped.

2.1.2. Flexibility

Flexibility is another test of fitness. The GFMT flexibility tests include a shoulder flexibility test and a right and left leg and middle split test. The shoulder flexibility test measures shoulder flexion flexibility. First the athlete holds her arms out straight in front of her at a ninety degree angle while holding a wooden dowel. Then a measurement of arm length is taken from her AC joint to the dowel. After this measurement the gymnast lies face down on the mat with her nose and chin touching the floor and with her arms extended in front of her. The gymnast is then asked to hold the dowel with an overhand grip and thumbs touching and then is told to lift her arms as high as she can without her nose coming off the mat and without changing her grip. When the athlete reaches the peak, a measurement is taken from the dowel to the floor. The split test assesses pelvis and lower extremity flexibility. Specific positioning of the body is determined and the gymnast is asked to slide into a split. Different measurements are taken from where the highest point of body clearance is from the floor mat. This test is done for both legs and a middle split.

2.2. Competitive Figure Skating

Competitive figure skating has evolved into a very physically demanding sport. Single skaters need to incorporate more double, triple and quadruple jumps into their routine, while pair skaters must execute more throws and lifts in order to impress judges and receive national recognition for their complex coordination skills. To accomplish such high levels of performance, in addition to training on ice, athletes are required to participate in weight training, dance and aerobic activities. This results in elite figure skaters training at least four to six hours per day, six times per week, for ten to eleven months per year. Such rigorous physical fitness demands leads to a number of acute and overuse injuries. Analyzing injuries in figure skating is a unique topic to evaluate because no other sport places the same diversity of forces on such a narrow base of support.

In both studies, *"Competitive Figure Skating Injuries"* and *"The Incidence of Injuries in Elite Junior Figure Skaters,"* most of the injuries observed with single competitive skaters were of the overuse type and occurred in the lower extremities. This suggests that there might be a relationship between the long hours of training and the injury itself. With pair skaters and ice dance skaters, on the other hand, most of the injuries were acute due to the fact that there are more lifts and throw jumps during their competitions. In general, pair skaters accounted for more injuries than single skaters and the injuries were mostly to the head and shoulder because of carries, throws and lifts. Women pair skaters were also more susceptible to injury because they are the ones being tossed in the air by their strong male counterparts. The research studies also concluded that single male skaters are more vulnerable to injury compared to single female skaters possibly because of the males' older age, greater body weight, and perhaps, the greater height, velocity, and impact of their jumps.

Since most competitive figure skaters begin training when they are physically immature, it is important for physicians, physiotherapists, and physical educators to advise, guide, and monitor the training of young skaters. Parents, coaches and administrators should also be aware of the overuse and acute injuries involved with competitive figure skating. As stated in the *"Incidence of Injuries in Elite Junior Figure Skaters,"* "Only through good postural alignment, adequate stretching and strengthening training programs, especially during the asynchronous development period of bone and soft tissue, can overuse syndromes be prevented and reduced." Educating parents, coaches and administrators about the susceptibility for injury in such a complex coordination sport can help prevent injuries. The findings from both studies suggest that more research is needed in evaluating the predisposing factors for overuse and acute injuries and methods of rehabilitation. The predominance of lower extremity injuries calls for investigation of footwear materials (fit, alignment and stability) as well as a detailed biomechanical analysis of propulsion and jump landing mechanics. Hopefully, with such knowledge about the injuries involved in figure skating, we can reduce the number of injuries.

2.3. Competitive Dancing

Competitive dancing is an activity that takes a lot of skill, physical strength, and flexibility in order to excel at it. Because of the specificity of the muscles that must be strong and the flexibility that is mandatory, acute and overuse injuries are a common factor during competitive dancing. Acute injury is defined as a sudden sharp pain or pop that a dancer could relate to a specific situation. Overuse injury is defined as a problem with insidious onset that continued to bother the dancer during a period of at least two weeks (Askling, 2002).

A study was done to see how prevalent injuries were in ninety-nine students at a dancing academy. All of these students were educated about the definitions of acute injury and overuse injury before answering a questionnaire about their injury history. Every third dancer had a history of acute injury in the last ten years, while every sixth dancer had a history of overuse injury in the last ten years. Eighty-eight percent of the students that defined their injury as acute stated that the injury occurred during slow activities in flexibility training, while the other twelve percent said it was caused by a powerful movement. All of the students with the acute injury said that it occurred in the proximal part of the hamstring muscle. Sixty-six percent of the injuries occurred while doing some form of the splits, which is a slow flexible activity. After injury most of the students stopped the physical activity for an extended period of time while some continued in the same training routine as before the injury. When comparing the two groups, there was no significant difference in continued problems after the injury (Askling, 2002).

Results from the study showed that extreme caution should be used while doing extensive stretching while in a competition or directly after. It was shown that a fatigued hamstring muscle is much more likely to be injured while stretching than a strong, rested muscle. The findings of the hamstring being injured during slow intense movements are contradictory to all other findings that showed hamstring injuries only occurred during high speed, high force activities. Many of the dancers in this study underestimated the time it would take for recovery. Most figured they would be back between a few weeks to a few months. The average recovery time was, however, was eight months with some lasting up to eighty months of problems after the initial injury. The study showed that dancers, and other participants in sports prone to acute hamstring injuries, should be more educated about the severity of the injury and the precautions that should be taken to prevent further injury. Educating athletes on how to treat the injury will minimize continuing problems (Askling, 2002).

2.4. Cheerleading

Cheerleading was nowhere close to being considered a sport when it began in 1884 at Princeton University. It began with a group of men displaying college spirit at athletic events. Now, cheerleading is highly recognized as a competitive sport with more than 400 North American teams at the collegiate level and 230 teams around the world. The sport has now introduced cheerleading gyms to the world's most famous cities, such as

Bangkok, London, Mexico City, Moscow, New York, Sydney, and Hong Kong (cheerleading.net). There are also more than 170 cheerleading associations and companies conducting training programs and cheerleading competitions around the world. With cheerleading evolving into a contact sport with high demands of agility, strength and gymnastic skills, cheerleading-related injuries are now on the rise. Between 1982 and 2002, the National Center for Catastrophic Sports Injury Research had forty-two direct catastrophic cheerleading-related injuries reported. The injuries varied from serious to fatal. They included traumatic head injuries such as skull fractures, cervical (C5) spine fractures, spinal cord injuries, coma, and quadriplegia.

Figure 1. Penn State Cheerleading team is getting ready for national competition

Most injuries result from pyramid building, basket tosses, and floor routine performances (Boden, 2003). Unfortunately, the National Collegiate Athletic Association injury surveillance does not include cheerleading injuries. Therefore, knowledge about injured cheerleaders at the collegiate level is limited. In 2005, the NCAA Insurance program stated that twenty-five percent of money spent on student-athlete injuries resulted from varsity cheerleading. After the monetary concern, the NCAA and Varsity Brands formed an alliance to improve safety in collegiate cheerleading. The American Association of Cheerleading Coaches and Administrators (AACCA) developed a safety program that mandatorily required coaches and team members to receive training and certification at Varsity Brand training camps (NCCSI, 2006).

2.4.1. Injury Assessment

Injuries experienced by athletes play influential roles in the their changes in behavior, self identity, performance, and adjustment to cognitive stress. However, injuries in competitive cheerleading can be perceived differently in the sense that they may have a greater influence on the concept of returning to play. Cheerleading has been determined as the sport with the highest average days lost to injury. Although football players experience the greatest number of injuries, cheerleaders lose more practice time than football players (Axe, 1991). This hard to believe phenomenon is due to the fact that all limbs are required to execute the required movements and stunts related to the sport. Many athletes of other sports experience injuries that do not limit their participation in practice. For example; a track runner with a strained bicep muscle still has the ability to run; however, a fractured arm or injured shoulder can easily prevent a cheerleader from participating for several weeks.

Full recovery after injury is also negatively affected by the time frame of the sport. Cheerleading is a year-round activity involving football season when most teams prepare for nationals and finally ending in the spring while committing to basketball season. This circulating and overlapping schedule of many collegiate teams allows limited time to fully recover from injuries. Shields and Smith (2006) conducted a study where cheerleading-related injuries of an estimated 208,800 injured-cheerleaders ages five to eighteen, were analyzed and categorized into injury and body part types. Data was retrieved from the National Electronic Surveillance System (NEISS). Results are shown in Table 1. The authors also concluded that injuries to the lower extremities were more common in cheerleaders twelve to eighteen years of age.

Table 1. Shows the categories and percentages of injury and types and body parts injured. *Abrasions, contusions, and hematoma **Foreign body, crushing injury, nerve damage, hemorrhage, dental injury, and anoxia. ***internal, pubic region, and (>25%) body.

Injury Type	Percentage	Body Part Type	Percentage
Fractures/Dislocation	16.4%	Upper Extremity	26.4%
Laceration/Avulsions	3.8%	Lower Extremity	37.2%
Soft tissue Injury*	18.4%	Head/Neck	3.5%
Concussion/ Closed head injury	3.5%	Trunk	16.8%
Strain/Sprain	54.4%	Other***	.8%
Other**	5.5%		

The Consumer Product Safety Commission reported an estimated 16,000 emergency room visits in 1986. In 1999, the number of injuries increased to approximately 21,916 and in 2004, the number increased to 28,414. A major factor playing a role in this increase has been the progression in athleticism in cheerleading, which now involves elite acrobatic type stunts. Mueller and Cantu (2005) also reported that cheerleading is responsible for more than fifty percent of all high school and collegiate direct catastrophic injuries experienced by female athletes.

A significant cause contributing to this increase in cheerleading injuries is coaching errors. A highly significant number of high school teams do not have qualified or certified coaching staff. Furthermore, athletics departments at the high school level do not consider cheerleading as a sport. Therefore, athletic trainers and medical staff are extremely limited. Other faults in coaching styles are inadequate training, overtraining, insufficient supervision, and practicing on inappropriate surfaces.

2.4.2. Fitness Assessment

Although research on cheerleading injuries is now increasing in publications, limited knowledge is known about the physiological status of a competitive cheerleader. For both men and women, there are demands of flexibility, strength, agility, and balance. Thomas et al. (2004) assessed the physical status of NCAA Division I collegiate cheerleaders. The authors determined fitness levels by measuring VO2max, maximum heart rate, push-ups, curl-ups, flexibility, quadriceps strength, and bench press. VO2max scores placed the men and women at the eightieth percentile of norms. Female cheerleaders' VO2max scores were similar to those reported for female basketball, gymnastics, swimming, and volleyball athletes. Male cheerleaders were similar to male basketball, football, and tennis players. In flexibility, the men were placed at the seventieth percentile and the women beyond the ninetieth percentile. Both men's and women's bench press scores were similar those of their basketball playing counterparts. Push-up scores placed the men at the ninetieth percentile and women at the seventy-fifth percentile. These demands for well above the norm of fitness levels demonstrate that collegiate cheerleaders match up to other collegiate athletes. Their high degree of fitness appears to reflect the demands of the sport.

3. FITNESS ASSESSMENT IN FOOTBALL

In athletics today in general, and in football in particular, there are countless methods of training. Some of them are effective and some of them can be damaging to the athletes. In order to prevent further maladaptive responses in the athletes of our chosen sport, our task was to look in depth into the three fields of fitness and create testing procedures for each of them. These testing procedures would help players, coaches, and the training staff evaluate the physical condition, and based on that, create a safe and effective training program for each player. Due to Penn State's rich history of success in the football program, the team sport that we chose to look at is football. In football, as in any sport, three major areas of physical fitness need to be focused on when training and testing. These three aspects are flexibility, endurance and, most importantly in football, strength. In football it is difficult to have a team-wide testing regimen due to the varying tasks that the many positions are required to do in competition. In order to overcome this we had to generalize testing procedures for the whole team. In the following paragraphs we will discuss within the three fields of fitness: the current trends in training, how the current trends could be made more efficient, and specific testing procedures for each of the three facets of fitness.

Flexibility is the first aspect of physical fitness that we have focused on. And unfortunately it seems to be the most overlooked in football, including in our team at Penn State. Flexibility should not be overlooked because it can be very beneficial in preventing injuries to joints that often plague football teams. One benefit of being flexible is that you have increased length in both your muscles and tendons. This leads to an increased range of movement, which helps your limbs and joints move further prior to an injury occurrence, theoretically making a potential injury less damaging.

A brief review of the situation discovered that not all football teams ignore flexibility as an aspect of fitness. The Denver Broncos employ a flexibility training program that includes three different types of stretching. The first, ballistic stretching, uses rapid bounce to stretch the muscle. However, these uncontrolled movements can easily result in excessive loading and can damage the connective tissue by extending it beyond its elastic capabilities. The second is static stretching, which is applying steady pressure at the extreme range of motion without bouncing. The final technique they use is passive stretching which is a slow, controlled stretch of the muscle. The technique of static stretching requires a slow, controlled elongation of the relaxed muscles; you feel a pull but no pain. This position is held for twenty seconds, and then the muscle is slowly allowed to shorten. We found an effective stretching regimen consists of a brief cardio-respiratory warm-up such as jogging, running in place, or jumping rope. Stretching should not be painful as you can easily damage soft tissues, ligaments, and tendons. To maximize flexibility, stretching exercises should be performed at least several times a day and four to five days a week.

Figure 2. Penn State Football team stretching prior to practice

For flexibility testing, there aren't extensive lists of testing procedures for football, or any other sport. Some general testing that could be used could be a sit and reach test. This would test the hamstring flexibility of the players. This could be useful since hamstring pulls are a common injury in football. Another simple way to test flexibility in football would be to have an athletic trainer passively stretch each player at various joints and gauge how good the player's range of motion is. As stated earlier, an increased range of motion from good flexibility will lead to a decreased chance of injury.

Endurance is the ability to maintain a high level of intensity in a specific activity for an extended period of time. Over the period of the game each player will need to perform his tasks repeatedly. This means a high level of endurance is required to compete at an elevated level throughout the game. Endurance, like all other aspects of fitness, is task specific. In other words, football does not require the same type of endurance that a marathon runner needs. Since we could not find any information on Penn State football's endurance training, we looked at how most football teams generally train endurance. There are two types of endurance that football players train. These are speed endurance and strength endurance.

Speed endurance is maintaining a high intensity anaerobic activity. This is done by developing coordination of slow and fast-twitch muscle fibers. To train speed endurance, football players perform a number of sprints at an intensity in excess of 85%. To test speed endurance, a coach could have players run several forty-yard sprints in a row. Then the coach could assess the player's speed endurance by comparing the player's best and worst times. The closer the times, the higher the level of speed endurance.

Strength endurance is a muscle's capacity to maintain the quality of the muscle's contractile force over a period of time. Some ways to train strength endurance, specific to football, are running with a parachute, hitting sleds, or basically any activity where muscle contractions last for an extended period of time. An easy way to test strength endurance would be to use a single man sled, and see how far each player can take it. These training and testing methods not only improve strength endurance, but also maximal strength can be improved simultaneously.

Maximal strength, or just strength as it is usually known, is the maximum amount of force that one can produce with a specific muscle. It is obvious that this is a very important feature of physical function to have in football. Every tackle, block, hit, and many other parts of the game call for the player to have as much force output as possible. Strength is arguably the most important aspect of fitness in football. Therefore, we focused on what Penn State's football program does for strength training, and how it could be improved upon, then finally some simple strength testing that could be employed by them.

The Penn State football team's lifting program only uses isolated muscle weightlifting machines, known as Hammer Strength machines, to increase strength. The players' common practice is to complete the maximum number of repetitions that they are able to accomplish for a given weight. Each week, either their repetitions or the amount of weight in which they are using is increased. Free weights are not utilized in their strength training regimen. This system of training poses many problems for the athletes for multiple reasons.

Using only Hammer Strength machines does not allow the athletes to realize the full benefits of weightlifting and develop their strength to the fullest. Focusing more on power movements, or exercises that recruit large groups of muscles, would allow each athlete to not only improve a larger group of primary muscles, but also it would allow them to strengthen their stabilizing muscles as well. For example, performing an exercise such as the leg press to strengthen the quadriceps and hamstrings does not allow for the strengthening of the athlete's core muscles such as the abdominals and paraspinals. Performing an exercise such as squats would not only strengthen the quadriceps and hamstrings but would also strengthen these core muscles along with other stabilizing muscles. Strengthening the stabilizer muscles is especially important for a football player. For instance, it is common to see a defender attempt to drag a ball carrier down by applying his weight to the ball carrier in a way that it would cause the ball carrier to lose his balance and fall to the ground. Increasing the strength of stabilizing muscles will increase his balance and ability to maintain an upright position when a load is added, causing a change in his center of mass and pressure. Not only can improved balance help performance but it can also help avoid many injuries of the lower body, including sprained ankles, torn knee ligaments, and hip injuries.

Power movements, such as a power clean, would not only additionally improve stabilizing muscles, but it is also a task more specific to football than the Hammer Strength machines. Frequently on a football field, many "explosive" movements are performed. Common football moves such as a first step toward a defender or the quick release of the football during a pass, rely on the quickness of an athlete. Using power movements improves this explosiveness which is so important in football. Training with power movements improves the interaction of the fast-twitch muscle fibers and the nerve endings in which innervate these fibers that are found throughout a player's body. In using an exercise such as a power clean these neuromuscular junctions are trained to perform at a much faster level through the use of certain strength training techniques and lifts. During this lift, quick contraction of the muscle fibers in the hamstrings, quadriceps, and calf muscles as well as the muscles found in the core are required to move the weight in the desired manner. Training these muscles to repeatedly fire at a quick and maximal rate will allow football players to reach maximum potential on the field through a newfound explosive ability. Using Hammer Strength machines alone would not allow an athlete to have the same benefits as can be acquired with power cleans. Strength can be easily assessed with any weight lifting exercise by having the athlete max out, or perform one repetition with the greatest possible weight. It is necessary that the players properly warm up and have spotters, professional instructors, and supervision, as one rep max tests can be somewhat dangerous due to the high exertion and probability of failure (coachr.org).

As said in the opening paragraph, it is difficult to have a team-wide testing regimen due to the varying tasks that the many positions are required to perform in competition. Therefore, coaches should adapt testing procedures for each of the various positions, and perhaps even each individual athlete. For example, a wide receiver does not require the same strength, flexibility, and endurance as a lineman, and should therefore be tested differently. All of these tests, for flexibility, endurance, and strength, need to be assessed not only at the beginning of the season, but at various points throughout the season. That way the coaches can adapt the periodization programs that they use to train their athletes, to focus on areas that need improvement as seen from the assessments.

4. FITNESS ASSESSMENT IN RUGBY

To be an elite athlete, one must posses certain specific skills. To make sure these skills are used to the highest ability possible, an athlete must perform fitness testing that include endurance, strength, and flexibility.

According to Professor Mike Stone, head of Sports Physiology with the United States Olympic Committee, strength is the ability of the neuromuscular system to produce force. Within this definition, aspects such as resistance, speed and power are included. While strength can be either dynamic or static, it can be measured in a few different ways, including isometric ally, concentrically, eccentrically and plyometrically. Endurance, which can be associated with strength also, is defined as the ability to sustain a prolonged stressful effort or activity before reaching fatigue (Webster). Two of the most common forms of endurance related to sports are cardiovascular and muscular endurance. While cardiovascular endurance deals with a person's ability to deliver oxygen to necessary tissues over a sustained period, muscular endurance involves a person being able to perform repeated contractions or perform contractions for a sustained period before reaching fatigue (McGlashan, 2003).

Tests for endurance include different mile tests (e.g. twelve minute, three mile run), doing sit-ups or push-ups until fatigue, treadmill test, or Maximal Oxygen Uptake (VO2max). Lastly, flexibility involves the "ability to move joints and use muscles through their full range of motion" (McGlashan 2003). One of the important reasons for having good flexibility is to reduce the risk of injury. A few ways to measure flexibility include the sit and reach test, trunk rotation, and using a flexometer (Fitness Testing, 2007). With these three aspects of fitness testing, I researched the assessment of them in the sport of rugby.

Even though rugby is thought to be football meets lacrosse, it is a high intensity sport. Professional rugby is an eighty minute game—two forty-minute halves with a ten minute break in between, so the average playing time is about thirty minutes a half. The other ten minutes are due to stoppages for injuries, conversions, penalty shots or ball out of play. Two teams play with fifteen players (eight forwards and seven backs) on the field each during the game. One interesting fact about the players of rugby is that each player's physical make up is based on his position. The front row, players 1-3, is more suited for strength and power since these players are required to gain possession of the ball. They also have frequent contact with opponents and are limited to running with the ball. Players 4 and 5 are called loose forwards. They have large body masses and power. The loose forwards, players 6-8, also have bodies specified for strength and power to gain and retain ball possessions. They also have to be mobile and powerful in open field play, have excellent speed, acceleration and endurance (Duthie, et al., 2003). Players 9 and 10, the half backs, are the endurance players since they control ball possession and have good speed. Players 12 and 13 are the midfield, and they too possess strength, power, and speed due to the increased frequency of contact. Lastly, players 11, 14 and 15 are called outsides, and they need only to specialize in speed since they provide run support, chase down kicks and cover defenses (Duthie, et al., 2003). Because of the differences in the specificity of the positions, forwards have a greater body mass than backs. This is due to forwards being exposed more to contact compared to the backs, who are more geared to running with the ball. Since rugby is a highly intensive sport, fitness testing is a must to help evaluate players.

With regards to *strength*, both muscular strength and power are tested in numerous ways. The vertical jump test is used to measure leg power. It can be measured by using measuring tape and a marking instrument (chalk or marker) or with a device called a Vertec or a jump mat. The most basic way to calculate the jump height is by finding the difference between the athlete's standing reach height (reaching up a wall, but keeping the feet on the ground) and the jump height (jumping vertically while away from the wall) (Fitness Testing 2007). Another test is the maximal strength test. The purpose of this is to measure the isotonic strength of different muscle groups. To perform this test, an athlete simply lifts a substantial amount of weight (barbell, dumbbell or any other type of free weight) and completes one repetition without fatigue. Usually this test can be repeated during one session since it might take a while for the athlete to meet maximal strength.

Another form of fitness testing that is crucial to professional rugby athletes is *endurance* testing. Since rugby is filled with bouts of sprinting that make it a high intensity sport, the measurement of VO2max is important to test (Kelton, 2004). This is usually measured with the aid of a treadmill where the athlete walks to the point of exhaustion, which would be his maximal oxygen uptake. Another method of testing endurance is the Multi-Stage Shuttle Run Test. This test can be performed as a team running continuously between cones that are twenty meters apart. The time it takes to make it from one cone to the next is recorded and can be used to figure out VO2max by using an equivalency chart (Fitness Testing, 2007).

One of the mostly overlooked aspects of fitness testing is *flexibility*. I must point out that while flexibility is a good thing to have in any sport, "ligamentous laxity or joint looseness, and athletes at either extreme of the flexibility continuum are probably at increased risk of injury" (Stewart, et al., 2004). With that being said, one popular test is the sit and reach test. This measures the hamstrings and lower back by the athlete sitting on the floor with his/her feet on the stretch and reach box, and then reaching forward as far as he/she can for a total of four times, with the fourth being held for a count of two seconds. The measurements are scored based on a scoring sheet that includes a positive for beyond the toes or a negative before the toes.

One more aspect of fitness testing that is very important for rugby players is *speed* testing. Just as endurance, strength, and flexibility help make an elite rugby player, speed is also a major contributor. Different sprint tests are performed by rugby players since the majority of the positions require speed. Tests are given at ten, twenty and forty meter distances, similar to tests performed by track athletes. The maximal running speed and acceleration are calculated and assessed. Consequently, by completing this test, an athlete could also perform an anaerobic test that measured fatigue (Kelton, 2004).

5. FITNESS ASSESSMENT IN CYCLIC SPORTS

Endurance sport and movement events that push the body to the limit are often overlooked when it comes to the amount of training these men and women go through. The amount of specific training that is designed for individual aspects of their performance is misunderstood by some amateur coaches, which can lead to injuries, lack of performance, and getting to peak performance at the wrong time. When it comes to cyclic sports such as swimming and endurance running, there are major differences in the way these athletes train, but in the end its almost all the same, as all are striving for the same goal by almost the same means, e.g. *periodization* and leveled training. Some types of training include, but are not limited to, running and swimming.

5.1. How to Measure:
a) **Strength**- weight training, plyo-metrics (training with one's own body weight, *i.e., pull-ups, push-ups, dips;*
b) **Flexibility**- measuring range of motion, sit-'n'-reach test, counting strides or measuring gait, stretching for a certain amount of time and then measuring range;
c) **Endurance**- timed mile, speed workouts (how long the athlete can hold the pace time), time the athlete for a certain distance or on a specific course (mile run indoors vs. mile run outside);

5.1.1. Strength
These three types of training lead to maximal performance, and this paper will tell you how. Strength training by athletes is important in any sport, especially for cyclic sports like running. It is important to understand how to increase training with certain athletes to avoid injury. Measuring the athlete's skill and progression can help to keep track of where the athlete needs to be in order to be able to perform to the best of his or her ability.

There are many ways to increase an athlete's ability and skill as a runner. The first step is to assess the athlete in order to understand where his or her abilities lie and what levels of force and challenges he or she can handle. This is very important for different training programs because not all athletes are at the same level. Coaches want to prevent any type of overtraining, but also want their athletes to be the best that they can be in that particular sport.

Different seasons call for different types of training. Pre-season consists of much lifting and endurance to ensure that the athlete will be strong enough to handle the seasonal workouts. In this particular season it is very important to have a particular workout plan and training session to prevent early injury and burnout during the season. The in-season regimen consists of endurance runs and workouts to ensure that the athlete can keep the physical condition that he or she gained in pre-season stable. Post- season is needed so that the athlete does not increase chance of injury. This is a time where athletes take some time off from their daily workout and rest their bodies. This ensures that athletes do not get hurt from wear and tear injuries and that their mind set will be fresh for another year of practice ad training. Post season also consists of an easy workout so that the athlete can stay in shape but not wear him or herself out.

Periodization is also applied to all types of training. This includes various workouts with changes in intensity and training volumes to achieve peak levels of fitness. This is done so that the athlete does not tire out and so that he or she reaches the optimal level at the correct time. There are many different strength-training techniques and starting in the weight room is one of them. First, help the athlete learn the machines, how to use them and how to increase weights if needed. Then devise a workout plan for each specific athlete so that he/she knows how much weight is needed and the specific number of repetitions and sets needed to achieve the optimal level of performance without injury. One can measure the increase of strength training through weights by monitoring the athlete's performance and the addition of weights to each workout. This can be done by the athletic trainer's observations or by a simple workout sheet that is used every time the athlete is in the weight room.

Other types of training can include cross training and ***plyo-metrics***. Cross training is need to prevent injury, but is also important because it targets the muscle group one would need doing the same repetitive movements. Some cross training exercises for running can include, but are not limited to, swimming, rock climbing, and kickboxing. All three of these exercises can target all parts of the body needed to strengthen running. According to Runner's World Magazine, "Because of the low impact nature of most cross-training activities, injury-prone runners can beef up their 'mileage' using this formula without increasing their risk of injury"

Plyo-metrics are also encouraged for all athletes during workouts and training. They consist of rapid eccentric movement followed by a short rest phase followed by an explosive concentric movement. This utilizes the stretch-reflex mechanism, allowing for much greater than normal force to be generated by pre-stretching a muscle...before it contracts" Some examples include drop jumping, standing jump, multiple jumps, single leg jumps, hops, and bounds. Neuromuscular firing patterns are developed to improve the muscle contractility of specific muscle groups (Mauro, 2005).

Many athletes can be hooked to machines and tested for specific increased amounts of certain muscle fibers. This observation can reassure trainers that athletes are getting proper conditioning techniques and results. An increase of movement and strength training can decrease or reduce the pain of "runner's knee" to as little as six to eight weeks. This can also relieve the recurrence of common injuries such as nagging hip and lower back pain. Overall strength training is highly important and encouraged in athletes especially among runners. Strength training is a key factor to prevent overtraining and injury.

Weight training can be used for many things including injury prevention, rehabilitation, strength gain, building general or specific fitness, or to cross-train to improve abilities in other sports. Many people would state that the best way to get better at swimming would be to swim. But in order to get additional gains once a swimmer has maximized swim time would be to add dry land work such as strength training. Swimmers need a lot of cross-training, especially strength training, in conjunction with their usual swim training. Swimming all day long would not be good for the body of a swimmer. In fact, many swimmers often get injured when they begin to purely swim and do not incorporate a strength training workout into their program. In order to optimize strength and power in the water, swimmers must practice strength training outside of the water in a gym setting.

For many swimmers, off season workouts are a lot less intense in strength training than pre-season and in-season strength training programs. Just prior to swim meets practices are often cut short in the water and strength training in the gym still occurs, but usually with lighter weights. A typical strength training routine for swimmers consists of a full body strength training workout three days a week for about half an hour. A full body strength workout is necessary because swimming requires the use of muscles throughout the entire body. There is no basic cookie cutter plan for strength training that applies to all swimmers. A swimmer's strength training program often is based on individual needs, individual goals, a season plan, and available equipment. However, as some general weight training rules, it is essential that swimmers give muscles time to rebuild by not lifting two days in a row and, as some may say, to prevent injury by not "lifting to failure". It is also essential to incorporate a warm up and cool down into any strength training program in order to avoid injury.

In conclusion, for swimmers, or any individual participating in a strength workout, it is important to start light and gradually increase. The best way to prevent injury during a strength training program is through slow progress. Swimmers should include cross training such as dry land workouts of strength training. Strength training can be a major contributor to faster swimming by increasing general fitness, speed, strength, coordination, balance, and body awareness. These increased abilities create positive benefits in a swimmer's swim training and performance.

5.2. Flexibility

Flexibility, which is an underrated aspect of sports, especially among males, makes a huge impact on the performance of athletes in cyclic sports. For some, running is an athlete's main sport, such as cross country or track contestants; for others, running is a key component of an athlete's training for another sport, such as football, basketball, or soccer. The main thing these two situations have in common is the ability to be flexible while running to protect your body.

Dr. Nicholas Romanov, a running guru from Russia who specializes in injury diagnosis, prevention, and exercise rehabilitation, among other disciplines, has done a great deal of research on the sport of running. Dr. Romanov states that the ability to bend in one's joints has three main components: mobility of joints; elasticity of ligaments and tendons; and muscle relaxation, providing their lengthening. All three parts require a separate approach and development, which varies with every individual. Romanov stresses that stretching and flexibility are not synonymous. He defines stretching as the forceful pull of muscles, tendons, and ligaments, which is not a body's natural position. However, a muscle relaxing is a normal state, as is flexibility (Romanov, 2007).

One factor to consider that directly affects flexibility is the running economy, or the submaximal energy cost of running at a given speed. Several recent research studies have suggested that trunk and lower limb flexibility are inversely related to running economy. After testing 100 males and females, the researchers determined that subjects who exhibited tightness in the trunk, which limited the turning of the leg, were the most economical at every speed test. The researchers hypothesize that they got these results because running occurs primarily in a forward direction, meaning that rotational motion is potentially energy-wasting and does not contribute to forward movement (Wilkinson & Williams 2007).

As one ages, flexibility decreases. Joint range of motion decreases due to muscle stiffness. If practiced, flexibility can contribute to anti-aging. The way to improve flexibility or maintain it is to perform stretches. It is recommended to stretch three times a week for ten to thirty seconds and do three to five repetitions. If flexibility and stretching techniques are taught at a young age, it will benefit the individual in the future.

To continue, flexibility is also important in athletes. It has been known to prevent injury and enhance performances. Flexibility should be practiced the same way athletes weight train, with specificity. While weight training, a springboard diver will not perform all the same exercises as a platform diver. Same is true with flexibility. For example, swimmers need flexible ankles to increase their range of kicking, while runners need stiff, strong calves and Achilles tendons to run on hard surfaces.

More specifically with swimmers, the shoulder joint is typically the most flexible joint. This is needed to perform all four strokes in the sport. However, too much flexibility can cause injury. It is important that swimmers maintain shoulder strength along with flexibility. In addition to the shoulder, flexibility in the lower extremities is also beneficial. To improve underwater kicking or kicking during a stroke, flexible ankles, knees, hips and lower back are needed. Flexibility allows swimmers to be efficient in the water and therefore improve racing.

As stated before, stretching increases flexibility. It is also important to note that stretches are beneficial when performed with warmed muscles. Stretching cold muscles could cause injury. Also if all limbs are used during an activity, the upper and lower body should be stretched equally. For instance, swimmers use both upper and lower extremities, so stretches for the triceps and latissimus dorsi are just as important as the quadriceps and hamstrings. Also it is important to know how a swimmer's most flexible joint is measured. A goniometer is used to determine a shoulder joint's range of motion. Different tools and techniques are used to measure other areas of the body, such as the standard sit-'n'-reach and V-sit tests to test the hamstrings.

5.3. Endurance

Endurance is an important component in performance for those individuals who participate in running. There are many different routines athletes can implement into their program to increase endurance. Several examples are: increasing the intensity of a workout over time, varying running speed during a workout, and training at the maximal level. There are also several ways to measure endurance, including heart rate and VO2 readings.

The first example of building endurance is increasing the intensity of a workout. This is achieved by starting out slowly and gradually increasing the intensity by walking faster, running, or walking on an incline. During the workout, the individual should continually make the exercise more difficult, which will in turn help to increase endurance. A second example is to vary running speed during a workout. In this routine, the individual runs at normal training speed, but incorporates short bursts of speed throughout the workout, lasting anywhere from a few seconds to a few minutes. By varying the speed times and sections, the individual is training his/her body to endure running at higher levels (endurance). The third example is to train at a maximum level. This increases endurance since athletes are training their body to tolerate a higher level of performance (Cosgrove, 2007).

Athletes can measure their endurance through a variety of methods such as checking heart rate and VO2max. As an athlete continues to build endurance, the resting heart rate (RHR) should continually decrease. By keeping a log of RHR from each training period and comparing the logs over time, an athlete can tell if his/her endurance has increased (Sinah, 2007). VO2max, or the maximal oxygen consumption during a workout, can also be measured to determine an increase in endurance. As endurance is increased, the VO2max of the individual increases as well (Wikipedia).

When dealing with a cyclic sport such as swimming, the way that has been proven best for measuring endurance is using Lactate Testing. Lactate testing can act as a superb educational tool, giving swimmers accurate information about what is actually happening in their muscles when they train and enabling them to understand the effects of different sets and the importance of the control of relationships among the various training parameters. Lactate testing shows the changing relationship between effort and speed. More effort means more lactate. At low intensities, speed increases faster than the lactate; at higher intensities, the lactate changes faster than speed. It also shows oxygen use where it counts—in the muscles at the cellular level. There are three major components for proper lactate testing and figuring out the true endurance of an athlete. They are Aerobic Work, Measures of Endurance, Maximum Lactate Value.

Many programs have used 4mM of lactate (4 milliliters of lactate at maximum effort) as their standardized measure of aerobic intensity. If one chooses a very low or very high intensity, results will not reflect the endurance capabilities accurately, so 4mM is a good choice and enables one to compare results with those of many other programs.

Measuring endurance capabilities in swimmers is one of the easiest coaching tasks. Any timed-distance swim will provide information about the endurance capabilities of the swimmer. You will know how fast he/she, but even if the swimmer gives maximum effort, you will not know how much energy he/she has used to produce the speed. If the swimmer does not give maximum effort, you will know very little. Lactate testing during one of these sets will tell you the effort precisely, even if the swimmer swims sub-maximally.

Lactic acid moving out of the muscles becomes lactate when it enters the bloodstream. As swimmers apply more effort to swim faster, we can measure more lactate. If they train to produce very high amounts of lactate and control their application of power, they will be able to swim even faster. The highest lactate result of a test is, therefore, a very important parameter— it corresponds to the maximum speed. Its abbreviation is LaMax.

All three vital components are interrelated. For example, training that is designed to produce changes in aerobic conditioning will, necessarily, also change the LaMax and the relationship between aerobic conditioning and maximum lactate. The results of lactic testing can be used to produce a training regimen that can positively affect the swimmer by increasing endurance and establishing at what stage of the race a swimmer produces the most lactic acid, and it pinpoints where improvement is needed.

These examples and real-world situations discussing the importance of training cannot be overlooked. The lack of knowledge of some coaches at the amateur level prevents youngsters from gaining the correct idea of not only how to train, but why they should train—not only to get better, but to avoid injuries that could be minor or catastrophic. Recently an Olympic runner from the United States collapsed and died while he was running. Not to say that the athlete did not train right, but it is an example to show that if that can happened to an athlete with the correct training and diet, then what could happen to immature bodies when they are incorrectly trained for strenuous physical activities?

Outside the physicality of sports, there is a huge part that is mental as well. Being put on a training program that is correct, and being tested to see levels of performance and health, helps the mental aspect of an athlete's ability. Mentally, when an athlete does the right thing, and is shown how to do the right thing when he/she is young, that athlete begins to have trust in the training which leads to him/her becoming a better athlete both physically and mentally. Some professional athletes feel they do not perform to their maximum potential because of the way their coaches run things, and it has nothing to do with their physical ability. Once an athlete believes in the program, it will have more benefits to lead that athlete to his or her greatest potential and most importantly will keep him or her on the field of play.

CONCLUSION

Gymnastics uses the test designed by Mark D. Sleeper, to assess gymnasts in flexibility and neuromuscular endurance through the use of rope climbing challenges, hanging pikes, over-grip pull-ups, the pushup test, and the handstand hold test. Competitive figure skating requires a variety of training techniques that involve more then just training on ice. These athletes are required to weight train as well as participate in dance and aerobic activities to improve their performance. Injuries that occur in this sport include overuse and lower extremity injuries. Competitive dancing requires physical strength and coordination, as a result the prevalent injuries are related to overuse. The demand on the body in competitive dancing requires extensive stretching, which can lead to even more injuries.

One of the underrepresented sports, cheerleading, showed clear signs of being similar to other sport related fitness testing while being tested using advanced methods that involved flexibility. This sport requires a lot of coordinated stunts that compromise the demands of flexibility, strength, agility, and balance. These sport-specific tests are important in reflecting and advancing the athlete's performance. Each sport places specific, concentrated demands on the body. If these tests are administered regularly, the prevalence of injuries may decrease.

Within the sport of rugby, fitness testing is very important. Unlike most sports, not all players are alike in rugby due to the fact that the different positions require different abilities. Flexibility is needed due to the high demands of running in the sport; therefore the hamstrings can't be tight. By performing the sit and reach test, an athlete can be assessed on flexibility, and greater emphasis on that aspect can be placed when training. Strength and power are important to test because many of the positions require these for ball possession or to guard against the opposition, especially since there is no equipment other than the body. Endurance is very important to test because rugby is filled with repeated sprints, usually without any form of break in play. Testing for endurance can ultimately mean the difference between winning and losing, especially if one team or player hasn't conditioned properly and has fatigued towards the end of a tight game. Last but not least, speed is just as important to test as the other three. Since most of the positions require speed while controlling ball possession or providing run support, speed is important. So the next time people say they are going to play rugby, stop and think about how skillful they are in these areas and you won't look down on this sport again.

If coaches were to focus on all of these aspects of physical fitness, not only in training, but also in testing their players for the specific tasks that are required during the game, they could adopt a more effective training *periodization* for their athletes. The end result would be that their athletes would function at a higher level with fewer injuries and better performance. Overall, as can be seen again from the previous text, one of the main reasons for injury is coaching errors, specifically with respect to the underestimation of fitness assessment throughout whole season. Hopefully, the aforementioned elaborations will gain popularity among young adults and make them realize that if they go into coaching, they hold the lives of those athletes in their hands and must make sure they don't over train and overwork athletes, and thereby avoid injury.

REFERENCES

Contos, A., Edlin, R., Collins, M. (2006). Aerobic fitness and concussion outcomes in high school football. In S. Slobounov and W. Sebastianelli, (Eds.), *Foundations of sport-related brain injuries,* pp. 315-340, Springer.

Askling, C., Lund, H., Saartok, T., Thorstensson, A. (2002). Self-reported hamstring injuries in student-dancers. *Scandinavian Journal of Medicine & Science in Sports, 12*(4), 11-17.

Boden, P.B., Tacchetti, R., Mueller, O.F. (2003). Catastrophic Cheerleading Injuries. *American Journal of Sports Medicine*, 31, 881.

Axe, M., Newcomb, W., Warner, D. (1991). Sports injuries and adolescent athletes. *Del Medical Journal*, 63, 359-63.

Shields B.J. and Smith G.A. (2006). Cheerleading-Related Injuries to Children 5 to 18 Years of Age: United State, 1990-2002. *NEISS, 117*; 122-1229.

Mueller F.O. and Cantu R.C. National Center for Catastrophic Sports Injury Research, Twenty-Fourth Annual Report Fall 1982-Spring, 2006.

Thomas D.Q., Seegmiller J.G., Cook T.L., and Young B.A. (2004). Physiological Profile of the Fitness Status of Collegiate Cheerleaders. *Journal of Strength and Conditioning Research, 18*, 252-254.

Webster's Dictionary. Merriam-Webster's Dictionary

McGlashan, L. (2003). Do you measure up?" Fitness Testing for Rugby.

Duthie, G, et al. (2003). Applied Physiology and Game Analysis of Rugby Union. *Sports Medicine* 33, 973-991.

"Fitness Testing for Rugby" 13 Nov 2007.
http://www.topendsports.com/sport/union/testing.htm.

Kelton, J. "Fitness Testing Assignment: Rugby" 27 Sept 2004. 12 Nov 2007.
http://physiotherapy.curtin.edu.au/resources/educational-resources/exphys/99/rugbye.cfm.

Mauro, P. (2005) – www.trainingsmartonline.com
http://www.trainingsmartonline.com/triathlon_plyometrics.php

Romanov, N. (2007). Flexibility in Running. November 14, 2007, cf: from Pose Tech.
http://www.posetech.com/training/archives/000375.html.

Wilkinson, M. & Williams, A. (2007). Running Economy. November 14, 2006, from Peak Performance Sport Excellence. http://www.pponline.co.uk/encyc/1007.htm.

Cosgrove, A. *Why "Endurance" Training Lacks Staying Power.* Retrieved November 8, 2007. http://www.alwyncosgrove.com/Endurance.html.

Sinah, A. *Heart Monitor Training.* Retrieved November 8, 2007.
Http://www.marathonguide.com/training/article/HeartMonitorTraining.cfm. *"VO2 max".* Retrieved November 8, 2007. http://en.wikipedia.org/wiki/VO2_max.

Sleeper, Mark D. "Gymnastics Functional Measurement Tool." 8 Jan. 2007. Department of Physical Therapy and Human Movement Sciences, Northwestern University. 5 Nov. 2007 <http://www.medschool.northwestern.edu/nupthms/research/sleeper/GFMT-Directions.pdf

PART III: PSYCHOLOGICAL TRAUMAS IN ATHLETES

CHAPTER 11

PSYCHOLOGICAL TRAUMA: UNFORTUNATE EXPERIENCE IN ATHLETICS

1. INTRODUCTION

Sport, a highly valued aspect of our culture, shapes the minds of athletes, organizers and spectators, as well as medical practitioners, partly because athletic injuries are an unfortunate part of modern sport today. Traumatic injury is defined as damage resulting in functional deficits and functional abnormalities at different levels of the CNS. Similarly, *psychological trauma* is defined as a type of damage to the *psyche* that occurs as a result of a traumatic event. A traumatic event involves a singular experience (i.e., a single episode of traumatic injury) or an enduring event (i.e., multiple traumas) or events that completely overwhelm the individual's ability to cope with or integrate the ideas and emotions involved with that experience. Psychological trauma in an athletic environment can be caused by a wide variety of events (e.g., previous traumatic injury, conflict with coaching staff etc.), but there are a few common aspects. It usually involves a whole complex of behavioral, cognitive and emotional sequelae, including a complete feeling of helplessness in the face of a real or subjective threat to life, bodily integrity, or sanity. Conventional wisdom is that mental problems in athletes are the direct consequences of physical trauma. However, it is important to note that psychological trauma may accompany physical trauma or exist independently of it.

Injury is an aspect of sport that has been highly evaluated and researched in recent years. Questions continuously arise as to whether or not traumatic injuries can be prevented? Or reduced? Or even predicted? These questions have plagued researchers, coaches, athletes, and doctors for some time. Unfortunately, we can say that injuries overall will never be 100% preventable because a combination of organism and environmental sources always poses uncontrollable threats to our bodies, no matter how much we redesign safety equipment and safety rules. The new question we are about to explore has to do with whether or not psychological trauma is linked to and caused by every injury, and whether or not this mental aspect of injury can be reduced, prevented or predicted.

However, before we are able to approach the issue of the prevention of psychological trauma in athletes, it is necessary at least to consider the following challenges: (a) to develop conceptual clarity and a framework about psychological trauma as a unique phenomenon in athletics; (b) to elaborate on both overt and covert essential components of psychological trauma; and (c) to define predisposing factors and causes leading to the development of psychological trauma. It is important to note that the concept of psychological trauma should not be viewed as a single and acute event, similar to those commonly observed acute traumatic injuries in athletics. Psychological trauma is rather a psychological abnormality that gradually develops at different rates in different athletes, influencing both the physical and mental status and the overall psychological well-being of an athlete.

To clarify this dissimilarity in terms of the evolution of traumatic events, let's consider for example a sport-related ACL tear and/or shoulder dislocation. These are complex phenomena influencing the entire person (not only physical disability and discomfort) with all of his/her emotional, cognitive and social life. However, most often, physical disabilities due to injury appear in the injury's acute stage, unlike the signs and symptoms of mental problems that develop as the injury and its recovery progress. Another point that should be made is that research into and clinical experience with the physical aspect of injury in athletes is more advanced, unlike with the psychological aspects of injury. About twenty years ago or so, knee meniscus damage was considered a *severe* injury, requiring surgery and use of crutches at least for one month as well as prolonged (up to six months) rehabilitation. Today, an athlete suffering from this or a similar type of injury is crutches free on the next day post-surgery and can return to full sport participation within two to three weeks post-injury. Unfortunately, advances in psychological research and clinical practices dealing with the mental aspects of injured athletes are less promising and optimistic. Indeed, partly due to the extremely subjective nature of mental problems with injury and its perception, it is a real challenge to conduct well-controlled studies on the psychology of injury. There is neither conceptual clarity or universally accepted in the clinical practice definition nor well-documented consistent symptoms of psychological trauma to-date.

It may be useful to think of all trauma symptoms as an athlete's adaptive responses. Specifically, symptoms may represent an athlete's unsuccessful attempt to cope *the best way he or she can* with irrational thoughts about the injury and with overwhelming feelings. For example, a preoccupation with the injury as evidenced by constantly talking about how it happened and whether it could have been prevented, not only manifest psychological symptoms of trauma but also an attempt to adapt to existing disabilities and cope with those the best way possible. When we observe some behavioral symptoms (avoidance rehabilitation sessions, social isolation, etc.) in athletes suffering from trauma, it is always significant to ask ourselves: "What purpose does this behavior serve?" We humans, especially elite athletes, are incredibly adaptive creatures. Often, injury is considered an opportunity to re-evaluate self-worth, who you are and what is the purpose and meaning of what you are doing. Limping at the acute stage of the injury and learning "new skills to run" on the crutches are other examples of athletes' adaptive responses to injury. Personally, I was so proud of myself to be able to speedily navigate all over the campus on my crutches after knee surgery requiring immobilization for about a month. Of course, we should also consider mal-adaptive responses, as evidenced by another set of symptoms, including various social problems and abusive behaviors commonly seen in injured elite athletes.

2. CONCEPT OF PSYCHOLOGICAL TRAUMA

Trauma is defined as damage resulting in functional deficits and functional abnormalities at different levels of the CNS. The concept of psychological trauma derived from clinical practice dealing with psychological symptoms in individuals suffering from exposure to traumatic events in life. Until rather recently, psychological trauma was noted only in soldiers with post-traumatic stress disorders (PTSD) after exposure to catastrophic war events. Now, because of advances in clinical practice and brain science, psychological trauma has further broadened its definition and conceptualization.

In lay terms "trauma" is often referred to as a highly stressful event, but the key to properly understanding the concept of trauma is that it should refer to express stress that overwhelms an individual's capacity to cope. Traumatic events create psychological trauma when a person is overwhelmed emotionally, cognitively and behaviorally, with a perceived inability to tolerate his/her fear of injury or illness, annihilation, mutilation or neuroticism. The more a person believes that s/he is endangered, the more traumatizing the event will be. If an athlete believes that an upcoming competition would endanger his/her well-being, more competitions would be perceived as traumatic. It is important to stress again that there may or may not be bodily injury involved, but psychological trauma coupled with physiological responses may develop as a chronic and long-term disability. *Hypervigilance, dissociation, avoidance and numbing* are examples of coping strategies that may have been effective at some time, but later interfere with the person's ability to live the life s/he wants (Allen, 1995).

There is no clear division between notions of stress and psychological trauma, which creates confusion among clinical professionals. Additional confusion stems from the fact that both psychological trauma and stress are considered to be a form of maladaptive response to life threatening events. Adaptive versus maladaptive responses to events based on an individual's subjective experiences determine whether these events are or are not perceived as traumatic. However, it was some progress to conceptualize differences between "stress" and "trauma." For example, one way to tell the difference between stress and psychological trauma is by looking at the outcome—how much residual effect are upsetting events having on individual lives, relationships, and overall functioning. Traumatic distress can be distinguished from routine stress by assessing the following: (cited from: http://www.helpguide.org/mental/emotional_psychological_trauma.htm)

- how quickly upset is triggered
- how frequently upset is triggered
- how intensely threatening the source of upset is
- how long upset lasts
- how long it takes to calm down

According to Pearlman & Saakvitne (1995), psychological trauma is a unique individual experience of an event or enduring conditions in which:

1. The individual's ability to integrate his/her emotional experience is overwhelmed;
2. The individual's experience is perceived as a threat to life, bodily integrity and sanity.

In a more narrow sense, the concept of "emotional trauma" was introduced based on recent research indicating that emotional trauma can result from such common occurrences as an auto accident, the breakup of a significant relationship, a humiliating or deeply disappointing experience, the discovery of a life-threatening illness or disabling condition due to traumatic injuries, or other similar situations. Traumatizing events can take a serious emotional toll on those involved, *even if the event did not cause physical damage*. This is an important point to consider in an athletic situation—emotional trauma could be present in athletes without any signs present of bodily structural damage.

Regardless of its source, whether serious illness, exposure to traumatic experiences or athletic injuries, an emotional trauma contains at least three common features:

- Emotional trauma <u>cannot be predicted</u> and/or anticipated and it is therefore an unexpected emotional reaction to harmful stimuli (e.g., injury in athletics that is always happening at the wrong time, prior to important competitions, at the peak of performance when an athlete is in great physical shape);
- An individual is <u>usually unprepared</u> for the traumatic event and therefore is unable, at least in the acute stage of trauma to cope with it. For example, acute injuries requiring termination of athletic activities are in conflict with pre-planned actions and are serious obstacles to athletes' goal setting. As a result, an athlete's behavioral reactions to injury are

unpredictable. There are common cases in clinical practice where someone who used to be hardy and in control of his/her emotions became easily embarrassed and/or confused and displayed severely debilitating behavior as a result of the injury;

- Nothing can be done to <u>prevent</u> it from happening. As discussed in the previous chapters, injury is an unfortunate and unavoidable part of athletics. Really it is difficult to prevent injury in athletics due to numerous extrinsic (i.e., growing demands of sports, all year around practices and competitions, travelling stresses and acclimatization, etc.) and intrinsic (i.e., personality predispositions, perceived demands, etc.) factors. Therefore, there is always uncertainty around sports inducing an enormous stress factor which ultimately leads to physical and emotional traumas;

- There is a high probability of long-term deficits and residual abnormalities affecting the life style and overall physical and psychological well-being of individuals suffering from emotional traumas.

With respect to psychological trauma directly linked to an athletic environment it should be noted that its severity and long term symptoms depend on numerous concomitant factors; specifically to those athletes who have been injured during combative, collision-related sports such as American football, hockey, boxing, and auto racing. The effects of head injuries in these sports linger long after the athlete has retired from the sport. The brain is in essence a ticking-time bomb that is ready to go off at any time. An injury suffered years before can remain dormant until it is one day triggered psychologically.

Athletes in contact sports can suffer from a cocktail of different debilitating problems ranging from cognitive problems, severe headaches, and depression to serious long-term neurological disorders like Alzheimer's disease, memory impairment, and dementia. Concussions and head injuries happen with a frightening frequency, yet nine out of ten cases go undiagnosed.

Brain injury is most likely cumulative. While an individual may recover from a concussion with little to no ill effect, which occurs in the vast majority of all concussion cases, an injury to the brain may constitute the first, second or third injury. As these injuries accumulate, difficulties that may begin the process for degenerative conditions like Alzheimer's accelerate. Because so many injuries go undocumented might be why some people experience the onset of post-concussion syndrome, as the neurological loss surpasses a volume that the brain can handle.

Culture plays a role in the way head injuries are handled. Ninety percent of the time, athletes just "shake off" what they are feeling because they are not educated enough in terms of neurological injury and treatment. They have been taught, moreover, to be tough and heroic, to fight through pain and adversity, and with that they will gain glory. The difference between the knowledge that doctors have about head injuries and that of athletes and coaches is vast. In America, football has become the king of sports. Children start playing the sport as early as five years of age. Because brain injuries are cumulative, and aren't necessarily seen until many years down the road, youth awareness and protection becomes an important issue. Youth sports are part of the American culture; children grow up emulating their idols and heroes, but the brain is most vulnerable when a child is young. The brain is still developing and is more prone to being injured. As a football player, or any athlete for that matter, gets older, the competition, speed, and force at which the game is played increases exponentially. According to doctors, by the time an athlete has progressed to the NFL and put in years of service, the trauma that his brain and head has endured would be equivalent to tens of thousands of car wrecks. One minute an athlete can be in the prime of his life, earning fame, glory and a handsome amount of money, and then the next a dormant brain condition can take that all away. Here are a few professional football players whose lives have been altered by debilitating brain injuries.

- <u>Mike Webster</u> - Hall of Fame Center for the Pittsburgh Steelers. He played for eighteen years and made it to nine Pro Bowls. Webster led a controversial personal life after his retirement in 1990. He is believed to have been legally disabled since 1996, and possibly before, suffering from amnesia, dementia, depression and acute bone and muscle pain. He lived out of his pickup truck or in train stations because he had become homeless. He attended his Hall of Fame ceremony in 1997, but disturbed some attendees with a rambling, twenty-minute long, incoherent acceptance speech. Webster's late-life sufferings, in which he would use a taser-stun gun as the only way he could get some rest, is credited to thirty-five years of playing football at various levels. The equipment of his day was inferior and the knowledge and education about brain injury was almost non-existent. Webster died in 2002, and in the autopsy it was found that Webster's brain showed neurofibrillary tangles and diffused amyloid plaque, similar to a brain affected by dementia pugilistica, or "punch-drunk syndrome." The doctors who performed the auropsy were the first to encounter an autopsy-proven chronic traumatic encephalopathy in a professional American football player.
- <u>Dr. Bennett Omalu</u>, a clinical instructor in the Department of Pathology at the University of Pittsburgh, who did the autopsy on Webster, said, "Repeated concussions over a long period of time may not actually be

innocuous. Over the years there are delayed cumulative and additive neuron-chemical sequelae just like in dementia pugilistica."

- John Mackey - Hall of Fame Tight End for the Balitmore Colts. A five-time Pro Bowl choice and member of two Super Bowl teams, Mackey was voted the tight end on the NFL's 50th Anniversary Team in 1969. In his ten NFL seasons, he recorded 331 receptions for 5,236 yards and 38 touchdowns. Mackey's illustrious career is almost forgotten, not by fans, but by himself. He now suffers from frontotemporal dementia, which makes him particularly protective of personal possessions and suspicious of anyone who tries to control his actions. In fact, Mackey became volatile while making an appearance at an Indianapolis Colts game, when he recognized that wide receiver Marvin Harrison was wearing his old No. 88 jersey. He has also had incidences at airports when he has to check his luggage in. Mackey's wife Sylvia never leaves his side; she is now his only means of control.
- Terry Long - Lineman for the Pittsburgh Steelers. Committed suicide in 2005, at the age of forty-five. Autopsy tests on his brain showed positive results for chronic traumatic encephalopathy, which the neuropathologist eventually began to refer to as "footballer's dementia," or neurofibrillary football-linked dementia.
- Larry Morris - Linebacker for the Chicago Bears. Morris began to show signs of dementia fifteen years ago, and now at age 73 the disease is full blown. He can't write or even remember his name, can't do basic hygiene tasks, and even struggles to dress himself. His neuropsychiatrist attributes his dementia to his football playing days, where he suffered four diagnosed concussions, and many more while playing college ball at Georgia Tech.
- Andre Waters - Defensive back for the Philadelphia Eagles. Waters committed suicide in 2006 at the age of forty-four. Depression was the original reason for the suicide but after Waters' family allowed tests to be done on his brain, doctors determined that his brain tissue had deteriorated to that of an 85-year-old man with similar characteristics to those of early-stage Alzheimer's victims. They attributed his condition to numerous concussions he suffered throughout his career, and doctors believed that he would have been completely incapacitated within ten years.

Football is not the only sport that causes life-long psychological ailments for its athletes. Boxing is another sport in which the head and brain take constant punishment. Boxers' main objective is to punch their opponents in the head in order to knock them out, or to win by judge's decision. Muhammad Ali, who many say is the greatest boxer ever, suffers from Parkinson's disease, a neurological syndrome characterized by tremors, rigidity of muscles and slowness of speech and movement, following the onset of which his motor functions began a slow decline. Although Ali's doctors initially disagreed about whether his symptoms were caused by boxing and whether or not his condition was degenerative, he was ultimately diagnosed with Pugilistic Parkinson's disease, which means that the disorder was brought on by an inordinate number of blows to the head.

Hockey also lends itself to cases of brain trauma injury. Players, including goaltenders, used to play without helmets or masks. Head injury was common and just perceived as part of the game, however, as research was done and technology improved, all major hockey organizations have required players to wear a protective covering on their heads. Even with helmets, head injuries still occur. While concussions and other brain injuries exist and are a concern at professional levels of hockey, the real concern lies with the youth hockey organizations of the world. It is known that injury in youth hockey is at the forefront of problems in hockey, it is also the area that experts and league organizers are least prepared to solve. In previous years the most publicized safety concern was minimizing the risk of catastrophic injury and potential sudden death from second impact syndrome, which is a rare outcome of two concussions in a short period of time.

Now the idea of limiting these long-term risks in active athletes is not simply about limiting the number of concussions they suffer, but also about how a concussion is managed in the minutes, days, and weeks following the event. In hockey, as with all youth sports, concussions go virtually unnoticed. The educational material and training just don't exist. The logical way to combat severe and long-term psychological effects on athletes is to protect the young kids. Victims of concussions and other brain injuries become more susceptible to more traumas after they have suffered one. If the problem is attacked at youth levels, then the next generation may not be as prone to multiple and potentially life-altering brain injuries. Further elaboration on causes, symptoms and consequences of concussion will be discussed later in the book.

3. COMPONENTS OF PSYCHOLOGICAL TRAUMA: SYMPTOMS

Psychological injury or trauma in a clinical setting is usually classified if victims of the trauma have developed a number of symptoms. Sometimes these symptoms can be delayed, for months or even years after the event. Often people do not initially associate their symptoms with the precipitating trauma. The following are symptoms that may result from a more **commonplace, unresolved trauma**, especially if there were earlier, overwhelming life experiences:

Physical

- Eating disturbances (more or less than usual)
- Sleep disturbances (more or less than usual)
- Sexual dysfunction
- Low energy
- Chronic, unexplained pain

Emotional

- Depression, spontaneous crying, despair and hopelessness
- Anxiety
- Panic attacks
- Fearfulness
- Compulsive and obsessive behaviors
- Feeling out of control
- Irritability, angry and resentment
- Emotional numbness
- Withdrawal from normal routine and relationships

Cognitive

- Memory lapses, especially about the trauma
- Difficulty making decisions
- Decreased ability to concentrate
- Feeling distracted

There are additional symptoms of emotional trauma commonly associated with a **severe precipitating event**. Extreme symptoms can also occur as a delayed reaction to the traumatic event.

Re-experiencing the trauma

- intrusive thoughts
- flashbacks or nightmares
- sudden floods of emotions or images related to the traumatic event

Emotional numbing and avoidance

- amnesia
- avoidance of situations that resemble the initial event
- detachment
- depression

- guilt feelings
- grief reactions
- an altered sense of time

Increased arousal

- hyper-vigilance, jumpiness, an extreme sense of being "on guard"
- overreactions, including sudden unprovoked anger
- general anxiety
- insomnia
- obsessions with death

Sometimes, emotional exhaustion may set in, leading to distraction, and clear thinking may be difficult. Emotional detachment, also known as dissociation or "numbing out", can frequently occur. Dissociating from the painful emotion includes numbing all emotion, and the person may seem emotionally flat, preoccupied or distant. The person can become confused in ordinary situations and have memory problems.

Some traumatized people may feel permanently damaged when trauma symptoms don't go away and they don't believe their situation will improve. This can lead to feelings of despair, loss of self-esteem, and frequently depression. If important aspects of the person's self and world understanding have been violated, the person may call their own identity into question.

These symptoms can lead to stress or anxiety disorders, or even post traumatic stress disorder, where the person experiences flashbacks and re-experiences the emotion of the trauma as if it is actually happening (cited from:http://en.wikipedia.org/wiki/Psychological_trauma#Symptoms_of_trauma).

Psychological traumas can also be divided into categories similar to any other traumatic injury. These classifications include behavioral, somato-sensory, cognitive, motivational, affective components, and self-efficacy evidence. Behavioral symptoms include limping, asymmetry, and avoidance behaviors such as the *bracing* effect. The *bracing* effect occurs when an athlete is afraid of re-injury and favors or protects the once damaged area either by avoiding using the area or by applying less force and impact to it. Behavioral symptoms also include an athlete's inability to regain pre-injury physical status and/or the inability to acquire new skills. Behavioral symptoms can easily be seen and diagnosed during observations of playing time.

For example, a baseball player who dove back to first base to avoid being out ended up dislocating his shoulder and impinging the auxiliary nerve. He slowly started to rebuild the function and strength in his arm but when it came time to play he would no longer dive for the ball because he was afraid of experiencing the pain again. This *bracing* and avoidance behavior caused him to avoid playing organized baseball until he was psychologically recovered. He then forced himself to replay the incident by returning to the same baseball field and running the scene back through his head. He used mental imagery to establish emotional ability that later helped him recover from his fear of re-injury.

Somato-sensory components primarily consist of anticipation of pain in conjunction with irrational thoughts that an initiated movement would be painful. As was discussed in the previous chapters, sensation and perception of pain not only depend on specifically dedicated activation of pain receptors (i.e., nociception) but also on numerous subjective factors that are difficult to control and assess. In fact, the overriding importance of pain in clinical practice as well as many aspects of pain neurobiology, psychology and pharmacology (including their interaction) remained an important and poorly understood area of athletic medicine to-date. Overall, the dissociation of "real" and "neurotic" pain symptoms both in acute and chronic forms requires the serious consideration of medical practitioners and coaches in terms of the assessment of injury and its consequences for development of psychological trauma.

Motivational symptoms include issues with self-efficacy and lack of confidence in the success of a treatment plan. Self-efficacy is the impression and belief that one does or does not have the capabilities to execute certain actions required for recovery. This can be detrimental in an athlete's recovery, especially if after a severe injury, because the athlete may sense that he or she will never be as good or at the same level as before. This leads to a lack of motivation, slow recovery and even termination of a treatment program.

Affective components of psychological trauma consist of fear, depression and anxiety. These issues may be hard to diagnose because in order to understand them the athlete must tell you how he/she is feeling and why he/she is having issues with recovery. Depression and anxiety could preoccupy an athlete's life and affect the recovery rate greatly. Imagine being on your high school's tennis team playing a few matches before the big championships. You are playing to the best of your ability with your partner and dive for the ball because you are confident in your skill. Instead of making the save you wake up on your back looking up at the sky feeling pain in your cheekbone after having missed the ball while falling face forward and hitting your face on the court. This happened, for example, to a senior tennis player a few matches before championships and the end of the season. She ended up with a severely fractured cheekbone and two black eyes. Completely depressed about her appearance and that she failed to complete the season she then lost her self-confidence, became fearful of re-injury, and favored her sensitive cheekbone. She experienced moodiness, trouble sleeping and neurotic pain until she talked to a sports psychologist who helped her understand what had happened and helped her to feel better about herself and her image. She was told that it was a case of bad luck and that it could have not have been prevented. The psychologist told her not to blame herself for her injury and this helped her to recover quickly.

Cognitive components consist overall of an athlete's mental attitude and self-awareness about the injury. The athlete may experience obsessive and preoccupied irrational mental thoughts about the injury. This can lead to the inability to focus, over-thinking, most often irrationally, and memory problems. Sometimes athletes who experience moderate to severe injuries which require some sort of *brace,* become too preoccupied with thoughts of the injury which causes them to have *bracing* or avoidance behaviors. For example, a defensive linebacker for the football team experiences nerve damage in his neck has to wear a neck brace. When he recovers he may exhibit *bracing* behavior and is he afraid of re-injury so he does not block as hard as he used to. He then has to overcome these symptoms by constant reinforcement and performing the blocks correctly at practice.

It is important to note that all of these symptoms or conditions relate or may lead to one another. Affective components such as fear or depression can lead to lack of motivation. If an athlete is depressed and his or her self-efficacy is low, he/she will lack the motivation to continue with the treatment plan. Many of these symptoms lead to not only recurrent physical injury, but psychological trauma as well. When this occurs, a physical therapist may not be enough and the help of a psychologist may become essential to preserve the well-being of an injured athlete.

4. PREDISPOSING FACTORS FOR TRAUMAS

4.1. Personality

The conceptual framework for studies of personality in sport in general, and in the personality-injury relationship in particular, is essential to understanding predictive values of personality characteristics in terms of the development of psychological traumas in athletes. As was mentioned by Singer, et al. (2001) in the past, research was based mainly on the long-standing conception of personality in terms of behavioral dispositions of traits that predispose individuals to activate relevant behaviors. Accordingly, the purpose of research based on this definition has been the identification of the personality factors/traits of elite versus averages athletes, similarly injured versus non-injured athletes. Within the framework of conception of personality, it was defined as "… a system of conscious and unconscious mediating processes whose interaction are manifested in predictable or coherent patterns of functioning in individuals" (Singer et al.,

2001, p.240). More specifically, personality research should address in a consistent manner the issues of how and why individuals think, feel, act and/or react as they do. The details of the conceptual framework for the study of personality are far beyond the scope of this chapter and can be found in Singer et al., (2001, Chapter: Personality and the Athlete, p.239-268). However, some early research and current trends in personality-injury relationships, with respect to the development of psychological trauma in athletes, will be discussed in the following text.

The current interest in investigating the relationship of psychological predispositions to athletic injuries was initiated back in the 1970s by sport psychologists with intention of extending findings in health psychology. One of the trends in this research initiative dealt with attempts to explore some commonalities and differences between the psychological responses of athletes to sport-related injuries and non-athletes' responses to illness and sickness. It was believed that some personality variables (e.g., Type A) may correlate with illness in non-athletic populations. Similarly, it was feasible to suggest that Type A personality may be a factor influencing exaggerated behavioral and psychological responses to injury in athletes. Over the next thirty years or so, the subsequent research conducted by sports psychologists all over the world using junior, collegiate and elite athletes created a solid foundation for future investigations into psychological substrates of athletic injuries.

From a historical perspective, one of the pioneering reports aimed at examining the personality-injury relationship was conducted by Brown (1971) using high school football players as the subjects. In this study the players completed the California Psychological Inventory (CPI), which consisted of eighteen factors ranging from self-acceptance to femininity, at the beginning of the season. At the end of the competitive season, the results of the testing between injured and non-injured athletes were compared for each of the psychological factors in this inventory. Surprisingly, no significant differences were observed between these two groups of subjects, suggesting that player personality has nothing to do with athletic injuries. Today, this controversial statement should be taken with great caution, since too many uncontrolled variables in addition to a possible lack of sensitivity and validity of the CPI may be present in this early study. However, this pioneering research triggered an interest among sports psychology researchers to further explore psychological causes and consequences of athletic injuries. Accordingly, several later studies have shown support for personality as a predictor to athletic injuries.

Specifically, Jackson, et al. (1978) have examined personality factors in high school football players using 16 PF Cattell's test and reported that injured athletes scored significantly higher on a factor indicating that they were more tender-minded, dependent, overprotected and sensitive than non-inured controls. Another factor indicated that injured players were more forthright. This initial report by Jackson using football players has been supported by another study by Valliant (1981) who observed the same tendency in injured male runners. However, one discrepancy that was noticed is that in this later study by Valliant, the injured athletes scored lower on the "forthright" factor than non-injured controls. Interestingly, a higher level of self-concept was found in injured high school basketball players (Young & Cohen, 1981), indicating that non-injured athletes, who scored lower on this factor, may play more cautiously because of lower self-concept values, and thus be less prone to becoming injured. Overall, this finding suggests that performance efficacy and an athlete's personality may be in conflict. These athletes avoid injury rather than seeking challenges which would put them in situations with a high risk of injury, are two possible scenarios to be considered in terms of personality-injury versus personality-performance dichotomy.

The current conceptualization of the personality-injury relationship stems from the popular model of "stress and athletic injury" (Anderson & Williams 1988). Most of the personality factors included in this model are: *hardiness, locus of control, sense of coherence, competitive trait anxiety* and *achievement motivation*.

Hardiness, earlier defined by Kobasa et al (1982) as "...a constellation of personality characteristics that function as a resistant resource in the encounter with stressful life events," may be linked to physical health in general, and to high risk for injury in the athletic population, in particular. The concept of hardiness with respect to injury was further elaborated on by Pargman (2007) who proposed that it had three interrelated components, such as commitment, challenge and control. The *commitment* component refers to a strong belief in one's own values and ability to achieve goals and involvement in sport activities. The *challenge* component refers to an individual's tendency to view difficulties and change as intermediate obstacles that should be overcome with some effort rather than considering these obstacles as a threat to self-worth and personal security. Not surprisingly, athletes with high scores on the challenge component of hardiness also possess a high degree of cognitive flexibility that permits effective appraisal of a situation, including potentially threatening events. That said, some anecdotal observations from clinical practice suggest that athletes who seek challenges more often put themselves at higher risk for injury than those who are prone to avoiding challenges. Finally, the *control* component involves a sense of personal power over the events in one's life. Athletes with strong "control" component usually assume responsibility for their actions, enabling them to avert a feeling of helplessness in a case of injury by using effective thinking, and by developing compensatory strategies to speed up the rehabilitation process.

Overall, *hardy* athletes are goal-oriented, committed and devoted to their actions, they view adaptation due to injury as an opportunity to take advantage of it and explore unknown experiences. The underlying mechanisms of connection between hardiness and health are not clear. However, it was suggested that both appraisal of the situation (realistic evaluation of the impact of an injury) and coping processes (development of a treatment plan), which interactively minimize the negative effect of trauma on psychological well-being, may contribute to this relationship (Williams et al., 1992). There is also indication in the literature that hardy people prone to rely on adaptive rather than on maladaptive coping strategies in the case of perceived stress (Wiebe, 1991).

There are several scales to assess hardiness. The original 71-item Hardiness Scale (HS), which was drawn from preexisting scales to measure psychological constructs such as commitment, challenge and control, was proposed by by Kobasa et al., in 1982. Later, the Revised Hardiness Scale (RHS) was developed and contained 36 items based on the original test. Most recently, the 50-item Personal Views Survey (PVS) was developed by the Hardiness Institute. It appears to possess reliable psychometric properties to assess hardiness as a predisposing factor for injury in athletes. In terms of its implication for practice, coaches should be aware that athletes with low scores on hardiness may perceive every new challenge (e.g., learning new skills or routines, changes in game plans, new rules etc.) as a personal threat. This can be observed via avoidance behavior (i.e., missing practices, making excuses of being sick, etc.). In the event of injury, medical practitioners should be aware that athletes with low scores on hardiness may constantly ruminate about their irrational and often exaggerated view of the injury, become tense and nervous when facing rehabilitation demands, get easily depressed and be inclined to isolate themselves from the team and coaches.

Locus of Control: Similarly to hardiness, the locus of control (Rotter, 1966) and sense of coherence (Antonovsky, 1985) were also considered to be contributing moderators of the personality-injury relationship. Locus of control is conceptualized as the degree to which individuals view their lives and their close surrounding environment as being under their control. Its internal (intrinsic) orientation is mostly characterized by a strong belief that one's own actions, rather than surrounding circumstances, regulate the behavioral outcome. The best way to describe an intrinsic orientation may be this: "… if you blame somebody else for your failure, you are no damn good…." If an athlete believes that the coach putting him/her in the field at the wrong time resulted in the injury, the athlete most likely possesses an external orientation. This is indicative of an individual who feels himself/ herself a victim of bad lack or circumstances. The lack of locus of control may prolong the recuperation of injured athletes who often attribute their failure to meet rehabilitation goals to lack of attention and/or social support from the team.

It is feasible to suggest that locus of control may influence a "positive state of mind," which may buffer the perceived level of threat and contribute to risk for injury. In fact, significant correlation between a high degree of tension/anxiety, anger/hostility, and a total negative mood state, as evidenced by psychological evaluation (POMS, Profile of Mood States), and high risk of injury was reported by Lavallee & Flint (1966). Athletes who constantly experience a "negative state of mind" may often exhaust their physical and mental resources and will respond inefficiently to the physical and psychological demands and strains of sports participation (Van Mechelen et al., 1996).

Competitive trait anxiety may be a serious predisposing factor for injury. This phenomenon was described as a general tendency to perceive a situation as threatening and to react to it with apprehension and anxiety (Spielberger, 1966). Accordingly, athletes with a high level of train anxiety may experience a high level of stress and an inability to cope with competitive pressure. As a result, these athletes would have a tendency to avoid tough competitive situations, modify their reaction to threat, perceive a neutral game situation as harmful and/or injury threatening leading ultimately to high risk for injury. In the case of previous injuries, athletes with high scores on train anxiety most likely will experience fear of re-injury, lack of confidence in the likelihood of gaining pre-injury status and higher risk for the development of chronic psychological trauma. In fact, several studies have supported this notion, indicating that athletes with higher scores on competitive trait anxiety had more injuries in general, and more severe injuries in particular (Blackwell & McCullagh, 1990; Petrie, 1993). Specifically, in Petrie's (1993) study, increases in train anxiety were positively correlated with a higher rate of injury in football starters. That said, it should be noted that like many other personality-injury relationship studies, the aforementioned report has failed to take into account numerous other factors, such as cognitive, affective and behavioral components of stress, which alter an athlete's vulnerability and/or resiliency to injury. Overall, it is feasible to speculate that since a high level of competitive anxiety may facilitate the detrimental effect of stress on performance, which was well-documented in numerous reports, it therefore could be considered a reliable predictor for the development of psychological trauma as a result of injury.

Achievement motivation addresses both the need to succeed and the need to avoid failure. With respect to injury, the need to succeed motivates athletes to achieve the required level of conditioning and the necessary skills prior to the season, to maintain this level over the whole season and to meet competition demands as reasonable challenges, but not obstacles to their goals. It should be noted that, surprisingly, no relationship between achievement motivation and the probability of injuries in athletics was reported by Van Mechelen et al., (1996). Again this may be due to the lack of control for many other concomitant factors predisposing high risk for injury. For example, such personality factors as "sensation seeing" can significantly moderate the effect of achievement motivation on injury risk. It should be remembered that Zuckerman (1979) defined sensation seeking as a biologically based predisposing factor that reflects individual differences in optical arousal level. Unlike "sensation avoiders," sensation seekers love an adrenaline rush, have high tolerance for arousal, care for change and unfamiliar situations, and enjoy risky activities. In fact, it was reported that only athletes who score lower in sensation seeking had a significantly positive correlation between major negative sports-specific life events and subsequent injury time-loss (Smith et al., 1992). Again, surprisingly there was no support for the assumption that high sensation seeking (i.e., more risk-taking behavior) would constitute an injury vulnerability factor in athletics.

In a more recent elaboration on the personality-injury relationship David Pargman (2007) came up with several additions and complementary personality constructs that will be briefly described in the following text.

Neuroticism is defined as a basic dimension of personality reflecting a general tendency towards emotional lability and negative affect (Eysenck & Eysenck, 1985). It is believed that those individuals high in neuroticism are proved to experience greater negative stress responses and severer reactivity to stress exposures. In combination with rumination associated with increased susceptibility to depression, individuals with high neuroticism may experience generalized anxiety disorders (Weinstock & Whisman, 2006) and overall are more susceptible to traumatic injuries and psychological traumas. Not surprisingly, the vast majority of personality-injury relationship studies aimed at linking neuroticism with cognitive (i.e., inability to focus on the task, memory problems, confusion, prolonged decision making mechanisms), emotional (i.e., exaggerated emotional responses to injury, fear of re-injury, irritability, hostility, anger and frustration), and behavioral (i.e., withdrawal and isolation, bracing behavior) responses to injury.

There are numerous theoretical models and associated scales to assess neuroticism in clinical practice. Although, the most commonly accepted measure is the 23-item neuroticism subscale from the Eysenck Personality Questionnaire (EPQ). There are also the Revised 48-item neuroticism subscale (Revised NEO PI-R) Personality Inventory and the shortened 12-item version of NEO PI-R, though less reliable it is still useful if time constraints are an issue. From a practical perspective, coaches should be aware that athletes with high scores on neuroticism (e.g., neurotic individuals) may express maladaptive responses to injury, exaggerate an injury's impact, severity and physical symptoms, and may express quick frustration and impulsive actions. Medical practitioners should expect that highly neurotic patients may experience problems with focused and goal-oriented treatment plans, and impatience with the rehabilitation outcome overall. It is highly recommended that "neurotic" injured athletes be referred for a comprehensive psychological evaluation and treatment along with the physical therapy.

Explanatory style is defined as the way athletes typically account for significant events in their life. It is a common and relatively stable tendency to explain things and life events in a certain way. For example, athletes with "defensive pessimism" in the case of injury usually blame coaches, teammates and medical staff as being responsible for the injury. The direction of thoughts in these individuals tend to explain a negative event, including injury, personally and in an exaggerated manner (e.g., I will never return to 100% after this bad injury; …this injury completely ruins my hope of getting drafted for the pros; etc.). In fact, early research clearly demonstrated that athletes who score highly on "defensive pessimism" usually experience a high degree of high stress and illness/injury symptoms (Perna & McDowell, 1993). Interestingly, "depressive pessimism" symptoms decrease significantly as the explanatory style becomes optimistic during successful treatment of the injury (Barber et al., 2005). This means that explanatory style is not a "stable" personality predisposition as was initially proposed, but rather a supportive-expressive dynamic process and success/failure dependent phenomenon.

A number of general explanatory style scales can be recommended for sports and medical practitioners to measure "defensive pessimism" in athletes both prior to and after traumatic injuries. These scales include several versions of the Attributional Style Questionnaire (ASQ), the Balance Attributrional Style Questionnaire (BASQ), the Attributional Style Assessment Test (ASAT), the Revised Causal Dimension Scale (CDSII). There are also more sport-specific tests which may be the instruments of choice in clinical settings. In terms of implication for practice, coaches should be aware that athletes with high scores on "defensive pessimism" may be predisposed to feeling of helplessness and depression, especially when facing more challenging athletic situations. In case of injury, medical practitioners should be aware that athletes with predominantly pessimistic attributions may fail to follow a recommended treatment plan and may exhibit a lack of effort and persistence in the face of perceived poor progress with rehabilitation.

Dispositional Optimism is defined as a general tendency and expectancy for good rather than bad events to occur. Athletes who possess this general tendency perceive a risky competitive situation as a challenge rather than a threat and strongly believe that a "smart game" is injury free. In the case of injury, these athletes believe that this is just "another annoying thing" common in athletics. Traditionally, *dispositional optimism* has been discussed within the scope of the general health-related consequences of personality. It has been shown that positive expectancies induce lower susceptibility to stress, depression and cardiovascular problems, and increase immunological functioning. Overall, dispositional optimism can be a good predictor of the possibility of fewer depressive symptoms in athletes suffering from injuries (Vickers & Vogeltanz, 2000).

Dispositional optimism may be assessed by the early Expected Balance Scale (EBS), the Optimism and Pessimism Scale OPS), and the more recent Life Orientation Test (LOT) instruments. The latest version of revised LOT contains just ten items allowing practitioners to calculate both "optimism" and "pessimism" and is highly regarded in current clinical practice.

Perfectionism is defined as a tendency by an individual to set extremely high and often unrealistic standards for his/her actions and/or to believe that others are setting the same high standards as well. This, in turn, may increase the negative stress response associated with injury due to the fact that perfectionists are prone to:

- constant striving
- self-doubt
- excessive concern about errors and mistakes
- all-or-none thinking
- overgeneralization of failure

Additional personality factors of perfectionism have been also linked to:

- elevated anger responses
- obsessive compulsive behavior
- depression
- psychosomatic distress (cited from Pargman, 2007, p.59)

With respect to predisposition for injury, a perfectionist-athlete may perceive (a) minor muscle soreness as an injury; (b) a minor setback due to fatigue as fear of losing control; (c) temporarily reduced strength and/or physical function as an indication of major performance deterioration. There is indication in the literature that a high level of perfectionism may be directly linked to injury, at least among dancers (Liederbach & Compagno, 2001). It is feasible to expect that the same tendency may be observed among athletes participating in complex coordination sports such as springboard diving, gymnastics, figure skating, etc., and to some extent among team sport players. However, no comprehensive research has been conducted regarding sport-related differences in terms of the effect of perfectionism on injury.

The most common instrument used to measure perfectionism in athletics is the Multidimensional Perfectionism Scale (MPS) developed by Frost, et al. in 1990. There are thirty-five statements in this test reflecting several aspects of perfectionism such as: personal standards, doubts about actions, concern over mistakes, need for organization, parental expectations, and parental criticism. Another test developed by Hewitt et al., (1991) consists of forty-five items aimed at assessing three dimensions of perfectionism, including self-oriented perfectionism, other-oriented perfectionism, and socially prescribed perfectionism. In terms of the implications for practice, it is feasible to expect that a high level of perfectionism presents a real challenge for coaches, especially when working with elite athletes. There is never "enough" practice time, rest time, number of repetitions, perfect skills, etc., among perfectionists—an attitude that could easily lead to overtraining injuries and injuries overall. Despite a high work ethic and a desire to achieve the highest standards in their sports, perfectionists are prone to overestimate their physical limits, and are therefore more susceptible to overuse injuries. There is a risk for lack of adherence to the treatment protocol in case of injury due to dissatisfaction with the short-term outcomes of the treatment procedures.

As a concluding note, most of the studies on personality-injury measuring personality characteristics associated with athletic injuries are inconclusive and often contradictory (Weinberg & Gould, 2003). The selective sensitivity of existing psychological personality tests is questionable when it comes to the issue of predictive values for injury in athletics. That said, it should be noted that recent evidence clearly suggests that personality factors such as optimism, self-esteem, hardiness and train anxiety indeed play a role in athletic injuries (Ford et al., 2000; Smith et al., 2000). However, these often studied personality constructs should not be examined in isolation but should be considered with respect to other injury predisposition factors, for example previous injuries, etc., in order to properly predict athletes at risk for both physical injury and psychological traumas.

4.2. Predictive Value of Previous Injuries

Previous injuries are an important predictor for recurrent injuries for a number of reasons. Numerous reports and anecdotal evidence suggest that athletes with a history of previous injuries are more susceptible to recurrent and often more severe injuries. With regards to orthopedic injuries, multiple traumatic injuries have been studied greatly. It has been shown that multiple orthopedic injuries may lead to repetitive strains (Mouhsine et al., 2004). By continuing to participate in athletic activity, which constantly exposes the injury to movement forces and impact forces, the athlete is likely to experience persistent pain (Mouhsine et al., 2004). Even with this pain athletes continue their participation in sport due to multiple pressures. It is well established that high endurance straining of prolonged periods may induce skeletal muscle damage (Grobler et al., 2004). Skeletal muscles have a high capacity to repair and adapt, but this capacity may be limited as a result of re-injury or multiple injuries (Grobler et al., 2004). Injured athletes may detect the signs of this limited capacity through physical pain. However, physical pain is often accepted as a part of athletics. Acceptance of this pain creates a particular tolerance in which the athletes simply learn to play with pain. Unfortunately, this acceptance is a misperception and can lead to severe health consequences. Athletes who have developed high levels of tolerance and continue high levels of endurance training despite re-injury show increased levels of skeletal muscle disruptions (Grobler et al., 2004).

Another aspect of orthopedic injuries that has become a concern is the psychological consequences of injury. Sports medicine specialists are beginning to see an increase in orthopedic injuries that are not limited to athletic activity. The psychological aspect of orthopedic injuries is also being studied in the work force due to the insurance and medical bills that corporations pay for their employees. Many adults experience chronic injuries. Studies that have targeted these populations have found that perceived stress and overall distress are very high among these individuals (Tjepkema, 2003). These findings are significant considering that the level of physical activity of corporate America's employees is significantly lower than that of athletes. Therefore, an increased level of activity with respect to re-injury and psychological stress poses a larger issue for athletes.

One of the specific injuries of concern is the cumulative effect of concussion. The current misconception is that multiple head injuries are unlikely in athletics. However, a recent study conducted among Ohio and Pennsylvania High School football players showed that more than 34% of participants had experienced multiple concussions (Langburt, 2001). Multiple brain injuries are likely to lead to cumulative neurological and cognitive deficits. In a study of amateur athletes, those who had suffered three or more concussions were 7.7 times more likely to exhibit drops in memory functioning (Iverson et al., 2004). If multiple head injuries are experienced within a short period of time, the effects could be fatal (MMWR, 1997). One sport in which we have observed the direct consequences of head trauma is boxing. This is a sport in which athletes are extremely susceptible to constant blows to the head. A study by Ravdin et al. (2003) was conducted to test the cognitive functioning of boxers. They found that boxers who fought in twelve or more bouts showed a significant decrease in cognitive functioning as a result of the number of blows suffered in competition. This suggests that there are cumulative effects related to multiple concussions (Ravdin et al., 2003). Another study examining the symptoms of concussion in relation to concussion history was conducted among high school football players (Collins et al., 2003). Specifically, this study was focused on symptoms such as loss of consciousness, anterograde amnesia, retrograde amnesia and confusion. It was shown that athletes who had suffered three or more concussions were 9.3 times more likely to exhibit these symptoms when suffering another concussion (Collins et al., 2003). DeRoss et al. (2002) conducted a study on rats in which differences were found between rats with one injury versus those with multiple head injuries. Eighty-five percent of the rats showed impairments in spatial recognition and deviations from baseline scores but motor control was not affected (DeRoss, 2002).

Similarly, Williams et al (1993) reported a positive correlation between prior and recurrent injuries. This report is consistent with the previous suggestions that physical education students with a history of previous injuries are more vulnerable to recurrent injuries. Moreover, Van Mechelen et al. (1996) proposed that previous injuries are more reliable predictor of future injuries than psychological, psychosocial and other factors. However, it should be mentioned that a contradictory notion that timing from the first injury is not related to subsequent injuries both in terms of frequency and severity.

Discussing the issue of previous injuries as a predisposing factor for recurrent injuries, *first*, it is important to understand the differences between return to sport participation versus return to competition. Athletes with a proper rehabilitation protocol may be physically and functionally ready for participating in practices and gradually acquire the pre-injury status. However, a premature return to competition, assuming "...ready to practice, therefore, ready for competition," may put athletes at higher risk for injury. In other words, additional competitive stress factors may induce alteration of anxiety, maladaptive appraisal of competitive situations, fear of injury, etc., if an injured athlete is not prepared psychologically for upcoming challenges. *Second*, it should be understood that return to pre-injury physical status is NOT indicative of fully functional recovery. In other words, an athlete may regain pre-injury strength, range of motion and stamina, but may fail to demonstrate "applicability" of these physical properties for real athletic situations such kicking or catching the ball, executing somersaults, etc. Finally, even though an athlete may regain 100% his own pre-injury status after months of rehabilitation, his or her injury free teammates have advanced to a new level of physical and psychological status. This would be a significant confusing factor upon return to sport participation that may lead to an athlete's loss of confidence in the ability "catch-up" with injury free athletes and regain his/her leading position.

5. MODEL OF PSYCHOLOGICAL TRAUMA

Considering controversial statements and confusing notions regarding the development of psychological trauma in an athletic environment it is feasible to propose very rough model underling its essential components for future research and applications for clinical practice. Indeed, the predictive values of cognitive, affective, motivational and behavioral symptoms cannot be fully appreciated until they are examined interactively within the scope of extrinsic and intrinsic properties of the dynamics of injury development. Clearly, the delayed effect of psychological trauma is one of the major concerns among medical practitioners requiring a proper assessment and monitoring of predisposing factors. A preliminary model of psychological trauma is shown below, requiring, of course, future research and empirical support.

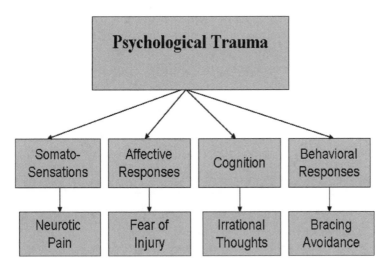

Figure 1. Model of psychological trauma in athletes

That said, further elaborations on predictive values and the classification of individuals at risk for the development of psychological traumas in an athletic environment are yet to be conceptualized and validated in clinical practice.

CONCLUSIONS

Psychological trauma overall is defined as an individual's subjective experience determining whether an event is or is not traumatic. "It is not the event that determines where something is traumatic to someone, but the individual experience of the event." (http://www.traumaresources.org/ emotional_trauma_overview.htm). The same traumatic event may or may not create psychological trauma. In the case that it does, a person would be overwhelmed emotionally, cognitively and behaviorally, with a perceived inability to tolerate his/her fears of injury or illness, annihilation, mutilation or neuroticism. As was mentioned before, the more a person believes that he/she is endangered, the more traumatizing the event will be. If an athlete believes that return to play after injury would endanger his/her well-being both physically (i.e., increased risk for re-injury) and psychologically (i.e., exaggerated irrational thoughts and misperceived high risk of danger), then more sports participation would be perceived as traumatic. It is important to stress again that there may or may not be bodily injury involved, but psychological trauma coupled with physiological responses may develop as a chronic and long-term disability. Therefore it is the responsibility of medical professionals to consider the importance of a timely assessment not only of the physical but also the psychological symptoms of injured athletes' disabilities, to determine an appropriate referral and coordination of care among the members of the core team to implement the best treatment plan.

REFERENCES

Allen, J. *Coping with Trauma: A Guide to Self-Understanding*. Washington, DC: American Psychiatric Press, 1995.

Pearlman, Laurie Anne, and Karen W. Saakvitne. *Trauma and the Therapist*. New York: Norton, 1995.

Singer, R., Hausenblas, H., & Janelle, C. (eds). *Handbook of sport psychology*. Second Edition. New York:John Wiley & Sons, Inc., 2001.

Andersen, M.B., & Williams, J.M. (1988). A model of stress and athletic injury: Prediction and prevention. *Journal of Sport & Exercise Psychology, 10*, 294-306.

Kobasa, S.C., Maddi, S.R., Kahn, S. (1982). Hardiness and health: A prospective study. *Journal of Personality and Social Psychology, 42*, 168-177.

Pargman, D. Psychological Bases of sport injuries. Third Edition. Morgantown: WV: FIT, 2007.

Williams, P.G., Wiebe, D.J., & Smith, T.M. (1992). Coping processes as mediators of the relationship between hardiness and health. *Journal of Behavioral Medicine, 15*, 237-255.

Wiebe, D.J.,(1991). Hardiness and stress moderation: A test of proposed mechanisms. *Journal of Personality and Social Psychology, 69*, 89-99.

Rotter, J.B. (1966). Generalized expectancies for internal versus external control. Psychological Monograph, 80 (Whole).

Antonovsky, A. (1985). The sense of coherence as a determinant of health. In J.D. Matarazzo, S.M. Weiss, J.A. Herd, & N.E. Miller (Eds.), *Behavioral Health: A handbook of health enhancement and disease prevention* (pp.37-50).New York: Wiley.

Lavallee, F., & Flint, L. (1966). The relationship of stress, competitive anxiety, mood state, and social support to athletic injury. *Research Quarterly for Exercise and Sport, 56*, 1222-130.

Van Mechelen, W., Twisk, J., Molendijk, A., Blom, B., Shel, J. & Kemper, N. (1996). Subject-related risk factor for sports injuries: A 1-yr prospective study in young adults. *Medicine and Science in Sport and Exercise, 28*, 1171-1179.

Spielberger, C.D. *Anxiety and behavior*. New York: Academic Press, 1966.

Blackwell, B., & McCullagh, P. (1990). The relationship of athletic injury to life stress, competitive anxiety and coping resources. *Athletic Training, 25*, 23-27

Petrie, T.A. (1993). Coping skills, competitive train anxiety, and playing status: Moderating effects of life stress-injury relationship. *Journal of Sport and Exercise Psychology, 15*, 261-274.

Zuckerman, M. *Sensation seeking: Beyond the optimal level of arousal*. Hillsdale, NJ: Erlbaum, 1979.

Smith, R. E., Ptacek, J.T., & Smoll, F. L. (1992). Sensation seeking, stress, and adolescent injuries: A test of stress-buffering, risk taking and coping skills hypothesis. *Journal of Personality and Social Psychology, 62*, 101-1024.

Eysenck, H.J., & Eysenck, M.W. *Personality and individual differences.* New York: Plenum, 1985.

Weinstock, L.M., & Whisman, M.A. (2006). Neuroticism as a common feature of the depressive and anxiety disorders: A test of the revised integrative hierarchical model in a national sample. *Journal of Abnormal Psychology, 115*, 64-74.

Perna, F., & McDowell, S. (1993). The association of stress and coping with illness and injury among elite athletes. Paper presented at the annual meeting of the Association for the Advancement of Applied Sport Psychology. Montreal, Canada.

Barber, J.P., Abrams, M.J., Connolly-Gibbons, M.B. et al. (2005). Explanatory style change in supportive-suppressive dynamic therapy. *Journal of Clinical Psychology, 61*, 257-268.

Vickers, K.S., Vogeltanz, N.D. (2000). Dispositional optimism as a predictor of depressive symptoms over time. *Personality and Individual Differences, 2*, 259-272.

Liederbach, M., & Compagno, J.M.(2001). Psychological aspects of fatigue-related injuries in dancers. *Journal of Dance Medicine & Science, 5*, 116-120.

Frost, R.O., Marten, P., Lahart, C. ,& Rosenblate, R. (1990). The dimensions of perfectionism. *Cognitive Therapy and Research, 14*, 449-468.

Hewitt, P.L., Flett, G.L., Turnbull_Donovan, W. & Mikail, S.F. (1991). The Multdimensional Perfectionism Scale: Reliability, validity, and psychometric properties in psychiatric samples. *Psychological Assessment: a Journal of Consulting and Clinical Psychology, 3*, 464-468.

Weinberg, R., Gould, D. *Foundations of Sport & Exercise Psychology.* 3rd Edition. Human Kinetics: Urbana-Champaign, 2003.

Ford, I.W., Eklund, R.C., & Gordon, S. (2000). An examination of psychosocial variables moderating the relationship between life stress and injury time-loss among athletes of a high standard. *Journal of Sport Science, 18(5)*, 301-321.

Mouhsine, E., Crevoisier, X., Leyvraz, P.F., Akiki, A., Dutoit, M., Garofalo, R. (2004). Post-traumatic Overload or Acute Syndrome of the os trigonum: a possible cause of posterior ankle impingement. *Knee Surgery Sports Traumatol Arthroscopy, 12*, 250-253.

Grobler, L.A., Collins, M., Lambert, M.I., Sinclair-Smith, C., Derman, W., Gibson, A., Noakes, T.D. (2004). Skeletal Muscle Pathology in Endurance Athletes with Acquired Training Intolerance. *British Journal of Sports Medicine, 38(6)*, 697-703.

Tjepkema, M. (2003). Repetitive Strain Injury. *Health Reports, 14 (4)*, 11-30.

Langburt, C., Cohen, B., Akhthar, N., O'neill, K., Lee, J.C. (2001). Incidence of Concussion in High School Football players of Ohio and Pennsylvania. *Journal of Child Neurology, 16(2)*, 83-85.

Iverson, G.L., Gaetz, M., Lovell, M.R., Collins, M.W. (2004). Cumulative effects of Concussion in Amateur Athletes. *Journal of Brain Injury, 18(5)*, 433-443.

Ravdin, L.D., Barr, W.B., Jordan, B., Lathan, W.E., Relkin, N.R. (2003). Assessment of Cognitive Recovery following sports Related Head Trauma in Boxers. *Clinical Journal of Sports Medicine, 13(1)*, 21-27.

Collins, MW., Field, M., Lovell, MR., Iverson, G., Johnston, KM., Maroon, J., Fu, FH. (2003). Relationship between postconcussion headache and neuropsychological test performance in high school athletes. *American Journal of Sports Med. Mar-Apr; 31*(2), 168-73.

DeRoss, A.L., Adams, J.E., Vane, D.W., Russell, S.J., Terrella, A.M., Wald, S.L. (2002). Multiple head Injuries in Rats: Effects on Behavior. *Journal of Trauma, 52(4)*, 708-714.

Williams, J.M., & Andersen, M.B. (1998). Psychosocial antecedents of sport injury: Review and critique of the stress and injury model. *Journal of Applied Sport Psychology, 10*, 5-25.

CHAPTER 12

FEAR AS ADAPTIVE OR MALADAPTIVE FORM OF EMOTIONAL RESPONSE

1. FEAR IS NEITHER ANXIETY NOR WORRY

As elaborated in the previous chapter, fear is one of the major components of *psychological trauma* that may or may not develop as a result of traumatic injury. It is a commonly accepted notion that fear as a form of emotion is a complex phenomenon that may be observed and manifested at different levels of analysis, such as overall behavioral responses (i.e., facial expression, muscular tension/freezing, trembling, etc.), cognition (i.e., a variety of rational and/or irrational decisions as result of the appraisal of stimuli), physiological processes (i.e., as reflected in alteration of heart rate, blood pressure, skill, galvanic reactions etc.), and activation of various brain structures which take part in the regulation of goal-oriented behaviors. According to Webster's Dictionary, fear is defined as "unpleasant, often strong emotion caused by expectation or awareness of danger." Although, current conceptualization of few trends to suggest that fear can be both adaptive response aimed at survival in face of life threatening circumstances and may create devastating psychiatric problems, including severe affective disorders. Specifically, according to Rosen & Schulkin (1998) fear responses (e.g., freezing, alarm, heart rate and blood pressure changes, and increased vigilance) are functionally adaptive behavioral and perceptual responses elicited during danger to facilitate appropriate defensive responses that can reduce danger or injury (e.g., escape and avoidance). However, pathologic anxiety, as a form of an exaggerated fear state, may develop from adaptive fear states as well. If that happens, the hyperexcitability of fear circuits that include several brain structures, such as amygdala, anterior cingulate cortex and prefrontal cortex (PFC), is expressed as hypervigilance and increased behavioral responsiveness towards fearful stimuli. Fear and associated behavioral/cognitive/physiological responses are closely linked, both functionally and structurally, as Williams James expressed in his famous quote more than 100 years ago:

"What kind of an emotion of fear would be left if the feeling neither of quick-ended heart-beats nor of shallow breathing, neither of trembling lips nor of weakened limb, neither of goose-flesh nor of visceral stirrings, were present, it is quite impossible for me to think...I say that for us emotion dissociated from all bodily feeling is inconceivable." **William James**, (*Psychology*, 1893, p.379).

Following are more quotes about fear, demonstrating that this human emotion is the most puzzling, complex and multimodal phenomenon ever seen.

- *"I stood stunned, my hair rose, the voice stuck in my throat."* **Virgil**
- *"The thing I fear most is fear."* **Michel de Montaigne**
- *"Then fear drives out all wisdom from my mind."* **Ennius, quoted by Cicero**
- *"Fear exceeds all other disorders in intensity."* **Michel de Montaigne**
- *"The only thing we have to fear is fear itself - nameless, unreasoning, unjustified terror, which paralyzes needed efforts to convert retreat into advance."* **FDR -*First Inaugural Address, March 4, 1933***
- *"One of the things which danger does to you after a time is -, well, to kill emotion. I don't think I shall ever feel anything again except fear. None of us can hate anymore - or love.* **Graham Greene -*The Confidential Agent (1939)***
- *"What are fears but voices airy? Whispering harm where harm is not. And deluding the unwary. Till the fatal bolt is shot!"* **Wordsworth**
- *"Fear - jealousy - money - revenge - and protecting someone you love."* **Frederick Knott - Max Halliday**
- *"listing the five important motives for murder...".* **Dial M for Murder (1952)**
- *"What potions have I drunk of Siren tears, Distill'd from limbecks foul as hell within, Applying fears to hopes, and hopes to fears, Still losing when I saw myself to win!"* **William Shakespeare– *Sonnets*.** *"Fear is a tyrant and a despot, more terrible than the rack, more potent than the snake."* **Edgar Wallace -*The Clue of the Twisted Candle* (1916)**
- *"Tush! Tush! Fear little boys with bugs."* **William Shakespeare-*The Taming of the Shrew***
- *"All of us are born with a set of instinctive fears - of falling, of the dark, of lobsters, of falling on lobsters in the dark, or speaking before a Rotary Club, and of the words "Some Assembly is Required."* **Dave Barry**
- *"Am I afraid of high notes? Of course I am afraid. What sane man is not?"* **Luciano Pavarotti**
- *"Fear makes the wolf bigger than he is."* ***German Proverb***
- *"Men fear death as children fear to go in the dark."* **Francis Bacon**
- *"The oldest and strongest emotion of mankind is fear."* **H.P. Lovecraft.**
- *"In politics, what begins in fear usually ends in folly."* **Coleridge**

- *"I must not fear. Fear is the mind-killer. Fear is the little-death that brings total obliteration. I will face my fear. I will permit it to pass over me and through me. And when it has gone past I will turn the inner eye to see its path. Where the fear has gone there will be nothing. Only I will remain."* **Frank Herbert, *Dune -Bene Gesserit Litany Against Fear***

- *"A man who has been in danger, when he comes out of it forgets his fears, and sometimes he forgets his promises."* **Euripides - *Iphigenia in Tauris (414-12 BC)***

- *"He either fears his fate too much, Or his deserts are small, That puts it not unto the touch To win or lose it all."* **James Graham - *Marquis of Montrose***

- *"Being frightened is an experience you can't buy."* **Anthony Price - *Sion Crossing* (1984)**

- *"What we fear comes to pass more speedily than what we hope."*

- **Publilius Syrus -*Moral Sayings (1st C B.C.).***

- *"Solitude scares me. It makes me think about love, death, and war. I need distraction from anxious, black thoughts."* **Brigitte Bardot**

- *"Why are we scared to die? Do any of us remember being scared when we were born?"* **Trevor Kay**

- *"A good scare is worth more to a man than good advice."* **Edgar Watson Howe -*Country Town Sayings* (1911)**

- *"Courage is not the lack of fear but the ability to face it."* **Lt. John B. Putnam Jr. (1921-1944)**

Symptoms and fear responses are often confused with those of anxiety. Although, according to Hackort & Schwenkmezger (1993), fear and anxiety are different phenomena. Fear is more specific, reflex-like defense and protection reaction. On the other hand, the sources of danger in anxiety are not always clearly understood. Stimulus recognition in fear is a key. Borkovec & Inz (1990) proposed that both fear and anxiety are focused on a threat of future harm, but they differ in the level of uncertainty, with fear associated more than anxiety to danger that is concrete and sudden. It has been also suggested that anxiety and fear differ in that fear is an active coping mechanism and anxiety results when threats are resistant to coping. If threats are resistant to coping, they might reasonably be expected to continue to influence cognitive appraisal of the threatening situations. In partial support of a connection between anxiety and persistent threat processing is a positive correlation between train anxiety scores and a measure of threat bias in implicit memory tasks (Mathews et al., 1989).

Similarly, fear is not the same as worry. For the most part, the evidence does not support a state of emotional arousal similar to fear during worry states. In fact, one prominent theory suggests that worry may actually be useful as an emotional avoidance strategy, similar to the notion of functional adaptation of fear responses (Borkovec & Inz, 1990). Individuals suffering from GAD (i.e., generalized anxiety disorder, and by definition chronic worry) are hypersensitive to threat cues in the environment and are especially distracted by emotional information when concurrently engaging in an unrelated cognitive task (Emotional Stroop test, Becker et al., 2001). Also, when individuals are asked to worry, they report approximately equal amount of anxiety and depressed mood. Indeed, these emotional reactions are not necessarily synonymous with bodily arousal as in a fear state. Even when worry increases, subjective reports of anxiety, physiological systems may not be preparing for fight or flight reactions. Thus, worry is a type of anxiety distinct from traditional fear responses.

From the cognitive-motivational-rational theory of emotion (Conroy et al., 2001), fear is considered a negative emotion, and will occur when the perceived or anticipated change is a decrease in the likelihood of achieving a goal. Fear is associated with the potential for failure and its resultant consequences (Birney et al., 1969). Specifically, at least three anticipated aversive experiences with respect to goal achievement were proposed, including: (a) fear of devaluation of the self-esteem—with respect to injury, following recovery and the return to sport participation, the athlete discovers that he/she is worse in terms of physical and psychological status in the team than previously hoped and anticipated; (b) "non-ego punishment" related to the anticipation of a loss of reward—with respect to injury, this would be losing the spot on the team. Accordingly, the higher the perceived risk of losing the spot, the higher the fear response; (c) fear of social devaluation, poor opinion of other. In terms of injury, the anticipated reduction of performance level may develop into fear of failure to meet the standards the sport demands. More recently, Conroy (2000) has further elaborated on the composite of fear of failure beliefs, associating "higher order fears" with cognitive "appraisal" as summarized in the following Figure 1.

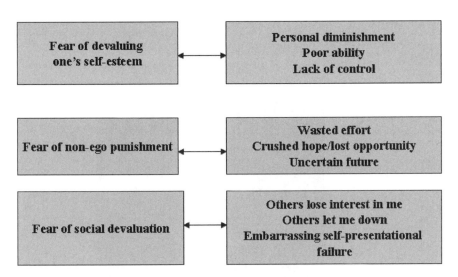

Figure 1. Higher order of fears associated with types of cognitive appraisal (Adopted from Conroy, 2000).

Clearly, this interesting model of fear of failure should be further considered and developed with respect to the alteration of the approach to goal achievement as a result of athletic injury. Specifically, single versus multiple injuries, mild versus severe injury, male versus female responses to injury, age and many other factors may differentially influence the possibility of the development of psychological trauma in general, and social-psychological composites of fear in particular.

2. NEUROBIOLOGY OF FEAR

One of the major ideas in the history of the neurobiology of emotions came in 1937 when Papez proposed the idea that certain structures of the brain form the anatomical basis of emotional reactions. Overall he suggested that the medial structures of the cerebral hemisphere acting on the hypothalamus produce emotional reactions. Later, in the 1950s, Paul Broca used the term "limbic lobe" to refer to the part of the cerebral cortex that forms a rim (*limbus* is Latin word for rim) around the corpus callosum and diencephalon of the medial part of both hemispheres. Two prominent components of this region are the **cingulate gyrus** and **parahippocampal gyrus**. It was also suggested that the cortex is necessary for transforming events produced by limbic structures into what we experience as "emotions." Specifically, the orbital and **prefrontal cortex,** along with the ventral part of the basal ganglia, anterior cingulated cortex, and a large nuclear mass in the temporal cortex called the **amygdala** now referred to as limbic system. The modern conception regarding the contribution of major brain structures in emotional experiences is shown in Figure 2.

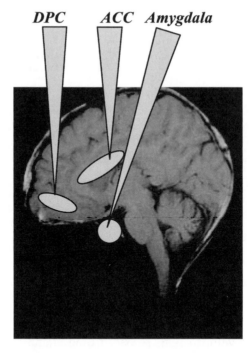

Figure 2. "Fear Triangle" including Amygdala, Prefrontal Cortex (PFC) and Anterior Cingulate Cortex (ACC) interaction as involved in fear processing response to a threat.

2.1. Animal Models of Conditioned Fear Responses

During the past twenty-five years or so, studies have been conducted examining the effect of lesions of the frontal cortex, paralimbic cortex, or amygdala – the structures that appear to be the most important forebrain areas involved in emotional experiences. Specifically, the earlier studies on nonhuman primates (experiments on rhesus monkeys), including amygdamectomy, clearly demonstrated that removal of the amygdala results in abnormal emotional behaviors, such as reduced aversion to biologically relevant stimuli that normal animals found threatening (Kluver & Bucy, 1339). This set of abnormal behaviors, such as bizarre oral behaviors, hyperactivity, hypersexuality, and changes in emotional behavior, including lack of hostility, unexpected fear responses and aggressiveness, is known as the Kluver-Bucy syndrome. They assumed that these changes in behavior were at least due in part to the interruption of the pathways earlier described by Papez.

Since this pioneering discovery, several models of emotional behavior have been developed. One of the most prominent approaches to the study of fear responses using an animal model is based on classic Pavlovian conditioned fear responses. Pavlovian fear conditioning is an associative learning task in which animals are trained to respond defensively to a neutral conditioned stimulus (CS) by pairing it with an aversive unconditioned stimulus (US). This type of learning is thought to be critically dependent on the amygdala, and evidence suggests that synaptic plasticity within the lateral nucleus of the amygdala (LA) may be responsible for storing memories of the CS-US association.

In a recent study by Blair et al., (2005) the researchers trained rats to fear an auditory CS by pairing it with a shock US delivered to one eyelid. Conditioning was assessed by measuring *freezing* responses evoked by the CS during a subsequent test session. The amygdala was unilaterally inactivated during either the training or the testing session by intracranial infusions of muscimol into the LA. It was found that both the acquisition and the expression of conditioned freezing to the CS depended on the amygdala contralateral, but not ipsilateral, from the eyelid where the shock US was delivered. Overall, the results demonstrate that the fear-learning circuitry of the amygdala is functionally lateralized according to the anatomical source of predicted threats. These results overall are consistent with early work described by LeDoux (1987) showing that projections from the central group of nuclei in the amygdala to the midbrain reticular formation are critical in the expression of freezing behavior, while other projections from this group to the hypothalamus control the rise in blood pressure. This may further justify the notion of differential neural bases for fear and anxiety, as discussed in the previous section of this chapter.

Another area of the brain related to the acquisition and extinction of conditioned fear in animals is the dorsal and ventral medial prefrontal cortex (mPFC). Specifically, the emotional reactivity of rats with a lesion of the dorsal portion of mPFC was examined by Morgan & LeDoux (1995) using a classical fear conditioning research paradigm. Behavioral fear, such as a *freezing* response, was measured during both the acquisition and extinction phases of the task. Interestingly, lesions enhanced fear reactivity to both conditioned and contextual stimuli, suggesting that dorsal mPFC lesions provide a general increase in fear reactivity in response to fear conditioning. This study is complementary to their previous report that lesions just ventral to the present lesion had no effect during acquisition of the same task and prolonged the fear response to conditioned stimuli during extinction (Morgan et al., 1993). It was concluded that both the dorsal and ventral areas of mPFC may be involved in the fear system, but each modulates different aspects of fear responsivity (i.e., fear acquisition versus fear extinction).

Based on these previous findings demonstrating that lesions of the ventral medial prefrontal cortex (mPFCv) disrupted performance during the extinction component of a classical fear conditioning task without affecting acquisition performance, Morgan et al. (2003) initiated another study aimed at demonstrating the importance of mPFC for the memory of prior extinction training. In other words, they examined the effect of mPFCv lesions made after training and compared the effects of pre-training and post-training lesions. The experimental findings suggest three conclusions. *First*, an intact mPFCv during acquisition may protect the animal from prolonged response during extinction trials following a brain insult. *Second*, changes in mPFCv may predispose animals toward enhanced fear reactions that are difficult to extinguish when re-exposed to fearful stimuli, due to a diminished capacity to benefit from the fear-reducing impact of prior extinction experience. *Third*, contextual cues processed by mPFCv may influence extinction performance. Overall, the findings that PFC may play a role in the regulation of fear in animals may help researchers to investigate the mechanisms and potentially the treatment protocols of uncontrolled fear, including anxiety, phobia, panic and post-traumatic stress disorders in humans.

2.2. Human Models of Neurobiology of Fear

There are several approaches to demonstrating the feasibility of the animal model of fear and its relevance to humans' fear responses. The early work by Dicks (1969) studied the loss of fear of humans and a general taming in animals with removed amygdala. A few interesting case studies were described in Neuroscience, 4th edition (Dale Purves et al., Chapter 29, *Emotions*), demonstrating the role of the amygdala in associating sensory stimuli with aversive consequences. For example, patients with bilateral damage to the amygdala without any motor or sensory impairment showed an impaired ability to recognize and experience fear, together with an impairment in rational decision making. Another patient with lesions in the orbital and prefrontal cortex also impaired emotional experience associated with influencing decision making processes. Considering that decision making entails the rapid evaluation of a set of possible outcomes with respect to future consequences (Damasio et al., 1994), it is feasible to suggest that patients with damage to the amygdala would suffer from impaired fight or flight mechanisms in face of danger.

Poor recognition of facial fear expressions as result of brain damage has been a focus of a Sprengelmeyer et al. (1999) study. They studied a person with bilateral amygdala damage and a left thalamic lesion. This patient showed an equivalent deficit affecting fear recognition from body postures and emotional sounds. His deficit of fear recognition was not linked to evidence of any problem in recognizing anger (a common feature in other reports), but for his everyday experience of emotion he reported reduced anger and fear compared with normal controls. These findings show a specific deficit compromising the recognition of the emotion of fear from a wide range of social signals, and suggest a possible relationship of this type of impairment with alterations of emotional experience.

This is consistent with a more recent study by Williams et al. (2006) suggesting that trauma to the amygdala and medial prefrontal cortex (mPFC) modulate consciously attended fear. They used fMRI methodology to elucidate the effect of trauma reactions on amygdala-mPFC function during overt fear perception tasks in normal controls and patients suffering from post-traumatic stress disorder (PTSD). PTSD patients showed a bilateral reduction in mPFC activity (in particular, the right anterior cingulate cortex, ACC) and a significant enhancement in the left amygdala. In other words, trauma was related to a distinct pattern of ACC and mPFC connections. More recently, Milad et al. (2007) directly examined a role of the ACC, specifically the dorsal anterior cingulate cortex, in fear responses using MRI. The skin conductance response (SCR) was the index of conditioned fear. They reported that: (1) cortical thickness within dACC is positively correlated with SCR during conditioning; (2) dACC is activated by a conditioned fear stimulus; and (3) this activation is positively correlated with differential SCR. Moreover, the dACC region implicated in this research corresponds to the target of anterior cingulotomy, an ablative surgical treatment for patients with mood and anxiety disorders. They concluded that convergent structural, functional, and lesion findings from separate groups of subjects suggest that dACC mediates or modulates fear expression in humans. Collectively, these data implicate this territory as a potential target for future anti-anxiety therapies. Overall, according to numerous facts, physically disconnecting the amygdala from other input via the temporal lobe may produce alterations in emotional behavior, which suggests that the amygdala may form part of the system for processing personally relevant information. Such a system would function in parallel to the object-recognition system of the anterior temporal cortex and hippocampus, and the spatial system of the posterior parietal cortex.

A goal of fear and anxiety research is to understand how to treat the potentially devastating effects of anxiety disorders in human. It should be noted that much of this research has utilized classical fear conditioning, a simple paradigm that has been extensively used in animals, aiming to outline the brain areas responsible for the acquisition, expression and extinction of fear. The findings from non-human subjects have recently been substantially extended in humans, using neuropsychological and imaging methodologies. For example, fMRI was used to study human brain activity during Pavlovian fear conditioning (Knight et al., 1999). In this brain imaging study subjects were exposed to lights that either signaled painful electrical stimulation (CS+), or that did not serve as a warning signal (CS-). Unique patterns of activation developed within the anterior cingulate and visual cortices as learning progressed. Training with the CS+ increased active tissue volume and shifted the timing of the peak fMRI signal toward CS onset within the anterior cingulated cortex. Within the visual cortex, active tissue volume increased with repeated CS+ presentations, while cross-correlation between the functional time course and CS- presentations decreased. This study clearly demonstrated the plasticity of the anterior cingulate and visual cortices as a function of learning, and implicated these regions as components of a functional circuit activated in human fear conditioning.

2.2.1. Amygdala

Considerable progress has been made over the past twenty years or so in relating specific circuits of the brain to emotional functions. Although much of this work has involved studies of Pavlovian or classical fear conditioning, behavioral procedures to couple meaningless stimuli to emotional (defense) responses incorporated with brain imaging research shed additional light on the problem. The major conclusion from studies of fear conditioning is that the amygdala plays a critical role in linking external stimuli to defense responses (LeDoux, 2003). To further support the notion that the same brain circuitry is involved in fear response in both animal and human populations, another recent report is worth mentioning. Specifically, Rosen and Donley (2006) reviewed research in both animals and humans and declared that considerable progress was made in elucidating a brain circuitry of fear, particularly the importance of the amygdala in fear conditioning. While there is considerable agreement about the participation of the amygdala in fear in both animals and humans, there are several issues about the function of the amygdala raised by the human research that have not been addressed by or may be answered by animal research. Three of these are addressed in this report: (1) Is the amygdala involved in or necessary for both fear learning and unconditioned fear? (2) Does the amygdala code for intensity of fear? (3) Is the amygdala preferentially involved in fear, or is it also activated when there is no overt fear or aversive stimuli, but where the situation can be described as uncertain? They present evidence indicating that the rodent amygdala is involved in some types of fear (conditioned fear), but not all types (unconditioned fear), and may therefore have significance for a differential neurobiology of certain anxiety disorders in humans. Further, similar to the human amygdala, the rodent amygdala responds to varying intensities of aversive stimulation. Finally, it is suggested that, similar to humans, the rodent amygdala is involved in the evaluation of uncertainty. The progress on elucidating the role of the amygdala in fear may be facilitated by the corroboration of findings from both animal and human research. Overall, animal models of fear are being translated to humans with the target of helping to facilitate the extinction or abolishment of fears, a trademark of anxiety disorders, by discussing the efficacy of modulating the brain circuitry involved in fear via pharmacological treatment or emotion regulation cognitive strategies (Delgado et al., 2006).

Empirical evidence obtained in human research suggests that the neural basis of fear is related with elevated activation of the amygdala, hypothalamus and brainstem, resulting in an elevated level of corticosteroid release together with behavioral and other physical responses (Davis, 1992). Seligman (190) explained that fear is easily acquired in a single trial as an adaptive response to threat when there is a high level of arousal (a condition in which humans are biologically prepared to be wary about their safety).

2.2.2. Prefrontal Cortex

The prefrontal cortex (PFC), interconnected with the amygdale, is also involved in organizing and planning future behavior, thus, the amygdala may provide the emotional input to fight or flight mechanisms in face of danger. A functional connectivity analysis using fMRI data clearly shows significant positive correlation between the PFC and hippocampus during extinction recall (Milad et al., 2007). Considering recent animal models of fear-learning that includes the amygdala vertrolateral PFC and hippocampus (Concoran & Quirk, 2007), it is feasible to suggest that the PFC is a site of neural plasticity that allows for the inhibition of fear during extinction recall. Also, the interaction of the amygdala with the neocortex and related subcortical areas most likely forms the neural basis for emotional experience: the highly subjective "emotional feeling." The emotional feeling may be conceived as the product of an emotional memory, and overall as an important substrate of "conscious awareness," which is the least understood topic of neurobiology to date.

2.2.3. Fear Perception: Brain Imaging Studies in Humans

The whole brain functional MRI studies of the neural substrates underlying the fear-related processing, traditionally involved viewing various pictures with different emotional content. For example, alternating blocks of pictures and captions evoking negative feelings versus irrelevant pictures versus picture-caption pairs evoking positive feeling induced different brain responses in an fMRI study by Teasdale et al. (1999). Specifically, compared with the reference picture-caption pairs, negative pairs activated the right medial and middle frontal gyri, right anterior cingulate gyrus, and right thalamus. In contrast, compared with the reference picture-caption pairs, positive pairs activated the right and left insula, right inferior frontal gyrus, left splenium, and left precuneus. Finally, compared with the negative picture-caption pairs, positive pairs activated the right and left medial frontal gyri, right anterior cingulate gyrus, right precentral gyrus, and left caudate. These authors concluded that (1) negative and reference picture-caption pairs and (2) positive and negative picture-caption pairs activated networks involving similar areas in the medial frontal gyrus (Brodmann's area 9) and right anterior cingu-late gyrus (areas 24 and 32). Activation of these same sites by a range of evoked affects is consistent with areas within the medial prefrontal cortex mediating the processing of affect-related meanings, a process common to many forms of emotion production, in including fear.

In addition, the pathway for fear perception may include interconnections between the amygdala and thalamo-cortical systems (Das et al., 2005). This was confirmed by the psychophysiological and physio-physiological interactions to examine the functional connectivity within the thalamus, amygdala and sensory (inferior occipital, fusiform) cortices, and the modulation of these networks by the anterior cingulate cortex (ACC). In this study, fMRI data were acquired for twenty-eight healthy control subjects during a fear perception task, with neutral as the 'baseline' control condition. The main effect analysis, using a region of interest (ROI) approach, confirmed that these regions are part of a distributed neural system for fear perception. Psychophysiological interactions revealed an inverse functional connectivity between occipito-temporal visual regions and the left amygdala, but a positive connectivity between these visual regions and the right amygdala, suggesting that there is a hemispheric specialization in the transfer of fear signals from sensory cortices to the amygdala. Moreover, physiophysiological interactions revealed a dorsal-ventral division in ACC modulation of the thalamus-sensory cortex pathway. While the dorsal ACC showed a positive modulation of this pathway, the ventral ACC exhibited an inverse relationship. In addition, both the dorsal and ventral ACC showed an inverse interaction with the direct thalamus-amygdala pathway. These findings suggest that thalamo-amygdala and cortical regions are involved in a dynamic interplay, with functional differentiation in both lateralized and ventral/dorsal gradients with respect to the processing of fear responses.

2.2.4. Facial perception

Facial perception design implemented with fMRI methodology may also shed light on the subconscious processing of emotional responses. Considering the known fact that the amygdala activates in response to fearful faces presented below the threshold of conscious visual perception, additional fMRI study has been recently conducted to examine the role of the amygdala in conjunction with the anterior cingulate gyrus (ACC) during pre-attentive presentations of sad versus happy faces (Killgore, 2004).

Masked happy faces in this study were associated with significant bilateral activation within the anterior cingulate gyrus (ACC) and the amygdala, whereas masked sadness yielded only limited activation within the left anterior cingulate gyrus. In a direct comparison, masked happy faces yielded significantly greater activation in the anterior cingulate and amygdala relative to identically masked sad faces. Conjunction analysis showed that masked affect perception, regardless of emotional valence, was associated with greater activation within the left amygdala and left anterior cingulate. Overall, these fMRI findings suggest that the amygdala and anterior cingulate are important components of a network involved in detecting and discriminating affective information presented below the normal threshold of conscious visual perception.

It should be noted, however, that the fear (signals for threat, fear versus anger) specific activation of the amygdala as a key neural substrate for facial displays of affect has been recently challenged by a more recent study by Fitzgerald et al. (2006). They performed a 4 Tesla fMRI study in which twenty subjects viewed a contemporary set of photographs displaying six different facial expressions (fearful, disgusted, angry, sad, neutral, happy, etc.) while performing a task with minimal cognitive demand. According to this study across subjects, the left amygdala was activated by each face condition separately, and its response was **not selective** for any particular emotional category. Thus, as mentioned before, these results may challenge the notion that the amygdala has a specialized role in processing certain emotions, such as only fear, and suggest that the amygdala may have a more general-purpose function in processing salient information from faces. Indeed, additional brain imaging studies using more sophisticated designs are needed to further explore the specialized role of the amygdala in conjunction with other brain structures in fear processing responses. In fact, new promising computational approaches, such as "network of effective amygdala connectivity" (Stain et al., 2007), using structural equation modeling (path analysis) may further clarify this issue. Specifically, Stain et al. (2007) developed an automated elaborative path analysis procedure guided by known anatomical connectivity in the macaque. They applied this technique to a large human fMRI data set acquired during perceptual processing of angry or fearful facial stimuli. Overall, they confirm and extend previous observations of amygdala regulation by an extended prefrontal network encompassing the cingulate, orbitofrontal, insular, and dorsolateral prefrontal cortex, as well as strong interactions between the amygdala and the parahippocampal gyrus. This validated model can potentially be used to study neurocognitive and affective correlates and functional interactions within the limbic system.

2.2.5. Body Actions as Threat Stimuli

Most recently, a whole body movement and gestures perception design was used in human fMRI studies to examine anatomical locations and neural substrates involved in the processing of fear responses. It was a well-known notion, according to Darwin's evolutionary perspectives, that whole body actions play a prominent role in expression of emotions. Accordingly, researchers in the field of social psychology and human development have long emphasized the importance of whole body motion as a reflection of emotional states, unlike cognitive neuroscientists, who mostly exclusively considered isolated facial expressions (fear, anger, disgust, shock, depression etc.) to study the neural bases of emotions. Using a short block of whole body expression of fear, Hadjikhani & de Gelder (2003), neural bases of emotions have been studies via high-field fMRI. Specifically, in this study subjects were presented with short blocks of body expressions of fear alternating with short blocks of emotionally neutral body gestures. All images had internal facial features blurred out to avoid confusion due to a face or facial expression. It was shown that exposure to body expressions of fear, as opposed to neutral body postures, activates the fusiform gyrus and the amygdala. The fact that these two areas have previously been associated with the processing of faces and facial expressions, as discussed in the previous section, suggests similarities between facial and body-action expressions of emotion. These findings may open a new area of investigation of the role of body expressions of emotion in adaptive behavior as well as the relation between processes of emotion recognition in the face and in the whole body.

Clearly, behavioral reactions such as putting the hands in front of the face, running for cover, avoidance of contact with a 300 pound football players running toward you, etc., may provide a strong fear signal to the observer who may or may not be aware of any danger. This was examined in humans by a recent fMRI study by Grezes et al., (2007). Specifically, they investigated how such dynamic (not static) fear signals from the whole body are perceived and processed in different brain regions. A factorial design allowed them to investigate brain activity induced by viewing bodies, bodily expressions of fear and the role of dynamic information in viewing them. Their critical findings are twofold. *First*, they found that viewing neutral and fearful body expressions enhances amygdala activity; moreover actions expressing fear activate the temporal pole and lateral orbital cortex more than neutral actions. And *second*, the differences in activations between static and dynamic bodily expressions were larger for actions expressing fear in the STS and premotor cortex compared to neutral actions. Current conceptualization with regard to brain imaging studies of the perception of fear may be summarized by the suggestion by Atkinson et al. (2007) that a full understanding of emotion perception and its neural substrate will require investigations that employ dynamic displays and means of expression other than the face. Indeed, additional research is needed to further explore the neural basis of fear responses induced by the perception of whole body threat stimuli. This would be of particular interest, for example, in terms of the examination of athletes' fear, induced by perceived collisions, of injury and re-injury.

2.2.6. Temporal Limitations of fMRI

Poor temporal resolution is a well known limitation of fMRI brain imaging methodology. The peak metabolic response (e.g., blood oxygen level dependent, BOLD signal) may occur on average within six seconds following the brain event (i.e., fear response). Therefore, there are attempts in the relevant literature to resoleve this fMRI limitation by employing alternative research methodologies such as electroencephalography (EEG), including event-related potentials (ERP) and magnetoencephalography (MEG). For example, in an attempt to address important empirical questions such as "when" and "where" the perceived signals of threat versus non-threat are processed in the brain, Williams et al. (2006) used the event-related potentials (ERP) procedure. They have elaborated on the ERP research based on their previously obtained fMRI findings suggesting that preverbal processing of fear may occur via a direct rostral-ventral amygdala pathway without the need for conscious surveillance, whereas the elaboration of consciously attended signals of fear may rely on higher-order processing within a dorsal cortico-amygdala pathway (Williams et al., 2006).

Overall their ERP study of the neural basis of fear processing was a traced temporal sequence ('when') and source localization ('where') of event-related potentials (ERPs) elicited by fearful and happy facial expressions, compared to neutral control expressions, using 219 healthy subjects. They scored ERPs over occipito-temporal sites (N80, 50-120 ms; P120, 80-180 ms; N170, 120-220 ms; P230, 180-290 ms; N250, 230-350 ms) and their polarity-reversed counterparts over medial sites (P80, 40-120 ms; N120, 80-150 ms; VPP, 120-220 ms; N200, 150-280 ms; P300, 280-450 ms). In addition to scoring peak amplitude and latency, the anatomical sources of activity were determined using low resolution brain electromagnetic tomography (LORETA). It was shown that fearful faces were distinguished by persistent increases in positivity, associated with a dynamic shift from temporo-frontal (first 120 ms) to more distributed cortical sources (120-220 ms) and back (220-450 ms). By contrast, expressions of happiness produced a discrete enhancement of negativity, later in the time course (230-350 ms) and localized to the fusiform region of the temporal cortex. In common, fear and happiness modulated the face-related N170, and produced generally greater right hemisphere activity.

These findings support the proposal that fear signals are given precedence in the neural processing systems, such that the processing of positive signals may be suppressed until vigilance for potential danger is completed. While fear may be processed via parallel pathways (one initiated prior to structural encoding), neural systems supporting positively valenced input may be more localized and rely on structural encoding.

The temporal dynamic of neural activity recorded from human amygdala during fear conditioning responses has been also addressed using magnetoencephalography (MEG) research methodology by Moses et al. (2007). In this study, the activation during conditioning training was compared to habituation and extinction sessions. Conditioned stimuli (CS) were visually presented geometric figures, and unconditioned stimuli (US) were aversive white-noise bursts. The CS+ was paired with the US on 50% of presentations and the CS- was never paired. The precise temporal resolution of MEG allowed the researchers to address the issue of whether the amygdala responds to the onset or offset of the CS+, and/or the expectation of the initiation or offset of an omitted auditory US. It was shown that fear conditioning elicited differential amygdala activation for the unpaired CS+ compared to the CS- extinction and habituation. This was especially robust in the right hemisphere at CS onset. The strongest peaks of amygdala activity occurred at an average of 270 ms in the right and 306 ms in the left hemisphere following unpaired CS+ onset, and following offset at 21 ms in the left and 161 ms in the right (corresponding to an interval of 108 ms and 248 ms after the anticipated onset of the US, respectively). However, the earliest peaks in this epoch preceded US onset in most subjects. Thus, the activity dynamics suggest that the amygdala both differentially responds to stimuli and anticipates the arrival of stimuli based on prior learning of contingencies. The amygdala also shows stimulus omission-related activation that could potentially provide feedback about experienced stimulus contingencies to modify future responding during learning and extinction.

CONCLUSION

First, fear as a form of emotion is a complex phenomenon that may be observed and manifested at different levels of analysis. Clearly, fear is not just a generalized anxiety and/or worry and should be properly conceptualized before proper fear managements interventions are proposed. *Second*, the **amygdala** emerges as a nodal point in the network that links together cortical and subcortical brain regions involved in emotional processing, including fear both in animals and humans. Most likely, the amygdala and its connections to the prefrontal cortex and basal ganglia influence the selection and initiation of behavior aimed at obtaining rewards and avoiding punishments.

As was noted in most recent report published in Nature, 2007, "Neural Mechanisms Mediating Optimism Bias," people usually expect positive events in the future even when there is no evidence to support such expectations. For example, people expect to live longer and be healthier than average. They underestimate their likelihood of getting a divorce, and overestimate their prospects for success on the job market. This report examined how the brain generates this pervasive optimism bias. Interestingly, it was shown that optimistic tendency, as provoked by positive versus negative imaging of life events, was related specifically to enhanced activation in the <u>amygdala</u> and in the rostral anterior <u>cingulate</u> cortex. These most recent results suggest a key role for the amygdala in monitoring emotional salience in mediating the optimism bias. The clinical implications of these well-documented brain imaging studies are awaiting further elaboration, including the development of therapeutic protocols of fear management in various patient populations, including athletes suffering from fear of injury.

REFERENCES

Rosen, J.B., Schulkin, J. (1998). From normal fear to pathological anxiety. *Psychological Review, 105*(2), 321-350.

Hackfort, D., & Schwenkmezger, P. Anxiety. In R. N Singer, M. Murphey, & L.K. Tennant (eds). *Handbook of research on sport psychology.* (pp. 328-364). New York: Macmillan, 1993.

Borkovec, T.D., & Inz, J. (1990). The nature of worry in generalized anxiety disorder: A predominance of thought activity. *Behavioral Research and Therapy, 28,* 153-158.

Mathews, A., Mogg, K., May, J., Eysenck, M. (1989). Implicit and explicit memory bias in anxiety. *Journal of Abnormal Psychology, 1,* 15-20.

Becker, E. S., RInck, M., Margraf, J., Roth, W.T. (2001). The emotional Stroop effect in anxiety disorders: General emotionality or disorder specificity? *Journal of Anxiety Disorders, 15,* 147-159.

Birney, R. C., Burdick, H., Teevan, R.C. *Fear of failure.* New York,: van Nostrand, 1969.

Conroy, D. E., Poczwardowski, A., Henschen, R.P. (2001). Evaluation criteria and consequences associated with failure and success for elite athletes and performing athletes. *Journal of Applied Sport Psychology, 13,* 300-322.

Kluver, H., Bucy, P. C. (1939). Preliminary analysis of the temporal lobes in monkeys . *Archives of Neurology and Psychiatry, 42,* 979-1000.

Blair, H.T., Huynh, V.K., Vaz, V.T., Van, J., Patel, R.R., Hiteshi, A.K., Lee, J.E., Tarpley, J.W. (2005). Unilateral storage of fear memories by the amygdala. *Journal of Neuroscience, 25* (16), 4198-205.

LeDoux J. 2003. The emotional brain, fear, and the amygdala. *Cell Mol Neurobiol, 23* (4-5), 727-38.

Morgan, M.A., LeDoux, J.E. (1995). Differential contribution of dorsal and ventral medial prefrontal cortex to the acquisition and extinction of conditioned fear in rats. *Behavioral Neuroscience, 109*(4), 681-8.

Morgan, M.A., Romanski, L.M., LeDoux, J.E. (1993). Extinction of emotional learning: contribution of medial prefrontal cortex. *Neuroscience Letters, 163*(1), 109-13.

Dicks, D., Myers, R., Kling, A. (1969). Uncus and amygdala lesions: Effects on social behavior in the free-ranging monkey. *Science, 165,* 69-17.

Purves, D., Augustine, G., Fitzpatric, D. Hall, W., LaMantia, A., McNamara, J., White, L. (eds.) Neuroscience. 4th edition.

Damasio, A. R. *Descartes Error: Emotion, Reason, and the Human Brain.* New York, Avon Books, 1994.

Sprengelmeyer, R., Young, A.W., Schroeder, U., Grossenbacher, P.G., Federlein, J., Büttner, T., Przuntek, H. (1999). Knowing no fear. *Proc Biological Sciences, 266* (1437), 2451-6.

Williams, L.M., Kemp, A.H., Felmingham, K., Barton, M., Olivieri, G., Peduto, A., Gordon, E., Bryant, R.A. (2006). Trauma modulates amygdala and medial prefrontal responses to consciously attended fear. *Neuroimage, 29* (2), 347-57.

Williams, L.M., Liddell, B.J., Kemp, A.H., Bryant, R.A., Meares, R.A., Peduto, A.S., Gordon, E. (2006). Amygdala-prefrontal dissociation of subliminal and supraliminal fear. *Human Brain Mapping, 27* (8), 652-61.

Milad, M.R., Quirk, G.J., Pitman, R.K., Orr, S.P., Fischl, B., Rauch, S.L. (2007). A Role for the Human Dorsal Anterior Cingulate Cortex in Fear Expression. *Biological Psychiatry, 62* (10), 1191-1194.

Milad, M.R., Rauch, S.L., Pitman, R.K., Quirk, G.J. (2007). Fear extinction in rats: implications for human brain imaging and anxiety disorders. *Biological Psychology, 73* (1), 67-71.

Milad, M.R., Wright, C.I., Orr, S.P., Pitman, R.K., Quirk, G.J., Rauch, S.L. (2007). Recall of fear extinction in humans activates the ventromedial prefrontal cortex and hippocampus in concert. *Biological Psychiatry, 62* (5), 446-54.

Knight, D.C., Smith, C.N., Stein, E.A., Helmstetter, F.J. (1999). Functional MRI of human Pavlovian fear conditioning: patterns of activation as a function of learning. *Neuroreport, 10*(17), 3665-70.

Rosen, J.B., Donley, M.P. (2006). Animal studies of amygdala function in fear and uncertainty: relevance to human research. *Biological Psychology, 73*(1), 49-60.

Deldago, M.R., Olsson, A., Phelps, E.A. (2006). Extending animal models of fear conditioning to humans. *Biological Psychology, 73*(1), 39-48.

Davis, M. (1992). The role of amygdala in fear and anxiety. *Annual Review of Neuroscience, 15,* 353-375

Seligman, M.E.P. (1990). Phobias and preparedness. *Behavioral Therapy, 2,* 307-320.

Corcoran, K.A., Quirk, G.J. (2007). Recalling safety: cooperative functions of the ventromedial prefrontal cortex and the hippocampus in extinction. *CNS Spectrum, 12* (3), 200-6.

Teasdale, J.D., Howard, R.J., Cox, S.G., Ha, Y., Brammer, M.J., Williams, S.C., Checkley, S.A. (1999). Functional MRI study of the cognitive generation of affect. *American Journal of Psychiatry, 156* (2), 209-15.

Das, P., Kemp, A.H., Liddell, B.J., Brown, K.J., Olivieri, G., Peduto, A., Gordon, E., Williams, L.M. (2005.) Pathways for fear perception: modulation of amygdale activity by thalamo-cortical systems. *Neuroimage, 26* (1), 141-8.

Killgore, W.D., Yurgelun-Todd, D.A. (2004). Activation of the amygdala and anterior cingulate during nonconscious processing of sad versus happy faces. *Neuroimage, 21*(4), 1215-23.

Fitzgerald, D.A., Angstadt, M., Jelsone, L.M., Nathan, P.J., Phan, K.L. (2006). Beyond threat: amygdale reactivity across multiple expressions of facial affect. *Neuroimage, 30*(4), 1441-8.

Stein, J.L., Wiedholz, L.M., Bassett, D.S., Weinberge, D.R., Zink, C.F., Matta, V.S., Meyer-Lindenberg, A. (2007). A validated network of effective amygdala connectivity., Neuroimage, *1*(36), 736-745.

Hadjikhani N, de Gelder B. (2003). Seeing fearful body expressions activates the fusiform cortex and amygdale. *Current Biology. 13*(24), 2201-5.

Grèzes J, Pichon S, de Gelder B. (2007). Perceiving fear in dynamic body expressions. *Neuroimage, 35*(2), 959-67.

Atkinson, A.P., Dittrich, W.H., Gemmell, A.J., Young, A.W. (2004). Emotion perception from dynamic and static body expressions in point-light and full-light displays. *Perception, 33*(6), 717-46.

Moses, S.N., Houck, J.M., Martin, T., Hanlon, F.M., Ryan, J.D., Thoma, R.J., Weisend, M.P., Jackson, E.M., Pekkonen, E., Tesche, C.D. (2007). Dynamic neural activity recorded from human amygdala during fear conditioning using magnetoencephalography. *Brain Research Bull, 71* (5), 452-60.

CHAPTER 13

FEAR OF INJURY, KINESIOPHOBIA & PERCEIVED RISK

1. INTRODUCTION

It is a well-documented fact that decreases in athletic performance after injury can be attributed to both psychological and physical factors (Dunn, 1999). The sports professionals have become aware of the integral role that psychological factors play in the injury occurrence and recovery processes. For this reason, such constructs like "perceiving fear" and "confidence in avoiding injury", "fear of pain" and "fear of injury" have become important factors to consider when dealing with injured athletes. The previous information paints an intricate picture of an athlete's psychological response to injury, but lacking information regarding emotional factors it is difficult to reliably predict athletes at high risk for injury. Although sports psychologists have contributed invaluable information to this concern, the emotional aspects of injury have not yet been fully addressed. The emotion of fear has not been excluded from research on general orthopedic injuries, but it has not been highly considered among athletes. Athletes are generally perceived as warrior type individuals who do not harbor emotions such as fear. This may be a major misperception considering that athletes face possible physical harm every time they step onto the field. When a non-athletic individual suffers an injury, he or she is faced with the difficulty of completing normal daily tasks due to pain and a loss of mobility. Once the person returns to pre-injury level, he or she is still only faced with the challenge of completing normal daily tasks. An athlete on the other hand, is not only faced with the challenges of daily functioning, but also faced with the challenge of returning to the field. Though he or she has overcome the injury, the athlete must deal with possibility of re-injury due to the high demands of sport activity. Therefore, the challenges for an injured athlete are quite complex because the act of returning to play forces an athlete to participate in the exact activity that caused the injury initially. Being faced with memories of pain and discomfort is likely to cause some level of fear. Given the complexity of the athlete's experience of injury, it seems erroneous to ignore fear as a possible component of re-injury.

2. FEAR OF INJURY UPON RETURN TO PLAY

Returning to sports participation following injury can be a difficult process for some athletes due to both physical (i.e., residual structural and functional abnormalities) and psychological (i.e., fear of re-injury; worry and anxiety; lost confidence) problems. Fear of injury manifests itself in several ways: being hesitant, holding back, not giving 100% effort, being wary of injury-provoking situations (particularly situations similar to the context of occurrence), and heavily strapping the injured body part. For those athletes with a history of injury to a particular body part, the fear of re-injury may be intensified because they know they already had a weakness there. Both anecdotal reports and empirical investigations indicate that a return to sport is often accompanied by fear of re-injury and fear of not performing up to pre-injury performance levels. For example, Crossman (1985) monitored the emotional responses to injury of thirty male athletes during four stages of recovery, including: (a) the day of the injury, (b) the following day, (c) halfway through rehabilitation, and (d) the day of return to practice. It was found that while 13% of injured athletes experienced fear during rehabilitation, a significantly higher number (40%) reported the same emotion upon return to competition. These researchers speculated that possible reasons for this fear included an uncertainty about the quality of future performances, a fear of failure, and the possibility of re-injury. In addition, as result of these fears and concerns, athletes may suffer heightened generalized anxiety due to uncertainty, an unnecessary focus on the injured area and preoccupation with anticipated pain, lowered confidence resulting in a temporary or long-term performance decrement, and a struggle to re-establish technical skills and impaired functions (e.g., strength, flexibility and endurance). Dealing with unrealistic expectations from others and existing pressure from coaches to return to competition prematurely may also be problematic for the returning injured athlete. Ultimately, multiple stressors may lead to a number of cognitive, emotional and physiological deficits that may increase the likelihood of re-injury.

Fear & Loss of Confidence: Much of the fear related to sport injury and rehabilitation emanates from not knowing what to expect and concern about what sensation and body reactions are appropriate (Flint, 1998). Therefore it is feasible to suggest that uncertainty and associated fears may link to lack of confidence in athletes suffering from injury. Indeed, there are several reports suggesting the link between fear of injury and loss of confidence. Specifically, the fear of re-injury may act as a factor in loss of confidence when returning to sports participation. For example, Macchi and Crossman (1995) questioned twenty-six professional ballet dancers to determine what impact injury had on their lives. The dancers initially reported feeling a host of negative emotions such as anger, fear, distress and depression. As a result of their injury, 42% of the dancers indicated that their attitude toward ballet had changed, that they were more careful when dancing, that they used better technique, and that they stretched more and modified exercises to avoid injury. Two of the ballet dancers expressed worry about re-injury.

Also, Magyar and Chase (1996) examined gymnasts and found that the fear of injury exists when athletes lack confidence in their ability to perform successfully in threatening or taxing competitive situations. Overall, this study confirmed that the fear of injury can be directly related to a loss of confidence. Accordingly, it is necessary to design and to assess the strategies that might be used by gymnasts to overcome their fear of injury. In other words, in order to perform at high competitive levels, one must learn to exercise control over fearful situations. This conclusion is consistent with their more recent research by Chase et al (2005) also examining the fear of injury, the sources of self-efficacy and the psychological strategies used to overcome fears in gymnastics. The participants were ten female gymnasts aged twelve to seventeen years. They had all taken part in competitive gymnastics and had experienced some type of injury during their careers. The results indicated that female gymnasts were most fearful of injuries because of the difficulty in returning from an injury and being unable to participate in practices and competitions while injured. Gymnasts described aspects of their past performance experience, such as success, consistency and communication with significant others, as important sources of self-efficacy. Some examples of psychological strategies used to overcome their fear of injury included healing imagery, muscle relaxation, etc.

Fear & Perceived Risk: Risk and danger are unavoidable attributes of any sports and recreational activities. However, it often happens that for some reasons an athlete may misperceive and/or exaggerate the risk of certain situations that in turn may increase the susceptibility for injury. Recently, there has been an interest in quantifying and determining athletes' perceptions of the risk of injury (Kontos et al., 2000). These researchers developed and statistically validated a general scale for the measurement of athletes' perceptions of risk of injury, called the Perceived Risk of Injury in Sport (RISSc). The RISSc contains six categories: (1) uncontrollable injuries, (2) controllable injuries, (3) overuse injuries, (4) surface related injuries, (5) upper body injury, and (6) re-injury. Each item uses the stem "what do you think are the chances you will..." Responses for each item are made on a scale between 1 (*very unlikely*) and 6 (*very likely*). As a follow-up, RISSc test has been used to examine gender differences in a sample of 501 adolescent team and individual sport athletes who participated in more than twenty different youth sport programs at the scholastic, club, and recreational level. Results showed that females scored higher on all subscales (i.e., controllable, surface-related, overuse, upper-body and re-injury) except for uncontrollable injuries where males reported higher scores. Results also revealed that males reported more previous injuries than females. Moreover, the type, location, severity, and repetitiveness of an injury may also influence the perceived risk of injury. For instance, an athlete who has suffered multiple, severe ankle sprains may perceive more risk in playing sports than an athlete who has had only one such injury. Also, injured athletes ($n = 111$) scored higher on all injury subscales when compared to uninjured athletes ($n = 241$). Overall, it is feasible to accept that the RISSc test is a viable measure, although it needs to be further studied using athletes of different sports and/or ages.

It should be noted that perceived risk as linked to injury is highly sport-specific phenomenon. For example, most common fears of competitive springboard and platform divers are the fear of "hitting" the board and fear of "being lost" in the air while executing multiple flips. Most often, platform diving is perceived as more dangerous than springboard diving due to "the fear factor of height." Not surprisingly, most of the divers who experienced "mental block," defined as inability to initiate the dive and being stuck, had a previous injury due to hitting the board and/or or "smacking." Similarly, baseball players perceived the most risk for controllable and uncontrollable injuries. Specifically, the risk for injury in baseball, as perceived by players, is associated with such actions as sliding into the base (more common in adults), over-exertion, falls, collision with another player, misjudged catches resulting in finger injuries and being hit by the baseball bat (more common in children). The above injuries are mostly due to uncontrollable factors. The author's numerous interviews with PSU gymnasts clearly indicate that a high perceived risk for injuries in gymnastics is associated with disorientation and improper landing at the end of the routine. Among female gymnasts, floor exercise is perceived as low risk, unlike the uneven bars and vault which are perceived as the most dangerous. Interestingly, no correlation between the perceived danger of specific routines and reported injuries during execution of these routines was found in gymnastics. There are several inconclusive reports aimed at establishing the relationship between perceived risk and confidence in ability to avoid injury. Due to numerous confounding factors influencing this relationship, it is not feasible to further elaborate on this issue in the present chapter.

3. FEAR OF MOVEMENT-KINESIOPHOBIA

3.1. Assessment: Tampa Scale of "Kinesiophobia" (TSK)

Over the past decades the construct of fear of movement/re-injury, currently known as "Kinesiophobia," has received increasing attention and scientific scrutiny as an important predictor of pain-related avoidance behavior and occupational disability (Vlaeyen et al., 2002; Severeijns et al., 2001; van den Hout et al., 2001; Crombez et al., 2002; Boersma et al., 2004; Roelofs et al., 2004; Peters et al., 2005). For athletic injuries, fear of movement is experienced in a different context. The fear not only originates from the exact event that caused the injury, but there is a fear of movement in general. The reason is that in athletics it is unlikely that re-injury will occur under the exact same circumstances, but rather that some type of movement will cause a secondary injury. Fear of movement due to anticipated pain and discomfort has been defined as "Kinesiophobia." More specifically, "Kinesiophobia" refers to "an excessive, irrational, and debilitating fear of physical movement and activity resulting from a feeling of vulnerability to painful injury or (re)injury" (Kori et al., 1990). The notion that the pain and (re)injury may be more disabling than the pain itself is contrary to the generally held notion that attributes disability solely to pain severity. Therefore it is feasible to suggest that fear of movement may serve as a potential predictor of observable and reported disabilities in addition to the pain severity index.

In 1991, Miller developed the Tampa Scale of Kinesiophobia (TSK). This is a 17-item questionnaire that is comprised of various questions concerning fear of movement (see below). All questions are based on a four point Likert scale ranging from strongly disagree (=1) to strongly agree (=4). The minimum quantification of the scale is 17 and the maximum is 68. Item numbers 4, 8, 12 and 16 are reverse scored. The reliability of the scale has been established as moderate to substantial (Cronbach's $\alpha= 0.70$ and $\alpha= 0.76$; Pearson's $r= 0.78$) (Swinkles-Meewisse et al., 2003).

ORIGINAL ITEMS OF THE TAMPA SCALE FOR KINESIOPHOBIA
(Miller et. al., 1991).

1. I'm afraid that I might injure myself if I exercise.
2. If I were to try to overcome it, my pain would increase.
3. My body is telling me I have something dangerously wrong.
4. My pain would probably be relieved if I were to exercise.
5. People aren't taking my medical condition seriously enough.
6. My accident has put my body at risk for the rest of my life.
7. Pain always means I have injured my body.
8. Just because something aggravates my pain does not mean it is dangerous.
9. I am afraid that I might injure myself accidentally.
10. Simply being careful that I do not make any unnecessary movements is the safest thing I can do to prevent my pain from worsening.
11. I wouldn't have this much pain if there weren't something potentially dangerous going on in my body.
12. Although my condition is painful, I would be better off if I were physically active.
13. Pain lets me know when to stop exercising so that I don't injure myself.
14. It's really not safe for a person with a condition like mine to be physically active.
15. I can't do all things normal people do because it's too easy for me to get injured.
16. Even though something is causing me a lot of pain, I don't think it's actually dangerous.
17. No one should have to exercise when he/she is in pain.

Research using the TSK has mostly been done on populations of patients suffering from orthopedic injuries. Several studies have specifically examined patients with lower back pain. In 1995, Vlaeyen et al. investigated fear of movement/re-injury and its relation to behavioral performance in lower back patients. By using the TSK they were able to find that fear of movement/re-injury is related to gender and depression but does not show high degrees of relation to pain coping and pain intensity. Swinkles-Meewisse et al. (2003) found that lower back patients with reduced levels of fear (Kinesiophobia) were more likely to participate in daily and social life activities. A year later an additional study was conducted on lower back patients, examining levels of fear associated with physical performance tests. Again correlations were found between fear and physical performance (Roelof et al., 2004). One study took the presence of fear a step further by counseling patients in order to increase their activity levels. Patients with lower back pain were exposed to information, via a physiotherapist and psychologist, regarding symptoms, beliefs and behaviors related to fear of injury. Patients who scored high in fear showed significant drops in scores and an increase in activity within three months of counseling intervention (Boersma et al., 2004).

Recently a study was conducted using the TSK among athletes. This study specifically investigated whether fear of re-injury due to movement is a component of return to play in athletes who have undergone anterior cruciate ligament (ACL) reconstruction. The TSK was used along with the Knee Injury and Osteoarthritis Outcome Score (KOOS). Fifty-three percent of the patients returned to activity and the athletes who did not return to previous activity levels possessed higher levels of fear. Fear was measured in the TSK. Additionally, there was a negative correlation found between fear of re-injury and knee-related quality of life. Based on findings from these studies, it was felt that the TSK was an appropriate measure of fear to use for the subject of the study. We feel that the TSK may provide us with additional insight regarding the psychological aspects of re-injury. Accordingly, a pilot study was conducted aimed at exploring the effect of different properties on injury on possibility of development of fear of movement among PSU students-athletes (Moss & Slobounov, 2006).

A total of eighty subjects were recruited for this study. The subjects were members of either a varsity or club sport at the university. There were forty men and forty women who were members of the rugby, football, basketball, track and field, and tennis teams. Subjects were assigned to either an injured (n=49, including fourteen subjects with mild traumatic brain injuries, (MTBI)) or non-injured (n=31) group based on their current injury status. The non-injured group served as a control group. These athletes were classified as controls based on the fact that they had not experienced any type of injury in the previous six months. The injured athletes were classified based upon whether they were currently injured (n=24) or had experienced an injury in the previous six months (n=24). Those athletes who were currently injured were then categorized by the severity of their injury in respect to recovery time before return to sports participation: mild injury (n=4), moderate injury (n=4), major injury (n=9), and concussed (n=14). Mild injury is classified as a minor physical hindrance that does not prevent an athlete from participating in sports for more than one week (e.g., mild ankle sprain). Moderate injuries are classified as those that may hinder an athlete from participating for two to seven weeks (e.g., extreme muscle pull or tendon damage). Major injuries are classified as those that hinder an athlete from participating in sporting activity for eight or more weeks (e.g., ACL tear, shoulder Labrum tear). Athletes who were categorized as concussed were assessed and diagnosed previously by a physician or athletic trainer.

For completion of the TSK, subjects attended experimental sessions in which the questionnaire was administered randomized between subjects. The TSK was administered by hand as a paper questionnaire. The data was collected using the original items of the TSK by Miller et al. (1991). Subjects rated their feelings on a Likert Scale from 1 to 4 on seventeen items of the questionnaire. Reliability of the scale has been established as moderate to substantial (Cronbach's $\alpha= 0.70$ and $\alpha= 0.76$; Pearson's $r= 0.78$) (Swinkles-Meewisse et al., 2003). Subjects were asked to write their answers directly on the questionnaire. Subjects were instructed to ask for assistance if there was a lack of understanding of any of the seventeen items. The Tampa Scale of Kinesiophobia is summed by adding the score of the seventeen items for each individual subject. The range of sums has a minimum of seventeen and a maximum of sixty-eight. Each subject's total score was added to a subject pool categorization based on injury status. The mean average of each category was calculated as a numerical comparison to various other categories. Total for each subject was imported into Microsoft Excel for statistical analysis.

Analysis of the TSK was conducted in order to identify differences in fear levels among different groups of athletes. As a measure of variance, ANOVA and the Tukey HSD Post Hoc test were performed to determine the significance of these differences. Tukey HSD is commonly used in psychological research tests. This test allows for computation of a single value that determines the minimum difference between group means that is necessary for significance. The value, called the honestly significant difference (HSD), is then used to compare any two group conditions. When the mean difference exceeds Tukey's HSD, it is concluded that there is a significant difference between groups (Gravetter et al., 2000).

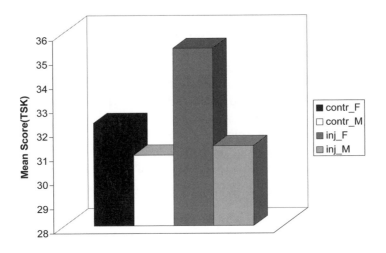

Figure 1. Effect of gender on fear of movement (TSC scores).

There are several major findings of interest from this study. First, female athletes reported higher levels of injury related-fear related due to movement than male subjects regardless of injury status (see Fig. 1). This observation is consistent with the previous TSK study, suggesting that females are significantly more fearful than males (Swinkles et al., 2003). Moreover, female athletes are more willing than males to disclose their concern to others and acknowledge their vulnerability (Martin, 2005). On the contrary, most male athletes gradually develop the belief that to be a man and an athlete requires them to learn to accept pain, physical risk, and injury in stoic silence (Messner, 1992; Nixon, 1996). The male gender role in most societies is to not discuss personal problems and/or admit vulnerability (Addis & Mahalic, 2003).

Second, the development of fear of re-injury due to movement may be influenced by the severity of previous injuries in relation to recovery time, regardless of gender (see Fig.2). Specifically, athletes with mild and moderate injuries reported the lowest levels of fear. Interestingly, athletes who suffered from mild traumatic brain injury experienced higher levels of fear of re-injury, similar to athletes with major orthopedic injuries.

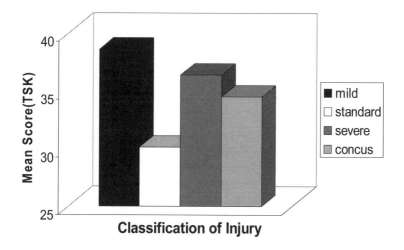

Figure 2. Effect of injury classification on TSC scores.

Third, the athletes who had suffered three or more injuries in the past experienced the highest level of fear of re-injury due to movement. Finally, the more severe the previous injuries (in addition to the number of injuries), the more likely that athletes would experience a higher level of fear of injury due to movement. In addition, athletes suffering from a single mild traumatic brain injury (MTBI) reported high levels of fear of re-injury due to movement, similar to the responses of athletes with major orthopedic injuries.

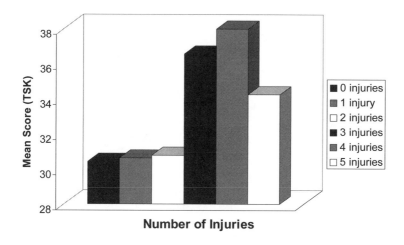

Figure 3. Effect of number of injuries on TSC scores.

Overall the results from this study indicated that there was a difference in fear levels between injured subjects based on severity of injury in relation to recovery time of the injury. Athletes with mild injuries and severe injuries reported the highest levels of fear. No significant differences were found when comparing severe injuries to concussion injuries. The third significant finding was related to fear levels and the number of injuries suffered in the past. There was a significant difference between subjects with two or fewer injuries and those with three or more injuries. The presence of a fear of injury-avoidance behavior relationship (Keefe, 1990) in athletes suffering from major orthopedic injuries was confirmed in this study. Our cursory analysis of field notes and what was observed without interference both clearly showed that most of the subjects under observation experienced various forms of *bracing* behavior. In the training room, this was evidenced by frequently restricted active range of motion at injured joint(s), abnormal asymmetry, sudden limping, abnormal speed and accuracy of prescribed exercise. On the football field, for example, a defensive player *braced* for an oncoming impact (tackle) by dropping his head too low. This is an extremely dangerous bracing technique frequently leading to serious injury. In basketball, a previously injured point guard closed his eyes when breaking down the defense out of fear of being hit again in the face. In track, hurdlers landed differently on their recently injured lead leg. Again, whether the observed *bracing* techniques are the consequence of uncompleted physical rehabilitation and response to physical pain, or fear of injury due to movement and response to anticipated pain is unclear and requires future research.

4. FEAR OF INJURY: OBSERVATION FORMATS

"Likewise in children, the tongue's speechless
Leads them to gesture what they express." **Lucretius**

The power of non-verbal communication including gestures and signs has been recognized for ages. Indeed, facial expression, hand gestures and head motions sometimes may deliver pure feeling free from deception and redundancy. This assumption was most beautifully described by Michel de Montaigne in his famous book: "Apology for Raymond Sebond" (1575-1580) as follows: "That of the hands? We beg, we promise, call, dismiss, threaten, play, entreat, deny, refuse, question, admire, count, confess, repent, fear, blush, doubt, instruct, command, incite, encourage, swear, testify, accuse, condemn, absolve, insult, despise, defy, vex, flatter, applaud, bless, humiliate, mock, reconcile, commend, exalt, entertain, rejoice, complain, grieve, mope, despair, wonder, exclaim, are salient, and what not, with a variation and multiplication that vie with the tongue. With the head: we invite, send away, avow, disavow, give the lie, welcome, honor, venerate, disdain, demand, show out, cheer, lament, caress, scold, submit, brave, exhort, menace, assure, inquire. What of the eyebrow? What of the shoulder? There is no movement that does not speak both a language intelligible without instruction, and a public language; which means, seeing the variety and particular use of other languages, what this one must rather be judged the one proper to human nature. I omit what necessity teaches privately and promptly to those who need it, and the finger alphabets, and the grammars in gestures, and the sciences which are practiced and expressed only by gestures, and the nations which Pliny says have no other language." (p.403)

As can be seen from the aforementioned quote, the emotional and cognitive substrate can be observable by trained clinicians. These substrates in injured athletes should be identified and observed in their natural environment (e.g., training room, practices and competitions). When they are, it is called *naturalistic observation* aimed at determining the frequency, strength and pervasiveness problem and behavior of the factors that are maintaining it (Phares, 1988, p.265). It is important to elaborate on more or less standard observational formats specifically designed for revealing the symptoms and signs of psychological trauma.

In the following text are some representative samples of observational formats of injury-related psychological trauma that can be recommended for medical practitioners.

Type of injury: 2nd degree sprained ankle
Consequences of injury: altered practice schedule and intensity
Sport: track/cross country
Site: track/training room
Duration: 1 hour
Dimension of intensity: light practice and jogging laps, ROM improvement exercises
Units/scale: 1 to 10, 1=normal activity/behavior and 10=highly abnormal activity/behavior

Behavioral Observation Chart	1	2	3	4	5	6	7	8	9	10
Vocal		X								
Posture				X						
Facial Expression					X					
Bracing Behavior				X						
Locomotion								X		

			X						
Gestures			X							
Specific Skillful Activity (ROM, Abnormal speed)							X			

Recast:

- Posture: bracing occurred as the intensity rose, and more noticeable compensation of weight to other ankle/leg
- Facial Expression: showed some signs of pain/grimacing during some of the workout
- Bracing Behavior: some points at which he favored injured ankle
- Locomotion: most noticeable out of all factors observed was the limp after run and slight compensation for ankle while running
- ROM: was not quite normal for a healthy ankle, ROM exercises helped
- No depression, abnormal social contact, excuses were detected during observation

Most of the aforementioned results ranged in the middle quadrant, which is a good sign showing he has no extreme psychological problems or outliers.

Subject: "Joe S", Wrestler for The Pennsylvania State University **Date of Observation: Monday, April 16, 2007** **Location: Athletic Training Room, Recreation Hall** **Injury: Anterior Cruciate Ligament of Right Knee**	
Injury History: Suffered an injury to ACL of right knee during 2004-2005 competitive season. After successful surgery and rehabilitation, returned to wrestle during 2005-2006 season. During team practice in late November 2006, injured ACL of left knee. Had surgery to repair ligament of left knee on December 26, 2006.	
REHABILITATION SCHEDLUE	
Monday, Thursday	"Hard Day" focusing on Strength Training
Tuesday, Friday	"Light Day" focusing on Balance and Flexibility
Wednesday	OFF

OBSERVABLE BEHAVIORS **(FOCUS ON PSYCHOLOGICAL TRAUMA RESULTING FROM INJURY)**		
Abnormal Behavior	**Behavior Present in Athlete?**	
	Yes	**No**
Bracing Behavior During Exercise (Such as asymmetric movements)		X
Evidence of Fear During Exercise		X

(Based on signs of an irregular heart beat, rapid breathing, nausea, overall feelings of dread, etc.)		
<u>Fear of Injury/Fear of Failure Beliefs</u> (Devaluing one's self esteem, fear of reduction in social value, crushed hope, uncertain future, etc.)	**X**	
<u>Evidence of Physical Pain</u> (Based on vocal and facial expressions, gestures, posture, and locomotion)		**X**
<u>Treatment Complications</u>	**X**	
<u>Compliance with Treatment</u>	**X**	

Further explanations of the Joe S's rehabilitation plan and observable behaviors can be found in the following chart:

Date	Physical Signs	Emotional Signs	Rehabilitation	Recommendations
1/15	Limping on left leg, wincing and pain during normal movement, unable to complete full practice	Signs of depression and frustration, vocally expressing pain when pressure on left leg, anger about injury and not being able to play	Work with physical therapist, imagery exercises, worked with athletic trainer, x-rays on ankle	Keep off leg, not ready to return to play, keep up with rehabilitation
1/16	Using crutches and brace to aid with walking/healing of injury, sat out of practice	Signs of depression and frustration still evident, work with physical therapist seemed to cheer him up	Work with athletic trainer, lifted weights and uninjured leg, fitted for a brace for ankle	Keep brace on during any type of physical activity, continue to train and attend practices
1/17	Easier for athlete to walk with crutches, still complaining of sore ankle and inflammation	Signs of depression, acceptance of injury, trying to deny pain to coaches	Work with physical therapist, imagery exercises, begin to return to	Apply minimal pressure to ankle, slowly return to practice if athlete feels up to it, if any signs of damage occur cease play

			practice slowly	
1/18	Bracing behavior on the court today, limping on leg even though using brace, may be due to failure to use crutches	Determined to return to play despite having serious injury – coach recommended sports psychologist	Continued work with athletic trainer and physical therapist, adjustments made to brace to improve fit and mobility of ankle	Athlete should slowly and carefully return to play, ankle seems to be taking a turn for the worse, athlete would benefit from seeing a sports psychologist
1/19	No physical activity today on court, limping when not using crutch, new brace seems to have helped ankle mobility	signs of frustration very evident – anger about slow healing process	Visit to sports psychologist Break from athletic trainer and physical therapist today at recommend-ation of coach	Meeting with psychologist on a regular basis will help athlete during recovery period
1/20	No physical activity on court, ankle sore and inflamed so mobility very limited	Anger and frustration – but seems to have subsided	Work with physical therapist, ankle iced to reduce swelling Visit to doctor for more x-rays	Minimal physical activity until swelling in ankle subsides
1/21	Swelling in ankle down, athlete observed practice and participated for minimal amount of time, limping minimal	Optimistic about improvement	Work with physical therapist and athletic trainer, meeting with coach to discuss healing process	Continue meeting with physical therapist and athletic trainer on a daily basis, slow return to sport advised, minimal pressure should be applied to ankle and brace should be worn at all times

Observational Results

Behaviors Examined	Description
Bracing Behavior	A self-protecting mechanism resulting in the athlete's improper execution of skill due to fear of re-injury.
Facial Expression	Any expression showing pain, fear, or fatigue due to fear of re-injury.

	Description of Exercise Session
Session 1	Practice from 2:30-5:00 p.m. Ran for approximately 40 minutes and completed strength exercises with weights.
Session 2	Cross training using bicycle for a total of 30 minutes.
Session 3	Practice from 2:30-5:00 p.m. Ran for approximately 40 minutes and completed strength exercises with weights and long sprints.

Behavior	Session 1	Session 2	Session 3
Bracing Behavior	None observed.	None observed.	None observed.
Facial Expression	Expressions were observed approximately two hours into duration of workout, mainly in the latter portion of subject's running exercises. Expressions showed exhaustion and were likely a result of normal workout fatigue.	None observed.	Expressions observed approximately one and a half hours into duration of workout. Expressions were similar to those observed in session one.

Dates / Times / Sites of Observations

Sat. March 10th	Sun. March 11th	Monday March 12th	Tuesday March 13th	Wed. March 14th	Thursday March 15th	Friday March 16th	Sat. March 17th	Sun. March 18th
Night Game Indoor Soccer Arena	**Rest**	**Practice** Dover Area High School	**Gym** Dover Area High School	**Practice** Dover Area High School	**Gym** Dover Area High School	**Rest**	**Night Game** Indoor Soccer Arena	**Rest** Public Observation

Observation Injury Chart

Name	Lower/ Upper Extrem-ities	Type and Loca-tion of Injury	Pain Assessment	Bracing Behaviors	Psychological Trauma
S.1 John	Lower Extremities	-Acute Patella Injury of the left knee	-Swelling in the knee joint -A sharp or stinging Acute Pain around the patella	-Slow when moving to a squatting position, i.e., using arms to guide down into a chair/car -Careful when making a downward step, i.e., using arms to brace when walking down a flight of stairs	**Emotional Trauma** -Depression, Irritability Anxiety, Anger and Resentment -Compulsive and Obsessive behaviors **Physical Trauma** -Sleep Disturbances (less than usual)
S.2 Matt	Upper Extremities	-Sprained Index Finger on the left hand	-Minor swelling over the joint -Sharp or stinging Acute pain when bending the finger and stressing the injured ligament	-Hiding/ Covering Hand, i.e., shields finger when ball is kicked at hand -Slow when grabbing or bracing with right hand, i.e., picking up a cup or opening a door	**Emotional Trauma** -Irritability

Name	Lower/ Upper Extrem- ities	Type and Location of Injury	Pain Assess- ment	Bracing Behaviors	Psycho- logical Trauma
S.3 **Tim**	Lower Extremities	-Lateral Cartilage Meniscus Inflammati on of the left knee	-Tenderness along the joint line of the knee -A Chronic Pain to the outside of the knee which is easily aggravated when running for long durations	-Careful when walking up an incline or down a declined surface, i.e., braces going up and down stairs; i.e., braces going up and down hills	**Emotional Trauma** -Mental Fatigue/ Anxiety **Physical Trauma** -Low Energy
S.4 **Just.**	Upper Extremities	- Concussion or Mild Traumatic Brain Injury	-Prolonged Pain in the area of the trauma which is usually associated with mild dizziness and headaches	-Hesitation to make certain plays, i.e., jumping to head the soccer ball when an opponent also jumps up for it simultaneously -Preparing for Impact i.e., closing eyes before heading the ball i.e., bracing/ freezing before impact with opponent	**Emotional Trauma** - Depression, Fearfulness **Cognitive Trauma** -Inability to follow coaching instructions (slight confusion) -Inability to make prompt decisions

Psychological discomfort due to persistent pain could be observed via:

Vocal	Sighing, yelling, making some noise at the signs of pain or discomfort.	5-6 times while kicking the ball to shoot, sighing while running after being on the foot for a while. 2-3 times while passing the ball. He made grunting noises and moaned a little after being on the foot for a while.
Facial	Grimacing, facial expressions that are not usual and seem to indicate pain or that the injured body part is still hurting.	Facial expressions of pain were present throughout the entire game. Every time he was running on it or kicked the ball, and once or twice when collisions occurred when he and another player were kicking the ball at the same time.
Gestures	Rubbing, touching, going through some type of movement or touching of the injured area in an effort to reduce the pain or discomfort.	Not really present during the scrimmage; after he was taken out, he took off his cleats, sat down and was rubbing the area on his foot and ankle. He did this and avoided moving.
Posture	Stiffness, compensations, bracing behaviors, guarding, anything done to prevent use of injured area or ease the pain and stress on that part of the body.	Many times throughout the scrimmage he avoided putting weight on his bad foot. He avoided collisions, and was running and walking with a limp to avoid putting more weight than necessary on that foot. Also went out of his way to use right foot instead of left when kicking the ball.

Upper Extremity	Type and location of injury	Rehabilitation Exercises	Pain Assessment	Bracing Behaviors	Psychological Trauma
Shoulder	Rotator Cuff	Passive range of motion exercises and active assisted range of motion exercises to repair	Pain Rating Index	Constant shrugged shoulder, stiffness	Depression

		tissue			
Shoulder	Rotator Cuff	Full passive motion; begin to strengthen the muscles that stabilize the shoulder blade	Pain Rating Index	Reluctant to lift arm above head during stretching	Anti-social and in denial of how bad injury is
Shoulder	Rotator Cuff	Passive stretching beyond the patient's own range of motion; more strengthening of the stabilizers of the shoulder blade	Pain Rating Index	Stiffness and facial expressions of pain during stretching exercises	Slowly understanding severity of injury; acceptance starting

There are numerous options to develop sport and injury specific observational formats. The most important issue is to find a way to reveal both overt and covert signs of fear of movement, perceived pain and/or discomfort that may debilitate the well-being of injured athletes. Another final note to consider is that there are possible errors that may jeopardize the validity and reliability of observation of injured athletes. Among them are: reactivity (e.g., athletes may react to the fact that they are being observed and modify their behavior accordingly); lack of ecological validity (e.g., truly representative samples of behavior); biases (e.g., anticipation of problem behavior that may not be present), to note just a few.

CONCLUSIONS

Overall, there two major types of fear in humans. One is the fear of physical harm and possible disability, even death, due to exposure to dangerous situations. Fear response associated with this type is usually processed at the "primitive" limbic system. There are numerous situations, especially in an athletic environment, in which people can get actually hurt. Not surprisingly, the physical risk associated with sports and recreational activities may build a "fear reflex" as a reaction to physical danger. This might be exaggerated when there is a previous injury resulted in sports participation—this most often leads to "Kinesiophobia," which is defined as a fear of movement due to anticipated pain and discomfort. The other type of human fear is the fear of looking foolish in the eyes of individual peers and being ridiculed because of it. One way or another, there are numerous links and consequences of fear including the development off loss of confidence, reduced self-esteem, and abnormal perceived risk of certain activities. In some cases the situation could even be worsened, leading to generalized and/or specific phobias with long-term residual psychological/psychiatric disorders. With special training and experience, the aforementioned signs and symptoms of fear of injury, which is most often exaggerated far beyond the actual threat and danger, can be accurately assessed, observed and properly treated.

REFERENCES

Dunn, G.H. (1999). A theoretical framework for structuring the content of competitive worry in ice hockey. *Journal of Sport & Exercise Psychology, 21,* 259-279.

Crossman (1985). Psychosocial factors and athletic injuries. *Journal of Sport Medicine and Physical Fitness, 25,* 151-154.

Flint, F.A. (1998). Integrating sport psychology and sport medicine in research: The dilemmas. *Journal of Applied Sport Psychology, 10,* 83-102.

Macchi, R., & Crossman, J. (1995). After the fall: Reflections of injured classical ballet dancers. *Journal of Sport Behavior, 19,* 221-234.

Magyar, M.T., & Chase, M.A. (1996). Psychological strategies used by competitive gymnasts to overcome the fear of injury. *Technique, 16,* Retrieved February 14, 2001.

Kontos, A. P., Feltz, D. L., & Malina, R. M. (2000). The perception of risk of injury in sports scale: Confirming adolescent athletes' concerns about injury. *Journal of Sport & Exercise Psychology, 22,* S12.

Vlaeyen, J.W., de Jong, J., Geilen, M., Heuts, .PH., van Breukelen, G. (2002). The treatment of fear of movement/(re)injury in chronic low back pain: further evidence on the effectiveness of exposure in vivo. *Clinical Journal of Pain, 18,* 251–61.

Severeijns, R., Vlaeyen, J.W., van den Hout, M.A., Weber, W.E. (2001). Pain catastrophizing predicts pain intensity, disability, and psychological distress independent of the level of physical impairment. *Clinical Journal of Pain, 17,* 165–72.

van den Hout, J.H., Vlaeyen, J.W., Houben, R.M., Soeters, A.P., Peters, M.L. (2001). The effects of failure feedback and pain-related fear on pain report, pain tolerance, and pain avoidance in chronic low back pain patients. *Pain, 92,* 247–57.

Crombez, G., Eccleston, C., Vlaeyen, J.W., Vansteenwegen, D., Lysens, R., Eelen, P. (2002). Exposure to physical movements in low back pain patients: restricted effects of generalization. *Health Psychology, 21,* 573–8.

Boersma, K., Linton, S., Overmeer, T., Jansson, M., Vlaeyen, J., de Jong, J. (2004). Lowering fear-avoidance and enhancing function through exposure in vivo. A multiple baseline study across six patients with back pain. *Pain, 108,* 8–16.

Roelofs, J., Goubert, L., Peters, M.L., Vlaeyen, J.W., Crombez, G. (2004). The Tampa scale for kinesiophobia: further examination of psychometric properties in patients with chronic low back pain and fibromyalgia. *European Journal of Pain, 8,* 495–502.

Peters, M.L., Vlaeyen, J.W., Weber, W.E. (2005). The joint contribution of physical pathology, pain-related fear and catastrophizing to chronic back pain disability. *Pain, 113,* 45–50.

Miller, R.P., Kori, S.H., Todd, D.D. (1991). The Tampa Scale of Kinesiophobia. Ref Type: Unpublished work.

Swinkels-Meewisse, I.E., Swinkels, R.A., verbeek, A. et al. (2003). Psychometric properties of the Tampa Scale for Kinesiophobia and fear avoidance beliefs questionnaire in acute low back pain. *Manual Therapy, 8*(1), 29-36.

Moss, R., Slobounov, S. (2006). Neural, behavioral and psychological effects of injury in athletes. In. S Slobounov & W. Sebastianelli (Eds.), Foundations of sport-related brain injuries, (pp.407-430). Springer.

Gravetter, F.J., Larry, B., Wallnau, B. (2000). Statistics for the behavioral sciences: a first course for students of psychology and education. 2nd ed. West Pub. Co. St. Paul. MN.

Martin, S.B. (2005). High school and college athletes' attitude toward sport psychology consulting. *Journal of Applied Sport Psychology, 17*(2), 127-140.

Messner, M.A. Power at play: Sports and the problem of masculinity. Boston: Beacon Press, 1992.

Nixon, H.L. (1996). Explaining pain and injury attitudes and experiences in sport in terms of gender, race, and sports status factor. Jounrnal of Sport and Social Science, 20, 33-44.

Addis, M.E., & Mihalic, J.R. (2003). Man, masculinity, and the contexts of help seeking. *American Psychologist, 58*(1), 5-14.

Keefe, F.J., Bradley, L.A., Crisson, J.E. (1990). Behavioral assessment of low back pain: identification of pain behavior subgroups. *Pain, 40*(2), 153-160.

De Mondaigne, M. The Complete works: Essays, Travel Journal, Letters. Alfred A. Knopf New York: Everyman's library, 2003.

Phares, J. Clinical Psychology: Concepts, Methods & Prodession. 3rd edition. The Dorsey Press, 1988.

CHAPTER 14

MULTIPLE FACETS OF PAIN DUE TO INJURY

1. INTRODUCTION

According to the International Association for the Study of Pain (IASP), pain is defined as an unpleasant sensory and emotional (conscious) experience associated with actual and potential tissue damage (www.isap-pain.org). Although unpleasant, pain plays an important role as a warning signal of harm giving rise to a number of physiological, physical, psychological and behavioral responses to injury, and initiating the processes necessary for recuperation and repair. Therefore, pain should be considered as a highly adaptive mechanism aimed at preventing further harm to the organisms and exaggerating the injury. Due to the highly subjective nature of pain, depending on individually perceived and psychologically appraised situations, pain can occur even in the absence of tissue damage, or when tissue damage is completely healed. This is the so-called "neurotic pain" phenomenon, which remains poorly understood and presents a real challenge for clinical professionals. There are numerous peripheral and central nervous system substrates clearly distinguishing "pure pain" and "affective-motivational" pathways. These well-documented and clearly indentified pain pathways are indeed in support of the notion that both physical and psychological pain are an adaptive and/or maladaptive response to harmful stimuli. Problems with "pain" studies is that there are multiple confounding factors such as age, gender, history of painful experiences and injuries, underlying disabilities that may or may not be associated with painful experiences, social and cultural norms about the acceptance of pain behavior, environmental circumstances, etc., that influence a person's pain responses.

The most striking feature of the pain phenomenon is that unlike other sensory modalities (i.e., visual, auditory, olfactory, etc.) pain is not subject to "habituation." In other words, in most sensory modalities the constant repetitive stimuli induce a reduced sensitivity resulting in diminished responses. No such habituation occurs in pain sensation/perception. In fact, "sensitization" and/or "*hyperalgesia*" may occur, a phenomenon that is characterized by increased sensitivity to noxious stimuli in the case of repeated stimulation. In the case of traumatic injury, a painful stimulus associated with tissue damage in the area of injury and the surrounding regions would be sensed and/or perceived as more painful, a phenomenon referred to as *hyperalgesia*. As an example of *hyperalgesia*: a slight mechanical pressure on the area surrounding the tissue damage, which used to be a neutral pain free stimulation, may induce significant pain response after injury. Pain is also "perceptual" in nature, referring to the conscious processing of noxious stimuli, involving both high order cognitive and affective/motivational structures. An example of altered "central sensitization" most likely involving high-order perceptual properties and induced by tissue damage is another pain phenomenon, called "*allodynia.*" Specifically, stimuli that under normal conditions (i.e., prior to traumatic injury) would be innocuous may activate *nociceptive* inputs, and give rise to a perception of pain. Therefore, pain as a phenomenon cannot be fully understood unless a multimodal approach is implemented.

Pain as a unique human experience has been a subject for scrutiny for ages. There are numerous quotes on pain demonstrating that this unfortunate human experience is one of the most puzzling, complex and multimodal phenomenon even seen:

- "Given the choice between the experience of pain and nothing, I would choose pain." **William Faulkner**
- "Pain is such an uncomfortable feeling that even a tiny amount of it is enough to ruin every enjoyment." **Will Rogers**
- "Pain is inevitable. Suffering is optional." **Anonymous**
- "We cannot learn without pain." **Aristotle**
- "Those who do not feel pain seldom think that it is felt." **Dr Samuel Johnson**
- "Pain is no evil unless it conquers us." - **George Eliot**
- "The art of life is the art of avoiding pain." - **Thomas Jefferson**
- "Nothing begins, and nothing ends, that is not paid with moan; for we are born in other's pain, and perish in our own." - **Francis Thompson**
- "Nobody is hurt. Hurt is in the mind. If you can walk, you can run." - **Lombardi, Vince (1913-1970)**
- "How much pain has cost us the evils which have never happened." **Thomas Jefferson**
- "What a distance runner learns to do is deal with pain. Learning to deal with pain is learning to deal with life." - **Unknown**
- "To banish cares, scare away sorrow and soothe pain is the business of the poet and singer." - **Bodenstedt**
- "It is easier to find men who will volunteer to die, than to find those who are willing to endure pain with patience." - **Julius Caesar**
- "All pain is either severe or slight; if slight, it is easily endured; if severe, it will without doubt be brief." - **Marcus T. Cicero**

- "He who makes a beast of himself gets rid of the pain of being a man." - **Samuel Johnson**
- "There is no coming to consciousness without pain." - **Carl Jung**
- "Even Pain pricks to livelier living." - **Amy Lowell**
- "The only antidote to mental suffering is physical pain." - **Karl Marx**
- "When there is pain, there are no words. All pain is the same." - **Toni Morrison**
- "Life is pain and the enjoyment of love is an anesthetic." - **Cesare Pavese**
- "No pain, no palm; no thorns, no throne; no gall, no glory; no cross, no crown." - **William Penn**
- "The aim of the wise is not to secure pleasure, but to avoid pain." **Aristotle**

2. NEUROANATOMY OF PAIN PROCESSING

Historically, pain as an essential attribute of acute injury has been usually examined in a manner of sensory channels similarly to other somatosensations derived from activation of mechanoreceptors, chemical and thermoreceptors. More recently, pain was conceptualized as a complex phenomenon and therefore needs to be considered from at least three mutually dependent and interacting dimensions including: (a) sensory-discriminative; (b) cognitive-evaluative; and (c) motivational-affective perspectives (Melzack & Casey, 1968).

There several structural units involved in the initial sensation of pain (e.g., nociceptors), transduction of nociceptive signals to the higher levels of the CNS and central processing of the pain signals terminated in the cortex and sub-cortical structures. An overall schematic of these structural unities is shown below.

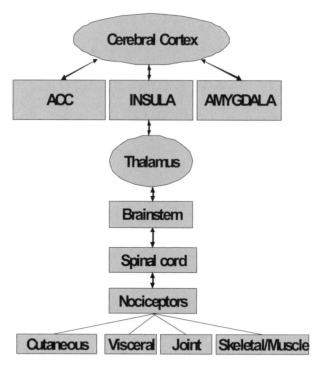

Figure 1. Pain processing at the different level of the CNS.

2.1. Nociceptors

From bottom-up, the specified nerve cell endings that are sensitive to tissue damage and pain are called "Nociceptors" (from the Latin word *nocere*, to injure). Like other receptors, *nociceptors* transduce a variety of stimuli into receptor potentials, which in turn trigger afferent action potentials. Nociceptors are present in the skin (i.e., cutaneous receptors) and in other body tissues, including bone, muscle, joint, visceta and blood vessels. In should be noted that nociceptors have NOT been found in brain and spinal cord tissues, indicating that brain is "painless," in a sense. There are two major nerve fibers known to transduce nociception to the dorsal horn of the spinal cord and then to super-spinal structures, including: (a) small-diameter myelinated relatively fast A fibers with a conduction velocity of 5-30 m/s; and (b) small-diameter unmyelinated relatively slow C fibers with a conduction velocity of about 2 m/s. Thus, even though the conduction of all nociceptive signals is relatively slow (for example, compared to thermal reception), there are fast and slow pain pathways. The activation of fast myelinated fibers is linked to a highly localized sensation of sharp pain or other dangerously intense mechanical and mechanothermal stimuli. The relatively slow unmyelinated fibers convey a diffuse pain sensation that is dull, in response to thermal, mechanical and chemical stimuli. These fibers are said to be *polymodal* in nature. There is a notion of two categories of pain perception due to activation of different A and C fibers, such as a sharp *first* pain and more delayed, diffused and longer-lasting *second* pain. The details regarding the mechanosensitive (A), mechanothermal (A), polymodal (C), as well as the classification of nociceptors with respect to the pain's location (cutaneous, joint, visceral, skeletal muscle) can be found elsewhere (see: Galea, 2002).

2.2. Pain Pathway

The somas of nociceptive afferent fibers are located in the dorsal root ganglion (**DRG**) of the spinal cord. The DRG is the first site for the processing of input from nociceptors. There are three different regions of the DRG and six laminae with different cytoarchitectonic criteria and chemical profiles. The incoming afferent fibers are differentially segregated and terminated in the DRG according to their size. The ascending pathway is further from the **dorsolateral track of Lissauer** and **spinothalamic track** (STT), which are the major systems conveying information about pain and temperature. This overall pathway is also referred to as the anterolateral system. Similarly, the mechanosensory pathway is referred to as the dorsal column-medial lemniscus system. In the brainstem, the STT passes dorsolateral to the inferior olivary mucleus in the **medulla**, then ascends dorsolateral to the medial lemniscus, via higher levels of the brainstem to the **thalamus**. After entering the pons, there are small myelinated and unmyelinated trigeminal fibers descending to the medulla and forming the **spinal trigeminal track** (e.g., it is commonly known as *spinal track of* *cranial nerve V*). Overall, somatic sensation of the head and oval cavity is conveyed by the trigeminal nerve, which is the largest nerve and consists of four nuclei: (a) the motor nucleus; (b) the main sensory nucleus; (c) the spinal nucleus; and (d) the mesencephalic nucleus. For example, the ascending pathway from the spinal nucleus mediates facial and dental pain.

The **spinoreticular track** (SRT) originates in the RDG and ascends to the ventral lateral column of the spinal cord and further to the brainstem. The final destination of the SRT is a number of nuclei of the reticular formation, including: (a) nucleus medullae oblangatae centralis; (b) lateral nucleus; (c) nucleus reticularis gigantocellularis; etc. The important role of the SRT underlying the affective-motivational aspect of pain processing is well documented both in animal studies and in humans. There are also well-defined and somatotopically organized tracks including: (a) the **spinomesencephallic track** (SMT); (b) the **spinocervical track** (SCT); and the **dorsal column-medial lemniscal track** consistent numerous afferent fibers and conveying various information (tactile, proprioceptive input, etc., in conjunction with nociceptive information) from different parts of the body (e.g., trunk, lower and upper extremities, etc.).

The **thalamus** represents the final link in conveying pain signals from the peripheral structures to the cerebral cortex. Specifically, there are six groups of thalamic nuclei involved in pain processing, including: (a) lateral (concern with motion functions); (b) medial (having links with the limbic system); (c) anterior (having links with ACC); (d) intrathalamic; (e) midline; and (f) reticular (receive inputs from reticular formation in the brainstem). It should be mentioned that an important role of the **hypothalamus** is to produce responses to emotional changes and needs, thereby providing necessary homeostasis. The hypothalamus is also known for its role in the control of sympathetic and parasympathetic functions. In addition, the hypothalamus receives inputs from the reticular formation and the amygdala. Therefore, it is reasonable to suggest the involvement of the hypothalamus in the affective-motivational pathway of pain processing.

Limbic structures have been previously discussed within the scope of fear (see Chapter 12 of this part). To reiterate, the "limbic system" consists of the cingulated gyrus (ACC), the parahyppocampal gyrus, hyppocampal formation in conjunction with prefrontal cortex and amygdale, and is sometimes called the "fear triangle." This notion came from the pioneering work of Papez (1937) suggesting that the limbic lobe formed a special neural circuit providing the anatomical substrate of effective responses (i.e., emotions). With associated with pain processing, it is relevant to note that ablation of the amygdala and the overlying cortex in animals and in humans causes not only reduction of fear responses but also reduction of responsiveness to noxious stimuli (Schreiner & Kling, 1953). Also, regarding the involvement of ACC in pain processing, it is worth mentioning that surgical section of the cingulum bundle was carried out to provide relief in cases of intractable pain, such as that from an inoperable carcinoma. Interestingly, patients were still able to perceive pain stimuli postoperatively, but were unconcerned about it (cited from Strong et al., 2002, p.29).

Cerebral Cortex, and specifically the primary somatosensory area of the cerebral cortex, is the major destination area of the thalamocortical projections from the ventral thalamus involved in the central circuit of pain processing. There are functionally different neurons in the somatosensoty cortex (Broadmann areas # 3, 1 and 2) receiving proprioceptive input as well as the signals from deep pressure and cutaneous receptors. In conjunction with nociceptive information, these inputs provide the major source of pain perception and active "fight-flight" responses. It should be noted that in addition to the primary somatosensory cortex, multiple other cortical areas are activated by painful stimuli, including the secondary somatosensory cortex (SII), the insula, the prefrontal cortex and the supplementary motor cortex. These and possibly other brain areas may underlie the neurobiological bases of both conscious and subconscious pain perception and processing. Overall, as can be concluded from the previous text, multiple cortical areas are activated by painful stimuli. Accordingly, pain signals may be considerably modified via both ascending (e.g., gate theory) and descending projections from the brainstem and cerebral cortex. The topographical organization and functional "responsibilities" of these distributed cerebral activations indeed reflect the complexity of the pain mechanism involving discriminative, affective, autonomic and sensory-motor components.

3. TYPES OF PAIN

There are lest four types of pain as result of injury commonly identified in clinical practice, including *acute* pain, *prolonged* pain, *chronic* pain and **neuropathic** pain. *Acute* pain is defined as an inherent biological function and a warning sign of actual and/or potential physiological harm (Melzack & Wall, 1988). This type of pain usually occurs in response to an injury or other painful stimulus, but goes away as soon as the injury heals or the stimulus is removed. Acute pain is commonly defined as being less than six weeks in duration. That said, it is important to note that patients may have recurrent attacks of acute pain beyond six weeks, especially when the treatment protocol is in violation. According to Melzack (1987), the properties of acute pain commonly expressed by victims of injury include but are not limited to:

- Stabbing;
- Sharp;
- Cramping;
- Shooting;

- Hot-burning;
- Gnawing;
- Aching, etc.

Some emotional components associated with patients' description of pain include:

- Fearful;
- Sickening;
- Tiring/exhausting;
- Tender;
- Splitting;
- Heavy;
- Light;
- Punishing;
- Horrible;
- Distressing;
- Annoying;

In case of tissue damage, if properly treated, acute pain usually diminishes long before healing is completed and may reoccur if the treatment protocol is insufficient. For example, sharp and prolonged pain may reappear if a injured knee joint was overloaded during a rehabilitation session. Thus, acute pain may serve as an index of "how much is too much" in terms of exercise duration, intensity and volume.

Most often, inefficient treatment protocol delays the healing process and is associated with a prolonged *inflammation* period. Accordingly, ***prolonged*** pain was identified as following the acute injury phase with the main function of preventing unnecessary disturbance during the healing process. ***Hyperalgesia*** (i.e., increase in pain elicited by sub-threshold mechanical, chemical, temperature noxious stimuli), and ***allodynia*** (i.e., pain evoked by normally innocuous stimuli, often reported after acute injury, defined as increased sensitivity to external stimuli – to protect damaged area and facilitate healing) are prominent features of the prolonged pain. There is a great risk for patients suffering from prolonged/acute pain beyond six weeks to develop a chronic pain and associated disability. It should be noted, however, that the distinction is NOT the duration of the pain, but rather the persistence of pain beyond the expected recovery time and the intractable nature of the pain (Waddell, 2004, p.33).

Unlike acute and/or prolonged pain, ***chronic*** pain was defined to describe pain that continues beyond what is normally expected for illness or injury healing, pain that occurs off and on over a period of months (three to six months post-injury) or years, and pain that is persistent, but has no identifiable cause. The examples of chronic pain include, but are not limited to, phantom limb pain and recurrent episodes of migraines. As Loeser (1980) mentioned in his perspective of pain, "acute and chronic pain have nothing in common but the four letter word pain." The clinical presentation and signs of chronic pain seem to be dissociated from particular tissue damage, most often with little evidence of nociception. Accordingly, attempts to continue to treat tissue damage with the intention of abolishing pain usually do not relieve the patient's suffering of chronic pain, but may actually reinforce acute pain and perpetuate the existing problem. As further concluded by Loeser & Melzack (1999, p.1609), "it is not the duration of pain that distinguishes acute from chronic pain but, more importantly, the inability of the body to restore its physiological functions to normal homeostatic levels."

An interesting approach to exploring the differences between acute and chronic pain was proposed by Sternbach (1977). Specifically, he linked acute pain to the sympathetic reaction of "fight versus flight." Accordingly, acute pain may have biological meaning of and value as a warning sing of tissue damage. Accordingly, acute pain and anxiety are closed linked in such a way that anxiety may help to reduce pain. When acute pain is prolonged for some reason, a pattern of "vegetative changes" may emerge, characterized by the appearance of numerous psychological/emotional residuals, including sleep and appetite disturbance, emotional lability, irritability, to name a few. In this stage of injury evolution, chronic pain loses its biological meaning to protect the organism from further tissue damage and becomes counterproductive. Not surprisingly, patients with chronic pain syndrome often experience feeling of helplessness and tend to be withdrawn from social activities. Overall, chronic pain is associated with suffering and may lead to considerable disability with numerous psychological, behavioral and environmental consequences. Thus, chronic pain is highly influenced by a patient's mental status and surrounding environment, therefore, psychological rehabilitation is a critical part of the clinical protocols for dealing with chronic pain syndrome.

Neuropathic pain is defined as "… any pain syndrome in which the predominating mechanism is a site of aberrant somatosensory processing in the peripheral or central nervous system." (Wright, 2002, p.352) It is a real challenge to understand the pathophysiological and psychological mechanisms contributing to the development of neuropathic pain syndrome. As a result, standard treatment protocols to control acute pain, including non-steroidal and anti-inflammatory drugs and opioids may fail, thus patients continue to experience pain and suffering. In the case of chronic neuropathic pain, the initial damage to the nociceptive system may be resolved, but pain continues or is even worsened (Braune & Schady, 1993). In terms of etiology, some neuropathic pain states may develop due to initial damage to the central nervous system (i.e., direct trauma) and are described as "central pain states." The others arise due to trauma to the peripheral nervous system and are defined as "peripheral pain states." Most common types of pain symptoms include allodonia to touch (i.e., cold stimuli may induce a sensation of burning pain), complete lack of sensitivity in the area of damage, but pain is experienced (i.e., anaesthesia dolorisa), to name just a few. Most often, in addition to pain, other forms of altered sensation are present in patients suffering from neuropathic pain syndrome. Several neuropathic pain mechanisms have been identified and described by Wright (2002), including:

1   *Neuroma formation*: Neuropathic pain associated with the development of a neuroma at the site of nerve injury. As a result, random activation arrived from the neuroma is perceived as a pain and consciously appraised as arising from the area of innervation;

2   *Phenotype change*: Nerve injury due to trauma may result in phenotype change in the peripheral nerve in such a way that myelinated A afferent fibers adopt properties that are similar to those of unmyelinated C fibers, and are therefore unable to generate central sensitization;

3   *Central sensitization*: Central sensitization describes the changes occurring at the cellular level within the framework of the notion of neural plasticity. This process is initiated by activation of peripheral nociceptors. In the case of neuropathic pain, central sensitization can be sustained in the absence of peripheral nociceptor input.

4   *Neuroanatomical reorganization*: In this situation, some cells, which previously received inputs from the denervated region, begin to respond to stimulation of other body parts. Patients may develop receptive fields in more proximal, innervated regions of the affected limb, or other regions of the body (i.e., patients may experience phantom pain, burning toe pain of amputated leg while talking).

5   *Disinhibition*: For example, peripheral nerve injury may result in disinhibition of spinal cord neurons, because firing of the dorsal horn projection neurons may be influenced not only by excitatory inputs from the periphery but also by input from the brain that may be both excitatory and inhibitory.

4. PAIN VERSUS DISABILITY

One of the important conceptual issues with a huge implication for clinical practice in dealing with injured individuals is that pain and disability often go together. However, pain should not be confused with disability (Loeser, 1980). Although, both of these phenomena possess four major attributes, including:

• *Nociception* refers to mechanical, chemical, thermal, etc., excessive stimuli causing tissue damage. These stimuli activate pain receptors to produce activation of nerve fibers that are further processed to the brain via two distinct pathways: (a) a pure sensation pathway conveying nociceptive information; (b) an affective-motivational pathway, signaling the unpleasant quality of pain (to be discussed later in this chapter);

- *Pain* per se refers to the perceived sensation of pain, meaning that both sensory information and high-order cognitive mechanisms should be involved in the appraisal of pain signals as a warning sign of possible harm to the organism;

- *Suffering* refers to an unpleasant emotional response generated in the higher nervous system (most likely in reticular formation, prefrontal cortex, amygdala and insula) by pure painful stimuli and emotional agents. Suffering, as often evidenced by grief, stress, depression, anxiety, may or may not be triggered by physical pain per se. Therefore, it is important to stress that pain and suffering, although connected, are quite different substrates of unfortunate human experiences. There are may be pain without suffering, and vice versa.

- *Pain behavior* refers to any covert and overt acts and expressions suggesting the presence of pain. These include both verbal (i.e., talking and moaning), and other forms of communication, such as facial expressions, limping, bracing, asymmetrical postures, taking medications, seeking health care, etc. Both voluntary (e.g., conscious awareness and appraisal) and involuntary (e.g., sub-conscious, dreams, etc.) processes may be indicative of pain behavior.

Pain and its cognitive interpretation of personal meaning and consequences of injury, along with a person's belief, emotional status and coping resources are integral part of the pain experience that may or may not lead to disability. **Disability** is referred to as restricted activity due to injury. According to the World Health Organization (WHO, 1980), disability is defined as "Any restriction or lack (relating from an impairment) of ability to perform an activity in the manner or within the range considered normal for a human being," not necessarily due to pure pain. Similarly, disability is defined as "...an alteration of an individual's capacity to meet personal, social or occupational demands because of impairment" (American Medical Association, 2000), not necessarily due to pain. The new International Classification of Functioning, Disability and Health (ICF) defined *activity* as "something a person does, ranging from vary basic elementary simple to complex," unlike *activity limitation*, which is defined by the ICF as "difficulty in the performance, accomplishment, or completion of an activity." Again, the core of these definitions is NOT pain and suffering but disability as a *restricted activity* that may or may not be due to pure pain. As was stressed by Waddel (2002), the simple model (see diagram below) of how most medical professionals and patients conceptualize pain and disability is too simplistic. Specifically, "pain is a symptom, not a diagnosis, not a disease".

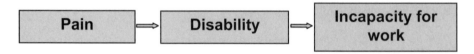

Some patients may have pain but can tolerate it and therefore have limited restricted activity or disability. Numerous examples in the athletic environment suggest that athletes were able to finish the game and then went to the doctor for treatment of bone fractures. Other patients have disability far beyond the proportion of their pain. The fundamental implication for clinical practice when dealing with the pain syndrome versus patients' suffering from disability is that:

- Pain and suffering resulting in restricted activity and possible disability may diverge greatly from peripheral nociception;
- Multiple other afferent impulses and neural activities at different levels of the CNS may modify the pain experience;
- Pain transmission may be modulated by various chemical substances and/or physical exercise which act as analgesics, e.g., opiates;
- Patient interpretation of pain and experienced restricted mobility due to pain may also reflect anticipation of harm, fear of pain and suffering; Therefore the way patients report pain will always be influenced by their mental status, direction of thoughts, belief and emotions.

5. PAIN MODULATION

5.1. Gate Control Theory

The conceptual distinction between pure pain and disability is based on well-documented empirical evidence (Melzack, 1996) and clinical observations that pain signals originated within the nociceptive system do not pass unaltered to the brain but are always modulated at the different levels of the CNS before reaching the conscious awareness. This notion was formulated within the scope of the famous ***gate control theory*** (Melzack & Walls, 1965). According to this theory, the stimulation of nociceptors produces signals in peripheral nerves entering the *dorsal root ganglion* in the spinal cord. Then, the dorsal horn ganglion acts as a "gate control" mechanism. Specifically, the interactive effect or balance between different types of fibers (e.g., myelinated A fibers and unmyelinated C fibers) may stimulate or inhibit the transmission of pain signals to the next ascending stage of pain information processing. In other words, the gate to the brain for processing pain can be closed or opened depending on the interactive activation of different nerve fibers. Consequently, sharp pain, aching pain, dull pain and/or no pain at all could be reported the by victim of injury even though tissue damage was present at the acute stage of injury. This type of modulation of pain is called up-regulation of nociceptive system functions.

5.2. Modulation of Nociceptive Input

Additional support for the notion that stimulation of peripheral receptors and activity within the neural pathway of pain processing is NOT the pain per se, can be further provided by numerous evidence that nociceptive inputs are modulated by the cortex and sub-cortical structures of the brain. It is well known since Reynolds' (1966) report that stimulation of discrete brain regions (i.e., periaqueductal gray area) could produce profound analgesia via the control of nociception. This type of modulation of pain is called down-regulation (or central regulation) of nociceptive system functions. This analgesic effect arises from activation of the descending pain-modulating pathway, particularly involving the rostral vertromedial medulla that projects to the dorsal horn of the spinal cord. Transmission of nociceptive information is also modulated by a number of brainstem sites, including the parabrachial nucleus, the dorsal raphe and the locus coeruleus, as well as the medullary reticular formation. These brain regions employ different neurotransmitters (e.g., serotonin, dopamine, histamine, noradrenalin, acetylcholine) whose interaction can exert both excitatory and inhibitory effects on the activity of neurons in the dorsal horn. The current notion is that these descending projections provide a balance of facilitatory and inhibitory influences that ultimately determine the efficacy of nociceptive system functioning. Also, a more recent development is that discovery the same areas in the brainstem may not only be responsible for pain inhibition, but also can facilitate pain and induce hyperalgesia when stimulated appropriately (Urban & Gebhart, 1999). This is known as a *bi-directional control* of pain perception mediated by a descending system from the brainstem. Another current development that may shed additional light on the nature of chronic and neuropathic pain is the capacity of the forebrain regions to control the processing of nociceptive inputs via descending bulbospinal tracks (Dubner & Ren, 1999). This observation may provide a basis for bi-directional modulation of pain perception and interpretation, depending on the patient's attention, cognitive and emotional states. In fact, numerous observations in clinical practice support the notion of bi-directional control and modulation of pain. For example, soldiers suffering from severe battle wounds often report little or no pain at all. The pain of the injured soldiers on the battlefield may presumably be mitigated by the rational thought system as a benefit of being removed from the danger of war. In contrast, the same type of injury and associated pain in a domestic setting would be interpreted and experienced quite differently. Similarly, injury and pain resulting from highly demanding athletic circumstances would be experienced quite differently by non-athletes suffering from the same type of injury.

5.3. Pain Modulation by Activities/Stress

Stress and activity induced reduction of pain sensitivity has been well-documented in clinical research as well as demonstrated by numerous anecdotal observations. It has been shown for example that profound pain reduction can be observed in animals by implementing different stressors including (from Kelly, 1986):

- Restraint;
- Cold;
- Hit;
- Predator exposure;
- Swimming;
- Forced exercise;
- Footshock;
- Metabolic stressors

Anecdotal evidence of athletes continuing to compete despite serious injury and experienced pain has been described by Stenrberg (2007). Athletes can be even unaware of the injury until competitions and practices are over. It was well-documented that competitions themselves can be considered a high stressor. Accordingly, in an explicit test of the hypothesis that athletic competition activates stress-related responses, experimental pain sensitivity was assessed in both male and females athletes immediately after sport events. The results of these data were compared with those obtained two days prior and two days after the sports events. It has been shown that pain sensitivity was significantly lower right after competitions, suggesting an analgesic effect of stress-related competitions (Sternberg et al., 1998). Interestingly, the effect of competition as a stressor may not be limited to athletic contests. Pain reduction can be induced by exercise as well (Sternberg et al., 2001).

5.4. Pain Modulation by Beliefs

Beliefs "…are the mental engine that drives human behavior, and may raise us to the skies or cast us down to the depths of hell" (Main & Waddel, 2002, p. 221). Overall, beliefs are relatively stable mental trends about the nature of events that are shaped by personal values and experience, learning, culture and many other factors. Therefore, the true meaning of personal beliefs is almost impossible to assess and consider when dealing with persons in pain. The most advanced knowledge about personal beliefs with respect to pain has been accumulated within the concept of "back pain" (see Main & Waddel, 2002 for details). However, the general schema of beliefs may also be applicable to any type of pain as a result of traumatic injury. As was stressed in the previous text, pain is a highly subjective experience, and this "subjectivity" itself is a product of personal beliefs about critical events (i.e., injury, pain and its consequences). In a more general sense, there appear to be four major components of beliefs that patients suffering from pain report in a clinical settings. These include:

- The nature of the injury/illness – beliefs about the cause of the injury and its symptoms;
- The future course and development of the injury/illness – beliefs about the duration and outcome of injury;
- Consequences - short and long-term effects and how this injury would impact future life;
- Cure or control – beliefs about how this injury can or cannot be handled;

Beliefs are not simply the product of pain but rather a personal appraisal of a harmful situation associated with injury, and decision making mechanisms forcing the system to choose fight or flight. Not surprisingly, it was an attempt to approach the concept of pain appraisal from the perspective of fear-avoidance beliefs (Lethen et al., 1983). As discussed in the previous chapters of this text, fear, although is a highly adaptive human response to possible harm and bodily damage, may cause us to experience stress and anticipated pain as well as force us to initiate avoidance reactions. Fear of pain may intensify actual pain, self-perception of pain and negative affective reactions leading to abnormal pain behaviors (i.e., avoidance reaction, bracing behavior etc.). This in turn may reinforce our appraisal of the harm and beliefs about the cause of the pain, and reward our efforts to avoid it. Our beliefs that initiated action would cause pain most likely to "block" our intention to move, the phenomenon early defined as "Kinesiophobia." Lethen et al., (1983) and later Troup et al. (1987) elaborated on these thoughts and developed a "fear avoidance model of exaggerated pain perception," specifically in application for patients suffering from low back pain. According to this model, fear of expectation of pain with the psychological context may trigger a chain of confrontation such as:

- Strong desire to return to normal activities and work (i.e., sport participation)
- Mobilize, exercise and confront personal pain;
- Increasing confrontation with pain experience;
- Accurate interpretation of pain sensation;
- Positive and active coping strategies aimed at minimizing illness

behavior and disability;
- Effective rehabilitation

On the other hand the adverse scenario can be observed when fear of pain may trigger avoidance behavior, including:

- Increased fear of pain and avoidance of physical and social activities;
- Physical consequences: loss of mobility and muscle strength; loss of fitness and weight gain;
- Psychological consequences: fear and avoidance of pain; abnormal interpretation of pain sensation; overcoming coping strategies; altered behavior; negative and positive reinforcement; and invalidity status;
- Exaggerated pain perception, increased disability and overall immobility;

It is feasible to suggest that fear avoidance beliefs may be linked to higher probability of re-injury due to the effect of the beliefs not only on psychological status but also overall on unhealthy behavior. In fact, this was clearly demonstrated by Vlaeyen et al (1995), who found that patients who developed fear of re-injury demonstrated numerous signs of avoidance behavior even when asked to produce a simple movement. There was no reason to believe that these simple motor acts would cause any pain, nevertheless, these patients anticipated possible harm and "braced" their movement. In addition, the patients under this study showed obvious signs of depression and catastrophizing much more than was warranted by the sensation of pain per se. A general "model of fear-avoidance behavior" with respect to two opposite paths of the evolution of injury and pain experience was well-illustrated by Main & Wadell (2002), adapted from Vlaeyen as follows.

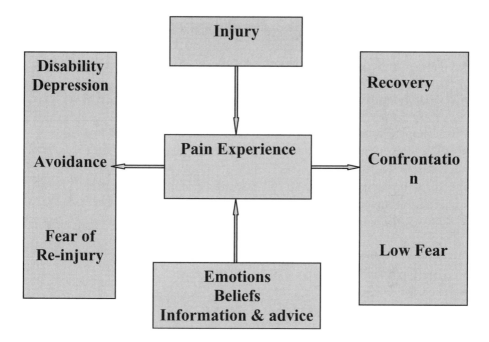

Figure 2. General model of fear avoidance behavior (adopted from Vlaeyen et al., 1995).

6. BIOPSYCHOLOGICAL MODEL

Pain and associated disability, although different phenomena, have been considered collectively to fully appreciate the consequences of injury. Moreover, pain is not only a physical problem that arises from tissue damage inducing a complex cascade of neurophysiologic processes, but also a psychosocial problem influencing subjective experiences when dealing with pain and disability. Therefore, from basic science and especially clinical perspectives, a number of critical components of pain experience should be considered. These were summarized within the scope of the "biopsychosocial model" (Engel, 1977; Waddel, 2002). This model was initially proposed with a specific emphasis on low back pain syndrome, although it might be applicable for any type of pain, especially for those with higher potential to evolve into chronic form.

The biopsychosocial model in its original content is constituted of three interactive components: (a) physical function/dysfunction; (b) psychological attributes; and (c) social elements. These three components evolve over time, develop differentially, and may alter a person's functional capabilities, overall behavior, heath status and life-style.

The physical functions/dysfunctions influencing overall pain behavior include but are not limited to:

- Reduced movability, range of motion and overall physical capabilities;
- Loss of strength and stamina;
- Loss of muscular coordination;
- Disrupted posture and gait patterns;
- Guarded movement and bracing behavior

Physiological function/dysfunction includes but is not limited to:
- Circulatory functions, tissue metabolic changes;
- Muscle wasting;
- Inflammation;
- Reduced muscle and joint position sensitivity;
- Muscle imbalance

Psychosocial substrates include but are not limited to:
- Irrational thought and beliefs;
- Fear of re-injury;
- Fear and anticipated pain;
- Lack of confidence;
- Fear of movement;
- Social withdrawal;
- Avoidance behavior

CONCLUSION

First, it is important to stress that pain is normally highly adaptive in nature, although lacking habituation, protective mechanisms and preventing further tissue damage due to injury. Pain is both a physical phenomenon and a highly subjective experience with a discrete neurobiological structure and distinct pathway. Originally induced by tissue damage, acute pain may evolve into prolonged, chronic and possibly neuropathic maladaptive forms most likely causing both short and long-term disabilities.

Pain experience is not a direct consequence of tissue damage and may be modulated by a variety of neurophysiologic mechanisms (e.g., gate control theory) and psychological processes (e.g., irrational thoughts, beliefs, fear, etc.) that significantly influence a person's pain experience. Overall, experienced pain may be enhanced as well as inhibited by the modulatory circuits within the CNS. Pain can also be modulated by stressful events, including athletic competitions and exercise. The complex interaction between the activation of nociceptors and the overall pain experience may have significant implication for clinical practice in dealing with injured individuals. Overall, it is not the PAIN that should be treated using various therapeutic modalities, but a PERSON suffering from pain who is in need of proper treatment.

REFERENCES

Melzack, R. & Casey, K.L. Sensory, motivational and central control determinant of pain. A new conceptual model. In: Kenshalo R. (Ed). *The Skin Senses*, (pp.423-443). Illinois: Thomas Springfiled, 1968.

Strong, J., Unruh, A., Wright, A., Baxter, G. *Pain: A textbook for Therapists*. Edinburgh: Churchill & Livingstone, 2002.

Papez, J.W. (1937). A proposed mechanism of emotion. *Archives of Neurology and Psychiatry, 38*, 725-743.

Schreiner, L., King, A. (1953). Behavioral changes following rhinercephalic injury in cat. *Journal of Neurophysiology, 15*, 643-659.

Melzack, R., & Wall, P.D. *The challenge of pain*. London: Penguin Books, 1988.

Melzack, R. (1987). The short-form McGill Questionnaire. *Pain, 33*, 191-197.

Waddell, G. *The Back Pain Revolution*. Edinburgh: Churchill & Livingstone, 2004.

Loeser, J.D. Perspectives of pain. In P. Turner (ed.), *Clinical pharmacy and therapeutics*, pp.313-316. London: Macmillan, 1980.

Loeser, J.D. & Melzack, R. (1999). Pain: an overview. *The Lancet, 353*(8), 1607-1609.

Sternbach, R.A. (1977). Psychologic aspects of chronic pain. Clinical *Orthopaedics and Related Research, 129*, 150-155.

Wright, A. Neuropathic pain. In Strong, J., Unruh, A., Wright, A., Baxter, G. *Pain: A textbook for Therapists*. pp.351-377. Edinburgh: Churchill & Livingstone, 2002.

Braune, S., Schady, W. (1993). Changes in sensation after nerve injury or amputation: the role of central factors. *Journal of Neurology, Neurosurgery, and Psychiatry, 56*, 393-399.

Melzack, R., & Wall, P.D. (1965). Pain mechanisms: a new theory. *Science, 150*, 971-979.

Reynolds, D.V. (1969). Surgery in the rat during electrical analgesia by focal brain stimulation. *Science, 164*, 444-445.

Urban, M.O., Gebhart, G.F. (1999). Supraspinal contributions to hyperalgesia. *Proceedings of the National Academy of Science, USA 96*, 7688-7692.

Dubner, R., Ren, K. (1999). Endogenous mechanisms of sensory modulation. *Pain, (Suppl. 6)*, S45-S53.

Kelly, D.D. (1986). Stress-induced analgesia. Annals of the New York Academy of Science. NY: New York Academy of Science.

Sternberg, W. (2007). Pain: Basic Concepts. In David Pargman. (ed.), *Psychological Bases of Sport Injuries*. Third Edition., pp.305-317. Morgantown: Fitness Information Technology.

Sternberg, W., Bailin, D., Grant, M., & Gracery, R. (1998). Competition alters the perception of noxious stimuli in male and female athletes. *Pain, 76*, 231-238.

Sternberg, W., Bokat, C., Kass, et al. (2001). Sex-dependent components of the analgesia induced by athletic competition. *Journal of Pain, 2* (1), 65-74.

Main, C., & Waddell, G. Beliefs about back pain. In: *The Back Pain Revolution*, (Ed), pp.221-239. Edinburgh: Churchill & Livingstone, 2004.

Lethem, J. Slade, P.D., Troup, J.D.G., Bentley, G. (1983). Outline of fear avoidance model of exaggerated pain perception. *Behavioral Research and Therapy, 21*, 4101-408.

Troup, J.D.G., Foreman, T.K., Baxter, C.E., Brown, D. (1987). The perception of back pain and the role of psychological tests of lifting capacity. *Spine, 12*, 645-657.

Vlayen, J.W.S., Kole-Snijders, A.M.J., Boeren, R.G.B., van Eek, H. (1995). Fear of movement/(re)injury in chronic low back pain and its relation to behavioral performance. *Pain, 62*, 363-372.

Engel, G.L. (1977). The need for a new medical model: a challenge for biomedicine. *Science, 196*, 129-136.

Main, C., & Waddell, G. (2004). Beliefs about back pain. In: *The Back Pain Revolution*, (Ed), pp.221-239; Edinburgh: Churchill & Livingstone.

Melzack, R. & Casey, K..L. (1968). Sensory, motivational and central control determinant of pain. A new conceptual model. In: Kenshalo R. (Ed). The Skin Senses, pp.423-443. Illinois: Thomas Springfiled.

Melzack, R. (1987). The short-form McGill Questionnaire. *Pain, 33*, 191-197.

Melzack, R., & Wall, P.D. (1988). *The challenge of pain*. London: Penguin Books.

Papez, J.W. (1937). A proposed mechanism of emotion. *Archives of Neurology and Psychiatry, 38*, 725-743.

Schreiner, L., King, A. (1953). Behavioral changes following rhinercephalic injury in cat. *Journal of Neurophysiology, 15*, 643-659.

Sternberg, W., Bokat, C., Kass, et al. (2001). Sex-dependent components of the analgesia induced by athletic competition. *Journal of Pain, 2* (1), 65-74.

Sternberg, W., Bailin, D., Grant, M., & Gracery, R. (1998). Competition alters the perception of noxious stimuli in male and female athletes. *Pain, 76*, 231-238.

Sternberg, W. (2007). Pain: Basic Concepts. In David Pargmen. (ed) Psychological Bases of Sport Injuries. Third Edition., pp. 305-317. Morgantown: Fitness Information Technology

Strong, J., Unruh, A., Wright, A., Baxter, G. (2002). Pain: A textbook for Therapists. Edinburgh: Churchill & Livingstone.

Waddell, G. (2004). *The Back Pain Revolution*. Edinburgh: Churchill & Livingstone

CHAPTER 15

PSYCHOLOGICAL TRAUMA: CASE STUDIES

1. INTRODUCTION

As was elaborated in the preceding chapters of this book, psychological trauma includes behavioral, motivational, affective, and cognitive evidence. Behavioral evidence consists of those aspects that can be seen physically during an observation, such as avoidance behavior like bracing. Motivational trauma evidence deals with self-efficacy issues such as a lack of confidence in the success of the treatment plan, and thinking treatment does nothing for recovery. Affective evidence becomes clear due to our primitive, limbic system, in which fear of and uncertainty about re-injury take over before our brain can actually assess the situation (rational fear system). The final type of evidence, cognitive evidence, deals with those aspects in which our thoughts become the main clue to trauma. This evidence includes obsessive thoughts, over-thinking, the inability to focus, and memory problems.

One common question is whether every single episode of physical injury may result in development of psychological trauma in athletes or if some additional factors contribute to it. Accordingly, is it feasible to advise that every injury requiring prolonged rehabilitation should also require psychological intervention, in order to avoid the risk for the development of chronic psychological discomfort and/or psychiatric disorders? Another important question to be answered is whether psychological trauma is always a side effect and/or a consequence of a previous physical injury. My personal clinical experience clearly suggests that this is not the case. First, I truly believe that athletes are strong enough both physically and mentally to overcome the psychological problems that accompany injuries. Of course, the type of injury is the most crucial factor to consider. That said, injured athletes may gradually develop an irrational attitude, abnormal sensations, obsessive thoughts about injury, fear and anxiety even without having been physically injured in the past. For example, a diver developed a career ending psychological trauma of panic and generalized anxiety disorder, after she witnessed another diver become seriously hurt while hitting the springboard board with the head. This injury was classified as a severe facial injury requiring reconstructive plastic surgeries. As she reported later, this image of the traumatic impact of the head on the board never left her mind when she was preparing for a dive. In clinical practice this case may be considered as *witnessing*. For example, seeing anyone being severely injured is stressful (you might be the next). In fact, the greater your attachment to the victim, the greater the stress and the greater the probability that you will develop psychological trauma.   In the case of the diver, psychological interventions failed to treat her trauma, so she quit the sport that she used to love.   This is not just an isolated case, but a common situation in a clinical practice.   With this in mind, in the following text are case studies of injuries categorized by severity, timing and intensity, and their relation to psychological trauma in athletics.

2. CATEGORIZATION IN TERMS OF SEVERITY OF INJURY

2.1. Mild to Moderate Injury

Acute anterior compartment syndrome: An eighteen-year-old cross-country runner had just started practice at a major university. The workouts are twice as hard as she had trained in high school, but she wanted to do all she could to become an All-American runner. She told her coach about the pain she had been experiencing for the last two weeks, since the workouts had progressively increased day by day. The coach just told her it was "shin splints" and that she should ice after practice. Since he was a college level coach, the runner believed him and she kept running with the pain. Finally, the pain increased until she was no longer able to run so she went to the emergency room. The doctor palpated the tender spots on the leg and ran a few tests to find out the runner had suffered from anterior compartment syndrome.

Unfortunately, this can be a highly neglected injury. Because athletes and coaches are not aware of the signs and symptoms, they shake their heads as if the pain indicates a minor injury.   If the athlete neglects anterior compartment syndrome long enough surgery will have to be performed to release the build up of pressure.   There are two kinds of compartment syndromes: acute compartment syndrome and chronic compartment syndrome. Acute anterior compartment syndrome arises from trauma like a direct blow to the lower leg (such as a soccer player getting kicked in the tibia). Chronic anterior compartment syndrome arises from overuse of the muscle (such as a runner putting on too many miles before proper training is performed).

A direct blow to the anterior lateral portion of the leg causes acute anterior compartment syndrome. The direct blow can cause an injury to the muscles of the anterior compartment causing severe bleeding and an increasing build up of pressure. Acute anterior compartment syndrome can also be cause by a rapid increase in activity (like my athlete in the opening scene). The rapid increase in activity would cause a muscle tear (from being under-trained) which would increase the bleeding, causing the increased pressure. The fascia surrounding the compartment doesn't give when rapid edema occurs, resulting in increased pressure on the tissues, arteries, and veins in the anterior compartment. The fascia does compensate for some edema, but the tibia and fibula can inhibit the anterior compartment from expanding further. When the compartment pressure exceeds the capillary perfusion pressure, the surrounding tissues may die from a loss of blood flow, which leads to a loss of oxygen. For the scope of this case, I will focus on acute anterior compartment syndrome.

The seven "Ps" are used to describe the signs and symptoms of compartment syndrome. The first is Pain. This would have to be severe pain different from just a blow to the tibia. The second is Pressure. The intra-compartmental pressure can cause a lot of problems with the lower leg down to the foot. The third is Pain with *passive* stretching. Stretching puts increased pressure on the already tight fascia. The fourth is Paresis of Paralysis. The muscle may not have enough room to move from the edema, causing foot drop. The fifth is Parathesia. Parathesia is caused by decreased sensation due to the compression of the nerves. The sixth is Pulse. The dorsal pedal pulse has to be checked periodically to make sure the anterior compartment isn't increasing in pressure. The last is Pallor. Pallor is defined as discoloration of the skin, due to the excess blood flow and no return.

An examined athlete has suffered from moderate acute compartment syndrome. As mentioned earlier, this injury was due to a rapid increase in activity when making the jump to collegiate athletics. She has not had compartment syndrome before so intervention with this injury will progress systematically. For her, the positive side of this injury is she decided to go against what her coach was telling her and got it checked out soon enough that she didn't need surgery. It was treated conservatively and eventually the swelling decreased to where she was able to regain full function of her lower leg and foot.

There have been some previous injuries she has had to deal with. She has suffered from stress fractures in her lower leg (as many of her teammates have as well). The stress fractures only kept her out of play for a week of no activity and they were able to resolve. She has also had a setback of a broken collar bone from a snowboarding accident. Although not running related, this injury also kept her out of running for a while because of the sling she had to wear. She was able to start biking after two weeks in the sling.

It is obvious that this particular athlete has suffered the compartment syndrome as a result of the increased workload of college sports. The coach has to have an adaptation period for the new athletes coming into the program so injuries are decreased. As I was talking with this athlete, she was saying that her first injury (the stress fractures in her lower leg) didn't bother her because she was able to get back to exercise in a week and the injury didn't recur. She was also familiar with this type of injury because her teammates suffered from them. However, the collar bone injury was more bothersome to her because it was not related to running. She had a hard time dealing with this because she felt like it was her fault that she went snowboarding in the off-season for some fun. It is important for the athletic trainer to work with the sports psychologist and keep the athlete's mind off previous injuries that are unrelated to the current injury. Because the most recent collar bone injury happened last year, it was still fresh in her mind. She was worried that if she had to sit out because of compartment syndrome this year, she would lose her spot on the relay team. She did not receive an intervention from a sports psychologist after her previous injuries because the stress fractures were minor, and the collar bone fracture was not an injury dealing with her legs so she could fight through the pain and absence from training for two weeks.

A good predictor of whether the athlete with compartment syndrome can return to participation is the girth measurements of the injured lower leg compared to the un-injured leg. My athlete did not have to have surgery, nor did she have a drop foot gait, or an absent dorsal pedal pulse. However, with that said, she still is classified as having a moderate injury because the compartment syndrome limited and altered her training.

This athlete did experience some psychological trauma in the form of fear. She didn't know if she would make the team if she sat out. She also showed some avoidance behavior because she was afraid to do anything besides walk to class and go to practice because the collar bone fracture happened during a leisure activity. There was also the fear related to the numbness she was feeling down her lower leg into her foot. She also had some fear of re-injury, and since she couldn't trust her coach anymore she didn't know what she would do if she got injured again.

This athlete did see a sports psychologist for a few sessions because she couldn't handle all the stress of college added to not being able to practice for a couple of weeks because of the injury. The sports psychologist broke down the sessions to deal with one problem at a time. First he talked to her about the previous injuries to get some background and see what type of personality she has. Then he talked each injury through with her, starting with the least traumatic and ending with the compartment syndrome. The stress fracture was overlooked, because it didn't cause negative thoughts in the athlete. Then the sports psychologist discussed the collar bone fracture with the athlete to see what she was still feeling from that injury. It was obvious that the compartment syndrome had rekindled some negative thoughts from the collar bone fracture because she didn't want to sit out any practices.

The athlete rested for a week, and slowly got back into some exercises under the close supervision of the sports medicine team. The sports psychologist spent three weeks with her and he was able to overcome some of the fears she had about this injury. The compartment syndrome eventually was gone and this athlete returned to full competition that year, and did not lose her spot on the relay team. She feels she has become a stronger person because of the injury. She believes the sports psychologist helped a great deal to get over her fears, but she also believes some of the psychological trauma was from being a freshman and getting overwhelmed with the college scene. No matter what the case, the athlete, athletic trainer, and sports psychologist all have to stay on top of injuries to reduce the possibility of developing psychological trauma after injury.

Whether acute or chronic, anterior compartment syndrome should not be taken lightly. The symptoms may seem like a different injury, such as shin splints, which is why athletes and coaches need to be aware of what is different about them. On the opposite side of the equation, the athletic trainer and orthopedic surgeon need to be active listeners to determine what signs and symptoms the athlete is experiencing. Anterior compartment syndrome can damage the lower leg if not treated properly. The bottom line with anterior compartment syndrome or any injury is to treat the injury as soon as possible to return the athlete to competition. There are cases (like this one) where compartment syndrome could have been avoided; however, some acute cases are just a result of bad luck. Education about the proper techniques to keep the body healthy and injury free is also important for both coaches and athletes.

Minor elbow injury: A former volleyball player "Mary K" reported an injury that took place during a volleyball competition in high school. The injury occurred during the game when she dove to save a ball that the team had hit. She was completely focused on the play at hand while she fully extended diving towards her right side, landing on her elbow. Since the incident was purely accidental and kind of "bad luck," not much could have been done differently to prevent the injury. She was immediately removed from the game and sent to the trainer's room where she learned that she had no broken bones or seriously damaged ligaments. Her injury was mild to moderate since it only lasted for four days, but it required some time away from practice and games. Even though she didn't see a doctor she healed easily on her own. She recalled having mild previous injuries, such as rolled ankles and bruises, which helped her quickly recover from the fall since they also healed quickly. The only treatment she could do was to sit out from practice and games and to ice and rest, which in turn, helped her regain strength and relief of pain in the elbow joint. After a few days she felt comfortable again diving and doing the things she did before the injury occurred.

She had very little indication of psychological trauma, if at all. Apart from a slightly slower reaction time and the slight pain there were no noticeable signs of diminished physical function following the injury. Within a week her entire physical discomfort and/or pain were gone and she returned to play, where she considered herself 100% with regards to the elbow bruise. Again, due to the fact that her injury was in the mild to moderate range, she suffered little to no psychological trauma at all, as evidenced by many observable characteristics. She neither experienced any fear of re-injury nor lack of confidence, nor any signs of avoidance and bracing behavior. Once she explained that she had had minor injuries before and knew how to take care of them, it was understood why she healed quickly and was ready to get back in the game without lingering trauma. This means that successful prior experience dealing with injuries is an important factor to consider when considering the possibility of psychological trauma. Another factor is that a mild to moderate injury only takes a relatively short time to heal. Because of this, athletes are not pressured to heal quickly or fake healthiness, and are not prone to re- injury due to the fact that they are almost completely 100% healed. The situation could be quite different if an athlete is rushed to play and a premature return to sport participation would cause another even mild recurrent injury. In this case, the risk for both multiple within-a-season injuries and psychological trauma will be significant.

2.2. Severe Injury

"**Juan T.**". Within the state of New York there has been a rising high school basketball star in the making. Born in Queens, New York City, "Juan T." has been a basketball prodigy ever since the age of ten. The now twenty year old had been recruited by every major university to play basketball for their esteemed school since the age of thirteen, the same age at which he was able to first dunk a basketball. North Carolina, Duke, Kentucky, Florida, and Maryland, just to name a few, all had their eyes on luring Juan to their school in the hope that he would lead the school to a national title. Standing 6'3" 195 pounds, this shooting guard from New York City has all the tools necessary to be a complete and tremendous basketball player. Sports analysts and commentators say that he was better than LeBron James in high school, as he had the uncanny ability to leap over and around any opponent in his way. Additionally, he had a shot that was already the equivalent of NBA range jump shots, leading many to believe he was already ahead of his time.

Nonetheless, unlike most high school star athletes, the then junior had plans for going to college to play basketball for Roy Williams and the North Carolina Tar Heels. Right now Juan is a junior averaging 19.6 points a game, 7 assists, and 5 rebounds for the Tar Heels. He is having a successful collegiate career and will enter the NBA draft after this season. However, it was always not glamorous for Juan, as the life he leads now pales in comparison to several years ago. Juan went through extraordinary circumstances to get where he is today, as a career ending injury almost took his life.

Coming off a brilliant high school junior year that saw Juan average 31.5 points, 9.7 assists, and 8.4 rebounds a game, Juan was looking forward to his senior year. His team had just won its second high school city basketball championship and all the players, especially Juan, were looking forward to repeating as champions for the first time in school history. It's no secret this lead Juan to work extra hard during the summer so that he could be better than ever for his senior year in high school. Things were going great for Juan as he had already committed to a full basketball scholarship to play basketball at Duke University in North Carolina. Coach Krzyzewski even said, "We are thrilled and excited to have a player of such great caliber come play for our university. I'm looking forward to working with Juan when he graduates from high school and comes here." Yes, things could not be any better for Juan, but it all came crashing down during the first game of his senior year.

Already dominating the competition in the first game Juan was having a great game and his team was playing tremendously. However, as Juan went up for a dunk in the third quarter an opponent reached out attempting to block his shot and Juan fell awkwardly towards to the ground. Writhing in pain, Juan screamed out that something was wrong with his arm and looking over at it he realized that his arm was completely contorted in the wrong direction. "Yeah it's broken," said the trainer as Juan was helped off the floor and back into the locker room. While still screaming in pain, tears began to fall down Juan's face as he started to realize that his season was probably over and he would not play basketball again with his teammates for months. It was a sad day for Juan and his Trojan teammates. Although the team won the game, every one knew that they had no chance at repeating without Juan in the lineup. As Juan went to sleep that night he kept visualizing all the scenarios that could happen with his broken arm. He started thinking maybe it wasn't a season-ending injury and maybe he could be back in a couple of months in time for the playoffs. Or maybe he could definitely be out for the season with months of rehab along the way. All of this was going to be found out the next day as Juan had a crucial doctor's appointment to let him know the status of his broken arm. Juan fell asleep nervously and optimistically wondering if his season could be salvaged and if he could make it back in time to play with his fellow seniors.

"You have broken your bone in half completely," the doctor said to Juan as he just sat there in awe. "You might not be able to play contact sports again, as another break like this could do serious damage to your arm to the point of amputation. Furthermore, you are looking at a year to a year and a half of rehab and recovery," said the doctor. Juan immediately started crying and screaming, "My life is over, basketball is all I have, I cannot believe this is happening to me." Juan went home in disbelief as he began to realize that his basketball career in the short term was over, and it might be done for good all together.

In the coming months Juan began to feel depressed as he watched his basketball team struggled to achieve a winning record. Sitting on the bench all Juan could do was watch while wishing he could be out there competing and playing with his teammates. The surgery to repair his broken arm was successful, but Juan knew that he could not start rehab for another four months as further damage could be done to the arm. Juan's depression nevertheless was growing even worse at home and within his social life; his depression was starting to affect not only him but the people around him. Constant arguments with his parents lead to Juan having anxiety and panic attacks on and off. He stopped eating properly, and his girlfriend broke up with him because he was being so negative all the time. As the school semester wound down for the summer Juan just kept falling into a deep despair of depression. His team did not make the playoffs, his grades were sinking even though he was going to graduate, and his friends and family seemed as if they did not want to be around him. Lying in his room with the blinds shut, alone and disheartened is the way Juan began his summer. He did not want to talk to or see anybody. His rehab program did not start for another month so Juan figured he would just lie around and do nothing till then. However Juan did not know things were about to get a lot worse as his whole world was going to get tangled upside down.

Two weeks before he was to start his rehab program Juan received a phone call from the athletic director of Duke University. The athletic director told Juan that they were pulling the basketball scholarship from Juan since he might not be able to play basketball again, and they were giving the scholarship to someone else. Juan immediately hung up the phone and began crying and started throwing and smashing things in his home. He then began to fell faint as his heart started to beat fast and he could not breathe. Having a panic attack, Juan began stumbling around his house until he finally passed out in his bedroom. Waking up a half-hour later Juan felt terrible and disgusted with himself—everything he wanted to do with his life was over. Juan went down stairs into the kitchen and began to drink his parents' Vodka to try and numb the pain that he was feeling. As the days went on Juan continued to drink heavily as the depression only got worse and he did not know what to do with his life.

Showing up drunk to his first rehab session Juan met his instructor Chad, who was excited to start working with him. During the interview process Chad could sense that Juan was drunk and asked him why he would do this. Juan immediately started crying, telling Chad everything that had happened within the last several months of his life. Chad came around his desk, gave Juan a hug and told him that "it's not the end of the world". Chad said, "Athletes have sports injuries every day and believe their life is over, but it's not. I'm going to recommend you start seeing a therapist before we start our rehab program so that you can be mentally stable when we're going through our workouts". Juan, who felt as if he did not have any alternative, agreed as he was determined to stop drinking, get rid of his depression and anxiety, and get his life back on track.

When Juan first began seeing his therapist he was quite optimistic about it as he had never experienced anything like this before. During his first visit he explained to the therapist his entire situation—losing his scholarship, drinking, panic attacks, and the broken arm that led to all of this happening. She told Juan that the healing process was not going to happen overnight, but he would need several sessions to help him with his problems. Moreover, to start the healing process Juan should take ten minutes every day to meditate and get rid of all the negative thoughts and the depression. In addition he should start a journal to keep track of his thoughts and ideas when they came to him. Juan accepted the advice and began meditating and keeping a journal while he was starting his rehab. Chad, his instructor, told Juan that it would take five to six grueling months to get his arm completely and totally healthy again. Chad instructed Juan that he was going to be put on a strict workout regimen that encompassed the physical and as well as the mental. He let Juan know that the physical part was easy; it was the mental part that would be the key in getting himself back to being a standout basketball player. Juan took a deep breath and said, "Lets go to work"

Juan continued this workout regimen for the entire six months, along with going to the therapist to help with his drinking and panic attacks. To keep his same workout instructor Juan went to a community college near his house so that he could stay around his friends and family for support. However things were not all good once his arm was healed. There was still a question of whether he was the same basketball player that he once was, and could he still do the same things he once could. Working with the coach at his community college Juan began to shoot and dribble slowly and surely. Juan began doing drills with basketballs so that he could compete with other players and see if his arm was able to take the contact. The day finally came when the coach let Juan scrimmage with the basketball team and it seemed as if Juan had not lost a step. Juan dominated the game by scoring 20 points and handing out 7 assists. After the game the coach said to Juan, "I thought for sure you would be shy and timid because this is your first time playing with contact, why weren't you?" Juan explained to the coach that during his rehab process every day he did visualization and imagery techniques about basketball so that he would be prepared for this day. The coach then invited Juan to join the team next year in hopes of helping them win their first ever community college championship. Juan joined the team the next year and did lead his team to its first ever community college championship while averaging 24 points, 6 assists, and 7 rebounds a game. The next year Juan transferred to North Carolina, where he led the Tar Heels in two beat downs of the Duke Blue Devils, averaging 36.7 points in the two games, and the rest is history. What a recovery!!!

"Patrick B", a nineteen year old soccer player, suffered this unfortunate trauma one fateful summer day and is still dealing with its scars today. Following his injury Patrick sought the advice of a physician whose analysis rendered a prognosis of a muscle tear. Prescribing an exercise regime along with the recommendation to pursue the expertise of a physical therapist, the doctor left Patrick with a false sense of security. With the development of an exercise routine, provided by the physical therapist, Patrick began the healing process.

A month had passed since the incident, and winter was slowly approaching. Feeling relief from his pain, Patrick began to relinquish his exercise program. Winter saw the decline of his routine workout along with the pain. Then, spring was at hand and a new season had begun. No longer hindered by the pain, Patrick once again took to the field. It wasn't long before playing time was limited due to interspersed bouts of discomfort. Once removed from play, return to the gamhe was impossible as his muscles contracted, making every movement pulse with pain. The season was once again short lived as the pains regained their hold on him. This time, however, more determined than ever, Patrick sought the help of an X-ray technician. The results of the X-rays revealed the presence of a vertebral stress fracture requiring further testing. To make a definitive diagnosis a CAT scan was performed. Once the fracture was confirmed, a regimen of rest and strengthening of basic core muscles was prescribed.

Three months had passed and once again with pain relief came a false sense of security. Summer was beginning to bud, bringing with it the bloom of soccer camp. Unable to resist its lure, Patrick once again took part in his favorite pastime. The relief from pain was short lived as was his tolerance of it. With the aid of a referral, he pursued an alternative source for injury identification through an MRI. Through his most recent tests, it was determined that the radiating pain was a direct result of a degenerative disc. The doctor confirmed that the degenerative disc was the outcome of continued injury to the original stress fracture due to continued play. With this news Patrick was faced with an important decision. The damage had already been done; no amount of rest and recovery could undo the affliction. Rather than succumb to the defeat of his situation, Patrick took this as an opportunity to strengthen himself as an athlete.

His drive and determination led him to try out that fall and with the burden of pain bearing heavy on his mind, he took the field once again ready to win. This now junior was proving to be unstoppable. His motivation stemmed from his team's cohesiveness that had made past successes possible and that made future seasons desirable. Game after game he fought through the brutal pain and anguish alongside his loyal teammates who proudly upheld the tradition of excellence this team had become accustomed to. Discouraged at times, especially when faced with sitting out of play, Patrick continued to press through the pain. As his muscles slowly tightened after their intense workout, searing waves of pain pulsed throughout his body, making sitting on the sideline a much feared and ill awaited event. Armed with friends and dreams for the future Patrick bravely fought his affliction.

Constantly in motion, Patrick maintained his muscle stretch, postponing the inevitable tightening. "I had to warm up and stretch a little longer then everyone else, and when I wasn't playing I couldn't sit down or it would tighten up, so I would always have to keep moving". Despite his increased need for proper stretching prior to performance, he failed to show any further signs of diminished function following his injury.

Patrick continues to pursue his favorite pastime, through intramurals in the spring and various leagues in the summer. Though the pain is much less severe than it had been, he continues to face the discomfort of his injury and the realization that he'll never fully recover. Doctors have predicted that Patrick will need to replace his damaged disc in about twenty to thirty years; until then he must endure the pain and tough it out.

Patrick's story describes a sad but very prevalent reality in the world of sports. Injury is no stranger to the many athletes who compete and train year round in all athletic activities. As sports professionals, who strive to properly prepare our athletes for performance in whatever form it may present, it is imperative that we train to avoid injury at all costs. This ideal raises an intriguing question: Can injury be predicted and/or prevented? If so, how could this scenario influence the development of chronic psychological trauma? Based on Patrick's story, it seems that the extent of his injury could have been prevented had doctors diagnosed the severity of his injury at an earlier stage. However, because injury is never planned for, it is hard to predict its appearance.

Another former Penn State varsity athlete, "Kyle P," was examined. After leaving his team he suffered a serious traumatic injury while playing recreation sports. He described the injury as "a freak accident," explaining that he believed his skills were exemplary since he played a variety of sports since he was young and his physical fitness level was very high. Kyle repeatedly said that he was always the "best athlete on every team" that he was on, which leads one to believe that he is overconfident in his skills and over-estimated both his skill and physical fitness, which could have contributed to the injury. After seeing a medical professional, it was determined that he had a complete tear of his anterior cruciate ligament (ACL) and a partial tear of his medial collateral ligament (MCL). Since he didn't seem to be recovering at the rate that he should be, the lack of attention given to his mental status may be directly linked to the following psychological and behavioral problems.

He seemed to have given up on his rehabilitation routine and stopped going to his physical therapist after a few months. This is a direct indication of avoidance reaction, giving him a chance to "attribute his disability to the lack of proper treatment". He only saw his doctor once every couple of months, and refused to see a sports psychologist. He also experienced favoring of his injured leg, as evidenced by accidental limping. Muscle soreness could be explained by lack of activity, resulting in being out of shape. The accidental "rubbing" of the injured knee and bracing behaviors indicate that he may or may not be aware of his psychological trauma. But it looks like he developed subconscious habits to avoid re-injury because he was so afraid of it. Kyle also expressed doubt about recovering and being able to play his best again. Since Kyle is someone who has been involved in and devoted to sports throughout his entire life, this injury that occurred over seven months ago obviously left some mental and physical effects.

Since the athlete was perceived to have an overly confident personality, it obvious that that he believed that he was the best at whatever he did. It was his false sense of security that led him to believe that he would gain possession of the ball and that he never could get seriously injured from such a trivial act. Even if he was highly skilled, in his mind it was just a matter of improper warming up and preparing for the contest. Ultimately, his mental state really contributed to this injury. Initially he exhibited denial, feelings of isolation and anger. He still harbors anxiety and is preoccupied with the injury because he hasn't fully recovered yet. Overall, his psychological trauma may be mostly attributed to his high level of athletic involvement prior to the injury and not being able to properly cope with the injury. Indeed, psychological intervention should be offered in this case to prevent the severe consequences of this common traumatic injury in athletics.

Here is the case of another former collegiate athlete, an ice hockey player from West Chester University, who also suffered the trauma of an ACL tear. "Josh M." tore his ACL, along with his MCL and a medial meniscus tear, at the age of seventeen, while at hockey camp. A second tear of his ACL occurred when he was playing basketball with friends on an uneven surface. Prior to this he neglected to tell the doctor that he would sometimes feel a "shift" while playing hockey. The third time he completely tore his ACL while skiing. He reported that he was able to push his prior injuries out of his mind after a few times down the slopes. He had two reconstructive surgeries and completed about eight months of rehabilitation under the supervision of an athletic trainer. Under the advisement of his coaches he did not return to ice hockey and turned to recreational sports instead. He experienced diminished function of his injured knees such as decreased knee flexion, swelling, posterior lateral pain, and a dull ache. This was occasionally accompanied by "sharp pain" that significantly reduced in intensity with immobility. He admitted he had feelings of depression, anxiety, and lack of motivation to return to play after his two surgeries.

Behavioral changes were also present. Josh exclaimed that he was not looking forward to participating in skiing or other sports as much as he usually did. He had less drive, intensity, and focus to be as good as he once was. His risk-taking personality, and his desire to meet challenges and risky situations were significantly altered, so he had developed a lowered sense of confidence in his ability. Josh also feared re-injury and had decreased ability to stick with a decision during sports activities. The lack of risk-taking occurs mostly in situations where not all aspects are perfect, thus making it more likely for injury to occur. By totally avoiding these situations, Josh put himself in a safe position to also avoid injury (i.e. avoidance behavior).

Interestingly, all of these athletes experienced a traumatic injury of the ACL tear while competing. It is significant that each athlete also encountered psychological trauma along with physical when suffering an injury classified as severe. It is well-known that all ACL injuries necessitate a long and dedicated rehabilitation process that requires commitment from the injured patient and regular visits with a physical therapist. It is difficult for athletes such as these two competitive ones above to be able to fully recover quickly enough in order to play at 100% for the season. Even if an athlete completes rehabilitation and is cleared to play again, he or she may not be mentally prepared while playing. That said, there are obviously two different outcomes and post-injury scenarios. They may depend on availability of coping resources, athletes' attitudes towards injury and many other psychological factors that should be seriously considered by medical practitioners while implementing the treatment protocols.

3. CATEGORIZATION IN TERMS OF TIME OF INJURY

3.1. Injury During Season

Penn State Football defensive linebacker "**Dan C.**" has been examined in relation to his 2004 season injury, which included nerve damage in the neck that was causing "stingers" down his arm. The cause of injury was attributed to the simple nature of the sport of football and his specific position as a linebacker, which consists of the constant act of hitting and tackling that put tons of pressure on the neck, shoulders, and back. In Dan's opinion, this injury was a moderate injury, because he had to be taken away from any contact for two weeks, which means he missed two games. The "stingers" down his arm from the nerve damage in his neck caused the arm to be continuously tender and he commented on having "very little strength" in his right arm. Although he started playing very quickly after his two games off, he needed a strength building program to get his right arm back into top shape. It was almost two to three months before it was back to optimal strength. There was an obvious decrease in performance due to this lack of strength training for so long. In addition, when exploring whether or not he suffered any psychological trauma from the injury, he informed me that he did. Dan went into great detail telling me that there was "a major mental aspect" to his injury when he returned to play. He said he found himself "bracing a lot" because of the still tender arm. More importantly, he added that the biggest mental factor contributing to the trauma was from the neck brace that he had to wear. He explained that the neck brace made him think more about the injury and instilled in him much more fear knowing that it was there, which caused the bracing avoidance behavior. Dan's evidence of psychological trauma came from both the affective, mental fear aspect, as well as the observable behavioral evidence from bracing and avoidance.

As far as personal strategies in overcoming the mental aspect are concerned, Dan explained that he really wanted to test himself and therefore forced himself to take as many hits as possible, and did so with as much strength as possible, because it restored his confidence knowing that he could produce the type of power he could without re-injury. He stressed that is was important to regain his confidence in his ability to perform optimally, because seeing that he could perform at his best released any prior fear he had. Dan believes that psychological trauma cannot be prevented because people are only human and have little control over the natural ability to have "fear," and also believes that trauma is completely sport specific as well as specific to personality types. This leads someone to believe that perhaps personality tests are a strong tool in predicting psychological trauma in sport. Lastly, he explained that he believes most psychological trauma cases occur in athletes recovering from a knee injury because of the importance of the knee in his sport. Although he doesn't believe trauma can be prevented completely, he does believe it can be reduced depending on the personality of the athlete and his or her mental state.

"**Kylene C.**" played tennis for her high school and suffered a moderate to severe injury during a match. As she dove for the ball she tripped and fell face first on to the court and blacked out. A CAT scan determined that she had a severely fractured cheekbone, and two black eyes that came along with the fall. Along with pain and suffering Kylene then experienced all the signs of emotional instability and depression. She felt like she failed and let down her team when they needed her the most, because she was unable to play out the rest of the season. She blamed herself constantly and thought that this injury could have been prevented. Kylene was ashamed to go to school because of her appearance. She lost her self confidence, went through withdrawal, and had a hard time being around her friends. Kylene experienced both *bracing behavior,* in which she favored her sensitive cheekbone, and avoidance reaction, in which she stayed away from all activities that could re-injure her face. She experienced moodiness, trouble sleeping and neurotic pain until she was able to talk to a sport psychologists who helped her understand what and why the injury occurred.

Since Kylene's injury occurred during the season she felt even more pressured to heal quickly. Since her injury was too severe to play again in the season she experienced much psychological trauma partly due to time pressure. Not only did she feel like she was a failure she also was ashamed of what she looked like because of the injury and had a hard time being around others. The severity and timing of the injury can have a great impact on the athlete's reaction to an injury. This case study is also indicative of gender differential responses to injury. If this trauma weren't related to a facial injury, psychological consequences could be less pronounced in this female athlete. Self-perception, body image and especially facial appearance are greater pressures for female than male athletes. The gender differential response to injury is a really important but poorly studied and understood phenomenon.

As a Penn State lacrosse player, "**Brian B.**" has suffered through various assortments of injuries. During a game in his sophomore year, he pushed off rapidly to accelerate towards the ball and felt his hamstring tighten up. As he exploded off the line during a drill, a ripping sensation tore across his upper hamstring. After seeing the trainer, he was diagnosed with a severe muscle pull of the left hamstring and was supposed to take five to seven weeks off for rehabilitation. After telling the trainer that he was experiencing very little pain they let him back to practice. That said, he knew he was not fully recovered and not surprisingly re-injured his hamstring.

Brian claimed to have been upset with what had happened because lacrosse was so important to his lifestyle. He became depressed, gained weight, lost confidence in himself and often feared re-injury after the second recovery. Every time he was thinking about accelerating towards the ball, he felt anxious and worried that rapid acceleration would cause re-injury. Clearly, this irrational thought of re-injury changed his perception of risk behavior. After the injury, and especially after the re-injury, this athlete associated body acceleration with high perceived risk of threat. Brian also explained that he had trouble concentrating on his studies and he withdrew from social activities since he wasn't happy with himself. He experienced the bracing behavior even outside of sports activities.

Because Brian was so intent on playing in the game against Notre Dame, he did not allow himself to fully recover from the injury. He even knew that he was not ready, but continued to press himself to assure himself that he was fine, so that he wouldn't have to let down his team. This poorly managed and irrationally speeded recovery led to even worse situations and re-injury. Overall, in Brian's case, psychological trauma was obviously present. This trauma required a serious psychological evaluation and systematic treatment in order to prevent long-term well-being deficits far beyond the athletics.

3.2. Injury During Off Season

In the following case studies two athletes were examined regarding their injuries during the off season with the intent to reveal the signs of psychological trauma following traumatic injury. The first athlete examined was "**Dan L.**", a fullback for the Penn State football team. The injury he experienced was a ruptured L4 disc in his lower back during his 2004 season. This was a very severe injury that he continued to play with until the end of the playing season. A "no pain no gain" and/or "injury is temporary, fame is forever…" attitude was the predominant factor in Dan's dealing with this injury. Once the season was over, he underwent two surgeries, and had his entire L4 disc removed from his back. He complained that the ruptured disc caused severe pain down his leg, and that as soon as it was removed via surgery he felt no pain. According to Dan, the cause of the injury was simply the nature of football. The severe hits he took as an offensive running back and blocker are common in football. Since he had the surgery right after the season ended, he spent the entire off season resting and did not enter into contact again until a year post-surgery.

He said that although he had no prior injury of this sort, both his father and his brother had the same L4 disc rupture while playing football, and both underwent the same surgery as he did. It was left to believe perhaps there is some genetic factor within his family. Since Dan spent so much time recovering during the off season, he experienced no diminished performance when he finally went back into contact after non-contact conditioning. He said, "I actually believed I performed better because of the lack of nerve pain shooting down my leg, which allowed me to move more freely and play with as much power as possible". Dan also added that he had no psychological problems, no reservations, no fear, and no lack of diminished confidence once he was allowed to return to play. He says, "I didn't think about the prior injury because I knew if I let it into my head, then it would have caused hesitation, which in turn causes more injuries." Looks like this athlete used "negative though stoppage technique," although nobody was teaching him its values. Dan also had a very long recovery period, and his body felt much better after the L4 ruptured disc that he had played with, was finally removed. Because of this, he had a greater feeling of well-being and release once he got back to competition without any of the nerve pain plaguing his performance.

Overall, Dan believes that he experienced no types of psychological trauma or discomfort in his personal injury, although he confessed that he has seen a lot of it from his experiences with his teammates through the years playing football. He says it depends a lot upon the personality of the athlete as well as the injury itself. His personal observation is that most mental trauma he witnesses in football is due to knee injuries. He believes that when athletes come back to play from a knee injury and have to play with a brace on their leg, it mentally scares the athlete and puts the injury in their head, and they do not compete at their best level. In fact, there is a common notion among coaching and medical staff advising that braces should be removed as soon as possible to prevent the development of "preoccupation with injury".

Another Penn State athlete examined was "**Adrienne S.**", the shooting guard for the varsity women's basketball team. Adrienne explained that her most recent injury occurred in the off season last spring and was diagnosed as rotator cuff tendonitis, bursitis, and impingement syndrome. Adrienne recalls her injury with these problems occurring simply because of the overuse of her shoulder due to basketball over the years, and also because of weight room training, which she was doing at the time she was diagnosed. After the first month only resting, she slowly got back into some exercises and conditioning, but could not return to play for another three months, a total of four months in all. Since Adrienne was lucky enough to suffer this injury during the off season, she saw no diminished performance when she returned to play, because she had adequate time to recover and get back into top shape. Adrienne explained, "I did not suffer any mental trauma with this injury because I knew it was an injury due from pure overuse. I didn't suffer the injury on the court, and it wasn't during the season, which I think was a big factor in recovering fast and healthy without any psychological issues". Interestingly, although Adrienne didn't suffer any mental trauma herself, she has seen a lot of cases of trauma within the women's basketball team. The specific case she told me about involved a teammate suffering from a knee injury. When her teammate returned to play, she was extremely hesitant from fear of re-injury which diminished her play on the court, and eventually forced to coach to start another player in her spot. That was another mental shock to this player, which was in fact more severe than the physical trauma.

Another interesting point is that according to Adrienne's observation, women are more prone to suffer from mental trauma after injury because they are more aware of their bodies than males are. Women take injury to their bodies much harder because their self-confidence and self-esteem has a lot to do with their appearance. As far as psychological trauma is concerned within both genders, Adrienne thinks that it has a lot to do with the personality of the athlete at hand.

As can be seen from these cases, athletes may not suffer any psychological drama despite the high severity of their injury. It is worth stressing the fact that both of these athletes had surgeries during the off season. Neither was rushed back into health in order to play and neither had the pressure of teammates and coaches wanting them to play prematurely.

Even though the injuries were of different severities and caused by different issues, each athlete recovered without any physical or psychological problems and safely returned to sports participation. It is believed that since each athlete had plenty of time to recover without any added pressure and with a good workout schedule to get back into shape slowly, there were no short and long-term signs of psychological trauma.

3.3. Injury Caused by Non-Intense Activity

"**Dana V.**", a Penn State student, and about nine others took a rigorous hiking trip through the French, Swiss, and Italian Alps led by a kinesiology instructor in summer 2004. They trekked though a variety of terrain including snow, ice, mud, rocks, and thick forests and the sheer face of Mont Blanc. By the fifth day of hiking she noticed a stinging pain on the outside of her left knee and ignored the pain until it became so intense that she had to put it in a brace. Being also a member of the collegiate ice hockey club team, Dana faced this misfortune as a bad lack. Once she returned home trying to rest her knee, she agitated it even more when she coached at a girls' ice hockey camp. Upon medical evaluation performed at Athletic Medicine Center, including X-rays and an MRI, she was diagnosed with iliotibial band friction syndrome. This consists of the IT band becoming tense and inflamed and unable to glide smoothly over muscle and bone, thereby causing pain outside the knee joint.

Dana's injury was labeled moderate and to reduce the inflammation she was told to refrain from any rigorous physical activity. She blames herself partially because she had inappropriate training, did not warm up before each hike, and had improper footwear. It should be noted that it is a responsibility of the instructor to inform students about possible health-related issued associated with this type of rigorous physical activity. Her boots caused her feet to ache, blisters to form, and Dana to be a little off balance due to an improper fit, which could have contributed to the misalignment of the IT band. In her athletic career, she had previous injuries such as a strained left hip flexor which she said may have contributed to the injury since it was weaker. Those who have had previous injuries are prone to re-injury, but this could be avoided if the athlete is properly prepared and trained.

There were some signs of psychological trauma in Dana's case. She definitely experienced behavioral (i.e., avoidance of activities causing perceived pain), cognitive (i.e., irrational thoughts about consequences of injury), and motivational (i.e., would never do it again) signs of trauma. Since there was a constant uncertainty of aching, she rarely trusted herself to move her own leg without inducing pain. She continued to wear the brace for a while and when she removed it she experienced a bracing effect (abnormal asymmetry even without feeling of pain or discomfort) and refused to walk without a limp. Dana then experienced learned helplessness (difficulty in learning new skills during rehabilitation sessions), lacking the motivation to attempt and speed-up the recovery, and she was constantly preoccupied with the thought of re-injury. In fact, this lasted until her graduation from Penn State in 2006. Dana's lesson is that even if the sport is not filled with high intensity competition, not only physical, but psychological trauma can occur. Dana did not experience any pressure to recover quickly, yet she still had lingering issues that dealt with her injury. There is no rule that says one will or will not go through psychological trauma depending on whether one performs a high or low intense activity.

II. LESSIONS LEARNED FROM CASE STUDIES

Traumatic injures in athletics can be classified at three different levels: *Mild*, *Moderate*, and *Severe*. Conventional wisdom is that each level of injury depends on the amount of time an athlete is away from practice or game. A mild injury may consist of an injury that would require some treatment, but would not interrupt training. Hyperextension of a limb could easily be classified as a mild injury. A moderate injury would be a little more severe in that it would interrupt training time and possibly games or performances. This level could consist of a severely sprained ankle or a partial muscle tear because it causes athletes to be away from the game for a certain amount of time. A severe injury would consist of a greater amount of time absent from training or play that could possible end an athlete's chance of play again. A head, neck, or spinal injury along with broken bones, surgeries, and hospitalization are a few examples of a severe injury.

2.1. Can Injuries be Predicted and/or Prevented?

Always in debate, people argue whether or not injuries could have been prevented and predicted once they occur. Many say that improper physical conditioning and planning may predict/anticipate injuries. The coach and the athlete should understand that a certain level of training is necessary for each athlete. Coaches should know what level an athlete is at so as to insure proper training techniques and intensities for each specific athlete so that over training and over loading do not jeopardize the athlete's ability to meet the sport's demands. A freshman cross country/track runner starts out on the varsity team in high school and has to prove not only to himself but to the team and spectators that he can handle the challenge of training with the faster, more experienced athletes. Since all of the runners have the same workout, he over trains and experiences the sensation of "burning out." If the coach does not have specific training programs for each athlete, the runner experiences pulled hip flexors and a twisted ankle to say the least. Since these injuries may be considered mild, the trainer lets him continue running with strength training as well, but he is still hesitant to run on it. This coaching and training error may result in the development of a chronic fear of injury and constant bracing behavior.

Although many injuries cannot be predicted, a greater number can be prevented. A cyclist doesn't predict a car accident, yet he or she reduces the risk of injury by always riding with a helmet. Athletes, like this cyclist, arm themselves with protective gear, knowledge and skill in the hope that their efforts will reduce their risk of injury. One of the rare instances in which physicians can actively participate in the prevention of injury is the preseason physical examination and preparation evaluation. According to *PubMed,* examinations of musculoskeletal, cardiovascular and psychological status are essentials of any evaluation. With clearly defined objectives that take into account the prospective athlete, the contemplated exercise program, and the athlete's personal motivation, evaluations allow physicians a greater opportunity to reduce risk. "Contents of a pre-exercise physical examination should include history, a physical examination, laboratory testing and additional specific screening evaluations". (*PubMed*)

Along with physical screening that occurs during the preseason, many "psychological researchers are continuing to show that thoughts, perceptions and aspects of personality may be linked to the incidence of injury" (Sports Injury Bulletin). Factors such as stress and personality characteristics can increase the risk of injury in sport. Stress that provokes attention narrowing, which in turn leads to missed peripheral cues to factors that pose potential danger, is common among sports that require multiple focus points. A football player whose only focus is the ball may miss the player charging at him from the sideline. Excessive muscle tension is another negative side effect of stress that increases the risk of injury. A figure skater who tenses prior to a jump is unable to complete the rotation and lands awkwardly on a delicate ankle. "Although the relationship between stress and injury is complex, one study that used a large sample size 452 athletes (3) showed that as predicted, athletes with more life stress, little social support and poor coping skills were associated with more days of non-participation due to athletic injury (Sport Injury Bulletin)."

Along with stress, attitude also plays a role in predicting injury in sports participation. Many high level athletes have a mentality that requires 110% effort 100% of the time. This attitude may push them to take undue risks in order to improve their performance. However, just as with stress, these traits do not exist outside the individual. Meaning, each person is unique in his or her interpretation of various situations—an event that may be viewed by one as highly stressful may be a walk in the park for another. Therefore, personality analysis and evaluations must be conducted prior to sports participation to eliminate the risk of injury due to psychological dispositions.

Perhaps more critical than any form of evaluation is the timing at which it occurs. Prevention requires that all evaluations take place prior to engagement in activity. Previous injuries must be properly addressed before future injuries can be prevented, therefore timing is vital.

Technique is often labored on by many coaches and trainers as a means for prevention of injuries. By teaching basic techniques, not jumping to higher and more complex skills, coaches can limit the occurrence of injury. This is why "proper practice makes perfect". If athletes do not use proper technique, they increase their risk of injury and their number of actual injuries. The reacquisition of proper technique after injury is another factor to consider in terms of rate of recovery and the possibility of returning to sports participation. The premature clearing of an athlete, despite existing problems with proper techniques, such as overcompensation, protecting and bracing, often results in recurrent and often more severe injuries. It is a very common observation that if an athlete is not 100% healed and recovered, the lack of function and physical properties, such as strength, flexibility and endurance, leads often to another and more severe injury.

Take a field hockey player, for example, who in years past suffered from a fractured nose while playing and now has a slight avoidance behavior. She sprains her ankle while playing one game when turning to avoid a near face collision with the ball and goes to the trainer for rehabilitation. She hurries through rehab and then decides to play with her ankle taped for a game a week later by telling the trainer she is feeling good enough to play. Not fully recovered or accustomed to the taping of her ankle she loses her balance and falls, breaking her collarbone. This injury could have been prevented had she had the patience to let her ankle heal with a little more time, but since she was so concerned with returning to the game she caused herself more injury, which could have been avoided.

The Personality of an athlete is another factor to consider in terms of prevention of injuries. Knowing the *personality* can help coaches with their training programs and help define certain types of athletes. This can also help predict psychological trauma in sports related activities and is also a good predictor of the recovery rate from injury. By issuing evaluations and personality tests to athletes coaches can better create different training programs for each athlete and thereby prevent injury. Such tests as the Sensation Seeking Scale (SSS), the Minnesota Multiphasic Personality Inventory (MMPI), the Profile of Mood States (POMS) and the State-Trait Anxiety Inventory (STAI) may help coaches and doctors better understand the athlete's predisposition for injury as well as the athletes' reaction to an injury in the event of it happening.

Usually an athlete who scores high on the SSS test is known to have a higher rate of injury due to his or her personality. An over-confident collegiate soccer player was tested with true/false, yes/no questions to indicate which type of personality he had and the particular rate at which injury would occur with him. As expected, he was on the higher range of the scale, which seemed to answer why he was having a slow recovery from an ACL tear. Since he was an athlete who was very confident about recovery he had a hard time with rehabilitation when results were not automatic. All he wanted to do was play and he became depressed, slowly stopped going to rehab when results where slow and started showing signs of bracing and avoidance behavior. He also experienced cognitive symptoms by constantly thinking of the injury and started to feel isolated, alone, and moody. With the help of a psychologist, he started using positive imagery, support from his friends and physical therapy activities and he slowly got better with time. The personality test helps further indicate that certain personality types act a certain way towards recovery.

Completion of rehabilitation is an important step toward full recovery and preventing recurrent injuries. This insures that an athlete returns to maximum function and physical capacities, and can perform once again to the best of his or her ability. However the clearing of a physical therapist or athletic trainer does not always mean that an athlete is at 100%. An athlete may tell and convince the physical therapist or trainer that his or her injury is healed, and that s/he has full range of motion and suffers no pain or discomfort. Athletes may do this due to anxiety to get back in the game to support the team. They do not want to be seen sitting on the sidelines when they know that they could contribute to the team's success. An unfortunate attitude: "Injured Athlete is Worthless," prevails in modern sports. This attitude may push an athlete with residual symptoms of previous injuries into an unfortunate and often dangerous situation to be harmed again at a higher level of severity.

One interesting observation also indicates that not all strength training programs provided by the physical therapist may result in full recovery. Constant and proper assessment of the rate of recovery of various diminished functions, including psychological status, is an essential attribute of the treatment plan. Otherwise, if doctors do not properly diagnose and assess the rate of recovery, an effective treatment plan may be jeopardized. The injury itself could not have been prevented, but the severity of the injury could have been far less. For example, a lacrosse player experiencing lower back pains goes to the physical trainer treating him for a slight muscle tear. The next season he returns to the trainer with more pain to discover a vertebral stress fracture, which is a result of a degenerative disc. Due to continued play and applied force, the injury became worse. If the coach had conducted preseason tests or diagnosed the injury correctly the athlete would have been stopped from playing. All that could be done was to constantly move stretch, relax, have a positive attitude and alleviate the pain until years down the road when a replacement of a vertebra would be needed.

The recovery time after the injury or surgery, and when the injury or surgery occurred, plays a major role in the prevention of psychological trauma among athletes. Many injuries in athletes occur during pre- and postseason. Those who experience injury in pre-season are constantly worrying about their future playing career, if they are going to recover in time for the season, how many games they are going to miss, and if they are going to be at the same level as they were before they were injured. This causes many athletes to rush recovery and rehabilitation which then results in re-injury. This subsequent injury is usually more severe than the first injury, putting the athlete out for the season. On the other hand, athletes who experience injuries during postseason tend to not worry as much since there is no pressured time against them. Take an athlete who plays basketball at a collegiate level who has rotator cuff tendonitis, bursitis, and impingent syndrome, for instance. The injury was caused by overuse of her shoulder during shots and passes so she was required to take off a few months from exercise. Since there was no pressure to recover there were no reported symptoms of psychological trauma. All she had to do was to relax, and after the inflammation was reduced she was allowed light exercises that kept her in shape. So, a gradual and consistent progression of recovery in a timely manner is another important attribute to consider when examining the probability of the development of psychological trauma in athletes suffering from injury.

Finally, not all injuries can be pinpointed to a certain precipitating event. Some injuries are just bad luck (i.e., wrong time, wrong place ...). An athlete cannot blame his/her coach or trainer for the result of an injury when there was no prevention possible to predict and implement. One example could include a wrestler who landed wrong while trying to pin an opponent, resulting in a broken thumb. A player in volleyball diving for the ball and landing on an odd spot on her elbow could be another example of just bad luck. Genetics is another factor that cannot prevent injury. Some people are just prone to certain disorders or injuries due to their genetic makeup. Take a fullback on a football team with a ruptured lumbar disc for example. Both his father and brother also had a ruptured lumbar disc when they played football. One may say this is coincidence, but the fact that all of them had the same injury to the same lumbar disc suggests that a genetic factor is a possible cause of "inherited injury". The player suffered from nerve pains shooting down his leg and had surgery in the off-season to remove the disc. He reported a greater feeling of well-being since there was no longer any pain and no fear of re-injury because both his father and brother talked him through what was going to happen.

CONCLUSION

Several conclusions about psychological trauma pertaining to sport-related injuries can be made. *First*, it seems quite clear that psychological trauma can not be prevented. Most of the examined athletes have either had trauma themselves, or seen it in others around them on a regular basis. If trauma could be prevented then we wouldn't see so many cases and it would not be as prevalent as it is. *Second,* it seems that most athletes find that the majority of trauma cases they see come from injuries to the knee. It can only be suspected that the importance of the knee in simple activities such as walking makes it a much bigger target for both physical and psychological trauma. Also, injuries that require a brace, which include most knee injuries, result in a bigger risk for mental trauma. It seems that braces, casts, or any other injury wear are items that cause the athlete to think about the injury more often and therefore cause more cases of fear of re-injury and bracing behavior.

Although it is evident that trauma cannot be prevented, we must evaluate whether or not it can be reduced. Most of the athletes under study also commented on how important the athlete's personality is in being able to predict or reduce trauma. Since personality seems like a huge indicator of how an athlete will mentally react to injury, perhaps personality tests are a very valuable tool for team doctors and coaches. There are several different inventory tests that can be used to predict injury itself, such as the Sensation Seeking Scale (SSS), the MMPI, the Profile of Mood States (POMS), and the State-Trait Anxiety Inventory (STAI), but perhaps having tests about being able to predict how athletes will respond to injuries is the most helpful in predicting whether or not an individual is at high risk or not for trauma.

Whether or not physical signs of injury are present has nothing to do with psychological trauma, since trauma may be completely independent of the physical aspects of injury. Because of this, we can not base return to play on symptoms resolution, because psychological problems can still be present. Therefore, it is imperative that athletes who are injured show functional readiness of psychological recovery before returning to play. Psychological readiness is reached when no behavioral, cognitive, motivational, or affective evidence of trauma is present. This means that an athlete must show no fear or uncertainty of re-injury, no avoidance behaviors, no confidence issues, no signs of anticipatory pain, and an overall well-being without depression and anxiety. It is crucial that when approaching an injured athlete's treatment plan, appropriate assessments of the athlete are made through interview, observation, and/or psychological tests. In addition, the assessment must be interpreted, with the formulation of a problem list followed by a set of goals established by both the athlete and the supervisor. Also, some great ideas to add into treatment plans to reduce the risk of pain and trauma are imagery techniques and relaxation techniques. Overall, it is clear that injury and the psychological aspects of injury will always be here. The most important way to approach the future is to learn how to reduce these factors affecting athletes, and to teach both athletes and coaches the correct ways to train as well as good techniques for treating injuries, to prevent, as much as possible, the onset of psychological trauma.

CHAPTER 16

PSYCHOLOGICAL TRAUMA: AGE & GENDER FACTORS

1. IMPACT OF ATHLETIC INJURIES IN YOUTH

It is understood that with any associated with any sports is the risk of injury. The primary motivation for the millions of the youth in America who participate in team sports is to have fun (Nelson et al., 2007). It is often asked, why are there so many injuries related to specific sports? One would believe that by now someone would have made efforts to reduce the risk of injuries. One might wonder what happened during the transition between youth sports, where everything is just simply play, to competitive sports, where injuries are more advertised and known. Many factors impact why a child is injured in the first place. Some of those factors include improper form, technique, susceptibility, gender, age, parents, and coaching errors.

With an increasing number of children participating in competitive and recreational sports, athletic injuries have become increasingly common. Youths are beginning sports at younger ages and tend to participate in more than one sport. Often one hears parents-to-be say things such as, "I can't wait to teach my little guy how to play baseball or basketball." This shows how popular and influential sports are in today's world. For youths, sports are often an integral part of their daily lives. The youth of today are not only involved in school-related sports teams but community-based leagues as well, and many of them will continue their sports into adulthood.

With the increased participation in youth athletic activities, it is important to note that there is a vast difference between youth in sports and adults in sports. With the difference between youth and adult athletes being pointed out, it is important to note that with any type of physical activity, especially sports, there is always an increased chance of injury (Kessler, 2007). Dr. Paul Stricker comments that with the marked increase in more intense exercise patterns and early specialization in sports, more injuries can occur (Stricker, 2002). With more than twenty million American youth participating in school or community sports there are vast numbers of injuries that will ultimately require one or more of the following: medical visits, surgery, missed school, time-loss from practices and/or decreased activity levels (Damore & Metzl, 2003).

Children with forceful coaches and parents are very susceptible to becoming injured due to the pressure put on them and the amount of time they spend on their sport. The fact that children are still growing and developing is a complex factor when training for sports. Just as different children pass certain developmental milestones, such as walking and talking, at different times, they also gain skills at different rates. These gained skills often make providing specific guidelines for training more difficult. Therefore, young athletes may have to postpone certain aspects of their sport until they have matured developmentally and attained proper coordination and a higher level of skill (Stricker, 2002).

1.1. Age Factor

The age of an athlete can have a great impact on how susceptible he or she is to injury, as well as how she/he responds and recovers. Younger people are more likely to sustain injuries, and these injuries may become more serious if not taken care of. The accidental injuries that youth experience have serious consequences. Once the body is fully matured, it is bigger, stronger, and better prepared for strenuous situations. Adults are still prone to injury, but the extent of their injuries is not as serious as it would be for younger athletes. For example, if a child falls on an outstretched arm, that person have a higher chance of injuring the arm or clavicle at a higher severity than a twenty-year old who has never injured his/her arm before.

The degree of injury inflicted on the body varies by context it is implied through. According to recent studies, accidental injuries occur less often in elderly individuals than in young ones (http://www.niams.nih.gov/hi/ topics/childsports/child_sports.htm). It seems that as people age, they are more prone to receiving injuries because of the wear and tear that their body has seen all their life. On the other hand, two high school athletes colliding in a sport are going to have a higher severity of injury than two elementary school athletes. This is because the younger kids are not able to move as fast or inflict as much damage upon each other. Young children are supposed to be able to play and have fun during sports, without taking games too seriously. However, within the past ten years, the innocence of youth sport has become irrelevant, as winning and pushing oneself to the maximum have taken over. There are thousands of young athletes who are putting their health at risk for the sake of winning. If a coach or parent makes it seem extremely important for their child to win all of their games, extra pressure is put on their shoulders. This may lead to the child feeling stressed and increasing the risk of obtaining an injury or returning to the sport too soon after an injury. It is no secret that the concept of winning has become more prevalent in our society as parents and coaches continually push their children. Our civilization today has taught young people to win at any cost. With the drive to win, athletes can become less focused on form and proper techniques, which can lead to serious injuries.

There are many reasons why young athletes experience injuries in sport, some of which include not being physically and mentally mature enough to gain or acquire the skills needed to perfect certain sports. Since the view on strength training and other "adult-type" physical activities such as weightlifting, triathlons, and long distance running, has changed in recent years, some coaches and parents view youth as merely mini adults who can tolerate the physical requirements needed to produce the intended advancements in their physiological training. The lack of knowledge may be responsible for the injuries accrued by youth because of overuse and even improper training techniques and intensity that are beyond the child's capabilities, but are rather based on length, actual intensity, and the duration of the activity. Clearly, an injury received through improper form may have negative consequences in the child's future.

Another possible cause for the injuries results from the failure to warm up, which is a direct consequence of coaching errors. Going into a competition with cold muscles and joints does not help the boy endure the stress imposed through the sport. Athletes should stretch before a practice or game to prevent pulling or straining any muscles during the activities. These results from the pressure imposed on young individuals to compete at a high level all the time. Additionally, some youth will experience the effects of unrealistic expectations from their parents, coaches, or even both which can pressure the young athlete to perform when he/she is not ready, or cause the denial of physical problems, among other issues. Controlling parents may push their child into participating in sports that the child is not interested in playing. Since the child is not enjoying the physical activity, he or she may get injured on purpose or fake an injury in order to avoid playing. Another factor is a discrepancy between the physician responsible for the young athlete and his or her parents. Often, parents will rely too much on their child's success and will be disappointed by the physician's instruction that their child must take time off to heal. Parents should be more supportive of their children by encouraging their children to participate in sports in which they are interested.

According to Michael C. Koester, the writer of "Youth Sports: A Pediatrician's Perspective on Coaching and Injury Prevention," the number of youth who participate in an array of non-scholastic organized sports between the ages of six and sixteen is nearly twenty million. Of these youth, it is estimated that young girls who play organized sports have an injury rate of 20 to 22 per 100 participants per season (Koester, 2000). The injury rate for youth boys is almost doubled that of the girls at 39 per 100 participants per season. These are indeed scary statistics. The data at the time this article was published indicate that sports and recreational activities account for 32.3% of all serious injuries in children ages five to seventeen years. According to Koester (2000), the potential causes for injury in youth sports participants are often stresses on the unique physiologic and biomechanical aspects of the growing body. The expanding literature concerning the effects of life stress on the occurrence of injury has yet to be scientifically investigated. However, anecdotal evidence implies that improper training methods and lack of proper sport-specific techniques also contribute to injury. It is recognized that injury results from a natural risk of participation in organized sport, although many of the injuries can be prevented, particularly those resulting from a lack of proper, well-supervised training and participation. Coaches play an imperative part in the young athletes' lives. According to Koester (2000), 85% of coaches are either parents or other interested persons who have no formal training in coaching. As the need for coaches has quadrupled over the last twenty years, there has consequently been a shortage of applicants who are qualified for the position. Overall, poor coaching may be directly linked to the growing number of injuries in youth sports.

1.2. Overuse Injuries in Youth Sports

As mentioned in the previous section, coaching errors significantly contribute to injuries in youth sports. This notion will be elaborated on in the following text. Athletic injuries such as fractures and sprains (macrotrauma) as well as enthesitis and tendonitis (microtrauma, which is secondary to overuse) are the most common injuries occurring in children. These injuries generally come from a combination of overuse and misuse. It was noted that many young athletes are never taught the proper conditioning methods or technique for their sports. As a result, youngsters and coaches alike often fall prey to the "more is better" philosophy of sports training, attempting to do "too much, too soon" while also using improper biomechanics (Koester, 2000).

Many of the coaches who teach youth sports simply do not have the proper knowledge base for instructing their players in good technique while performing major drills. The volunteers, who have not been formally trained, often base their coaching style and techniques on their own experiences. These experiences are often outdated and not sufficient for the current development of sports. Nonetheless, even trained coaches will likely have learned many of the technical aspects of their job by observing and listening to other coaches. Both styles of coaching knowledge are prone to misinformation and improper theory being perpetuated for years, setting the stage for injury. Thus, injury in youth sports should be considered an evolving process, rather than individual accident due to bad luck. Therefore, when dealing with youth sports, it is very important that the athletes are well supervised and taught by a qualified instructor. Forceful and aggressive coaches can have a negative impact on the youth on their teams. Overuse can be a result of too many or too intense practices. If a child cannot handle all of the activity expected of him/her, he/she may be prone to injuries.

Another factor that is often overlooked is the environment in which the youth actively practice and participate in their sports. Many times, there is inadequate supervision from coaches, or youth practice in overly hot environments. It is noted by Stricker (2002) that heat stroke is the third most common cause of death in high school athletes. It is very important for children to learn the proper forms and rules for the different sports that they are participating in. Improper form while playing a sport only increases the child's risk of injury. For example, having children play in a peewee football league and tackle an opponent with the wrong form, risks possible injury to them and their opponents. Another example would be a child heading a ball in soccer, there is a very specific way to do this, and the children need to learn that you have to hit the ball a little above the eyes in the center of your forehead to prevent any brain damage.

As mentioned with the coaching styles, overuse is quite possible in youth sports. If a particular action is repeated too many times, certain parts of the body can become worn out and then physically injured. The overriding factor for risk of injury is overuse, concludes orthopedic surgeon James R. Andrews, MD, who says about this issue: "I see hundreds of young athletes in our clinic whose tremendous athletic ability has been compromised, or even cut short, because their musculoskeletal systems have not had a chance to mature" (Andrews, 2007). This can be attributed to the fact that, with a higher level of intensity of play being needed, the rise in overuse injuries spike in the young athletes who participate. Moreover, the Academy of Pediatrics states that significant sport specialization at an early age carries the risk of potential negative outcomes. Guidelines to protect young athletes from overuse injuries have been developed based on numerous scientific studies that prove kids are being damaged by asking them for too much too soon.

A common overuse injury in sports that requires throwing or swimming is rotator cuff tendonitis, in which the muscles and tendons in the shoulder area are overused, causing the tendons to develop tears (Kessler, 2007). This injury can prove to be problematic since the shoulder is very mobile in everyday tasks, but especially in sports activities. In order to manage and treat this injury, the athlete must decrease the activity that initially caused it for usually two to six weeks, depending on the actual severity of the injury. Constant injury and muscle overuse can lead to problems down the road for the joints, tendons and muscles, causing pain when using these areas later on.

In addition, musculoskeletal overuse injuries are the most common injuries seen in adolescent athletes. There are different grades of overuse injuries ranging from Grade 1 to Grade 4. The article "*Sports Injuries in*

Adolescents" (Nelson, & Patel, 2000) gives a chart of each grade and what is usually associated with that particular level:

Grade 1: Pain follows hours after activity; diffuse pain and soreness; mild tenderness not well localized; usually <2 weeks' duration and resolves rapidly with rest.

Grade 2: Pain immediately after activity-may localize; usually <2-3 weeks' duration; poorly localized tenderness; most athletes present at this stage.

Grade 3: Pain during activity; localizes; recurs with activity; increase in severity-localized tenderness; may limit activity; performance deteriorates.

Grade 4: Pain at rest; continuous; also with daily activity; localized severe tenderness; functional disability usually >4 weeks' duration; consider stress fracture or other cause.

There are a plethora of different sports that youth participate in, each requiring a different set of skills and movements. There are many possible injuries for youth sports. Some youth may specialize in one particular sport, while others may be involved in all kinds of sports depending on the season. Different sports often have distinct sport-related injuries that are associated with them. Overuse injuries are serious issues that should be carefully considered by parents, coaches and medical professionals.

1.3. Epidemiology

A study published in the Journal of Athletic Training explores the most common sport-related injuries, ankle injuries, using national data. This descriptive epidemiological study investigated the rates of ankle injuries by sex, type of exposure, and sport in one hundred United State high schools (Nelson et al., 2007). The study reviewed data collected during the 2005-2006 school year on ankle injury, from athletes participating in nine sports. It was noted that during that school year, more than 7.1 million students in the United States (US) participated in high school athletics. This represents an all-time high of 53.5% of enrolled students and demonstrates the seventeenth straight year of growth in high school athletic participation. This nationally representative sample was obtained by an injury surveillance system called High School RIO. The sports studied included boys' football, soccer, basketball, wrestling, and baseball. The girl sports included soccer, volleyball, basketball, and softball.

During the 2005-2006 school years, the national number of ankle injuries was estimated at 326,396. This accounted for 22.6% of all injuries and yielded an ankle injury rate of 5.23 ankle injuries per 10,000 athlete-exposures. During competition, there was a significantly higher rate of occurrence (9.35 per 10 000 athlete-exposures) compared to during practice (3.63), (risk ratio = 2.58; 95% confidence interval = 2.26, 2.94; P < .001). The results of the study showed that the highest rate of ankle injuries was seen in four sports, specifically football, basketball, soccer, and volleyball, each which involved jumping in close proximity to other players, as well as swift changes of direction while running. Boys' basketball had the highest rate of ankle injury (7.74 per 10 000 athlete-exposures), it was followed by girls' basketball (6.93) and boys' football (6.52). With the exception of girls' volleyball, all the investigated sports had ankle injury rates higher in competition than during practice. This is an interesting finding requiring further investigation.

Ligament sprains with incomplete tears accounted for 83.4% of the diagnosed injuries. Athletes most commonly missed less than seven days of activity (51.7%) as a result of their ankle injuries. Athletes who lost seven to twenty-one days of activity accounted for 33.9% while those who lost more than twenty-two days of activity accounted for 10.5%. The conclusion from this comprehensive study indicated that sports combining jumping in close proximity to other players and swift changes of direction while running are most often associated with ankle injuries. However, future research on ankle injuries is needed to drive the development and implementation of more effective preventive interventions.

Other types of injuries commonly associated with youth sports are strains, sprains, fractures, brain injuries, and injuries to the upper and lower extremities. Another injury that is important to take care of is damage to the growth plate, which can happen at the tissues at the end of the bones. If an injury is sustained by the tissues before the growth plate is fully developed, then the child needs to see an orthopedic surgeon to fix the problem before it causes a long-term effect (http://www.hughston.com/hha/a.soccer.htm). Injuries also include contusions; spinal cord injuries although rare can still happen to young athletes and have a tremendous effect on their nervous system; and other skeletal injuries that include fractures and other strains that can affect bone growth. A strain is a serious injury to consider the consequences of because it can lead to the weakening of the tendon. If there is a complete tear of the tendon, the athlete would require surgery and perhaps would not be able to compete again.

Overall, skeletal traumatic injuries can have a damaging effect on the growth of adolescents. Problems occur when the athlete sustains an injury that affects his or her growth plates, which includes the tissues that is growing at the ends of the bones and develops into bones and get larger while a person is maturing. Even though some of these injuries are unlikely to happen to most athletes, any injury if not treated correctly or noticed and corrected can lead to problems that develop down the road.

1.4. Psychological Impact

No matter how severely a child is injured, he/she is psychologically impacted by it in some manner. If a child is treated differently by parents or coaches after an injury, he/she is psychologically and emotionally impacted. For example, a star athlete would miss the attention he or she received when playing and being positively reinforced. Many children feel like they can never fully recover, which is upsetting. Injuries can result in anxiety, depression, fatigue, pressure, rehabilitation problems, and fear of re-injury. When athletes endure an injury and are going through the recovery process, it is almost certain that they will experience negative thoughts of re-injury. Psychological trauma is an individual's subjective experience that determines whether or not an event is traumatic. The psychological trauma that comes with overusing and overworking kids far outweighs winning games on grade or high school teams.

It was suggested that negative emotional reactions to a sports injury rend to diminish with age (Brewer, 2003). As an example, a ten year old soccer player who is injured on the field is more likely to express concern about the physical pain he or she experienced, rather than the severity of the injury and recovery time involved. A high school athlete with ambitions to play at the collegiate level would express the most concern about severity of the injury and whether or not he/she can continue to participate without rest and rehabilitation. Most athletes in this circumstance express little concern about the initial pain they are sensing. Although collegiate athletes are much closer to each other in age, a similar response can be noted. Freshman football players who are injured on the field have been noted to express concern about the severity of the injury and immediately seek help from the team trainer to take care of the injury. Senior players who are injured on the field are more likely to express less concern about the severity of injury and, in many cases, attempt to hide the injury from coaches, trainers and other players (Nelson, 2006).

Age plays an important role in post-injury psychological consequences. Age maturity should be considered in psychological consequences, as it is in the physical consequences of a sports injury, because the different levels of maturity affect the road to recovery. For example, adolescents report more pain and anxiety than adults within twenty-four hours post-surgery. Also, pre-surgical cognitive, emotional and behavioral variables differ between adolescents and adults. Adults who are more self-motivated and perceive a high level of social support will recover quicker than younger patients who are not eager to recover (Brewer, 2003). The reverse is also true. Young patients who are highly invested in the athlete role as a source of self-worth will recover faster than adults who are not self-motivated.

It was documented that the interactive effect of age and intelligence is a significant determinant of injury and the possibility for the development psychological trauma as a result of injury. According to JAMA, more intelligent six year olds are less likely to experience a traumatic injury by age seventeen and if they do, are less likely to develop symptoms of psychological trauma. Children with an IQ above 115 at age six are less likely to have psychological trauma after suffering a traumatic injury by age seventeen, than those who have an IQ less than 115. Also, children are less likely to have psychological trauma after a serious injury because children tend to heal much quicker than adults do. Adults who have experienced a serious injury will take a longer time heal and may have a harder time dealing with injury and recovery. Overall, the injury has a higher probability of affecting the adult's everyday life than it does a child who does not have as many responsibilities.

Another important psychological factor that may indicate further need for intervention as a result of youth injuries is using the injury as an excuse to quit the sport or activity in exchange for not embarrassing one's self, parent, or coach. When a pitcher gets injured, especially a child, it is hard for that athlete to determine whether the event is distressing or not. This type of pressure put on a young person could have lasting effects, as the child could grow up always feeling nervous and depressed. Additionally, he or she might not know how to cope with life, especially when something negative like an injury happens. Despite fundamental procedures to rehabilitate youth athletes physically, there are also psychological factors that are important and quite different from the psychological traumas that adult athletes may experience as a result of sport-related injury. There are differences in the way youth and adult athletes respond to physical injuries. As Nelson and Patel (2000) state:

"The athlete may go through a series of reactions after an injury similar to those described in the context of death or other loss—disbelief, denial, and isolation; anger; bargaining; depression; and acceptance and resignation with hope, in that order of progression. This pattern is believed to be seen far less commonly in adolescents than in adults, and young adolescents seem to proceed rather rapidly from anger and frustration to acceptance."

Many doctors and even coaches are unaware of the differences in training, treating injuries, and dealing with family issues that may arise when injuries are brought into the life of a young athlete. Young athletes should be able to recover quickly from injuries if listening to the right advice. Proper rest and treatments, along with a positive imagery of recovery, help these athletes to recover quicker. With most minor injuries to younger children, they just need rest, ice, elevation, and compression to get better, so with proper time and treatment most young athletes will be able to recover and play again. Indeed, well-documented research on the susceptibility of children to psychological trauma as a result of injury is lacking and deserves special consideration.

1.5. Prevention

In order to prevent youth athletic injuries from occurring in the first place, many actions could be taken. Rules and regulations could be implemented for particular sports in order to avoid overuse. Coaches would have to regulate how often their teams practice, as well as how intense those practices would be. There are several steps that can be taken in order to decrease the number of overuse injuries and harmful psychological problems that come along with them. Speaking to the young athletes about how winning is not everything and having fun is the main goal would initiate the steps. When young athletes are playing the game for fun, instead of worrying about winning, there is a reduction in the pressure of success. Positive and encouraging environments would help avoid kids' resorting to negative solutions when they get hurt or injured. Parents and coaches can help develop a plan so that rehabilitation is successful once completed. A great emphasis should be placed on teaching children the proper way to participate in an activity. Coaches and instructors also have to make sure that the children are getting a proper amount of rest and are not overusing their muscles and tendons because this can lead to injuries.

A former teacher of physical education, Willem van Mechelen, now a professor of Occupational and Sports Medicine, investigated preventative measures that may reduce the risk of sports injuries in youth in his essay entitled *Injury prevention in young people- time to accept responsibility*. Dr. Mechelen speaks of the great potential health gains that result from prevention of sport injuries, specifically in youth. "In the short-term, the absolute number of sports injuries falls, and, in the longer term, the risk of injury recurrences and chronic damage is prevented. Prevention of injury also promotes a physically active lifestyle from childhood into adulthood." (van Mechelen & Verhagen, 2005)

In youth, the measures to prevent sports related injuries should be based on knowledge of the incidence and likely severity of the particular injury, causal factors, and mechanisms that contribute to the risk of sustaining sport-related injuries. Once identified, preventive measures should be implemented and assessed, preferably in randomized trials. Although not much is published and known with respect to sports injuries in young people, there is one crucial factor that is known to have an important role in injury prevention: an understanding of the behavior of children, by their parents and sports instructors (van Mechelen & Verhagen, 2005). As children grow they continue to learn and develop skills. From six to twelve years of age the nature of physical activity evolves from joyful play into competitive sport. Everything a child of this age range learns about ways to avoid injury, the capabilities of their body, safe play, and fair play will be carried forward and will affect the way they take part in sports for the rest of his or her life. At this age, the coaches and parents think of the children as "mini adults and potential Olympic athletes, and enforce the children to value performance over enjoyment" (van Mechelen & Verhagen, 2005). The added pressure encourages the children to mimic professional role models. During the growth phase, in team sports such emulation can lead to risky situations in which there are great differences between children in, for example, body size, degree of skill, and strength. In individual sports, excessive training and incomplete rehabilitation after injury can lead to injuries associated with overuse and subsequent growth problems.

The main cause of injuries between the ages of twelve and eighteen years is behavior. Most children take part in competitive sports, and need to be made aware of their behavior and whether it is putting themselves or others at risk. For instance, proper protection in inline skating is rare in this age group, since wearing a helmet is not thought cool. Another example comes from football, where children do not always see the danger of certain forms of tackling. If taught good values at a younger age, while sports is still more about play than competition, fair play in adolescence will represent a natural progression. If children are better educated by the adults who influence them, injury prevention could improve in youth sports. The improved knowledge would then pass on to their later years and less injury could be seen in collegiate athletics.

CONCLUSION

Despite the alarming statistics of youth-related sport injuries, there are many benefits to youth sport participation. Some of the physical benefits are "motor skill development, foundational hand-eye coordination, and improved cardio respiratory fitness, promotion of physical growth, increases in total daily energy expenditure and positive effects of body composition." (Stricker, 2002). There are also numerous psychological benefits such as improved self-esteem, social skills, and motivation to continue a healthy lifestyle throughout puberty and adulthood. There is hope for the future in terms of research and interventions for youth injuries. Properly understanding why children get injured and how the injuries impact them is important for athletics in years to come. Among all the controversy that surrounds overuse injuries and the psychological trauma that comes with them, there is the hope that we can help young athletes. That said, it is feasible to propose that with regard to age and psychological trauma, a greater degree of psychological trauma would be experienced by those of a younger age. It is a growing concern among medical practitioners that children may suffer from psychological trauma sometimes resulting in post-traumatic stress disorder, if not properly treated.

It should be noted that teenagers are more prone to "sloughing off" traumatic injuries because of the love of the sport and the drive to continue to play. But if a child was affected by psychological trauma at a young age, this trauma may have residual effect for years afterward. Although adolescent athletes may have better success in dealing with physical trauma and heal faster, their psychological responses to injury may be highly maladaptive, leading to long-term consequences and residual effects.

In youth sports, there are different factors that come into play that impact injuries as well as their psychological effect. Parents and coaches' interaction with their aspiring youth athlete is crucial. If both parents and coaches were educated about the harm they put their children through just because they want them to win, there might just be a reduction in injuries. In sports in which ankle injures are prevalent, if more research is done to take preventative measures and analyze the way individuals play the sports, there might just even more reduction of growing youth. The psychological impacts of injuries need to be considered when a child is injured. A severe injury could greatly impact a person for the rest of his or her life. It is believed that with proper recovery, the child is able to overcome the negative psychological effects after injury.

2. GENDER EFFECT IN RESPONSES TO INJURY

For a long time, sports were viewed as strictly masculine activities. However, the involvement of women in athletic activities became one of the most revolutionary schools of thought that have evolved to change how women are viewed in society. Originally, women were prevented from participating in sports because they were thought of as fragile, frail and physiologically weak. This dates back to and even before the games of ancient Greece, were women were not even allowed to watch athletic competitions. Eventually, they began to compete in their own Olympic games, known as the games of Hera. One of the most important legislative improvements in the U.S. was the Educational Amendment Act, Title IX, which stated that "federal money could not be given to public school programs that discriminated against girls."

For the most part, people feared that women were physiologically different from men and this could affect them if they participated in sports. For the most part, this idea was unfounded. However, as more research is done, a difference in the injury rate between men and women, depending on the sport and the type of injury, is being revealed. For example, women are two to eight times more likely to suffer from injury to the ACL, compared to men (Harmon, 2000). In addition, body composition is a major factor in the types and number of injuries sustained by each gender. It has been shown that females sustain a higher number of lower body injuries, while males sustain a higher number of upper body injuries (Schenck, 1999). It was also reported in this study that male athletes tend to become more insecure after an injury, while females demonstrate increased anxiety and tension. Also females are more likely to conceal anger and become less assertive than males.

This topic is important because knowing the specific mechanisms that lead to injury in men and women can be used to design training programs to alter motor behavior in men and women for different types of sports. Furthermore, knowing how men and women are affected psychologically by injury can lead to more well designed rehabilitation programs. Overall this information can be used through varied programs to prevent forms of injury that are physical or psychological, and for a safer and faster recovery or return to play. It should be noted that gender differences in terms of causes, consequences and individuated responses to injury are poorly studied.

Psychological trauma can have different effects based on gender. While females may have greater degree of pain tolerance than males, women allow pain to be visibly seen more than males. This is partly because males have more of a complex with their pride. If they show they are in pain, they might fear that they are being seen as weak. Men have an easier time hiding their emotions, and consequently are able to mask any signs of psychological trauma they may be suffering from. Females, on the other hand, may not have as easy a time masking their emotions and mental pain, but they also are more likely to admit they are suffering the psychological impact of injury (Zlotnic et al., 2001). Females are much more likely to reach out and ask for help when dealing with psychological trauma than males. The psychological ramifications of injury, including the disruption of social support networks, a compromised relationship with coaches, and possible change in playing position and team hierarchy, weigh heavily on the minds of injured female athletes. It was known that females have the tendency to more frequently re-experience and re-live the trauma they suffered through. It should be noted, however, that the results of this report may be biased since females are more likely to come forward and admit they have a problem and seek help. Whereas males might suffer just as much if not more than females, yet because they do not seek help as often or are less open to admitting they are suffering, there is a level of uncertainty and room for possible errors of generalization.

There are several well-documented studies that are worth considering. Accordingly, in the next few paragraphs, the discrepancy in injury rate as a result of *intrinsic* (physical or biomechanical) and *extrinsic* (social & environmental) factors will be discussed (a) in general; and (b) in relation to ACL injuries, which are most common in the athletic environment, as well as psychological factors and their effect on recovery.

Gender differences can be traced to several specific areas. ***First*** is that the relationship between athletes and coaches serves as the backbone of an athlete's career. This is well-documented in sports science literature. For example, female athletes are usually unhappy with the relationship with their coaches following sustained injury. Specifically, 94% of studied females reported negative feeling as compared to only 20% of males under this study. Many female athletes felt that they were ignored by coaches after they suffered from an injury. The females interpreted their coach's actions as negative feelings. Males, on the other hand, expressed a much different perspective. Many of the males interpreted the coach's actions as part of their job. "I did not hear from my coach much, but that's O.K. Their job is to put the best tram out on the field. Their job can't stop just because I'm hurt." This quotation was from a male football player.

Second, females also noted the lack of sympathy from their coaches when they were injured. Many female athletes felt that the coaches failed to recognize the psychological impact the injury has on the athlete. Unlike males, who reported the opposite, saying that coaches are very understanding and responded to injured athletes in very positive way. Along with ignorance and lack of sympathy, female athletes also noted negative social behavior toward the injured athletes. Examples included blaming the athlete for getting hurt, poor team performance, insisting that the athlete faked the injury and hassling the athlete to rush back onto the field. Male athletes did not mention negative behavior from their coaches and mentioned that the coaches would not push injured athletes back onto the field, for fear of recurrent injuries.

Third, gender differences can also be noted in the relationship with a significant other during the time of suffering from injury. The majority of male athletes mentioned positive support from their girlfriend during the injury recovery process. None of the studied females remarked on support of any kind from their boyfriend. The reasons for this discrepancy deserve to be fully elaborated and further studied. Fourth, a very significant difference in genders involved the future health of the athlete following athletic injuries. Female athletes had much more concern about their physical health later in life in apposed to their athletic career. The majority of males expressed less concern about their future and mainly focused on getting back on the field.

Overall, research in psychosocial responses to athletic injury has shown much diversity among athletes. Differences in gender appear to be quite disturbing. Male athletes seem to be careless with their future well-being and believe that playing with injuries is a sign of leadership and courage. Females expressed much more concern fro their future health. This careless behavior in males can result in short athletic careers and lifetime disabilities, psychological and physical. Differences in the relationship with significant others came of no surprise. Many male athletes in sports such as football or basketball in high school and at the collegiate levels typically have supportive girlfriends. As for elite female athletes, many are so involved with their sport and are put under more stress from coaches, and parents, that they do not have time to devote to an additional relationship. It is quite opposite that most female athletes mentioned were not involved with a significant other.

Gender-specific psychological responses to injury have been addressed in several studies. Specifically, psychological trauma has been examined with the "Impact Event Scale," commonly used in individuals who have experienced natural disasters. Athletes were graded upon their answers to the intrusion and avoidance sub-scales. The intrusion scale examined the "unbidden thoughts and images" while the avoidance scale examined the "avoidance of situations that remind the person of the traumatic events." The results showed that the natural disaster victims had higher intrusion scores than the athletes, but the athletes' avoidance scores were significantly higher. The group of female athletes was found to have much higher avoidance scores than their male counterparts, indicating that females tend to be more psychologically distraught after a major injury (Shuer & Deitrich, 1997).

The most common sports that result in injury for females are: soccer, basketball, track and volleyball. Most athletes experience psychological trauma after the first injury. The most common injury where psychological trauma occurred is anterior cruciate ligament (ACL) injuries (Manuel et al., 2002). This is consistent with previous research. For example, according to the article: "*A comparison of knee joint motion patterns between men and women in selected athletic tasks*" (Harmon, 2000), ACL disruptions are common knee injuries, with 70% of them resulting from athletics. Most of them are non contact in nature, meaning that there was no direct contact at the time of injury. In this study, twenty recreational athletes (eleven men, nine women) with relatively average knees were randomly recruited from Duke University. Their movements were analyzed using Maximal Voluntary Contraction (MVC) tests. The results showed that the knee flexion angle was smaller for women by about 8. Furthermore, women tend to have more knee valgus angles, greater quadriceps activation, and lower hamstring activation during the tests, in comparison to men. Literature suggests these alternated knee motion patterns of women tend to increase the load on the ACL, which could lead to a greater instance of injury. Moreover, women were found to be more lax in the ligaments, which may be related to higher levels of estrogen and relaxin. Women also rely less on hamstring activation, which protects the ACL (Malinzak et al., 2001). It should be noted that this preliminary research is inconsistent and should be further elaborated with respect to numerous uncontrolled variables.

Since Title IX was passed in 1972, women may have been disadvantaged athletically. The article states "With the passing of Title IX in 1972, many women began to compete in organized sports with little or no previous experience, which was contrasted with their male counterparts, who typically had many years of athletic experience by the time they entered high school. There was concern that this lack of experience would translate to decreased skill and coordination and increased injury rates. Lack of experience has been proposed to contribute to the increased rate of ACL injury in women." Some methods of prevention to reduce injury rates include early intervention for proper techniques and learning how to land safely and properly, but more research needs to be dedicated to this understudied topic.

When it comes to gender related psychological causes or emotional responses following athletic injuries, some sources report that there is no statistically significant data to support this conclusion, while others agree that there may be some differences among genders. For example in a study called *"Emotional Responses of Athletes to Injury,"* conducted by Smith et al (1990), no gender differences were identified but the article did state that the most seriously injured group experienced significantly more tension, depression, and anger and less vigor than college norms, a mood disturbance that lasted one month. Furthermore, the study *"Positive States of Mind and Athletic Injury Risk"* discusses a model in which in stressful situations such as demanding practice or important competitions, the athlete's history of stressors (e.g., life event stress, daily hassles), personality characteristics (e.g., trait anxiety,) and coping resources (e.g., social support, coping skills) contribute interactively or in isolation to the stress response (Williams et al., 1993). The central hypothesis of the model states that individuals with high stress and personality characteristics with few coping mechanisms and responses when placed in a stressful situation are more likely to view the situation as stressful or maybe even threatening. They will exhibit greater muscle tension and attentional disruptions, leading to greater risk of injury compared with individuals with the opposite profile.

Gender differences exist in emotional responses; however, they may become more significant after the injury. Therefore, following injury, it is very important to be aware of emotional responses to injury in order to be certain that the athlete is ready to return to competition. As stated, emotional responses can manifest themselves as fear, lack of confidence, and frustration, among others, and can manifest during the competition. In complex coordination events like gymnastics or diving, these emotional responses can lead to severe injury of the spinal cord, for example. One of the reasons for the difficulty in finding emotional gender differences and their contributions to injury is that female athletes are more likely to admit to their feelings and fears. The male's ego or need to be viewed as strong and masculine may cause him to want to hide an injury or hide the fear. While males and females both have fear of pain, the openness of females in admitting it may help the coach incrementally return them to practice and raise and observe the athlete's confidence levels as she works toward return to competition. According to *"Practical Guideline for Treatment of Patients with Acute Stress Disorder and Posttraumatic Stress Disorder"* (Ursano, 2004), positive encouragement and attention may help in preventing negative emotional response to physical trauma, which in turn reduces dissociative tendencies in female athletes. One of the implications for this information is that since female athletes are usually more open to admitting fear of injury or failure, it can be easier to determine when a female athlete, as opposed to a male athlete, is ready to return to competition following injury. Furthermore, emotional responses such as depression or fear can be counseled to allow the female-athlete to safely return to play. At this time, the degree of susceptibility of female athletes to psychological trauma and its long-term consequences is unknown.

CONCLUSION

The limited literature showed that there are differences between men and women in the way that they are built physically and the way that they perform biomechanically. For example, women have smaller knee flexion angles and recruit hamstrings, which are important to support the ACL, less. This may contribute to a greater instance of ACL injury in women. However, women may be ready to return to play in a safer manner because they are more likely to talk about their feelings, and admit theirs fears and preoccupations.

According to the literature, while males and females are prone to different types of injury because of their different body composition, females did tend to express a higher degree of psychological trauma when compared to their male counterparts. Overall, females experience increased anxiety and tension, therefore creating more psychologically stressful situations. While male athletes also deal with psychological trauma, it appears to be less prevalent and/or long lasting.

Psychological trauma often goes unnoticed by athletes and coaches alike. Even though the athletes may be completely rehabilitated, this does not mean that they are at their full potential to perform in the same way they did prior to the injury. That said, with counseling, the root of the problem can be discovered and professionals can work toward healing injured athletes holistically from both physical and psychological perspectives. However, the current reality is that athletes are often reluctant to admit they have psychological problems, and coaches continue to preach that injury is not allowed to be discussed and that pain is unacceptable and should be ignored.

REFERENCES

Nelson, E. (2007). Complaints of injury Cover-up. Oakland Tribune.

Nelson, A., Collins, C.L., Yard, E.E., et al. (2007). Ankle Injuries Among United States High School Sports Athletes, 2005-2006.*Journal of Athletic Training,42*(3), 381-387.

Stricker, P. R. (2002). Sports training issues for the pediatric athlete. *Pediatric Clinics of North America, 49*. Retrieved May 6, 2007, from MD Consult database.

Damore, D. T. & Metzl, J. et al. (2003). Patterns in Childhood Sports Injury. *Pediatric Emergency Care 19*(2), 65-67.

Michael, C. Koester, F. (2000). Youth sports: A pediatrician's perspective on coaching and injury prevention. *Journal of Athletic Training,35*(4),466-470.

Andrews, J.R. MD. Medical Director, Orthopaedic Surgeon. American Sports Medicine Institute. Birmingham, AL. Clinical Professor of Orthopaedic Sports Medicine, University of Virginia Medical School (2007).

Nelson, T. L., & Patel, D. R. (2000). Sports injuries in adolescents. *Medical Clinics of North America, 84*. Retrieved May 6, 2007, from MD Consult database.

Brewer, S., & Britton, W. (2003). Developmental differences in psychological aspects of sport-injury rehabilitation. *Journal of Athletic Training, 38*(2), 152-153.

van Mechelen, W. & Verhagen, E. (2005). Essay: Injury prevention in young people-time to accept responsibility. *The Lancet:Medicine and Sport, 366*, S46

Schenck, R.C. (1999). Athletic training and sport medicine, Rosemount. Il American Academy of Orthopaedic Surgeons.

Zlotnick, C., Zimmerman, M., Wolfsdorf, B., Mattia, J. (2001). Gender and trauma-related predictors of use of mental health treatment services among primary care patients. *Psychiatic Services, 57*(10), 1505-9.

Manuel, A., Janeen, C. et al. (2002). Coping with sport injuries: An examination of the Adolescent athlete. *Journal of Adolescent Health, 31*, 391-393.

Malinzak, R.A., Colby S.M., *et al.* (June 2001). A comparison of knee joint motion patterns between men and women in selected athletic tasks. *Clinical Biomechanics. 16*(5), 438-445.

Williams, J.M., Hogan, T.D., Andersen, M.B. (1993). Positive states of mind and athletic injury risk. *Psychosomatic medicine, 55*(5), 468-72.

Williams, J.M., & Roepke, N. (1993). Psychology of injury and injury rehabilitation. In R. Singer, M. Murphey, L. Tennant (Eds.), Handbook of research on sport psychology, (pp. 825-839). New York: Macmillan.

Ursano, R.J., Bell, C. *et al.* (2004). Practical Guideline for Treatment of Patients with Acute Stress Disorder and Posttraumatic Stress Disorder. *American Journal of Psychiatry*, (September),1-96.

PART IV: CONCUSSION IN ATHLETES

CHAPTER 17

CONCUSSION: WHY BOTHER?

1. INTRODUCTION

The generally accepted definition of mild traumatic brain injury (commonly known as concussion) is as follows: "… a complex pathophysiological process affecting the brain, induced by traumatic biomechanical forces (McCrory et al., 2005). There are several additional features that may help in the definition of concussion. Specifically, the severity and duration of symptoms may dissociate mild, moderate and severe forms of traumatic brain injuries (TBI). The severity of TBI is typically categorized during the acute stages of injury on the basis of the presence and duration of three post-injury signs: (1) loss of consciousness (LOC), (2) retrograde amnesia (i.e., lack of memory and forgetting information about events that occurred prior to injury) and anterograde amnesia (i.e., lack of memory and forgetting information about events that occurred after the injury), and (3) presence/absence of brain structural and functional alterations as a result of injury (i.e., brain lesion, abnormal EEG/MRI etc.). Overall, based on the presence and duration/severity of these aforementioned signs and symptoms, a TBI is classified as mild (concussion), moderate or severe. Moderate and severe TBI typically involve prolonged (e.g., several hours/days) LOC and amnesia and possibly varying sizes of brain structural abnormalities.

Over the past decade, the scientific information about traumatic brain injury has increased considerably. A number of models, theories and hypotheses about traumatic brain injury have been elaborated (see Shaw, 2002, for review). For example, a search of the Internet resource *PubMed* (National Library of Medicine) for the term "brain injury" returned 1,990 articles published between the years of 1994-2003, compared to 930 for the years 1966-1993. However, despite dramatic advances in this field of medicine, traumatic brain injury, including mild traumatic brain injury (MTBI), commonly known as concussion, is still one of the most puzzling neurological disorders and least understood injuries facing the sports medicine world today (Walker, 1994; Cantu, 2003).

Historically, LOC and amnesia have been considered the primary features of concussion, however, as evidenced by the definition provided by McCrory et al. (2005) consensus committee, mild TBI is rarely accompanied by LOC and/or prolonged amnesia. Rather, signs and symptoms such as disorientation, confusion, dizziness, headache, nausea, excessive sensitivity to light or noise, balance problems and a host of others are listed on the Post-concussion Symptom Scale and are sufficient to warrant an accurate diagnosis of concussion. Overall, the presence and duration of a variety of signs and symptoms are important factors to consider when assessing TBI.

Definitions of concussion are almost always qualified by the statement that loss of consciousness (LOC) can occur in the absence of any gross damage or injury visible by light microscopy to the brain (Shaw, 2002). According to a recent *NIH Consensus Statement*, mild traumatic brain injury is an evolving dynamic process that involves multiple interrelated components exerting primary and secondary effects at the level of individual nerve cells (neuron), the level of connected networks of such neurons (neural networks), and the level of human thoughts or cognition (NIH, 1998).

The need for multidisciplinary research on mild brain injury arises from recent evidence identifying long-lasting residual disabilities that are often overlooked using current research methods. The notion of transient and rapid symptoms resolution is misleading since symptoms resolution is not indicative of injury resolution. No two traumatic brain injuries are alike in mechanism, symptomology, or symptoms resolution. Most grading scales are based on loss of consciousness (LOC), and post-traumatic amnesia, both of which occur infrequently in MTBI (Guskiewick et al. 2001; Guskiewick, 2001). There is still no agreement upon diagnosis (Christopher & Amann, 2000) and there is no known treatment for this injury besides the passage of time. LOC for instance, occurs in only 8% of concussion cases (Oliaro et al., 2001). Overall, recent research has shown the many shortcomings of current MTBI assessment rating scales (Maddocks & Saling, 1996; Wojtys et al., 1999; Guskiewicz et al., 2001), neuropsychological assessments (Randolph, 2001; Shaw, 2002; Warden et al., 2001) and brain imaging techniques (CT, conventional MRI and EEG, Thatcher et al., 1989, 1998, 2001; Barth et al., 2001; Guskiewicz, 2001; Kushner, 1998; Shaw, 2002).

The clinical significance for further research on mild traumatic brain injury stems from the fact that injuries to the brain are the most common cause of death in athletes (Mueller & Cantu, 1990). It has been estimated that in high school football alone, there are more than 250,000 incidents of mild traumatic brain injury each season, which translates into approximately 20% of all boys who participate in this sport (LeBlanc, 1994, 1999). It is conventional wisdom that athletes with uncomplicated and single mild traumatic brain injuries experience rapid resolution of symptoms within 1-6 weeks after the incident with minimal prolonged sequelae (Echemendia et al., 2001; Lowell, 2003; Lowell et al., 2003; Macciocchi et al., 1996; Maddocks & Saling, 1996). However, there is a growing body of knowledge indicating long-term disabilities that may persist up to 10 years post injury. Recent brain imaging studies (MRS, magnetic resonance spectroscopy) have clearly demonstrated the signs of cellular damage and diffuse axonal injury in subjects suffering from MTBI, not previously recognized by conventional imaging (Garnett et al., 2000). It is important to stress that progressive neuronal loss in these subjects, as evidenced by abnormal brain metabolites, may persist for up to 35 days post-injury. Therefore, athletes who prematurely return to play are highly susceptible to future and often more severe brain injuries. In fact, concussed athletes often experience a second TBI within one year post injury. Every athlete with a history of a single MTBI who returns to competition upon symptoms resolution still has a risk of developing a post-concussive syndrome (Cantu & Roy, 1995; Cantu, 2003; Kushner, 1998; Randolph, 2001), a syndrome with potentially fatal consequences (Barth et al., 2001).

Post-concussive syndrome (PCS) is described as the emergence and variable persistence of a cluster of symptoms following an episode of concussion, including, but not limited to: impaired cognitive functions such as attention, concentration, memory and information processing; irritability; depression; headache; disturbance of sleep (Hugenholtz et al., 1988; Thatcher et al., 1989; Macciocchi et al., 1996; Wojtys et al., 1999; Barth et al., 2001; Powell, 2001); nausea and emotional problems (Wright, 1998). Other signs of PCS are disorientation in space, impaired balance and postural control (Guskiewicz, 2001), altered sensation, photophobia, lack of motor coordination (Slobounov et al., 2002) and slowed motor responses (Goldberg, 1988). It is not known, however, how these symptoms relate to damage in specific brain structures or brain pathways (Macciocchi et al., 1996), thus making accurate diagnosis based on these criteria almost impossible. Symptoms may resolve due to the brain's amazing plasticity (Hallett, 2001).

Humans are able to compensate for mild neuronal loss because of redundancies in the brain structures that allow for the reallocation of resources such that undamaged pathways and neurons are used to perform cognitive and motor tasks. This functional reserve gives the appearance that the subject has returned to pre-injury health while in actuality the injury is still present (Randolph, 2001). In this context, Thatcher (1997; 2001) was able to detect EEG residual abnormalities in MTBI patients up to eight years post injury. This may also increase the risk of *second impact syndrome* and multiple concussions in athletes who return to play based solely on symptom resolution criteria (Barth et al., 2001; Kushner, 2001; Randolph, 2001).

2. INCIDENCE OF CONCUSSION IN ATHLETICS

The exact incidence of concussion related to sports activities is not well-defined. This is primarily because of lack of recognition of concussion by the player, coaches, or trainers, and underreporting by players. Indeed, in many cases, a player may not even realize that he or she has suffered a concussion. Extensive education has been required to emphasize the fact that a concussion can occur in the absence of loss of consciousness. Indeed, more likely than not, the vast majority of concussions in sports fall into this category. The incidence of concussion seems to be rising in virtually all sports, but the higher numbers may reflect an increase in recognition and reporting by team physicians. The National Athletic Injury/Illness Reporting System began following all injuries in various sports in 1975. Their statistics indicate that the risk of a concussion is 2% to 6% in football and generally less than 2% in other sports. For the sake of comparison, the risk of minor head injury for the general population in the United States is 0.1% or a rate of approximately 131 per 100,000 per year. The true incidence of concussion in sports may be significantly higher than these estimates suggest (Buckley, 1988). Gerberich et al. (1983) reported that 19% of high school athletes had experienced at least one concussion during their career. A prospective study of 2,500 college football players found the risk of a concussion to be approximately 10% (Macciochi et al., 1996). Overall, when the epidemiology of concussion in high school athletes is considered, the rates appear to be relatively high. The self-reported incidence rates of concussion were between 15.3% and 47.2% (Langburt et al., 2004).

An equally concerning issue is the occurrence of subsequent concussions in the same player over time. Various authors have reported a four to six fold increased risk for subsequent concussion in athletes who have suffered a prior concussion. Guskiewicz et al. (2000) reported that an athlete who has sustained a single concussion is three times more likely to sustain a second concussion in the same season. In the same article, he reported that there was an increased severity of symptoms with subsequent concussions. Powell & Barber-Foss (1999) reported that 10.3% of high school athletes who had sustained a concussion at some point during their high school career had a second concussion before ending their high school career, with 63.3% of the second concussions occurring in the same season and 19.4% in the following season. During a six-year period, 1996 to 2001, 787 concussions were reported in National Football League games – an incidence of 0.41 concussions per game. The highest risk was from helmet impacts, while 21% occurred from contact with other players' body parts, and 11% occurred from contact with the ground. In 91% of the cases, the concussion was not associated with a loss of consciousness (Pellman et al., 2004).

3. MECHANISMS OF CONCUSSION

From a historical perspective, a concussion has been defined as a short lasting disturbance of neural functions provoked by a sudden acceleration and/or deceleration of the head usually without skull fracture (Denny-Brown and Russell, 1941; cf: Shaw, 2002). Sudden loss of consciousness and profound paralysis of neuronal functions which happen (although not always) in concussion accidents are at odds with the fact that no obvious sign of demonstrable lesion, including "laceration, edema, hemorrhage, or direct injury to the neuron ..." (Symonds, 1974; cf. Shaw, 2002) can account for the observed symptoms. Cerebral concussion refers to the disturbance or shock of impact (*Oxford English Dictionary*) and is still the most puzzling and controversial phenomenon in sports medicine today. Currently the most common, accepted definition of a concussion is an immediate and transient impairment of neural function, such as alterations in consciousness, disturbance of vision, memory, equilibrium, or other similar symptoms, caused by a direct or indirect (e.g. rotation) force transmitted to the head. Delayed symptoms of concussion may include chronic fatigue, tinnitus, sleep and eating irregularities, irritability, depression, inability to perform daily activities, and academic problems (Wojtys et al., 1999).

4. BIOMECHANICS OF CONCUSSION

The biomechanics of concussion have been extensively studied and found to be primarily related to acceleration, deceleration, and translation or rotation of the head. It should be noted, however, that an attempt to quantify the biomechanics of a concussion in general or of any comparatively simple type of brain injury is an extremely difficult task. Numerous factors need to be taken into consideration, including the shape of the skull, its size and geometry, density and mass of neural tissue, thickness of scalp and skull, extent/type and direction of the concussive blow, head–body relationships and mobility of the head and neck (Shaw, 2002). Recently, however, a sophisticated finite element analysis was undertaken using a detailed anatomic model of the brain to determine brain responses from concussive impacts occurring in National Football League games. With deceleration and rotation, a variety of "hot spots" were defined, indicating progressive areas of brain deformation subsequent to the impact, leading to the signs and symptoms of a concussion. There is an early response of the brain directly adjacent to the impact site (*coup* injury). In this linear case, a sufficient force in the form of an opposite velocity vector may cause the brain to strike against the inner skull in the direction it was initially traveling. Since the majority of NFL concussion impacts are oblique and lateral, the earliest signs of brain deformation or strain were in the temporal lobe. Subsequently, a slightly delayed response was seen on the opposite side of the brain from the impact (*contrecoup* injury) – typically, the temporal lobe opposite the site of the impact, though, any brain area can be affected. In other words, the brain may be "rebounding" from the direction of the deceleration and hit the inner lining of the skull in the opposite direction. When rotational force is applied, the sites of brain contact with the skull can be manifold. It should be noted however, that there is a notion that no true coup or *contrecoup* brain injury may exist, and the magnitude of the brain tissue alteration (i.e., diffuse axonal injury, DAI) can be significantly larger when excessive rotational forces are applied (Barth et al., 2001).

Late in response to the concussive impact, deformations are seen in the midbrain above the brainstem. The study by Viano and colleagues concluded that concussive injuries occur from rapid displacement and rotation of the cranium after peak acceleration and momentum transfer in helmet impacts, and that various regions of the brain are serially affected by deformational strains as a result of this momentum transfer (Viano et al., 2005).

5. GLOBAL METABOLIC CASCADE OF CONCUSSION

At present, there is a lack of complete understanding of the pathophysiology of cerebral concussions and explanations as to why after even mild concussions, the brain may become extremely vulnerable to secondary injury. It was initially felt that deformational strains produced by concussive forces would result in only a temporary disturbance of brain function related to neuronal, neurochemical, or metabolic function without associated structural brain injury. In recent years, however, it has been recognized that structural derangements may indeed occur, and that there may be a period of selective vulnerability to additional insults (e.g., second impact syndrome) or prolonged vulnerability to cumulative concussions and their long-term effects (i.e. *dementia pugilistica*). Indeed, during the minutes to few days after a concussion blow, brain cells that are not irreversibly damaged remain alive but exist in a vulnerable state (Woytys et al., 1999). Some patients suffering from a mild form of concussion may be extremely susceptible to the consequences of even minor changes in cerebral blood flow, as well as slight increases in intracranial pressure and apnea (Hovda, 1995). Metabolic dysfunctions during acute post-concussive events may be responsible for maintaining a state of brain vulnerability, characterized by increase in the demand for glucose and an inexplicable reduction in cerebral blood flow (CBF). In healthy controls, the CBF is tightly coupled to neuronal activity and glucose metabolism. This coupling may be disrupted as a result of acute brain injury. There are some supporting evidences for this. Experimentally induced fluid percussion following brain injury may significantly reduce CBF up to 50% of normal (Doberstein et al., 1992). In a setting of increased glucose use (i.e., hyperglucolysis), this mismatch in supply and demand may result in a potentially damaging energy crisis (Giza & Hovda, 2001). An acute brain injury-induced increase in glucose utilization has been shown in the presence of low CBF in a number of animal studies (Pfenninger et al., 1989; Yanakami & McIntosh, 1989), and in humans with severe head injuries

(Bergsneider et al., 1997). After the initial period of increased glucose unitization, the injured brain transitions into a period of depressed metabolism that may lead to long-lasting and worsening energy crisis. Specifically, in relation to evidence from experimental animals, following the initial stage of hyperglycolysis the CBF was found to be diminished by twenty-four hours post-injury and remained low for the next five to ten days (Ballanyi et al., 1987). During the prolonged metabolic depression after traumatic brain injury (TBI), neurons are less able to respond metabolically to peripheral stimulation (Ip et al., 2003). The results of lateral fluid percussion injury (LFPI) clearly indicate that stimulation with impulses capable of inducing a vibrissa twitch resulted in an increase in the cerebral metabolic rate for glucose (CMR(glc) within one hour and was maintained up to day seven post-injury. However, on day one LFPI stimulation induced a 161% increase in CMR (glc) and a 35% decrease in metabolic activation volume. Extracellular lactate concentrations during stimulation significantly increased from 23% to 55% and to 63% on day one and day seven, respectively, post-injury. Extracellular glucose concentrations during stimulation remained unchanged on day seven but decreased 17% on day one post-injury. The extent of cortical degeneration around the stimulating electrode on day one post-injury nearly doubled when compared with controls. In humans suffering from severe brain injury decreased glucose utilization may last up to four weeks post-injury (Bergsneider et al., 2000, PET study). Reduced cerebral glucose utilization was also found in comatose patients (Bergsneider et al., 2000), though it is still unclear how depressed metabolism may be related with acute neuropsychological and behavioral symptoms of traumatic brain injury. These neurobiological evidences may be at odds with common practices involving the clearing of brain injured athletes for sport participation within few days post-injury.

The early findings regarding depressed metabolism in acute brain injury have been supported by a number of recent studies. Specifically, it has been clearly shown that brain trauma is accompanied by regional alterations of brain metabolism, reduction in metabolic rates and possible energy crisis (Vespa et al., 2005). In this study, microdialysis markers of energy crisis were found during the critical period of intensive care despite the absence of brain ischemia. Patients underwent combined positron emission tomography (PET) for the metabolism of glucose (CMRglu) and oxygen (CMRO(2)) and cerebral microdialysis (MD) at a mean time of thirty-six hours after injury. Microdialysis values were compared with the regional mean PET values adjacent to the probe. The data revealed a 25% incidence rate of metabolic crisis (elevated lactate/pyruvate ratio (LPR) > 40) but only a 2.4% incidence rate of ischemia. Positron emission tomography imaging revealed a 1% incidence of ischemia across all voxels as measured by oxygen extraction fraction (OEF) and cerebral venous oxygen content (CvO(2)). In the region of the MD probe, PET imaging revealed ischemia in a single patient despite increased LPR in other patients. Lactate/pyruvate ratio correlated negatively with CMRO(2), but not with OEF or CvO(2). It was concluded that traumatic brain injury leads to a state of persistent metabolic crisis as reflected by abnormal cerebral microdialysis LPR that is not related to ischemia. In another recent study, the course of cerebral blood flow (CBF) and metabolism in traumatic brain injury (TBI) patients was examined with special focus on changes in lactate and glucose indices in the acute post-traumatic period (Soustiel et al., 2005). Global CBF, cerebral metabolic rates of oxygen (CMRO2), glucose (CMRGlc), and lactate (CMRLct) were calculated. In all patients, CBF was moderately decreased during the first twenty-four hours in comparison with normal controls. Both CMRO2 and CMRGlc were significantly depressed and correlated to the outcome of Glasgow Coma Scale (GCS) gradings. Moreover, CMRLct analysis revealed positive values (lactate uptake) during the first forty-eight hours, especially in patients with a favorable outcome. Both CMRO2 and CMRLct correlated with GCS gradings. These findings emphasize the clinical significance of monitoring the CBF and metabolic changes in TBI and provide evidence for metabolic coupling between astrocytes and neurons. These findings are also consistent with a more recent animal study examining the effects of traumatic brain injury (TBI) on brain chemistry and metabolism in three groups of rats using high-resolution (1)H NMR metabolomics of brain tissue extracts and plasma (Viant et al., 2005). Evidence was found of oxidative stress (e.g., a decrease in ascorbate of 16.4% in the cortex and 29.7% cortex and hippocampus combined in TBI rats versus the untreated control group). Also there were indicators of excitotoxic damage (e.g., a decrease in glutamate of 14.7% and 12.3% in the cortex and hippocampus, respectively), membrane disruption (e.g., a decrease in the total level of phosphocholine and glycerophosphocholine of 23.0% and 19.0% in the cortex and hippocampus, respectively), and neuronal injury (e.g., decreases in N-acetylaspartate of 15.3% and 9.7% in the cortex and hippocampus, respectively). Significant changes in the overall pattern of NMR-observable metabolites using principal components analysis were also observed in TBI animals.

It is important to note that the pathophysiology of peri-lesion boundary zones in acute brain injury is highly dynamic, and it is now clear that spreading-depression-like events occur frequently in areas of cerebral cortex adjacent to contusions in the injured human brain (Parkin et al., 2005). In this study, an automated method to assay microdialysate from a peri-lesion cerebral cortex for the assay of glucose and lactate in eleven patients with intracranial haematomas combined with electrocorticogram (ECoG) revealed several patterns of changes in metabolites. The number of transient lactate events was significantly correlated with the number of glucose events. In addition, progressive reduction in dialysate glucose was very closely correlated with the aggregate number of ECoG events. The authors suggested that adverse impact of low dialysate glucose on clinical outcome may be because of recurrent, spontaneous spreading-depression-like events in the perilesion cortex. Interestingly, abnormal metabolic cascades may be present at a remote site of brain injury, including subcortical brain regions. Specifically, a positron emission tomographic study examined the nature, extent, and degree of metabolic abnormalities in subcortical brain regions remote from hemorrhagic lesions using sixteen normal controls and ten TBI patients (Wu et al., 2004a). Sixteen normal volunteers and ten TBI patients (Glasgow Coma Scale score, 4-10) participated in this study. Data from gray matter and white matter (WM) remote from hemorrhagic lesions, plus the whole brain, were analyzed. There was a significant reduction in the subcortical WM oxygen-to-glucose utilization ratio after TBI compared with normal values, whereas the mean cortical gray matter and whole-brain values remained unchanged. WM metabolic changes, which were diffuse throughout the hemispheres, were characterized by a reduction in the metabolic rate of oxygen without a concomitant drop in the metabolic rate of glucose. This finding suggests that the extent and degree of subcortical WM metabolic abnormalities after moderate and severe TBI are clearly diffuse. Moreover, this pervasive finding may indicate that the concept of focal traumatic injury, although valid from a computed tomographic imaging standpoint, may be misleading when considering metabolic derangements associated with TBI. Moreover, the apparent loss of overall gray-white matter contrast (GM/WM) may be seen in TBI patients on FDG-PET imaging reflecting the differential changes of glucose metabolic rate (CMRglc) in cortical gray mater (GM) and subcortical white mater (WM) (Wu et al., 2004,b). In this study, the stabilities of the global and regional FDG lumped constants (LC) were examined. Parametric images (pixel unit: mg/min/100g) of FDG uptake rate (CURFDG) and CMRglc were generated and changes of CMR (glc) in the whole brain; GM and WM were studied separately by using a MRI-segmentation-based technique. The GM-to-WM ratios of both CURFDG and CMRglc images were significantly decreased (>31%) in TBI patients that was highly correlated with the initial Glasgow Coma Scale score (GCS). The patients with higher CMRglc GM-to-WM ratios (>1.54) showed good recovery twelve months after TBI. There was also a selective CMRglc reduction in cortical GM following TBI. However, the pathophysiological basis for the reduction in GM-to-WM CMRglc ratio seen on FDG-PET imaging following TBI remains unknown.

Abnormal metabolic cascades, in acute TBI patients, as evidenced by significant alteration of glucose utilization, may be also present in the thalamus, brain stem, and cerebellum (Hattori et al., 2003). In this particular study, the regional cerebral metabolic rate of glucose (CMRglc) of the cortical areas (remote from hemorrhagic lesions), striatum, thalamus, brain stem, cerebellar cortex, and the whole brain was compared with the severity of the injury and the level of consciousness. This was evaluated using

GCSini (full form: Glasgow Coma Scale initial) and the Glasgow Coma Scale score at the time of PET (GCSpet). It was shown that regional CMRglc of the brain stem is relatively unaffected by the TBI. Compared with healthy volunteers, TBI patients exhibited significantly depressed CMRglc in the striatum and thalamus. CMRglc levels were not statistically lower in the cerebellum and brain stem. However, comparison of noncomatose with comatose patients showed that CMRglc values in the thalamus, brain stem and cerebellar cortex were significantly lower in noncomatose than in comatose patients. It should be noted that CT or MRI findings were normal for the analyzed structures except for three patients with diffuse axonal injury of the brain stem. The presence of shear injury was associated with poor GCSini. The metabolic rate of glucose utilization in these regions significantly correlated with the level of consciousness at the time of PET. It is feasible to assume that after traumatic brain injury (TBI), subcortical white matter damage may induce a functional disconnection leading to a dissociation of the regional cerebral metabolic rate of glucose (CMRglc) between the cerebral cortex and deeper brain regions, including thalamus, brain stem and cerebellum. Not surprisingly, patients suffering from TBI may experience long term behavioral deficits including abnormal balance and postural control.

6. NEUROCHEMICAL CASCADE OF CONCUSSION

Direct evidence of diffuse abnormal neuronal excitation/inhibition in acute concussion can be obtained by examining the depolarization of nerve cells soon after traumatic brain injury. This involves directly measuring ionic fluxes, in particular, the concentration of extracellular potassium (K^+) and the release of excitatory amino acid (EAA) neurotransmitters. The most common technique is the insertion of ion-sensitive microelectrodes into the animal brain immediately after the induction of experimental concussion (Takahashi et al., 1981). In a set of elegant experiments, microdialysis techniques were used to measure K^+ concentration in the hippocampus of an experimental rat right after mild or moderate induced concussions (Katayama et al., 1990). On a cellular level, several acute ionic changes may occur in concussed brains, such as the disruption of neuronal membranes, axonal stretching, and the opening of a voltage-dependent potassium channel leading to five fold increases in extracellular potassium (K^+). In addition, nonspecific membrane depolarization may disrupt neural transmission by depolarizing neurons, leading to the excessive release of excitatory amino acids (EAA) such as glutamate, which exacerbates the potassium flux into the extracellular space by activating kainite, NMDA, and D-amino-3 hydroxy-5-methyl-4-soxazole-propionic acid (AMPA) receptors (Giza & Hovda, 2001). It seems that only when extracellular potassium reaches a critical threshold, does this trigger the release of EAA and other neurotransmitters from nerve terminals. Normally, excessive concentration of extracellular potassium is neutralized by surrounding glial cells, allowing the brain to maintain physiological equilibrium of K^+ following mild disturbances. In fact, the EAA inhibitor drugs (i.e., kynurenic acid) may significantly reduce the post-traumatic potassium efflux in rats (Katayama et al., 1990). Moreover, EAA release may be unaffected by administration of TTX modified by cobalt, implying a role for neurotransmitter release. Within the scope of this pathological cascade the excessive extracellular K+ may trigger neuronal depolarization, release of EAAs and ultimately even greater concentration of extracellular K^+. Post-synaptic EAA receptors subsequently activate the opening of associated ligand-gated ion channels, therefore permitting the rapid outflow of a large amount of K^+ accompanied by an influx of extracellular calcium (Nilsson et al., 1993). In other words, initially there is a massive excitatory process (due to excessive concentration of potassium) and this is ultimately followed by an abrupt wave of relative neuronal deactivation (outflow of K^+), and this phenomenon is known as *spreading depression*. There is a notion that acute loss of consciousness, memory loss and cognitive abnormalities are direct manifestations of post-traumatic *spreading depression* (Giza & Hovda, 2001).

Through a phenomenon known as excitotoxicity, glutamate activation of NMDA receptors opens calcium (Ca^{2+}) channels and allows an influx of calcium into cells. Excessive accumulation of Ca^{2+} can damage intracellular organelles, especially the mitochondria, resulting in aberrant oxidadative metabolism and ultimately energy crisis or failure. In fact, a potent N-type calcium channel blocker, SNX-111, may significantly reduce post-concussive calcium accumulation and may be suggested as a treatment with NMDA receptor antagonists (Samii et al., 1999; Giza & Hovda, 2001). Another trigger for influx of Ca^{2+} may be mechanical stretching of axons resulting in membrane disruption and mitochondrial swelling (Maxwell et al., 1995). Increased Ca^{2+} has been shown to lead to microtubule breakdown up to twenty-four hours post-injury, and along with focal axonal swelling may lead to secondary axotomy and formation of axonal bulbs. These are other intra-axonal cytoskeletal pathologies that are commonly considered within the scope of diffuse axonal injury as a result of head trauma. It is important to note that post-traumatic increase in Ca^{2+} may not necessarily lead to immediate cell death, although this may possibly lead to an impairment of mitochondrial metabolism. The detailed discussion of neurochemical cascades associated with excessive accumulation of intracellular Ca^{2+} triggering the cell death is beyond the scope of this chapter. Again, electrolyte homeostasis is usually restored within minutes to hours post acute traumatic brain injury. However, long-term perturbations may occur, resulting in neuronal vulnerability to further insults and/or be responsible for post-concussive symptoms (Katayama, 1990). See Fig. 1 for details.

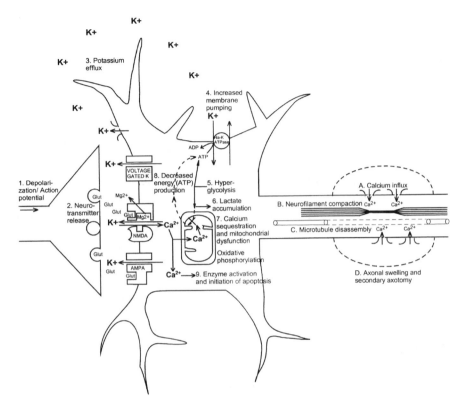

Figure 1. Neurometabolic cascade following traumatic brain injury. (1) Nonspecific depolarization and initiation of action potentials. (2). Release of excitatory neurotransmitters (EAAs). (3). Massive influx of potassium. (4) Increased activity of membrane ionic pumps to restore homeostasis. (5) Hyperglucolysis to generate more adenosine triphosphate (ATP). (6) Lactate accumulation. (7) Calcium influx and sequestration in mitochondria leading to impaired oxidative metabolism. (8). Decreased energy (ATP) production. (9) Calpain activation and initiation of apoptosis. A, Axolemmal disruption and calcium influx. B, Neurofilament compaction via phosphorylation or sidearm cleavage. C, Microtubule disassembly and accumulation of axonally transported organelles. D, Axonal swelling and eventual axotomy.
K+, potassium; Na+, sodium; Glut, glutamate; Mg2+, magnesium; Ca2+, calcium; NMDA, N-methyl-D-aspartate, AMPA, d-amino-3-hydroxy-5-methyl4-isoxazole-propionic acid.
(Re-printed with permissions from: Giza, C., & Hovda, D. (2001). The neurometabolic cascade of concussion. (*Journal of Athletic Training, 36(3)*, 228-235.)

7. NEURAL BASIS OF COGNITIVE DISABILITIES IN CONCUSSION

There is a considerable debate in the literature regarding the extent to which mild traumatic brain injury results in permanent neurological damage

(Levin et al., 1987; Johnston et al., 2001), psychological distress (Lishman, 1988) or a combination of both (McClelland et al., 1994; Bryant & Harvey, 1999). Lishman's (1988) review of the literature suggested that physiological factors contributed mainly to the onset of the MTBI while psychological factors contributed to the duration of its symptoms. As a result, causation of MTBI remains unclear because objective anatomic pathology is rare and the interaction among cognitive, behavioral and emotional factors can produce enormous subjective symptoms in an unspecified manner (Goldberg, 1988).

To-date, a growing body of neuroimaging studies in normal subjects has documented involvement of the fronto-parietal network in spatial attentional modulations during object recognition or discrimination of cognitive tasks (Buchel & Friston, 2001; Cabeza et al., 2003). This is consistent with previous fMRI research suggesting a supra-modal role of the prefrontal cortex in attention selection within both the sensori-motor and mnemonic domains (Friston et al., 1996, 1999). Taken together, these neuroimaging studies suggest the distributed interaction between modality-specific posterior visual and frontal-parietal areas servicing the cognitive tasks of visual attention and object discrimination (Rees & Lavie, 2001). Research on the cognitive aspects in MTBI patients indicates a classic pattern of abnormalities in information processing and executive functioning that correspond to the frontal lobe damage (Stuss & Knight, 2002).

The frontal areas of the brain, including the prefrontal cortex, are highly vulnerable to damage after traumatic brain injury, leading to commonly observed long-term cognitive impairments (Levin et al., 2002; Echemendia et al., 2001; Lowell et al., 2003). A significant percentage of mild traumatic brain injuries will result in structural lesions (Johnston et al., 2001), mainly due to diffuse axonal injury (DAI), which are not always detected by MRI (Gentry et al., 1988; Liu et al., 1999). Recent dynamic imaging studies have finally revealed that persistent post-concussive brain dysfunction exists even in patients who sustained a relatively mild brain injury (Hofman et al., 2002; Umile et al., 2002).

Striking evidence for DAI most commonly involving the white matter of the frontal lobe (Gentry et al., 1998) and cellular damage after mild TBI was revealed by magnetic resonance spectroscopy (MRS). Specifically, MRS studies have demonstrated impaired neuronal integrity and associated cognitive impairment in patients suffering from mild TBI. For example, a number of MRS studies showed reduced NAA/creatine ratio and increased choline/creatine ratio in the white matter, which can be observed from three to thirty-nine days post-injury (Mittl et al., 1994; Garnett et al., 2000; Ross & Bluml, 2001). The ratios are highly correlated with head injury severity. More importantly, abnormal MR spectra were acquired from frontal white matter that appeared to be normal on conventional MRI images. Predictive values of MRS in the assessment of a second concussion are high, because of the frequent occurrence of DAI with second impact syndrome (Ross & Bluml, 2001). The language, memory and perceptual tasks sensitive to frontal lobe functions have been developed because a disruption in frontal-limbic-reticular activation system following closed head injury has been hypothesized (Johnston, 2001). Patients with MTBI performed poorly in these tasks. Long-term functional abnormalities, as evidenced by fMRI, have been documented in concussed individuals with normal structural imaging results (Schubert & Szameitat, 2003; Chen et al., 2003). Overall, abnormal brain metabolism may present between one-and-a-half to three months post-injury, indicating continuing neuronal dysfunction and long-term molecular pathology following diffuse axonal brain injury.

8. SECOND IMPACT SYNDROME

In 1984, three athletes died from massive brain swelling after a minor concussion. In all three cases there had been an antecedent concussion from which they were still symptomatic – this typifies the second impact syndrome (Saunders & Harbaugh, 1984). Since then, more than fifty such occurrences have been reported (Cantu, 1992; Cantu & Voy, 1995). The postulated pathophysiology is a defect in cerebrovascular autoregulation initiated by the initial concussion. Ongoing cerebrovascular vulnerability at the time of the second concussion triggers massive vasodilatation and subsequent lethal brain swelling due to a marked increase in cerebral blood volume. This notion has been called into question by autopsy findings of acute subdural hematomas in 15% to 20% of cases (Cantu & Voy, 1995). However, recent studies using transcranial Doppler ultrasonography to access cerebrovascular resistance and cerebral blood flow after concussion have demonstrated poor or absent cerebrovascular autoregulation in up to 30% of patients (Junger et al., 1997).

There are several potential pathophysiologies for the second impact syndrome. First, acute abnormal glucose metabolism and energy crisis shortly after traumatic brain injury indicate a window for potential vulnerability in the traumatized brain. Moreover, after the initial period of increased glucose unitization, the injured brain transits into a period of depressed metabolism (e.g., post-traumatic *spreading depression* (Giza & Hovda, 2001) that may lead to a long-lasting and worsening energy crisis. Thus, a secondary blow to the head delivered shortly after the first one may further exacerbate the energy crisis due to further demand in energy and previously impaired blood flow. Thus, an acutely injured brain may be capable of recovering after the fist blow, but a second blow during energy failure can lead to irreversible neuronal injury and massive cell death. Another potential candidate for the second impact syndrome is that excessive Ca^{2+} accumulation may irreversibly impair mitochondrial metabolism during the second blow to the head, inducing massive cell death. Further development of and elaborations on pathological mechanisms of second impact syndrome require additional empirical evidence and research.

9. CUMULATIVE EFFECT OF CONCUSSION

Increasing evidence suggests that repeated concussions may have the potential for long term neurologic and cognitive sequelae. Several high profile athletes in recent years have ended their careers as a result of such concerns. The punch drunk syndrome or dementia puglistica was first described in 1928 and involved a spectrum of dementia, personality disturbances and cerebellar or Parkinson-like symptoms (Martland, 1928). Subsequent retrospective studies of a number of former boxers with no neurologic or cognitive impairments found high rates of abnormalities on CT scans, electroencephalography and neuropsychological testing (Casson et al., 1984; 1982). One autopsy study found a high rate of neurofibrillary tangles, amyloid angiopathy and neuritic plaques (all markers of dementia) in fifteen former boxers (Corsellis, 1973). Four of the fifteen boxers' brains with clinically active Parkinson's disease demonstrated substantial substantia nigra depigmentation. There is also a line of evidence, primarily from experience with boxers, that there may be a genetic predisposition to both the severity of as well as the susceptibility to recurrent concussions. Jordan, et al. (1997) reported that the presence of a specific allele of the apolipopritein E (APOE) gene was associated with an increased likelihood of severe cumulative effects in a study of thirty active and retired boxers.

Soccer players may also be susceptible to cumulative effects on cognition and neuropsychological functioning, ostensibly from long term head to ball contact. Matser et al. (1998) found evidence of chronic neurocongitive impairments in fifty-three active European professional players. Retired soccer players have been found to have a variety of neuropsychological, CT scan and electroencephalographic abnormalities (Sortland & Tysvaer, 1989; Tysvaer & Lochen, 1991). In football, there is considerable ongoing debate regarding the cumulative effect of repeated concussions. Most studies, while finding significant neuropsychological impairment after single or multiple concussions, have also found a resolution of these abnormalities within one to two weeks (Macciochi, 1996).

In a prospective cohort study of 2,905 football players from twenty-five US colleges, Guskiewicz, et al. (2003) looked at the incidence and effects of repeat concussions over the course of three seasons. During this time concussions occurred in 184 players (6.3%), with twelve (6.5%) of these having a second concussion in the same season. Of players reporting three or more concussions, the only substantive finding was a prolongation of post-concussive symptoms as compared to those with one prior concussion -

- 30% with > one week of symptoms compared to 14.6% of these having the first concussion. Iverson, et al. (2004) also studied amateur athletes with three or more concussions, comparing them to a matched group whose members had never suffered a concussion, utilizing the *ImPACT* computerized neuropsychological test battery. On testing two days post-injury, significantly lower memory performance was manifested by athletes with multiple concussions compared to athletes with none. However there was no long term follow up provided.

Over five seasons, information was collected on concussions reported by thirty National Football League (NFL) teams, with a 6.3% incidence of single and 6.5% incidence of multiple concussions (Pellman et al., 2004). Slower recovery from post concussive symptoms was seen more frequently in players who had sustained multiple concussions. Of those with three or more concussions, 30% had symptoms lasting more than one week, compared to 14.6% with a history of only one concussion. However, at least one study has demonstrated the persistence of neuropsychological abnormalities up to six months after multiple concussions (Wilberger, 1989). Serial neuropsychological testing of players for up to six months showed a correlation between not only the number of concussions but also the duration and severity of neurocognitive abnormalities. The long-term significance of these findings, if any, is yet to be known. A comprehensive health survey of former NFL players found a correlation between the frequency of concussions and depression, but not with the incidence of dementia or Alzheimer's disease. (Cantu's personal communication, Bailes, J.B., see also, Bailes & Hudson, 2001). Thus, further research is necessary to define the true significance of the possible cumulative effects of concussion and its underlying pathology. However, it is clear that repeated brain injuries incurred within a short time frame can lead to much larger neuroanatomical, cognitive and behavioral impairments than isolated brain injuries.

CONCLUSION

There is still considerable debate in the literature as to whether mild traumatic brain injury (mTBI) results in permanent neurological damage or in transient behavioral and cognitive malfunctions. One of the reasons for this controversy is that there are several critical weaknesses in the existing conceptualization of and/or research into the behavioral, neural and cognitive and neuropsychological consequences of mTBI. *First*, most previous research has failed to provide the pre-injury status of mTBI subjects that may lead to misdiagnosis of the persistent or new neurological and behavioral deficits that occur after injury. Rarely are pure baseline data available to medical practitioners in order to consider the patient's pre-injury status. *Second*, previous research unfortunately has focused selectively on pathophysiology, cognitive, neuropsychological and/or behavioral sequelae of MTBI in isolation. Thus, real symptoms of concussion that are beyond the expertise of the researchers may be overlooked. *Third*, previous research has focused primarily on single concussion cases and failed to examine the subjects who experienced a second concussion at a later time. However, there may be a cumulative effect of previous concussive episodes that dramatically change the perception of concussion. Finally, previous research has failed to provide analyses of biomechanical events and the severity of a concussive blow at the moment of the accident. Biomechanical events set up by the concussive blow (i.e. amount of head movement about the axis of the neck at the time of impact, the site of impact etc.) ultimately result in concussion, and their analysis may contribute to a more accurate assessment of the degree of damage and potential for recovery. Overall, a multimodal approach using advanced technologies and assessment tools may dramatically enhance our understanding of this puzzling neurological disorder facing the sports medicine world today.

The currently accepted conventional notion of transient and rapid symptoms resolution in athletes suffering from even mild traumatic brain injury is misleading. There are obvious short-term and long lasting structural and functional abnormalities as a result of mild TBI that may be revealed using advanced technologies. There is a need for the development of a conceptual framework for examining how behavioral, cognitive, neuropsychological and underlying neural mechanisms (EEG and MRI) are interactively affected by single or multiple mTBI. A set of tools and advanced scales for the accurate assessment of mild traumatic brain injury must be elaborated, including the computer graphics and virtual reality (VR) technologies incorporated with modern human movement analysis and brain imaging (EEG, fMRI and MRS) techniques. Semi-quantitative estimates of biomechanical events set up by a concussive blow should be developed using videotape analysis of the accident, so they may be correlated with other assessment tools. Numerous current research studies of athletes prior to and after brain injury have provided strong evidence for the feasibility of the proposed approach of utilizing technology in examining both short-term and long-lasting neurological dysfunction in the brain as a result of mild TBI.

RERERENCES

McCrory, P., Johnston, K., Meeuwisse, W., Aubry, M., Cantu, R., Dvorak, J., Graf-Baumann, T., Kelly, J., Lovell, M., Schamasch, P. (2005. Summary and agreement on the 2nd International Conference on Concussion in Sport, Prague 2004. *Clinical Journal of Sports Medicine, 15*(2), 48-55.

Shaw, N. (2002). The neurophysiology of concussion. *Progress in Neurobiology, 67*, 281-344.

Walker, A. E. (1994). The physiological basis of concussion: 50 years later. *Journal of Neurosurgery, 81*, 493-494.

Cantu, R. (2003). Neurotrauma and sport medicine review, 3rd annual seminar, Orlando, Fl.

National Institute of Health. NIH Consens Statement, v.16. Bethesda, MD: NIH, 1998.

Guskiewicz, K.M., Ross, S.E., Marshall, S.W. (2001). Postural Stability and Neuropsychological Deficits After Concussion in Collegiate Athletes. *Journal of Athletic Training, 36*(3), 263-273.

Guskiewicz, K.M. (2001). Postural Stability Assessment Following Concusion: One Piece of the Puzzle. *Clinical Journal of Sport Medicine, 11*, 82-189.

Christopher, M., & Amann, M. (2000). Office management of trauma. *Clinic in Family Practice, 2*(3), 24-33.

Oliaro, S., Anderson, S., Hooker, D. (2001). Management of Cerebral Concussion in Sports: The Athletic Trainer's Perspective. Journal of Athletic Training, 36(3):257-262.

Maddocks, D., & Saling, M. (1966). Neuropsychological deficits following concussion. *Brain Injury, 10*, 99-103.

Wojtys, E., Hovda, D., Landry, G., Boland, A., Lovell, M., McCrea, M., Minkoff, J. (1999). Concussion in Sports. *American Journal of Sports Medicine, 27*(5), 676-687.

Randolph, C. (2001). Implementation of neuropsychological testing models for the high school, collegiate and professional sport setting. *Journal of Athletic Training, 36*(3), 288-296.

Warden, D.L., Bleiberg, J., Cameron, K.L., Ecklund, J., Walter, J., Sparling, M.B., Reeves, D., Reynolds, K.Y., Arciero, R. (2001). Persistent Prolongation of Simple Reaction Time in Sports Concussion. *Neurology, 57*(3), 22-39.

Thatcher, R.W., Walker, R. A., Gerson, I., & Geisler, F. H. (1989). EEG discriminant analyses of mild head injury. *EEG and Clinical Neurophysiology, 73,* 94-106.

Thatcher, R.W., Biver, C., McAlister, R., Camacho, M., Salazar, A. (1998). Biophysical linkage between MRI and EEG amplitude in closed head injury. *Neuroimage, 7,* 352-367.

Thatcher, R.W., Biver, C., Gomez, J., North, D., Curtin, R., Walker, R., Salazar, A. (2001). Estimation of the EEG power spectrum using MTI T2 relaxation time in traumatic brain injury. *Clinical Neurophysiology, 112*, 1729-1745.

Barth, J.T., Freeman, J.R., Boshek, D.K., Varney, R.N. (2001). Acceleration-Deceleration Sport-Related Concussion: The Gravity of It All. *Journal of Athletic Training, 36*(3), 253-256.

Kushner, D. (1998). Mild traumatic brain injury: Toward understanding manifestations and treatment. *Archive of Internal Medicine, 158*, 10-24.

Mueller, F.O., & Cantu, R. C. (1990). Catastrophic injuries and fatalities in high school and college sport. Fall 1982 – spring 1988. *Medicine and Science in Sport and Exercise, 22*, 737-741.

LeBlanc, K. E. (1994). Concussion in sport: guidelines for return to competition. *American Family Physician, 50*, 801-808.

LeBlanc, K.E. (1999). Concussion in sport: Diagnosis, management, return to competition. *Comprehensive Therapy, 25*, 39-44.

Echemendia, R.J., Putukien, M., Mackin, R.S., Julian, L., Shoss, N. (2001). Neuropsychological Test Performance Prior To and Following Sports-Related Mild Traumatic Brain Injury. *Clinical Journal of Sports Medicine, 11*, 23-31.

Lowell, M., Collins, M., Iverson, G., Field, M., Maroon, J., Cantu, R., Rodell, K., & Powell, J., & Fu, F. (2003). Recovery from concussion in high school athletes. *Journal of Neurosurgery, 98*, 296-301.

Lowell, M. (2003). Ancillary test for concussion. *Neurotrauma and sport medicine review.* 3rd annual seminar, Orlando,Fl.

Macciocchi, S.T., Barth, J.T., Alves, W., Rimel, R.W., & Jane, J. (1966). Neuropsychological functioning and recovery after mind head injury in collegiate athletes. *Neurosurgery, 3*, 510-513.

Garnett, M., Blamir, A., Rajagopalan, B., Styles, P., Cadoux_Hudson, T. (2000). Evidence of cellular damage in normal-appearing white matter correlates with injury severity in patients following traumatic brain injury: A magnetic resonance spectroscopy study. *Brain, 123(7)*, 1403-1409.

Cantu, R.C., & Roy, R. (1995). Second impact syndrome: a risk in any sport. *Physical Sport Medicine, 23*, 27-36.

Hugenholtz, H., Stuss, D. T., Stethen, L. L, & Richards, M. T. (1988). How long does it take to recover from a mild concussion? *Neurosurgery, 22(5)*, 853-857.

Powell, J. (2001). Cerebral Concussion. Causes, Effects, and Risks in Sports. *Journal of Athletic Training, 36(3)*, 307-311.

Wright, S.C. (1998). Case report: postconcussion syndrome after minor head injury. *Aviation, Space Environmental Medicine, 69(10)*, 999-1000.

Slobounov, S., Sebastianelli, W., Simon, R. (2002). Neurophysiological and behavioral Concomitants of Mild Brain Injury in College Athletes. *Clinical Neurophysiology, 113*, 185-193.

Goldberg, G. (1988). What happens after brain injury? You may be surprised at how rehabilitation can help your patients. *Brain injury, 104(2)*, 91-105.

Hallett, M. (2001). Plasticity of the human motor cortex and recovery from stroke. *Brain Research Review, 36*, 169-174.

Buckley, W.E. (1988). Concussions in College Football: A Multivariate Analysis. *American Journal of Sports Medicine,16*, 51-56.

Gelberich, S.G., Priest, J.D., Boen, J.R., Straub, C.P., & Maxwell, R.E. (1983). Concussion incidences and severity in secondary school varsity football players. *American Journal of Public Health, 73*(2), 1370-1375.

Macciochi, S.N., Barth, J.T., Alves, W., et al. (1996). Neuropsychological Recovery and Functioning after Mild Head Injury in Collegiate Athletes. *Neurosurgery, 39*, 510-514.

Langburt, W., Cohen, A., Arhthar, N., O'Neill, K., & Lee, J.C. (2001). Incidence of concussion in high school football players of Ohio and Pennsylvania. *Journal of Child Neurology, 16*(2), 83-85.

Guskiewicz, K., Weaver, N., Padua, D., Garrett, W. (2000). Epidemiology of Concussion in Collegiate and High School Football Players. *American Journal of Sports Medicine, 28*, 643-650.

Powell, J, Barber-Foss, K. Traumatic Brain Injury in High School Athletes. JAMA 282:958-963, 1999.

Pellman, E.J., Viano, D.C., Casson, I.R., Affken, D., & Powell, J. (2004). Concussion in professional football: Injuries involving 7 or more days out – part 5. *Neurosurgery, 55*(5), 1100-1119.

Viano, D.C, Casson, I.R., Pellman, E.J.. (2005). Concussion in Professional Football: Brain Responses by Finite Element Analysis- Part 9. *Neurosurgery, 57*,891-916.

Hovda, D.A. (1995). Metabolic dysfunction. In: Narayan R.K., Wilberger, J.E., Povlishock J.T.(Eds). *Neurotrauma*. pp.1459-1478. McGraw Hill, NY.

Doberstein, C., Velarde, F., Babie, H., Vovda, D.A. (1992). Changes in local cerebral blood flow following concussive brain injury. *Society for Neuroscience, Abstract 18*, 175. Anaheim, CA.

Giza, C., & Hovda, D. (2001). The neurometabolic cascade of concussion. *Journal of Athletic Training, 36(3)*, 228-235.

Pfenninger, E.G., Reith, A., Breitig, D., et al. (1989). Early changes of intracranial pressure, perfusion pressure, and blood flow after acute head injury. Par 1. Journal of *Neurosurgery, 70*, 774-779.

Yamakashi, I., & McIntosh, T.K. (1989). Effects of traumatic brain injury on regional cerebral blood flow in rats as measured with radiolabeled microspheres. *Journal of Cerebral Blood Flow Metabolism, 9*, 117-124.

Bergsneider, M., Hovda, D.A., Shalman, E., et al., (1997). Cerebral hyperglucolysis following severe traumatic brain injury in humans: A positron emission tomography study. *Journal of Neurosurgery, 86,* 241-251.

Ballanyi, K., Grafe, R., ten Bruggencate, G. (1987).   Ion activities and potassium uptake mechanisms of glial cells in guinea-pig olfactory cortex slices. *Journal of Physiology, 382,* 159-174.

Ip, E.Y., Zanier, E.R., Moore, A.H., Lee, S.M., Hovda, D.A. (2003).   Metabolic, neurochemical, and histologic responses to vibrissa motor cortex stimulation after traumatic brain injury. *Journal of Cerebral Blood Flow Metabolism, 23(8),* 900-910.

Bergsneider, M., Hovda, D.A., Lee, S.M., et al., (2000).   Dissociation of cerebral glucose metabolism and level of consciousness during the period of metabolic depression following human traumatic brain injury. *Journal of Neurotrauma, 17,* 389-401.

Vespa, P., Bergsneider, M., Hattori, N., Wu, H.M., Huang, S.C., Martin, N.A., Glenn, T.C., McArthur, D.L., Hovda, D.A. (2005).   Metabolic crisis without brain ischemia is common after traumatic brain injury: a combined microdialysis and positron emission tomography study. *Journal of Cerebral Blood Flow Metabolism, 25(6),* 663-774.

Soustiel, J.F., Glenn, T.C., Shik, V., Bascardin, J., Mahamid, E., Zaaroor, M. (2005). Monitoring of cerebral blood flow and metabolism in traumatic brain injury. *Neurotrauma, 22(9),* 955-965.

Viant, M.R., Lyeth, B.C., Miller, M.G., Berman, R.F. (2005).   An NMR metabolomic investigation of early metabolic disturbances following traumatic brain injury in a mammalian model. *NMR Biomedicine, 18(8),* 507-571.

Parkin M, Hopwood S, Jones DA, Hashemi P, Landolt H, Fabricius M, Lauritzen M, Boutelle MG, Strong AJ. (2005).   Dynamic changes in brain glucose and lactate in pericontusional areas of the human cerebral cortex, monitored with rapid sampling on-line microdialysis: relationship with depolarisation-like events. *Journal of Cerebral Blood Flow Metabolism, 25(3),* 402-413.

Wu, H.M., Huang, S.C., Hattori, N., Glenn, T.C., Vespa, P.M., Hovda, D.A., Bergsneider, M. (2004a).   Subcortical white matter metabolic changes remote from focal hemorrhagic lesions suggest diffuse injury after human traumatic brain injury. *Neurosurgery, 55(6),* 1306-1315.

Wu, H.M., Huang, S.C., Hattori, N., Glenn, T.C., Vespa, P.M., Yu, C.L., Hovda, D.A., Phelps, M.E., Bergsneider, M. (2004b).   Selective metabolic reduction in gray matter acutely following human traumatic brain injury. *Journal of Neurotrauma, 21(2),* 149-61.

Hattori, N., Huang, S.C., Wu, H.M., Yeh, E., Glenn, T.C., Vespa, P.M., McArthur, D., Phelps, M.E., Hovda, D.A., Bergsneider, M. (2003).   Alteration of glucose utilization correlates with glasgow coma scale after traumatic brain injury. *Journal of Nuclear Medicine, 44(11),* 1709-1716.

Takahashi, et al. 1981. H. Takahashi, H., S. Manaka, S., & S. Keiji, S. (1981).   Changes in extracellular potassium concentration in cortex and brainstem during the acute phase of experimental closed head injury. *Journal of Neurosurgery, 55,* 708-717.

Katayama, Y., Becker., D., Tamura, T., Hovda, D. (1990).   Massive Increases in Extracellular Potassium and the Indiscriminate Release of Glutamate Following Concussive Brain Injury. *Journal of Neurosurgery, 73,* 889-900.

Nilsson,P., Nilsson, P., L. Hillered, L., Y. Olsson, Y., M. Sheardown, M., A.J. Hansen, A.J. (1993).   Regional changes of interstitial K^+ and Ca^{2+} levels following cortical compression contusion trauma in rats. *Journal of Cerebral Blood Flow Metabolism, 13,* 183-192.

Samii, A., Badie, H., Fu, K., Lusher, R.R., Hovda, D.A. (1999).   Effect of an N-type calcium channel antagonist (SNX 1111 Ziconotide) on calcium-45 accumulation following fluid perfusion injury. *Journal of Neurotrauma, 16,* 879-892.

Maxwell, W.L, McCreath, B.J., Graham, D.I. Gennarelli, T.A., (1995).   Cytochemical evidence for redistribution of membrane pump calcium-ATPase and ecto-Ca-ATPase activity, and calcium influx in myelinated nerve fibres of theoptic nerve after stretch injury. *Journal of Neurocytology, 24,* 925-942.

Levin, N.S., Mattis, S., Raff, R.M., Eisenberg, H.M., Marshall, L.F., & Tabaddor, K. (1987). Neurobehavioral outcome following minor head injury: a three center study. *Journal of Neurosurgery, 66*, 234-243.

Johnston, K, Ptito, A., Chsnkowsky, J., Chen, J. (2001). New frontiers in diagnostic imaging in concussive head injury. *Clinical Journal of Sport Medicine, 11(3)*, 166-175.

Lishman, W.A. (1988). Physiogenesis and psychogenesis in the post-concussional syndrome. *Biological Journal of Psychiatry, 153*, 460-469.

McClelland, R.J., Fenton, G.W. , Rutherford, W. (1994). The postconcussional syndrome revisited. *Journal of the Royal Society of Medicine, 87*, 508-510.

Bryant R., & Harvey, A. (1999). Postconcussive symptoms and posttraumatic stress disorder after mind traumatic brain injury. *Journal of Nervous Mental Disease, 187*, 302-305.

Buchel, C. & Friston, K. (2001). Extracting brain connectivity. In *Function MRI: an introduction to methods*. Jezzard, P. Matthews, P.M., & Smith, S.M. (Eds). pp.295-308. Oxford University Press: N.Y.

Cabeza, R., Dolcos, F., Prince S.E., Rice, H.J., Weissman, D.H., Nyberg, L. (2003). *Neuropsychologia, 41(3)*, 390-399.

Friston, K.J., Holmes, A., Poline, J.B., Price, C.J., & Frith, C.D. (1996). Detecting activations in PET and fMRI: Levels of inference and power. *Neuroimage 40*, 223-235.

Friston, K.J., Holmes, A,P., & Worsley K.J. (1999). How many subjects constitute a study? *NeuroImage, 10*, 1-5.

Rees, G. & Lavie, N. (2001). What can functional imaging reveal about the role of attention in visual awareness? *Neuropsyschologia, 39(12)*, 1343-1353.

Stuss, D., & Knight, R. (2002). Principles of frontal lobe function. Oxford, University Press

Levin, B., Katz, D., Dade, L., Black, S. (2002). Novel approach to the assessment of frontal damage and executive deficits in traumatic brain injury. In: Principles of frontal lobe function Stuss & Knight (Eds.) pp. 448-465.

Gentry, L., Godersky, J., Thompson, B., Dunn, V. (1988). Prospective comparative study of intermediate-field MR and CT in the evaluation of closed head trauma. *American Journal of Radiology, 150*, 673-682.

Liu, A., Maldjian, J., Bagley, L., (1999). Traumatic brain injury: diffusion-weighted MR imaging findings. *AJNR, 20*, 1636-1641

Hofman, P.,Verhey, F., Wilmink, J., Rozendaal, N., & Jolles, J. (2002). Brain lesions in patients visiting a memory clinic with postconcussional sequelae after mild to moderate brain injury. *Journal of Neuropsychiatry and Clinical Neuroscience, 14(2)*, 176-184.

Umile, E., Sandel, M., Alavi, A., Terry, C., Plotkin, R. Dynamic imaging in mild traumatic brain injury: support for the theory of medial temporal vulnerability. *Archive of Physical Medical Rehabilitation, 83(11)*, 1506-1513.

Mittl, R., Grossman, R., Hiehle, J., Hurst, R., Kauder, D., Gennarelli, T., Alburger, G. (1994). Prevalence of MR evidence of diffuse axonal injury in patients with mild head injury and normal head CT findings. *American Journal of Neuroradiology, 15(8)*, 1583-1589.

Ross, B., Bluml, S. (2001). Magnetic Resonance spectroscopy of the human brain. *The American Records (New Anat), 265*, 54-84.

Schubert, T., Szameitat, A. (2003). Functional neuroanatomy of interference in overlapping dual tasks: fMRI study. *Cognitive Brain Research, 23*, 334-348.

Chen, J-K., Johnston, Frey, S., Petrides, K., Worsley, K., Ptito, A. (2003). Functional abnormalities in symptomatic concussed athletes: an fMRI study. *Neuroimage, 22*, 68-82.

Saunders, R.L., Harbaugh,, R.E. (1984). The second impact in catastrophic contact sports head trauma. *JAMA: 252*, 538-539.

Cantu, R.C. (1992). Second Impact Syndrome: Immediate Management. *Physician and Sports Medicine, 20*, 55-66.

Cantu, R.C., & Voy, R. (1995). Second Impact Syndrome: A Risk in any *Sport. Physician and Sports Medicine 23(6)*, 91-96.

Junger, E.C., Newell., D.W., Grant., G.A., et al. (1997). Cerebral Autoregulation Following Minor Head Injury. *Journal of Neurosurgery, 86*, 425-432.

Martland, H.S. (1928). Punch Drunk. *JAMA 91*, 1103-1107.

Casson, I.R., Siegel, O., Sharm, R., et al. (1984). Brain Damage in Modern Boxers. *JAMA 251*, 2663-2667.

Casson, I.R., Sharon., R., Campbell., E.A., et al. (1982). Neurological and CT Evaluation of Knocked-Out Boxers. *Journal of Neurology, Neurosurgery, 45*, 170-174.

Corsellis, J.A.N., Bruton, C.J., Freeman-Brown, D. (1973). The Aftermath of Boxing. *Psychological Medicine, 3*, 270-273.

Jordan, B.D., Relkin., N.R., Ravdin, L.D., et al. (1997). Apolipoprotein E e4 Associated with Chronic Traumatic Brain Injury in Boxing. *JAMA 278*, 136-140.

Matser, J.T., Kessels, A.G., Jordan, B.D., et al. (1998). Chronic Traumatic Brain Injury in Professional Soccer Players. *Neurology, 51*, 791-796.

Sortland, O., Tysvaer., A.T. (1989. Brain Damage in Former Association Soccer Players. An Evaluation by Cerebral Computed Tomography. *Neuroradiology, 31*, 44-48.

Tysvaer, A.T., Lochen, E.A. (1991). Soccer Injuries to the Brain: A neuropsychological study of Former Soccer Players. *American Journal of Sports Medicine, 19*, 56-60.

Guskiewicz, K.M., McCrea, M., Marshall, S.W., et al. (2003). Cumulative Effects Associated with Recurrent Concussion in Collegiate Football Players: The NCAA Concussion Study. *JAMA 19*, 2604-2605.

Iverson, G.L., Gaetz, M., Lovell, M.R, Collins, M.W. (20040. Cumulative Effects of Concussion in Amateur Athletes. *Brain Injury, 18*, 433-443.

Wilberger, J.E., Maroon, J.C. (1989). Head Injury in Athletes. Clinical Sports Medicine, 8, 1-9.

Bailes, J., & Hudson, V. (2001). Classification of sport-related head trauma: a spectrum of mild to severe injury. *Journal of Athletic Training, 36(3)*, 236-243.

CHAPTER 18

CONCUSSION CLASSIFICATION: HISTORICAL PERSPECTIVES AND CURRENT TRENDS

1. GRADING SYSTEMS

There is still enormous controversy both in the literature and in clinical practice as to how to classify concussed individuals. There is still NO universally accepted concussion grading system. The overall problem with concussion classification is that with the exception of an unconscious athlete (which is a relatively rare case in athletics) or someone who is severely dazed, it is often very difficult to identify who has sustained a concussion and who has not (Cantu, 2006). Despite some advances in clinical practice and research, NO consensus has been reached on how to classify concussion cases nor on proposed universal criteria for a return to sport participation. That said, there are currently three more or less conventionally accepted grading systems which serve to classify concussion based on four criteria: presence/absence of loss of consciousness (LOC), presence/absence and length of injury-related memory problems (amnesia), presence/absence of disorientation/confusion at the time of injury, and the duration of post-concussive symptoms.

A consensus committee of experts released a report in 2002 which recommended that grading systems and their corresponding return to play (RTP) guidelines be abandoned largely because there did not seem to be widespread compliance with RTP due to its conservative nature. Moreover, there still little empirical support for any of the existing return to play criteria. A more recently convened consensus committee that assembled in Prague to revise the recommendations from the 2002 report concluded that a two-grade classification should be used to assist in the management of concussion.

Although the Prague report recommended that a two-grade system should be widely implemented in the clinical management of concussion, other authorities, including some of the Prague committee members, recommend that the Cantu Evidence-Based System should be used to classify the severity of concussion. Overall, based on the consensus report outlining the current opinions of experts in the field, it is apparent that there is still enormous controversy about which concussion grading systems should be utilized in clinical practice. To fully appreciate the whole complexity of the issue, in the following text, some historical perspectives and currently accepted trends of concussion classification will be briefly discussed.

From a historical perspective, the first grading system of concussion was the Nelson grading system, published in *The Physician and Sportsmedicine* journal in 1984. Along with Dr. Nelson, there was some pretty "heavy horsepower" in this group: Dr. John Jane. This was a five-level concussion grading system going from zero to four. Most mild forms of concussion were defined as when an individual experienced a headache or some difficulties concentrating after a blow to the head but a) did not sustain any loss of consciousness (LOC), and b) had some clinical evidence of amnesia. More severe forms of concussion involved amnesia and/or progressive periods of LOC. If the subject was unconscious for less than one minute this subject was given a grade three of concussion. If someone was unconscious for more than one minute, this case was classified as a grade four concussion. (see also Table 1 for details).

Table 1. Nelson grading system for concussion

Grade 0	Head struck or moved rapidly; not stunned or dazed initially; subsequently complains of headache and difficulty in concentrating
Grade 1	Stunned or dazed initially; no loss of consciousness or amnesia; sensorium clears in less than 1 minute 2
Grade 2	Headache; cloudy sensorium longer than 1 minute in duration; no loss of consciousness; may have tinnitus or amnesia; may be irritable, hyperexcitable, confused, or dizzy
Grade 3	Loss of consciousness less than 1 minute in duration; no coma (arousal with noxious stimuli); demonstrates grade 2 symptoms during recovery
Grade 4	Loss of consciousness for more than 1 minute; no coma; demonstrates grade 2 symptoms during recovery

From:Nelson WE, Jane JA, Gieck JH. Minor head injury in sports: a new system of classification and management. Physician Sportsmed. 1984;12(3):103-107.

What is important historically about this grading system is that it was focused on primarily whether or not there was loss of consciousness (LOC) and its duration and/or whether or not there was short or long-term amnesia. At the time this grading system was proposed, there were no prospective studies of concussion and very little objective data were available. It was really the best grading system available at its time.

A year later, a more or less an experimental grading system proposed by Ommaya and colleagues was published based on empirical studies using primate animals (See Table 2). It should be noted that in reality, the last three grades in this classification schema should be considered more similar to the state of a coma rather than a concussion per se. Therefore, although it was called a concussion grading system it was really a grading system of different degrees of diffuse axonal injury. The first three grades of this six-tier system were later adopted by a Colorado Grading System. In the mild grade concussion assessment there is no amnesia. If amnesia is present, it becomes a grade two, if there is brief loss of consciousness (LOC), it becomes a grade three. This is historically what was used in the Colorado Grading System as shown in Table 4.

Table 2. Ommaya grading system

Grade 1	Confusion without amnesia (stunned)
Grade 2	Amnesia without coma.
Grade 3	Coma lasting less than 6 hours (includes classic cerebral concussion, minor and moderate head injuries)
Grade 4	Coma lasting 6-24 hours (severe head injuries)
Grade 5	Comas lasting more than 24 hours (sever head injuries)
Grade 6	Coma, death within 24 hours (fatal injuries)

From: Ommaya AK. Biomechanics of Head Injury: Experimental Aspects . In Nahum AM, Melvin J (eds): Biomedics of Trauma. Appleton & Lange, 1985, pp 245-269

As can be seen from Table 2, the grade two in the Ommaya's system is characterized by presence of amnesia without any LOC. It should be noted that any LOC in this grading system is indicative of the grade 3 concussion. Since this system was developed using animal studies, it was relatively easier to know the duration that the subjects were unconscious. In fact, this grade three LOC could be observed in any animal model of experimental concussions. Not surprisingly however, it is impossible to know if "concussed" animal suffered from amnesia, tenderness or lightheadedness.

A year later, without any prospective studies, a new grading system was proposed by Cantu, weighted heavily on whether or not the individual was unconscious and whether or not the individual had post-traumatic amnesia (Cantu, 1986). This grading system has subsequently been changed over the years. However, just for historical reasons, it is appropriate to look at it. However, one should be aware that the current body of knowledge has surpassed what was known at the advent of that scale. According to grade one, way back in 1986, there should not be LOC and/or post-traumatic amnesia within thirty minutes post-injury. And if the LOC occurred and this lasted less than five minutes, then one could consider the presence of a grade two concussion. If patients experienced post-traumatic amnesia for longer than thirty minutes, but less than twenty-four hours, this may also be indicative of grade two concussions. Finally, a grade three is assigned to the patient if LOC is present for more than five minutes or post-traumatic amnesia persists for more than twenty-four hours. This Cantu's classification scale has been revised since, as experience has been gained over the last fifteen years. The bottom line is that there are almost NO mild concussions on athletic fields that last longer than a minute. It is very rare that concussed athletes experience LOC for more than five minutes. In the great majority of concussion accidents athletes are unconscious for a few seconds. Therefore, most likely a lot of concussion cases are missed where athletes were unconscious briefly. The time frame in which athletes usually lose consciousness is on the order of five minutes or less.

Table 3. Cantu grading system

Grade 1	No loss of consciousness; posttraumatic amnesia less than 30 minutes.
Grade 2	Loss of consciousness less than 5 minutes in duration or posttraumatic amnesia lasting longer than 30 minutes but less than 24 hours in duration.
Grade 3	Loss of consciousness for more than 5 minutes or posttraumatic amnesia for more than 24 hours.

Cantu RC. Guidelines for return to contact sports after a cerebral concussion. Phys Sportsmed 1986;14:76-79. Used with permission of McGraw-Hill, Inc.

In 1986, the Cantu Grading System was proposed which relied on the presence or absence of LOC and the duration of post traumatic amnesia. The most mild, Grade one, was assigned to concussion episodes in which there was no LOC and post traumatic amnesia was brief, usually less than 30 minutes. In the Grade 2, there could be LOC of less than 5 minutes, post traumatic amnesia greater than 30 minutes and less than 24 hours. In Grade 3 there was LOC for more than 5 minutes or post traumatic amnesia for greater than 24 hours. Overall, this grading system is most popular among medical practitioners.

Now there will be a brief focus on another system, the Colorado Grading System. According to this system, the presence of confusion without amnesia is an indication of a grade one concussion. The presence of confusion with amnesia is an indication of grade two. Any sign or presence of LOC is indicative of a grade three concussion. This suggests that if patient did not lose consciousness at the time of injury, there was mild TBI rather than a severe grade three concussion. However, we should not forget the cases of Steve Young, Meryl Hodges, Al Tune, Pat LaFontaine and numerous other professional athletes who were never rendered unconscious at the time of injury, but had to retire because of persistent and long lasting post-concussive symptoms. It is inappropriate to consider these cases within the category of a moderate grade, because it costs these people their athletic career. To reiterate, LOC at the time of injury should NOT be considered as a primary predictor of concussion grade (Cantu, 2006).

Table 4. Colorado medical society grading system for concussion

Grade 1	Confusion without amnesia; no loss of consciousness
Grade 2	Confusion with amnesia; no loss of consciousness
Grade 3	Loss of consciousness

From Report of the Sports Medicine Committee. Guidelines for management of concussion in sports. Colorado Medical Society, 1990 (revised May 1991). Class III

In 1989 Barry Jordan, who has made significant contributions in the concussion field, especially as it relates to boxers, proposed his own grading system which is kind of a hybrid of what had preceded it. This system is outlined in Table 5. According to this classification system, confusion without amnesia and no LOC should be considered as grade one. This is quite similar to many other proposed classification systems of concussion. The presence of confusion with amnesia lasting less than twenty-four hours with no LOC is an indication of grade two.

Table 5. Jordan grading system for concussion

Grade 1	Confusion without amnesia; no loss of consciousness
Grade 2	Confusion with amnesia lasting less than 24 hours; no loss of consciousness
Grade 3	Loss of consciousness with an altered level of consciousness not exceeding 2-3 minutes; posttraumatic amnesia lasting more than 24 hours
Grade 4	Loss of consciousness with an altered level of consciousness exceeding 2-3 minutes.

From: Jordan BJ, Tsairis PT, Warren RF (eds): Head Injury in Sports. In Sports Neurology. Aspen Publications, 1989, p 227.

This system is also similar to a Myers' grade two concussion classification, which is characterized by confusion with amnesia, but without LOC. When LOC with altered levels of consciousness not exceeding two to three minutes, or post-traumatic amnesia lasting more than twenty four hours is present, this situation should be considered as grade three. Grade four assumes LOC for a longer period of time (usually more than few minutes). Thus, according to this system, the hallmarks of concussion are the presence of post-traumatic amnesia and LOC. Both LOC and its duration are important determinants of concussion grade.

In 1991 Joe Torg, who is much better known for his cervical spine axial load compression injuries causing quadriplegia, proposed a six-tiered concussion grading system. This system was published in the textbook titled *Athletic Injuries to the Head, Face and Neck*. The major themes of this system can be found in Table 6. It should be noted that head injuries were only partly discussed and a major part of the book did indeed deal with cervical spine and neck injuries. As it relates to concussion, his grading system has mainly focused on short-term confusion and the presence of amnesia at the time of injury or shortly after the incidence. He also introduced the "bell rung" term, referring to possible excessive noise sensitivity following mild TBI. It is important to note that duration of transient LOC was also considered an important feature in this classification system.

Table 6. Torg grading system for concussion

Grade 1	"Bell rung"; short-term confusion; unsteady gait; dazed appearance; no amnesia
Grade 2	Posttraumatic amnesia only; vertigo; no loss of consciousness
Grade 3	Posttraumatic retrograde amnesia; vertigo; no loss of consciousness
Grade 4	Immediate transient loss of consciousness
Grade 5	Paralytic coma; cardiorespiratory arrest
Grade 6	Death

From: Torg JS. Athletic Injuries to the Head, Neck, and Face. St. Louis, MO:Mosby-Year Book; 1991, p226.

As can been seen from Table 6, the most prominent symptoms of mild grades of concussion include some confusion, but no amnesia. Within grade two, there was presence of some post-traumatic amnesia. By definition, Torg considered post-traumatic *antrograde amnesia* as synonymous with post-traumatic amnesia (Torg, 1991). In other words, difficulties with cognitive abilities must be present from the moment of concussion and continue as the concussion progresses. According to this system, anterograde amnesia was considered in conjunction with LOC. However, it is a real problem to dissociate LOC from amnesia.

On the other hand, retrograde amnesia, if it was present without LOC may be an indication of grade three. Grade three may also be assigned if any sign of LOC was present at the time of injury. Grades five and six are really not concussion any more, but should be considered as severe TBI. Here, two basic categories are considered (i.e., amnesia and LOC). The LOC, regardless of how long it was present at the site of injury was considered to be severe TBI.

One year later, Dr. Roberts (1992), a primary care sports medicine specialist in Minnesota and a past president of the American College of Sports Medicine, proposed his grading system (see Table 7). Both LOC and post-traumatic amnesia were critical indices of this system. The lack of symptoms such as LOC or amnesia, and the presence of other symptoms for less than ten minutes were classified as either so-called "Bell Ringer" or grade zero. Grade one was defined when post-traumatic amnesia lasted less than thirty minutes, but more than ten minutes with no LOC. Grade two was defined when LOC was present for less than five minutes, and post-traumatic amnesia was present for longer than thirty minutes. The most severe case of TBI was defined when LOC lasted longer than five minutes and post-traumatic amnesia lasted longer than twenty four hours.

Table 7. Roberts grading system of concussion

Bell Ringer	No loss of consciousness; no posttraumatic amnesia; symptoms less than 10 minutes
Grade 1	No loss of consciousness; posttraumatic amnesia less than 30 minutes; symptoms greater than 10 minutes
Grade 2	Loss of consciousness less than 5 minutes; posttraumatic amnesia greater than 30 minutes
Grade 3	Loss of consciousness greater than 5 minutes; posttraumatic amnesia greater than 24 hours

From: Roberts WO. Who plays? Who sits? Managing concussion on the sidelines. Phys Sportsmed 1992; 20:66-76.

The same individuals who wrote the Colorado Grading System subsequently came up with the American Academy of Neurology Grading System which was published in *Neurology* in 1997 (Kelly & Rosenberg, 1997). It is basically similar to the Colorado Grading System except that there were some time limits placed on the grade one concussion. Specifically, by definition there should be transient confusion with no LOC and symptoms that all resolving within fifteen minutes. If these did not resolve within fifteen minutes it became a grade two. Grade three was defined if any LOC was present no matter how brief.

Based on recent collective work with Drs. Guskiewicz, Lovell and Kohn as well as with the founder of Hiplounder, Dave Erlounder, Dr. Cantu came up with a revision of his original grading system. They fully realized that five minutes of being unconscious is not what happens on athletic fields. A one minute duration of symptoms became the cut-off point between a moderate and a severe concussion, provided that there was LOC. Moreover, all post concussion symptoms, not only LOC, should be considered as important features of concussion and must be used as classifiers. These are all important features in terms of managing individuals with brain injury. It is not acceptable that people return to play while they still have post concussion symptoms. Also, symptoms such as post-traumatic amnesia, headaches, balance problems that lasted less than thirty minutes were considered as primary classifiers of grade one.

Table 8. ANN practice parameter (Kelly and Rosenberg) grading system

Grade 1	Transient confusion; no loss of consciousness; concussion symptoms or mental status abnormalities on examination resolve in less than 15 minutes
Grade 2	Transient confusion; no loss of consciousness; concussion symptoms or mental status abnormalities on examination last more than 15 minutes
Grade 3	Any loss of consciousness; either brief (seconds) or prolonged (minutes)

From: Kelly JP, Rosenberg JM. The diagnosis and management of concussion in sports. Neurology 1997; 48:575-580

Starting with the work of Lovell as well as that of Erlanger, subsequent prospective studies, the Vienna and Prague consensus statements and the American College of Sports Medicine Team Physician Statement on concussion, all refute the notion that brief LOC represents a serious concussion episode.

Table 9. Data driven Cantu revised concussion grading system

Grade 1	No LOC* PTA‡/PCSS‡‡ < 30 min (Mild)
Grade 2	LOC <1 min or PTA > 30 min <24hrs, other (Moderate), PCSS >30 min <7days
Grade 3	LOC $\geq$ 1 min or PTA $\geq$ 24 hrs, PCSS > 7 days (Severe)

*Loss of consciousness
‡Post-traumatic amnesia (anterograde/retrograde)
‡‡Post-concussion sign/symptoms
Cantu, RC Post-tramatic (retrograde and anterograde) amnesia: pathophysiology and implications in grading and safe return to play. J of Athletic Training 36(3)244-248, 2001

The final grading system that worth discussing is Cantu's based on prospective studies: "Data Driven Cantu Revised Concussion Grading Guidelines." This current grading system was published in the *Journal of Athletic Training* in 2001 and shown in Table 9. This is the only concussion grading scale where all post concussion symptoms are taken into consideration with extra weight given to post traumatic amnesia. In this grading scale the mild grade one involves no LOC, brief post traumatic amnesia and/or other post concussion symptoms, all of which are less than thirty minutes in duration. The intermediate grade two involves brief LOC of less than one minute, post traumatic amnesia longer than thirty minutes but less than twenty-four hours, or other post concussion symptoms longer than thirty minutes but less than one week in duration. Finally, grade three under that guideline involves LOC longer than one minute or post-traumatic amnesia greater than twenty-four hours and/or other post concussion symptoms lasting longer than seven days. It is reasonable to suggest that one should neither grade concussed athletes nor treat these athletes until all the symptoms have cleared.

There is a growing agreement among professionals that the duration of the post concussion symptoms is a very significant component of the severity of the head injury experienced by an athlete. Again, in the previous grading systems the emphasis was given to the presence and duration of post-traumatic amnesia and LOC. All concussion symptoms should be carefully considered and included in a classification model of concussion in athletics (Cantu, 2006). These post concussion symptoms can be found in the Post Concussion Signs and Symptoms Checklist in Table 10.

Table 10. Post Concussion Signs/Symptoms Checklist

Bell rung	**Memory deficits**
Depression	**Nausea**
Dinged	**Nervousness**
Dizziness	**Numbness/tingling**
Drowsiness	**Poor balance**
Excess sleep	**Poor coordination**
Fatigue	**Poor concentration**
Feel "in a fog"	**Easy distraction**
Feel "slowed down"	**Ringing in the ear(s)**
Headache	**Sadness**
Inappropriate emotions	**Light intolerance**
Personality change	**Noise intolerance**
Irritability	**Anxiety**
LOC	**Confusion**
Loss of orientation	**Stupor**

Note: A PCSS checklist is used not only for the initial evaluation but for each subsequent follow-up assessment, which is periodically repeated until all PCSS have cleared at rest and exertion.

It should be stressed that there are a number of other grading systems that might be considered from both historical and conceptual perspectives. Existing lack of consistency as well as the controversies in assessing athletes suffering from TBI is obvious. More sophisticated scientific data-driven assessment scales and numerical categories of concussion need to be developed in the future.

2. RETURN-TO-PLAY AFTER CONCUSSION

A number of guidelines for return-to-play (RTP) criteria after a first concussion are shown in Tables 11-14, reflecting those developed by Dr. Cantu and published by the Colorado Medical Society. These guidelines are currently most widely used in clinical practice dealing with brain-injured athletes. It is important to note, however, that none of these guidelines are founded on prospective data but rather reflect their authors' personal experience and common practice. They are best estimates of a way to manage concussion in athletics. We definitely should face the reality that many athletes suffering from concussion are not recognized and properly treated. And, even if a concussion is recognized at the time of injury, premature return to play solely based on some symptoms resolution may create further problems, including recurrent brain injuries, with many consequences, including fatality.

The major problem with return-to-play criteria is related to the lack of prospective experimental data on individual responses to concussion. For instance, neuropsychological data (such as memory, attention, information processing speed and other cognitive scores) and subjective symptoms reported by injured athletes are often uncorrelated and/or inconsistently present. Moreover, symptom reports document variable patient responses, even though they experience a similar type of concussive blow. A more complex situation is observed when athletes have experienced two or more concussive blows within a short time frame. Following multiple brain injuries, symptoms resolution and recovery from injury are quite individual and most often unpredictable. Therefore, it is a real challenge for medical practitioners to make a decision regarding the time at which injured athletes are ready for a safe return to sport participation. Another important factor to consider is that a return to sport participation versus a return to competitions should not be confused. This important distinction is discussed elsewhere in this book.

There are many factors that should be taken into consideration before allowing an athlete to return to play after a concussion. Among these are: clinical history, number of previous concussions, severity of previous brain injuries, athlete's personality, athlete's individual responses to injury, pain tolerance, current results of clinical evaluation, etc. In general, if an athlete has any symptoms on the field or outside the sport environment, this athlete should not be allowed to resume athletic practices, regardless of contact versus not-contact sports. Another general rule that should be followed is that criteria for return to sport participation in asymptomatic athletes should be the same for all sports, regardless of the degree of contact or use of protective devices. If neuropsychological evaluation is used, special caution should be exercised by medical practitioners, because the athlete's motivation, peer pressure, other financial issues, and/or pressure from coaching staff may be a serious confounding factor. The issue of athletes' motivation during neuropsychological assessment both at baseline (before the occurrence of any head trauma) and particularly after concussion is crucially important. Finally, it cannot be assumed that an athlete is asymptomatic when he or she "feels fine" based on subjective reports, since subjective feeling may not correlate with objectively obtained clinical evaluation and testing.

Table 11. Cantu guidelines for return to play after a first concussion

Grade	Recommendations
1	May return to play if asymptomatic* for one week
2	May return to play if asymptomatic* for one week
3	Should not be allowed to play for at least one month. May then return to play if asymptomatic* for one week
* rest and exertion (Cantu et al., 1986).	

It is commonly accepted now that athletes suffering from mTBI should not return to sport participation until neuropsychological, balance and other testing are done and this athlete is reported as asymptomatic for at least one week post-injury. It should be noted that recently a number of studies have clearly demonstrated the diagnostic value of balance testing. Accordingly, several guidelines include balance testing in addition to standard neurological examinations for the evaluation of athletes who have sustained brain injuries. Unfortunately, currently there are no neuroanatomic, physiological or diagnostic neuroimaging data that can be used to precisely determine the extent of brain injury in concussed individuals, the severity of metabolic dysfunction, or the precise moment it has cleared. "Normal" neuroimaging data would not warrant the clearance of post-concussive symptoms. Therefore, a clinical decision as to when to allow an athlete to return to play after a concussion should not be made solely on the basis of the results of neuroimaging or other tests. Rather, a clinical decision should be made based on the presence of other symptoms such as dizziness, disorientation, slowness in responding to questions, evidence of difficulty concentrating, physical sluggishness, and memory deficits, especially retrograde and anterograde amnesia. It is important to note that athletes who experience retrograde amnesia do not usually fully recover during the athletic contest in which they experienced the injury. This sign may be a strong predictor of TBI classification.

Table 12. Colorado medical Society Guidelines for return to play after first concussion

Grade	Recommendations
1	May return to play if asymptomatic* after at least 20 minutes observation.
2	May return to play if asymptomatic* for 1 week.
3	Should not be allowed to play for at least 1 month. May return to play if asymptomatic* for 2 weeks.

*rest and exertion

From: Kelly JP, Nicholas JS, Filley CM, et al. Concussion in sports: Guidelines for the prevention of catastrophic outcomes. JAMA 1991;266-2867; Report of the Sports Medicine Committee for the management of concussion in sports. Colorado Medical Society, 1990 (revised May 1991). Class III

Whether an athlete has been unconscious is, of course, important in terms of the return to play decision. It is generally believed that the degree of TBI is indicated by the depth and duration of the unconscious state. While not diminishing the importance of being rendered unconscious, it is inappropriate to make a decision of return to play solely on this symptom. It is illogical to assess as less severe the concussion occurring without LOC that produces post-concussive symptoms which last months or years, than the concussion which results in brief LOC and a resolution of all post-concussive symptoms within a few minutes or hours.

Regarding the clinical history, medical practitioners should be aware that athletes who have had a concussion will more than likely have further concussions due to possible cumulative brain trauma. According to recent evidence, once a player has incurred an initial cerebral concussion, his or her chances of incurring a second one are three to six times greater than for an athlete who has never sustained a concussion. One cause for this is a premature return to play based solely on clinical symptoms resolution. A sobering reality is that the ability to process information may be reduced after a concussion, and the severity and duration of functional impairment may be greater with repeated concussions. The damaging effects of the shearing injury to nerve fibers and neurons are proportional to the degree to which the head is accelerated and these changes may be cumulative. Taking into consideration the differential responses to single versus multiple concussions in terms of symptoms duration and symptoms resolution, special guidelines for return to play following multiple concussion were proposed.

Table 13. Cantu Guidelines for Return to Play After a Second or Third Concussion

Grade	Second Concussion	Third Concussion
1	Return to play in 2 weeks if asymptomatic	Terminate season
2	Minimum of 1 month	Terminate season
3	Terminate season; may return to play next season if asymptomatic*.	

From Cantu RC, Guidelines for return to contact sports after a cerebral concussion. Phys Sportsmed 1986;14:76-79. Used with permission of McGraw-Hill, Inc.

There is still debate in the literature regarding the possibility of the cumulative effect of previous concussion episodes. However, there is some evidence which suggests that high school athletes who suffer more than three concussions may experience more concussion symptoms. Moreover, these symptoms may be more severe and last longer. It is also important to document a history of previous concussions in terms of severity. There is definitely an interaction effect between the number and severity of previous concussions that should be considered in terms of return to play criteria. A few anecdotal facts suggest that even extremely mild multiple concussions may lead to a career ending catastrophic injury. Brett Lingros, Al Toon, Jim Miller, Steve Young, and Merill Hodge are professional athletes whose carriers were ended by numerous mild concussions without LOC, which produced sustained long-term post-concussion symptoms.

Table 14. Colorado Guidelines for Return to Play After a Second or Third Concussion

Grade	Second Concussion	Third Concussion
1	Terminate contest or practice play if without symptoms for at least 1 week.	Terminate season; may return to play in 3 months if without symptoms.
2	Consider terminating season; may return to play in 1 month if without symptoms.	Terminate season; may return to play next season if without symptoms.
3	Terminated season; may return to play next season if without symptoms.	Terminated season; strongly discourage return to contact or collision sports.

From: Kelly JP, Nicholas JS, Filley CM, et al. Concussion in sports: Guidelines for the prevention of catastrophic outcomes. JAMA 1991;266-2867; Report of the Sports Medicine Committee for the management of concussion in sports. Colorado Medical Society, 1990 (revised May 1991). Class III

Overall, universal agreement cannot be reached regarding concussion grading and return-to-play criteria. However, there is unanimous agreement that an athlete still suffering post-concussion symptoms at rest and exertion should not return to sport participation. There can be significant pressure placed on both athletes as well as medical practitioners to return the athlete to practice and play as soon as possible after the brain injury. However, returning to play may be delayed because of concern about the susceptibility to a second and even recurrent multiple TBI. Partial returning to practice may be a reasonable means of maintaining physical conditioning while awaiting full recovery. This decision should be made on individual basis. Again, every concussion episode should be considered as a unique case, thus, generalization in terms of RTP should not be encouraged.

CONCLUSION

It is important to stress that all of the guidelines are in agreement with the notion that athletes should NOT return to sport participation/competition until they have a normal neurologic examination, and they are asymptomatic at rest and exertion. Moreover, a neuropsychological test battery should indicate at least baseline level or above, and if done, a CT or EEG/MRI of the head should show no functional deficits and/or intracranial lesions that place the athlete at increased risk for re-injury. Most importantly one must understand that an athlete, while still symptomatic at either rest or exertion should NOT be allowed to return to competition. No athlete who has experienced LOC or amnesia should be allowed to go back into the event that same day. The general tenor is *"if in doubt, sit them out"*.

Additional factors that need to be considered include the athlete's total concussion history; including the number and the severity of those prior concussions. Moreover, the temporal proximity of concussions and the severity of the blow causing the concussion need to be assessed. Minor blows causing serious concussions should make a physician more hesitant to return an athlete to sport participation. The exact mechanisms of both short term and long lasting abnormalities in the brain's functional, behavioral, and cognitive abilities, and many other overseen abnormalities as a result of concussion in athletes still remain to be elucidated in future research.

Acknowledgements

The earliest, currently edited version of this chapter was published by Cantu (2006). Concussion classification: Ongoing Controversies. in Slobounov & Sebastianelli (Eds): *Foundations of Sport-related Brain Injuries* Springer, (obtained with permission from Springer), and also presented by Dr. Cantu at the Conference on Concussion in Athletics at Pennsylvania State University, University Park, April 29-30, 2004.

REFERENCES

Cantu, R.C. (2002). *Neurologic Athletic Head and Spine Injuries* Philadelphia, W.B. Saunders Company.

Entire July 2001 issue *Clinical Journal of Sports Medicine, 11(3)*, 131-209.

Entire October 2001 issue *Journal Athletic Training, 36(3),* 213-348.

National Athletic Trainers' Association Position Statement: Sport-Related Concussion. (2004). *Journal of Athletic Training, 39*, 280-295.

Summary and Agreement Statement of the 1st International Conference on Concussion in Sport. Vienna – November 2-3, 2001. (2002). Published simultaneously in *British Journal of Sports Medicine, and Physician and Sports Medicine.*

Summary and Agreement Statement of the 2nd International Conference on Concussion in Sport. Prague 2004. (2005). *British Journal of Sports Medicine, 39(4)*, 196-205, *Clinical Journal of Sport Medicine,* (2005), *15(2)*, 48-56, *Physician and Sports Medicine,* (2005) *33(4),* 29-44.

Nelson, W.E., Jane, J.A., Gieck, J.H. (1984). Minor head injury in sports: a new system of classification and management. *Physician Sportsmedicine, 12(3)*, 103-10.

Ommaya, A.K. Biomechanics of Head Injury: Experimental Aspects. In Nahum AM, Melvin J (eds), *Biomedics of Trauma*. Appleton & Lange, pp 245-269. 1985.

Cantu, R.C. (1986). Guidelines for return to contact sports after a cerebral concussion. *Physician Sportsmedicine, 14*, 76-79.

From Report of the Sports Medicine Committee. (1990). Guidelines for management of concussion in sports. Colorado Medical Society, (revised May 1991). Class III.

Jordan, B.J., Tsairis, P.T., Warren, R.F. (eds), *Head Injury in Sports: Sports Neurology*. Aspen Publications, p 227. 1989.

Torg, J.S. *Athletic Injuries to the Head, Neck, and Face*. St. Louis, MO:Mosby-Year, 1991.

Roberts, W.O. (1992). Who plays? Who sits? Managing concussion on the sidelines. *Physician Sportsmedicine, 20*, 66-76.

Kelly, J.P., Rosenberg, J.M. (1997). The diagnosis and management of concussion in sports. *Neurology, 48*, 575-580.

Cantu, R.C. (2001). Post-tramatic (retrograde and anterograde) amnesia: pathophysiology and implications in grading and safe return to play. *Journal of Athletic Training 36(3)*, 244-248.

Kelly JP, Nicholas JS, Filley CM, et al. Concussion in sports: Guidelines for the prevention of catastrophic outcomes. JAMA 1991; 266-2867; Report of the Sports Medicine Committee for the management of concussion in sports. Colorado Medical Society, 1990 (revised May 1991). Class III

CHAPTER 19

EVALUATION OF CONCUSSION: SIGNS AND SYMPTOMS

1. SIGNS AND SYMPTOMS OF CONCUSSION

Clinically observable signs and symptoms of concussion are the result of the pathophysiological processes that ensue immediately after the introduction of traumatic external forces. Although, additional information is necessary to further clarify the most common signs and self-reported sequelae of concussion: post-concussive symptoms, amnesia, and LOC. Proper concussion management begins as soon as an injury is suspected, often on the playing field, court, or sidelines. Once more serious injury is ruled out (e.g., open head injury, airways obstructed, etc.) and concussion is the working diagnosis, a thorough assessment of signs and symptoms should be completed. Often this assessment is performed by the team physician, athletic trainer, or emergency medical personnel. These professionals are often faced with the decision of whether the injured athlete should be sent to the emergency room for further evaluation, including a brain imaging examination. A proper assessment of concussion should include a thorough assessment of signs and symptoms, with documentation of the presence and duration of each.

Apart from self-reported symptoms, cognitive impairments are less observable by medical staff, teammates, or coaches, less transient than other signs of concussion (e.g., LOC, vomiting, etc.), and less susceptible to dissimulation and underreporting from an athlete who is motivated to return to play. Additionally, in the absence of signs and symptoms, lingering cognitive deficits which may be undetectable to the athlete may be apparent on neuropsychological tests, thus serving as a valuable indicator of concussion severity.

1.1. Cognitive Impairment After Concussion

Several cognitive domains have been found to be negatively affected by concussion, as evidenced by deficits identified with neuropsychological instruments: attention/ concentration (Gronwall & Wrightson, 1974; Bohnen et al., 1992; Echemendia et al., 2001), working memory (Echemendia et al., 2001; Lovell et al., 2003), verbal memory (Field et al., 2003; Echemendia et al., 2001; Bruce & Echemendia, 2004), visuospatial memory (Lovell et al., 2003), verbal learning (Echemendia et al., 2001; Field et al., 2003), speed of information processing (Barth et al., 1989; Macciocchi, et al., 1996), reaction time (Barth et al., 1983; Voller et al., 1999), and executive functions such as set shifting (Macciocchi et al., 1996; McCrea et al., 2003). It should be noted that deficits in all of these domains do not occur for every athlete. The data do, however, consistently show that one or more of these deficits typically persist for five to ten days post-concussion (Field et al., 2003; Echemendia et al., 2001; Barth et al., 1989; McCrea et al., 2003). Although it is a physical sign of concussion, postural stability may also be impaired after a concussion occurs. Postural stability impairments tend to resolve one to five days post-injury (Guskiewicz et al., 2001; McCrea et al., 2003). However, recent studies clearly indicate that proper assessment using advanced methodologies could reveal signs of balance problems far beyond seven days post-injury (Slobounov & Sebastianelli, 2006).

1.2. Common Signs of Concussion

Although it was once thought to be the hallmark sign of concussion (Gasquoine, 1998), LOC is no longer considered necessary for the diagnosis of concussion to be made, and only 6.3% to 9.3% of concussions involve LOC (Guskiewicz et al., 2000; Guskiewicz et al., 2003; Pellman et al., 2004). Another commonly identified symptom of concussion is amnesia, which occurs in 13% to 24.1% of concussions (Pellman et al., 2004; Guskiewicz et al., 2003; Erlanger et al., 2003). Perhaps the most commonly exhibited signs of alterations in mental status are confusion/disorientation. These are present in anywhere from 45% to 95% of athletes with concussion (Erlanger et al., 2003; Macciocchi et al., 1996).

1.3. Post-concussive Symptoms

The outward signs and symptoms of the neurochemical cascade and the subsequent recovery from injury have been briefly reviewed. As noted, LOC and amnesia are not necessary for a concussion to occur, but rather concussions are necessarily marked by other symptoms. These symptoms can be broken into three different categories: physical, cognitive, and affective. Characteristic physical symptoms are headache, dizziness, nausea, vomiting, sensitivity to light, sensitivity to noise, and drowsiness, among others. Typical cognitive symptoms are difficulty concentrating, difficulty remembering, feeling in a "fog," and feeling slowed down. The affective symptoms that are common following a concussion are irritability, sadness, and anxiety (Kelly & Rosenberg, 1997).

Headache, dizziness, difficulty concentrating, fatigue, attentional difficulties, and memory problems tend to manifest immediately after a concussion occurs (Alves et al., 1993; Cantu, 2001). These signs and symptoms often resolve within three to fourteen days, with a mean duration of about five to seven days (Echemendia & Julian, 2001; Macchiocchi et al., 1996, Lovell et al., 2004).

1.4. Signs and Indicators of Concussion Severity

Each of the aforementioned signs and sequelae of MTBI are important in assessing the severity of concussion. However, the relatively low prevalence of LOC and amnesia after concussion precludes the exclusive use of these sequelae as indicators of concussion severity. Rather, the length of post-concussive symptoms and duration of cognitive deficits may provide more useful data for assessing concussion severity, because either, and, more frequently, both sequelae are commonly experienced after a concussion. Moreover, the resolution of these sequelae roughly corresponds to the resumption of normal physiological brain functioning (Giza & Hovda, 2001).

1.5. Loss of Consciousness

As indicated, LOC was once considered by many to be the primary indicator of concussion severity, especially in the early concussion literature (Symonds, 1962; Gasquoine, 1998). However, more recent literature suggests that an LOC duration of no longer than twenty to thirty minutes tends to produce cognitive deficits at levels that are similar to individuals without LOC (Erlanger et al., 2003; Collins et al., 2003). It should be noted that the mean LOC durations reported in these studies were well below twenty minutes). Thus, it appears that the presence rather than the duration of LOC (within limits) in MTBI serves as the marker of severity.

1.6. Amnesia

Reports from experts and findings from empirical studies indicate that amnesia (either retrograde or anterograde) can serve as a fair indicator of concussion severity. Anterograde amnesia represents the period following injury where a disruption in continuous memory function has occurred for information and experiences occurring after the injury. Conversely, retrograde amnesia represents memory dysfunction for information occurring before the injury. Both types of amnesia typically vary in duration from a few seconds to a few days, depending on the injury. Interestingly, Collins and others found that amnesia was a much better predictor of outcome and recovery from concussion than was loss of consciousness (Collins et al., 2003). This is evidenced by the data showing that individuals with amnesia tend to show slower cognitive recovery than athletes without amnesia (Cantu, 2001; Lovell et al., 2003; Collins et al., 2003).

1.7. Confusion and Disorientation

The specific indices of confusion and disorientation are common and considered to be indicators of alterations in mental status. Lovell et al. (2003) found that longer periods of confusion/disorientation led to a longer duration of recovery of cognitive functioning after concussion. A similar pattern was reported by Collins et al. (2003), and the researchers also found that longer periods of disorientation tended to lead to more post-concussive symptoms about two days post-injury.

1.8. Balance and Postural Stability

Human postural stability is a product of an extremely complex dynamical system that relies heavily on integration of input from multimodal sensory sources, including those of vision, vestibular and proprioceptive systems (Mergner et al., 2003, for review). Several previous studies have identified a negative effect of mTBI on postural stability (Lishman, 1988; Ingelsoll & Armstrong, 1992; Wober et al., 1993). Recently, Geurts et al. (1999) showed an increased velocity of the center of pressure and overall weight-shifting speed, indicating both static and dynamic instability in concussed subjects. Interestingly, this study also indicated the association between postural instability and abnormal mental functioning after MTBI. The use of postural stability testing for the management of sport-related concussion is gradually becoming more common among sport medicine clinicians. A growing body of controlled studies has demonstrated postural stability deficits, as measured by the Balance Error Scoring System (BESS) on post-injury day one (Guskiewicz et al., 2003; Rieman et al., 2002; Volovich et al., 2003; Peterson et al., 2003). The BESS is a clinical test that uses modified Romberg stances on different surfaces to assess postural stability. The recovery of balance occurred between day one and day three post-injury for most of the brain injured subjects (Peterson et al., 2003). It appeared that the initial two days after MTBI are the most problematic for most subjects standing on the foam surfaces, which was attributed to a sensory interaction problem using visual, vestibular and somatosensory systems (Valovich et al., 2003; Guskiewicz, 2003).

However, more recent studies by Cavanaugh et al., (2005 a, b; 2006) have shown that advanced methods from non-linear dynamics (i.e. Approximate Entropy, ApEn) may detect changes in postural control in athletes with "normal" postural stability measures based upon conventional balance testing, longer than three to four days after cerebral concussion. Our own recent study has also shown alteration of virtual-time-to-contact (VTC) measures in the absence of traditional measures of postural instability in concussed individuals (Slobounov et al., 2008). These new findings are in agreement with our original hypothesis that residual abnormalities in concussed individuals may be undetected using isolated unimodal and/or traditional research methodology. Despite the general recognition of motor abnormalities (Kushner, 1998; Povlishock et al., 1992; Heitger et al., 2006) and postural stability specifically, resulting from neurological dysfunction in the concussed brain, no systematic research exists on how both dynamic balance and underlying neural mechanisms and neuropsychological measures are interactively affected by single and multiple MTBI. A database search of PubMed using a combination of EEG + balance (posture) + neuropsychological + concussion (or MTBI) returned only two articles and these are from our group at PSU.

1.9. EEG Signs and Symptoms

There are numerous EEG studies of concussion. Early EEG research in 300 patients clearly demonstrated the slowing of major frequency bands and focal abnormalities within forty-eight hours post-injury (Geets & Louette, 1985). A study by McClelland et al. (1994) has shown that EEG recordings performed during the immediate post-concussion period demonstrated a large amount of "diffusely distributed slow-wave potentials," which were markedly reduced when recordings were performed six weeks later. A shift in the mean frequency in the alpha (8-10 Hz) band toward lower power and an overall decrease of beta (14-18Hz) power in patients suffering from MTBI was observed by Tebano et al. (1988). The reduction of theta power (Montgomery et al., 1991) accompanying a transient increase of alpha-theta ratios (Pratar-Chand, et al., 1988; Watson et al., 1995) were identified as residual symptoms in MTBI patients. The most comprehensive EEG study using a database of 608 MTBI subjects up to eight years post-injury revealed (a) increased coherence in frontal-temporal regions; (b) decreased power differences between anterior and posterior cortical regions; and (c) reduced alpha power in the posterior cortical region, which was attributed to mechanical head injury (Thatcher et al., 1989). A study by Thornton (1999) has shown a similar data trend in addition to demonstrating the attenuation of EEG within the high frequency gamma cluster (32-64 Hz) in MTBI patients. Recently, the usefulness and high sensitivity of EEG in the assessment of concussion have been demonstrated. In our work, significant reduction of the cortical potentials amplitude and the concomitant alteration of gamma activity (40 Hz) was observed in MTBI subjects performing force production tasks three years post-injury (Slobounov et al., 2002). More recently, we showed a significant reduction of EEG power within theta and delta frequency bands during standing postures in subjects with single and multiple concussions up to three years post-injury (Thompson, Sebastianelli & Slobounov, 2005) and reduced amplitude of cortical potentials (MRCP) up to thirty days post-injury (Slobounov et al., 2005). It should be noted, however, there is no systematic EEG research available in the literature on subjects suffering and recovering from multiple concussions. Specifically, EEG abnormal signs and symptoms should be further supported by empirical studies to be used in clinical practice.

2. DURATION OF POST-CONCUSSION SYMPTOMS AND COGNITIVE DEFICITS

Unlike PTA and LOC, which may serve as initial indicators of concussion severity, post-concussive signs and symptoms and cognitive deficits can be considered "real time" indicators of concussion severity. These sequelae serve as markers that something pathological is happening in the brain which is likely due to the persisting effects of concussion. There is evidence to suggest that the duration of post-concussive signs and symptoms and cognitive deficiencies may be influenced by the age of the injured individual. Research findings indicate that symptoms resolve later in high school athletes relative to college athletes (McClincy et al., 2006; Field et al., 2003). Specifically, it appears that high school athletes tend to report persisting symptoms for about eight to fourteen days (Field et al., 2003; McClincy et al., 2006), whereas college athletes tend to report symptoms for less than one week (Guskiewicz et al., 1997; Echemendia et al., 2001).

The results of post-concussion neuropsychological testing also reveal a similar pattern of prolonged concussion sequelae. In particular, cognitive deficits were present at one week post-injury (Lovell et al., 2003; Field et al., 2003) and even at two weeks post-injury (McClincy et al., 2006). It is important to note that the research groups who completed these three studies did not examine their concussed athlete samples until cognitive deficiencies fully resolved. Rather, the athletes received serial neuropsychological testing until either one week (Field et al., 2003; Lovell et al., 2003) or two weeks (McClincy et al., 2006) post-concussion. Thus, it is not clear when cognitive deficits tend to resolve after concussion in high school athletes. However, it is apparent that cognitive deficits tend to last for more than one week in high school athletes and may potentially persist for longer than 14 days in high school athletes.

In college athletes, cognitive deficits tend to persist for five to ten days (Echemendia et al., 2001; Macciocchi et al., 1996; McCrea et al., 2003), and the average duration of cognitive deficits tends to be shorter (i.e., about five to seven days) than that of the high school athletes. Therefore, the findings of studies examining symptoms and cognitive deficits in high school athletes suggest these samples of adolescents seem to be experiencing sequelae of concussion for much longer than athletes who are just a few years older. These results suggest that age may be a moderating factor in the brain's response to concussion, and there is physiological research which will be reviewed later that provides more evidence for this assertion.

It should be noted that signs and symptoms following concussion will certainly vary greatly across athletes. Athletes may report or exhibit only one symptom of injury, or report several. Proper evaluation of signs and symptoms depends heavily on the professional's knowledge of common symptoms. There are a number of assessment instruments that include a checklist of signs and symptoms. The Post-Concussion Symptom Scale (PCS; Lovell et al., 1998) evaluates the presence and severity of twenty-one of the most commonly reported post-concussion symptoms, indicated on a seven-point Likert-type scale. This inventory can be administered at baseline (pre-injury), and serially following injury in order to track an athlete's recovery. Table 1 shows the paper version of the PCS, though this data may also be collected electronically through the ImPACT neuropsychological test battery or ImPACT sideline.

Table1. Post-Concussion Scale Rate your symptoms over the past two days.

Symptom	None	Mild		Moderate		Severe	
Headache	0	1	2	3	4	5	6
Nausea	0	1	2	3	4	5	6
Vomiting	0	1	2	3	4	5	6
Balance Problems	0	1	2	3	4	5	6
Dizziness	0	1	2	3	4	5	6
Fatigue	0	1	2	3	4	5	6
Trouble Falling Asleep	0	1	2	3	4	5	6
Sleeping More Than Usual	0	1	2	3	4	5	6
Sleeping Less Than Usual	0	1	2	3	4	5	6
Drowsiness	0	1	2	3	4	5	6
Sensitivity to Light	0	1	2	3	4	5	6
Sensitivity to Noise	0	1	2	3	4	5	6
Irritability	0	1	2	3	4	5	6
Sadness	0	1	2	3	4	5	6
Nervousness	0	1	2	3	4	5	6
Feeling More Emotional	0	1	2	3	4	5	6
Numbness or Tingling	0	1	2	3	4	5	6
Feeling Slowed Down	0	1	2	3	4	5	6
Feeling Mentally "Foggy"	0	1	2	3	4	5	6
Difficulty Concentrating	0	1	2	3	4	5	6
Difficulty Remembering	0	1	2	3	4	5	6
Visual Problems	0	1	2	3	4	5	6

Listed in Table 1 are results of a study examining the frequencies of the most common symptoms presenting acutely following concussion in high school and collegiate athletes.

Table 2. Frequencies of symptom endorsements (in percentage) for the Post-Concussion Symptom Scale in concussed athletes (N=52).

Symptoms	%
Headache	88.5
Difficulty Concentrating	82.7
Feeling Slowed Down	78.8
Dizziness	78.8
Nausea	77.3
Fatigue	76.9
Feeling Mentally "Foggy"	75.0
Drowsiness	73.1
Difficulty Remembering	69.2
Sensitivity to Light	57.7
Balance Problems	55.8
Sensitivity to Noise	50.0
Trouble Falling Asleep	45.0
Irritability	38.5
Sleeping More Than Usual	34.6
Visual Problems	32.7
Sleeping Less Than Usual	30.8
Nervousness	30.8
Feeling More Emotional	19.2
Sadness	19.2
Numbness or Tingling	15.4
Vomiting	11.5

Adopted from Lovell & Pardini (Cf: Slobounov & Sebastianelli, 2006).

A review of the table two above suggests that, although concussed athletes will not display a consistent pattern of deficits from case to case, it is reasonable to expect that most will experience headache, dizziness, nausea, and cognitive problems. Given that there appear to be many common symptoms among concussed athletes, it may be beneficial from a clinical standpoint to conceptualize symptoms based upon the "type" of symptoms being reported by the athlete. Recent factor analysis by Lovell and colleagues revealed four distinct symptom clusters emerging from the PCS in acutely concussed athletes. This analysis revealed somatic, cognitive, emotional, and insomnia factors. Somatic symptoms include visual disturbance (such as blurring or tracking problems), dizziness, balance problems, sensitivity to light and noise, nausea, etc. Symptoms of emotional disturbance include sadness, emotionality, and nervousness. Insomnia symptoms are trouble falling asleep and sleeping fewer hours overall. Cognitive symptoms are reported problems with concentration, memory, mental fogginess, fatigue, and cognitive slowing. Currently, the relation of these specific factors to concussion recovery is under investigation, though using the categories as descriptors when discussing issues with concussed athletes is often conceptually helpful.

Assessment of all post-concussion symptoms should occur as quickly as possible following injury, as well as multiple times during the recovery process. An athlete's report of symptoms will be a helpful indication of his or her perception of current health status, recovery, and readiness to return to sport participation.

3. EVALUATING CONCUSSION: COGNITIVE TESTING

Clearly, past concussion management strategies were based upon subjective information and dependent upon patient report. With return to play guidelines based only on presence and duration of mental status changes and self-reported symptoms, some injured athletes—especially those with mild injuries—were "falling through the cracks" in the concussion management system. Of interest, concussed athletes were subjectively reporting cognitive problems such as memory disturbance, inattention, and slowed thinking in addition to somatic symptoms such as headache and nausea. In the 1980s, researchers such as Alves (1987) and Barth (1989) began exploring these symptom complaints in a more objective manner. Since the 1980s neuropsychological testing has been a valuable tool in concussion research and practice. At first, neuropsychological testing was used in sports concussion for research purposes. From over two decades of research, important findings have consistently emerged, which has led to the clinical use of neuropsychological testing to manage the injury.

Professional sports organizations were some of the first to recognize and put into practice a neuropsychological testing program, which typically included baseline cognitive assessments and follow-up testing of concussed athletes. In 1994, the National Football League (NFL) established a committee on mild traumatic brain injuries in response to concerns about repetitive sports concussions. The committee included team physicians, athletic trainers, and equipment managers from the NFL, as well as a variety of experts in brain injury, basic science, and epidemiology. In 1996, the NFL began to record details of each concussion and chart the recovery process. This research has laid the foundation for a series of papers on concussion in professional football, published in *Neurosurgery* on a regular basis. In 1998, the commissioner of the NFL supported baseline neuropsychological testing and data-based concussion management programs throughout the league. Although each NFL club is an individual organization, most have decided to adopt a concussion management program that includes baseline and post-injury neuropsychological testing. At present, many clubs are beginning to transition from paper and pencil based testing to computerized testing.

Another professional sports organization that has led the way in concussion management is the National Hockey League. This organization became increasingly concerned about understanding concussion and its short- and long-term effects in ice hockey athletes. In the mid 1990s, Gary Bettman, the Commissioner of the NHL, initiated a rule which mandated neuropsychological testing (pre-injury and post-injury) for all of its players. The players' association, Commissioner Bettman, and many other staff members in this organization continue to be extremely supportive of this league-wide concussion program. At this point in time, more than 4,000 NHL players have undergone baseline testing and more than 800 concussed athletes have been studied. The data from eight years of testing and research are now under evaluation and are undergoing detailed analysis in our laboratories.

Although we have chosen to highlight these two pioneering programs in concussion management, there are many sports organizations, professional and amateur alike, that have adopted some sort of concussion program for contact sports. For example, the UPMC Sports Medicine Concussion Program and ImPACT have recently formed alliances with the USA Ski teams, including the aerial, downhill, and snowboarding teams. Also, given the risk of head injuries in motorsports, the NASCAR, Indianapolis, and Formula One racing leagues have adopted a neuropsychological testing program to ensure their drivers are recovered before returning to racing. Drs. Olvey and Cantu are in the process of collecting data on racing accidents, which should provide very interesting and useful information.

3.1. Measuring Change & Neuropsychological Testing

Practice effects on neuropsychological measures are well documented (Chelune et al., 1993; Iverson and Gaetz, 2004) and occur because exposure to test procedures or stimuli facilitates improved performance on subsequent testing. The improved performance may be due to procedural practice effects because of familiarity with the test procedure, or content practice effects that occur because test stimuli are remembered from one test administration to the next. Practice effects occur in both traditional "paper and pencil" tests and computerized test batteries. The use of alternate forms helps to mitigate content practice effects but not procedural practice effects.

These practice effects can be problematic in test interpretation because improvement due solely to practice effects may be confused with improved neurocognitive functioning. Practice effects are typically measured using a non-injured group of participants tested at least twice. Practice effects vary with the number and time interval of testing, with those tests occurring in close temporal proximity having the greatest practice effects. It is also worth noting that practice effects have shown an asymptote pattern and may reach a ceiling after two administrations. There are relatively few studies that have reported on practice effects in competitive athletes with concussion. Competitive athletes tend to be young (<35 years of age), motivated to perform well due to a desire to return to competition, and are susceptible to brain injury. Such susceptibility necessitates neuropsychological testing to monitor injuries and prevent further injuries from occurring before the athlete fully recovers (Echemendia & Julian, 2001). To provide greater precision in identifying real cognitive change post-concussion, it would be valuable if more information concerning practice effects on commonly used neuropsychological instruments were available for competitive athletes. The present chapter will present data to address practice effects in a group of elite college athletes who are at risk for or who have sustained a concussion. This chapter expands upon and refines previously presented data from the Penn State Concussion project (Mackin et al., 1997).

Conventional concussion monitoring and management programs have evolved using the baseline test paradigm pioneered by Barth and his colleagues at the University of Virginia (Barth et al., 1989). Using this paradigm, athletes involved with sports that are at risk for concussion are usually tested when they first join a team. If athletes sustain a concussion, they are tested serially post-injury using the same battery of tests with alternate forms, if available. Post-injury test data are compared to baseline data in order to identify any decrements in performance. If significant declines are found, the athlete is generally not returned to play until baseline levels of functioning are achieved. There is wide variability in the number and timing of post-injury tests. Some programs require a fixed testing interval such as two to seven days post-concussion, while others advocate that players should not be tested until asymptomatic (McCrea et al., 2005). Generally, it is recommended that players should be returned to competition when neuropsychological test performance returns to baseline and overt symptoms are no longer reported (Echemendia & Cantu, 2004).

Because of the relatively high prevalence of concussion and the potentially serious consequences of premature return to competition, neuropsychological measures have provided useful information to assist in the monitoring of concussion and in return to play decisions. Furthermore, because of frequent post-injury retest intervals, it is also important to consider the likely influence of practice effects. Several methods have been developed to evaluate the magnitude of practice effects and clinically significant change. These include effect size (Dikmen et al., 1999), reliable change index (RCI; (Jacobson & Truax, 1991)), and a reliable change index adjusted for practice effects (RCI practice; (Chelune et al., 1993)). There have only been a few studies where the RCI method has been applied to the neuropsychological performance of athletes with or without concussion (Barr and McCrea, 2001; Hinton-Bayre et al., 1999; Iverson et al., 2003; Iverson and Gaetz, 2004). More typically, statistically significant change has been examined through the use of conventional statistical analyses. These analyses are not intended to measure clinically significant change in individuals; instead, these methods provide general group information about mean changes. The use of RCI and RCI practice help to accurately capture "error variance" in test scores and thereby produce a more clinically useful method for identifying when a clinically significant change has occurred. A central goal for this chapter is to provide a comprehensive analysis of several commonly used neuropsychological measures with concussed athletes using three of the most commonly employed measures of change. In particular, the Hopkins Verbal Learning Test – Revised (HVLT-R; Brandt & Benedict, 2001), Stroop tests (Trennery et al., 1989), Trailmaking tests A and B (Reitan, 1958), and the Symbol Digit Modalities Test (SDMT; (Smith, 1982)) will be reviewed based on data obtained from the Penn State Concussion project (Echemendia, 1997). What follows is a review of some data addressing the practice effects in and reliability of these measures. The Appendix of this chapter includes a description of each way of measuring change discussed, and formulas for calculating them.

Although our focus in this chapter is practice effects in non-injured control athletes who are tested at time intervals similar to concussed athletes, we also chose to examine the possibility of practice effects in injured athletes. At first glance, the use of the term "practice effects" in injured athletes appears to be a misnomer, because the change in score for injured athletes contains both practice effects and true change because of neurocognitive improvement. However, we chose to examine possible practice effects in injured athletes as well, especially by one-week post concussion, because if we observe improvement of injured athletes that goes beyond their original baseline, then this cannot simply be due to cognitive recovery. Some practice effect must be present in such a scenario. Regardless, for ease of exposition in this chapter, we will refer to both practice effects as they are observed in non-injured athletes and practice effects plus cognitive recovery observed in injured athletes simply as "practice effects."

3.2. Measuring Change on the Trailmaking Tests

Several studies have examined practice effects on Trails A and B in non-athlete populations (Basso et al., 1999; Bornstein et al., 1987; Craddick and Stern, 1963; desRosiers and Kavanagh, 1987; Dikmen et al., 1999; Dikmen et al., 1983; Dye, 1979; Matarazzo et al., 1974; Mitrushina and Satz, 1991). However, most of these studies have examined practice effects at only one retest interval (Basso et al., 1999; Bornstein et al., 1987; desRosiers and Kavanagh, 1987; Dikmen et al., 1999; Matarazzo et al., 1974). Two studies on non-athletes have examined practice effects at two or more retest intervals (Craddick and Stern, 1963; Mitrushina and Satz, 1991). Results from both studies revealed statistically significant practice effects on Trails A. On Trails B Mitrushina and Satz (1991) compared mean differences between time points and found statistically significant practice effects at two, one-year intervals. Craddick and Stern (1963) used the same approach and found statistically significant practice effects on Trails B after a third one-month interval. In the studies using only one retest interval, significant practice effects were found on both Trails A and B (desRosiers & Kavanagh, 1987; Dye, 1979). In contrast, Dikmen and colleagues (Dikmen et al., 1999) used the effect size method and Basso and colleagues (1999) used the RCI practice method and found no evidence of significant change on either Trailmaking test.

Practice effects on Trails A and B have also been examined in athletes. In a study conducted by Macciocchi and colleagues (2001), MANOVA was used to examine differences among twenty-four concussed athletes on Trails A and B, the SDMT, and several other neuropsychological tests. Athletes were tested at twenty-four hours, five days, and ten days post-injury following a first concussion. These investigators found that even these concussed athletes showed significant improvement from pre-season baseline scores to five days post-injury on Trails A, however, they did not show significant improvement at twenty-four hours post-injury. On Trails B, these athletes showed a large improvement in performance from baseline to twenty-four hours post-injury that persisted to five days post-injury. For both Trails A and B, there was no further improvement in performance from five to ten days post-injury.

Macciocchi and colleagues (1996) reported on practice effects for Trails A and B in 183 athletes with one concussion and matched non-injured, non-athlete student controls. Practice effects were reported for both athletes and controls at twenty-four hours, five days, and ten days post-injury. Conventional statistical analyses were used to compare mean differences between injured athletes and controls. These investigators found that both injured athletes and their non-injured matched controls showed improvements in performance at each testing interval, improvement that increased as the number of testing intervals increased on both Trails A and B. No RCI or RCI Practice comparisons were reported.

Guskiewicz and colleagues (2001) also examined practice effects for Trails A and B. These researchers used repeated measures ANOVAs to examine practice effects relative to baseline at twenty-four hours, three days, and five days post-injury in a sample of thirty-six injured athletes and thirty-six, non-injured matched control athletes. In contrast to Macciocchi and colleagues' studies, they found that injured athletes showed worse performance relative to controls compared with baseline at all time points post-injury on Trails B. Injured athletes also displayed a significant decline from their baseline scores at twenty-four hours post-injury. The groups were not differentiated on Trails A at any post-injury time point, but examination of mean scores for the groups reveals that the concussed group showed evidence of notable improvement from baseline to day five post-injury. The control sample showed effectively static performance from baseline to the twenty-four hour time point, and then notable improvements at three and five days on Trails A.

Iverson and Gaetz (2004) compared 126 non-concussed collegiate football players at pre-season and then post-season on Trails A and B. They found that, overall, the athletes displayed significant practice effects between time points for both measures. Applying the RCI methodology to this same sample and using 90% confidence intervals (as we do in our sample below), these investigators found that only about 5% of both football and soccer players improved from pre-season to post-season on Trails A. These values were similar for football players when the RCI Practice method was applied, whereas practice effects for the soccer players were reduced to about 3%. On Trails B using the RCI methodology, about 8% of both groups of players improved from pre-season to post-season. When the RCI Practice methodology was applied, this value was reduced to about 5% to 6%. Several researchers have reported test-retest reliability coefficients for Trails A and B. The estimates for Trails A have ranged from 0.46 to 0.79 and from 0.44 to 0.90 for Trails B (Basso et al., 1999; desRosiers & Kavanagh, 1987; Matarazzo et al., 1974; Mitrushina and Satz, 1991).

3.3. Measuring Change on the Stroop Tests

Several studies have examined practice effects on the Stroop tests. Because there are several different versions of the test, however, the conclusions that can be drawn from the results cannot necessarily be applied to the version of the test that we have used in our research. Nonetheless, the results of these studies are fairly consistent. Statistically significant practice effects have been demonstrated using a computerized Stroop test (Davidson, 2003), an abbreviated (forty item) version (Houx et al., 2002), the Stroop

(1935) version (Dikmen et al., 1999), the Golden (1978) version (Connor et al., 1988), and in the normative sample of the Trenerry (1989) version. With the exception of Dikmen and colleagues' (1999) study, which used the RCI practice method, statistical calculations comparing mean differences between administrations were used for each of the other studies to identify practice effects. Furthermore, these studies used between one and five retest intervals, but the intervals were not comparable to those typically used in neuropsychological testing for concussed athletes.

Hinton-Bayre and Geffen (2004) describe a study of thirteen concussed and thirteen control Australian rules football players that used the Stroop, among other tests, at baseline and then two days post-concussion. On the Stroop A, where examinees were simply required to read color words on a page, concussed athletes displayed significantly slower performance at two days post-concussion compared with baseline. No significant change was noted for the control group. In contrast, on Stroop B, the color-word version of the test, no change post-concussion was observed in the concussed group, but the controls displayed a significant practice effect.

The test-retest reliability estimates of Stroop tests in previous research range from 0.79 to 0.84 on the color only version (Franzen et al., 1987; Dikmen et al., 1999) and from .77 to .90 on the color-word trial (Houx et al., 2002; Franzen et al., 1987; Trenerry et al., 1989).

3.4. Measuring Change on the Symbol Digit Modalities Test (SDMT)

A few studies have examined the nature and extent of practice effects on the SDMT. The data collected from the original normative sample (Smith, 1982) only included the mean scores of the participants at two testing intervals; no statistical analyses were reported to identify whether the difference between the mean values was statistically significant. Uchiyama and colleagues (Uchiyama et al., 1994) compared mean values at different retest intervals. The results of their analyses showed that the differences between the baseline scores and the scores at the retest intervals did not reach statistical significance.

In the Hinton-Bayre and Geffen (2004) study of Australian rules football players described above, they also used the SDMT. They found that the concussed group declined significantly in performance from baseline, but the performance of the control athletes was comparable to baseline. So, like the Stroop 1, the SDMT showed no evidence of a practice effect in the control group between two testing points.

In the study described earlier using twenty-four concussed collegiate athletes tested at baseline and then at post-concussion intervals of twenty-four hours, five days, and ten days, Macciocchi and colleagues (2001) also examined the SDMT. Athletes who had sustained one concussion displayed notable improvement in performance from baseline at five days post-concussion, but especially by ten days. Test-retest reliability coefficients for the SDMT have been reported. Uchiyama and colleagues (1994) reported a 0.79 coefficient and Smith (1982) reported a coefficient of 0.80.

3.5. Measuring Change on the Hopkins Verbal Learning Test – Revised (HVLT-R)

In Guskiewicz and colleagues' (2001) study described earlier, repeated measures ANOVAs were conducted to examine practice effects at twenty-four hours, three days, and five days post-injury in a sample of injured athletes and non-injured control athletes on the HVLT-R. They reported a significant group by day interaction for the HVLT-R, but examination of the group means of the concussed and control groups reveals very little change from baseline to any of the post-testing intervals. For example, the largest raw word increase for total immediate memory across the three HVLT-R learning trials was less than two words (from day three post-injury to day five post-injury time points in controls). Most of the other changes for both groups were approximately one word or less, suggesting that practice effects are not very significant at a clinical level. The use of alternate forms may have significantly attenuated practice effects in this case. The HVLT-R-revised manual (Brandt & Benedict, 2001) indicated that the test-retest reliability coefficient for this measure is 0.74.

3.6. Summary and Conclusions Regarding Measuring Change with the Trailmaking Tests, the Stroop Test, the SDMT, and the HVLT-R

As presented above, some data have been published regarding practice effects on the SDMT, HVLT-R, Stroop tests, and Trails A and B in samples of injured and control athletes. Also, some investigators have considered the impact of practice effects on the interpretation of data after serial neuropsychological testing in athletes (e.g., Bohnen et al., 1992; Echemendia & Julian, 2001); however, most studies have compared mean differences between athletes and control groups rather than using methods that provide information about the magnitude of the change (e.g., effect size) or about clinical significance (e.g., RCI and RCI practice) and the number of participants exhibiting significant practice effects. A few studies have examined RCI and RCI practice, but to our knowledge there are no published data in the sports neuropsychology research literature comparing the two methods. Although some studies have compared methods for the measurement of change using computerized measures (Erlanger et al., 2003), we have not found published research that has focused on examining practice effects on each of the four highlighted commonly used paper-and-pencil clinical neuropsychological tests presented in this paper. That said, McCrea and colleagues (2005) recently published a study on collegiate athletes using a standardized regression-based method to evaluate change in a group of concussed and control athletes on some of the same tests we examine in this chapter, and at the same time intervals post-concussion; however, the focus was not on practice effects, per se, but change following concussion. The data we outline below address some of the limitations in this literature.

4. NEUROPSYCHOLOGICAL RESERCH METHODOLOGY

4.1. Paper and Pencil Neuropsychological Assessment

In the beginning, traditional paper-and-pencil neuropsychological testing was used to evaluate the acute and more distant effects of sports-related concussion. Various test batteries have been used, though most tap similar abilities that are most often affected by concussion, including memory, attention, processing speed, and reaction time. In the 1989 Barth study of more than 2,350 collegiate football athletes, cognitive testing was administered at preseason, one, five, and ten days post injury, then at the end of the season. These authors used well-normed neuropsychological tests such as Trail Making Tests A and B (Reitan & Davison, 1974), Paced Auditory Serial Addition Task (Gronwall & Wrightson, 1980), and the Symbol Digit Test (Smith, 1973). On these measures, significant deficits were observed at twenty-four hours post-injury, with most athletes demonstrating a return to baseline by day ten. When Macciocchi and others (1996) compared the group of concussed athletes from Barth's large study to a control group on the above measures, most athletes evidenced generally equivalent performance to non-injured controls by day five.

In a study of professional rugby players, Hinton-Bayre and colleagues performed a baseline assessment of eighty-six players, thirteen of whom sustained concussions over the course of the ensuing season. Neuropsychological tests measuring verbal learning, verbal fluency, executive functioning, attention, processing speed, and psychomotor speed were used. When compared to their own baseline scores, as well as to a matched uninjured control group, concussed athletes demonstrated difficulty with psychomotor speed, processing speed, and aspects of executive function at twenty-four to forty-eight hours post-injury (Hinton-Bayre, et al., 2004). A study of concussion in the National Football League revealed no significant decline in performance on commonly used neuropsychological tests when athletes were tested, on average 1.4 days after injury. However, group differences emerged when comparing the post-injury performance of athletes who evidenced on-field memory dysfunction noted on a standard NFL physician form, to those who did not demonstrate memory dysfunction on the field. This comparison showed attenuated immediate and delayed memory functioning on visual memory, and a trend toward increased dysfunction on verbal memory tasks. Processing speed and psychomotor speed-based tasks did not show these differences (Pellman, et al., 2004).

These and other studies demonstrated that sport-related concussion did produce at least mild, though short-lived cognitive difficulties in most athletes. However, many researchers and practitioners began wondering if traditional neuropsychological testing, where response times could be measured accurately only to the nearest second and where practice effects were unavoidable, was sensitive enough to identify milder forms of cognitive impairment (Maroon et al., 2000; Pellman et al., 2004).

4.2. Computerized Testing

Through the 1980s and 1990s, research by a variety of scientists (neuropsychologists, neurosurgeons, athletic trainers, physicians, etc.) revealed that concussion not only caused the subjective experience of cognitive problems in athletes, but also caused verifiable declines in cognitive functioning in the areas of memory, attention, and/or processing speed (depending on the study). Thus, neuropsychological testing was becoming a useful adjunct, and an objective data point, through which to characterize and manage concussion. However, there were concerns about the more traditional model of completing a neuropsychological test battery.

First, in order to properly evaluate the primary cognitive functions that can be impaired by concussion, a lengthy battery of paper and pencil tests was required. Baseline testing has always been recommended as a "gold standard" of comparison to have available when an athlete becomes concussed. However, sports teams and their larger governing bodies often found the process of providing baseline evaluations to all players time consuming and expensive. A neuropsychologist or trained psychometrician was required to administer and interpret the tests, as well as determine the validity of each test. In addition, depending upon the battery, these measures could take more than two hours to administer per athlete. Secondly, many paper and pencil neuropsychological tests are subject to practice effects, which create difficulties in reliably assessing change over multiple administrations (Podell, 2004).

In the late 1990s, computerized neuropsychological batteries emerged in response to the need for concussion management programs in contact sports that were less expensive, more time efficient, and better able to detect small changes in response times. The ability to administer valid and reliable cognitive measures, which could be administered to groups and automatically scored, would ease the financial and personnel demands of traditional neuropsychological testing in this novel environment of athletics. Prior to 1998, no computerized neuropsychological test was available that was designed or normed for use with athletes. However, early computerized testing of concussed athletes was accomplished through using tests designed for use in other populations (such as the elderly or military personnel).

The development of cognitive assessment tools designed to assess sports-related concussion not only benefited professional sports organizations who were beginning to better protect the health of their injured athletes through adding an objective data point to assist in return to play decisions, but also benefited the group largest in number, and perhaps most at-risk for concussion—children and teenaged athletes. More than 1.25 million athletes compete in sports at the high school level each year (Bailes & Cantu, 2001), and more than 60,000 cases of MTBI occur at that level each year (Powell & Barber-Foss, 1999). Clearly, when considering that many young athletes begin contact sport participation such as football or soccer when they enter elementary school, the number of at-risk athletes is even higher. Thus the ultimate challenge for the sports medicine practitioner or sports neuropsychologist, as well as team coaches, parents, and athletic trainers, is to provide the same level of care for our younger athletes as we do for professional athletes. This is especially important given that research has indicated that children may be more vulnerable to injury.

4.3. Computerized Testing: ImPACT

Although other tests were used to assess sports-related concussions, ImPACT (Immediate Post-concussion Assessment and Cognitive Testing) was the first designed specifically with the athletic population in mind (Maroon et al., 2000). This test was developed at the University of Pittsburgh Medical Center by Drs. Mark Lovell, Joseph Maroon, and Micky Collins, and remains the foundation of our concussion management program. Thus, we will review this test in this chapter. There are other concussion assessment and management tools available, and the reader is encouraged to research these platforms as well. Recent research indicates that ImPACT is a sensitive and specific instrument with adequate reliability and validity (Iverson, Lovell, & Collins, 2005). ImPACT measures many cognitive processes that are often affected by concussion, including memory, attention, reaction time, and processing speed. (See Table 3 for a listing and description of the tasks comprising ImPACT.)

To address the issue of practice effects, the test contains multiple forms along with random generation of some stimuli. The test can also be easily administered to large groups with minimal supervision, in a school or team computer laboratory; thus providing an efficient and cost-effective way to provide baseline (pre-injury) evaluations. Like many traditional neuropsychological tests, ImPACT has validity indicators built into the system. In addition, normative data are provided for a wide range of student athletes, specific to age and gender. As previously mentioned, the test also contains an electronic version of the Post-Concussion Symptom Scale (PCS). Given that contact sports are popular across the globe, ImPACT has been developed for other language groups as well. The ability to track cognitive and symptom data across multiple time points in the recovery period allows the sports medicine practitioner to effectively track the athlete's recovery, and determine his or her deviation from baseline, increasing the fund of information upon which a return to play decision can be based.

Table 3. ImPACT Neurocognitive Test Battery.

Test Name	**Neurocognitive Domain Measured**
Word Memory	Verbal recognition memory (learning and retention)
Design Memory	Spatial recognition memory (learning and retention)
X's and O's	Visual working memory and cognitive speed
Symbol Match	Memory and visual-motor speed
Color Match	Impulse inhibition and visual-motor speed
Three Letter Memory	Verbal working memory and cognitive speed
Symptom Scale	Rating of individual self-reported symptoms

Composite Scores	**Contributing Scores**
Verbal Memory	**Word Memory** (learning and delayed), **Symbol Match** memory score, **Three Letters** memory score
Visual Memory	**Design Memory** (learning and delayed), **X's and O's** percent correct
Reaction Time (RT)	**X's and O's** (average counted correct RT, **Symbol Match** (average weighted RT for correct responses), **Color Match** (average RT for correct response)
Visual Motor Processing Speed	**X's and O's** (average correct distracters), **Symbol Match** (average correct), **Three Letters** (number correctly counted)

5. CONCUSSION MANAGEMENT USING COMPUTERIZED TESTING

Whether one is managing sport-related concussion in a professional league or seeing a thirteen year old recreational league soccer player in clinic, it is important to have in place a solid and consistent plan for managing concussion and making return to play decisions. For maximum benefit to both the athlete and the care provider, any athlete participating in contact sports should receive a baseline evaluation of his or her cognitive functions as well as baseline symptom reporting. This will provide a foundation around which the care provider can evaluate deviations in performance following injury. In addition, establishing a baseline program in a school or organization provides an excellent opportunity for the education of athletes and their parents about head injury in sports. Through completing a symptom inventory or interview, athletes can become familiar with many signs and symptoms of injury. Many teams use the beginning of a baseline session to allow athletes to view a presentation, film, or both about sports-related concussion. Educating a team about the injury, even if baseline testing is not performed, will communicate the seriousness of the injury and the importance of reporting an injury to a medical professional immediately.

Once an athlete sustains a concussion, the attending professionals should begin collecting data about the mechanism and sequelae of the injury. This will assist in follow-up evaluations. There are many sideline assessment tools available for recording this information, as well as for performing a brief evaluation of mental status. The Standardized Assessment of Concussion (McCrea et al., 1998), the UPMC Sports Medicine Concussion Program Concussion Card (see Table 3a), the McGill Abbreviated Concussion Evaluation, and Sideline ImPACT are just a few examples.

Table 3a. UPMC Sports Medicine Concussion Program Concussion Card: Side 1

Signs Observed by Staff	Symptoms Reported by Athlete
• Appears dazed or stunned	• Headache
• Is confused about assignment	• Nausea or vomiting
• Forgets plays	• Balance problems or dizziness
• Is unsure of game, score, or opponent	• Double or fuzzy vision
• Moves clumsily	• Sensitivity to light or noise
• Answers questions slowly	• Feeling "foggy"
• Loses consciousness	• Changes in sleep patterns
• Shows behavior or personality change	• Concentration or memory problems
• Forgets events prior to hit	• Irritability, emotionality, sadness
• Forgets events after hit	

Table 3b. UPMC Sports Medicine Concussion Program Concussion Card: Side 2

Orientation	Retrograde Amnesia
What city is this?	What happened in the prior quarter/period?
What stadium is this?	What do you remember just before the hit?
What is today's date (month/day/year)?	What was the score of the game prior to the hit?
Who is the opposing team?	Do you remember the hit?
What is the score?	**Concentration**
	Repeat the days of the week backward, starting with today.
Anterograde Amnesia	Repeat these numbers backward: **63; 419**
Repeat these words and try to remember them: **Girl, dog, green**	**Word List Memory**
	Repeat the three words from earlier.

In addition to collecting symptom and mental status data, the attending professional should determine when the athlete received the first concussive blow, if he or she played through the injury for any given time period, and if there were subsequent blows sustained during injured play. There are some instances in which the athlete will not be the person who reports a concussion to medical professionals. His or her team members, coaches, athletic trainers, or even parents in the stands, may be the first to observe changes in behavior, play, or coordination. Thus, third party information will be helpful in reconstructing the event and early symptoms.

Once the sideline evaluation is complete, the attending professional may be faced with making a decision about return to play during the same contest. Previous concussion grading scale-based guidelines have allowed return to play during the same game or practice if the athlete's symptoms last fewer than fifteen minutes, and the concussion did not involve loss of consciousness (LOC). In the case of high school students and younger, recent research and expert recommendations from international meetings of the Concussion in Sport group (Aubrey et al., 2002; McCrory et al., 2005), state that any young athlete should NOT be returned to play following any concussion, no matter how mild it may seem. These conclusions are based upon a growing body of literature suggesting that symptoms may abate prior to cognitive deficits, that athletes do not always accurately report symptoms, and that there may be a delayed onset of symptoms, even in mild concussions.

One to two days after injury, the concussed athlete should be reassessed for both symptom presentation and cognitive functioning. This will allow the care provider to determine the severity of the injury based upon the departure from baseline, and begin to determine the trajectory of recovery by comparing sideline symptom and mental status data to more thorough symptom and cognitive assessments. The follow-up assessment may be completed by a private or institution-based neuropsychologist, a properly supervised psychology technician or community psychologist, or a properly trained physician. In addition, if a computerized test battery is used, the program may be administered in the high school the following day by an athletic trainer or coach, then sent electronically to a consulting neuropsychologist for interpretation and case management. Preferably, a concussed athlete will receive at least one face-to-face consultation with a neuropsychologist or other professional trained in head-injury management. We are of the strong opinion that face-to-face evaluation by a concussion specialist is a must, and that utilizing cognitive and symptom data without a true knowledge of the injury may lead to false negatives and false positives.

While the athlete is recovering (e.g., experiencing symptoms and cognitive deficit), the care provider may wish to obtain follow-up assessments on a serial basis as the athlete recovers. The frequency at which one should administer assessments has been debated (see McCrory et al., 2005), though this is ultimately a decision that rests with the practitioner, often in cooperation with the sports organization. Given that recovery times can significantly vary, we believe it is helpful to conduct serial assessments. Once the athlete reports being symptom-free at rest, and he or she has achieved baseline or expected levels of functioning, the athlete may begin what is essentially exertional testing, which must precede safe return to play. The following description of exertional testing is derived in part from recommendations of the Concussion in Sport group (Aubrey et al., 2002).

A graduated return to exertion involves having the athlete begin with light non-contact forms of physical exertion (e.g., walking, stationary biking) after he or she no longer experiences any post-concussion symptoms. If the athlete is able to tolerate light physical exertion without the return of symptoms, he or she may then try moderate non-contact physical exertion, which usually involves sport-specific activity (such as running in soccer or skating in hockey). Once asymptomatic with moderate exertion, the athlete may proceed to heavy, non-contact physical exertion, which usually involves training drills, heavier running or weight lifting, etc. Generally, there should be at least twenty-four hours between steps, and many practitioners are more comfortable with extending the time between steps. In the case of some contact sports, it is possible to introduce light contact drills prior to a return to full contact practice or game play. Some practitioners may suggest that a soccer athlete try supervised ball heading, or a football player attempt light contact through supervised hitting prior to a full return to sport. Regardless, each of these steps is designed to ensure that concussion symptoms do not re-emerge when the athlete increases his or her levels of physical activity, which would be a sign of incomplete recovery. If an athlete experiences a return of ANY concussion symptom during physical exertion, he or she must be returned to the previous level at which there were no symptoms, and should begin the stepwise process from that point.

CONCLUSION

The science and clinical practice of concussion assessment and management has grown and improved in recent years. Current research using neuropsychological assessment tools incorporated with balance and brain imaging testing has dramatically improved our understanding of concussion. Surely, we are learning the importance of managing each concussion episode as an individual event, thus a multi-dimensional approach should be undertaken to properly assess the severity of a concussion and predict possible consequences. As mentioned by Lovell & Pardini (2006), the current state of concussion research should communicate to any reader that there really is no such thing as a simple concussion. From a clinical perspective, there are cases of concussions that seem mild at first and unpredictably lead to prolonged symptoms, and may interfere with not only athletic activities but also daily functioning. Therefore, it is the responsibility of medical practitioners, coaches and other interested parties to protect our athletes by exercising updated concussion assessment and management.

Acknowledgements

The information for this chapter was partly acquired from Lovell and Pardini' as well as Rosenbaum et al., chapters published in Slobounov & Sebastianelli (Eds): "Foundations of Sport-related Brain Injuries," Springer, 2006 (obtained with permission from Springer publishing company). Dr. Rosenbaum contribution to this chapter is especially appreciated.

REFERENCES

Gronwall, D., & Wrightson, P. (1974). Delayed recovery of intellectual function after minor head injury. *Lancet, 2*(7881), 605-609.

Bohnen, N., Twijnstra, A, & Jolles, J. (1992). Post-traumatic and emotional symptoms in different subgroups of patients with mild head injury. *Brain Injury, 6*(6), 481-487.

Echemendia, R. J., Putukian, M., Mackin, R.S., Julian, L.J., & Shoss, N. (2001). Neuropsychological test performance prior to and following sports-related mild traumatic brain injury. *Clinical Journal of Sports Medicine, 11*, 23-31.

Lovell, M. R., Collins, M. W., Iverson, G. L., Field, M., Maroon, J. C., Cantu, R., Podell, K., Powell, J. W., Belza, M., & Fu, F. H. (2003). Recovery from mild concussion in high school athletes. *Neurosurgery, 98*(2), 296-301.

Field, M., Collins, M. W., Lovell, M. R., & Maroon, J. (2003). Does age play a role in recovery from sports-related concussion? A comparison of high school and collegiate athletes. *The Journal of Pediatrics, 142*(5), 546-553.

Bruce, J. M., & Echemendia, R. J. (2004). Delayed-onset deficits in verbal encoding strategies among patients with mild traumatic brain injury, *Neuropsychology, 17*(4), 622-629.

Barth, J. T., Alves, W. M., Ryan, T. V., Macciocchi, S. N., Rimel, R. W., Jane, J. A., Nelson, W. E. (1989). Mild head injury in sports: Neuropsychological sequelae and recovery of function. In H. S. Levin, H. M. Eisenberg, & A. L. Benton (Eds.), *Mild head injury* (pp. 257-275). New York: Oxford University Press.

Macciocchi, S. N., Barth, J. T., Alves, W., Rimel, R. W., & Jane, J. A. (1996). Neuropsychological functioning and recovery after mild head injury in collegiate athletes. *Neurosurgery, 39*(3), 510-514.

Barth, J. T., Macciocchi, S. N., Giordani, B., Rimel, R., Jane, J. A., & Boll, T. J. (1983). Neuropsychological sequelae of minor head injury. *Neurosurgery, 13*(5), 529-533.

Voller, B., Benke, T., Benedetto, K., Schnider, P., Auff, E., & Aichner, F. (1999). Neuropsychological, MRI, and EEG findings after very mild traumatic brain injury. *Brain Injury, 13*(10), 821-827.

McCrea, M., Guskiewicz, K. M., Marshall, S. W., Barr, W., Randolph, C., Cantu, R. C., Onate, J. A., Yang, J., & Kelly, J. P. (2003). Acute effects and recovery time following concussion in collegiate football players: The NCAA Concussion Study. *Journal of the American Medical Association, 290*(19), 2556-2563.

Guskiewicz, K. M., Ross, S. E., & Marshall, S. W. (2001). Postural stability and Postural stability and neuropsychological deficits after concussion in collegiate athletes. *Journal of Athletic Training, 36*(3), 263-273.

Slobounov, S. M., & Sebastianelli, W. (2006). *Foundation of Sport-Related Brain Injuries* (pp. 445-477). New York: Springer.

Gasquoine, P. G. (1998). Historical perspectives on postconcussion symptoms. *Clinical Neuropsychologist, 12*(3), 315-324.

Guskiewicz, K. M., Weaver, N. L., Padua, D. A., & Garrett, W. E. (2000). Epidemiology of concussion in college and high school football players. *American Journal of Sports Medicine, 28*, 643-650.

Guskiewicz, K. M., McCrea, M., Marshall, S. W., Cantu, R. C., Randolph, C., Barr, W., Onate, J. A., & Kelly, J. P. (2003). Cumulative effects associated with recurrent concussion in collegiate football players: the NCAA Concussion Study. *Journal of the American Medical Association, 290*(19), 2549-2549.

Pellman, E. J., Viano, D. C., Casson, I. R., Arfken, C., & Powell, J. (2004). Concussion in professional football: Injuries involving 7 or more days out—Part 5. *Neurosurgery, 55*(5), 1100-1119.

Erlanger, D., Feldman, D., Kutner, K., Kaushik, T., Kroger, H., Festa, J., Barth, J., Freeman, J., & Broshek, D. (2003). Development and validation of a web-based neuropsychological test protocol for sports-related return-to-play decision-making. *Archives of Clinical Neuropsychology, 18*(3), 293-316.

Kelly, J. P., & Rosenberg, J. H. (1997). Diagnosis and management of concussion in sports. *Neurology, 48*(3), 575-580.

Alves, W., Macciocchi, S. N., & Barth, J. T. (1993). Postconcussive symptoms after uncomplicated mild head injury. *Journal of Head Trauma Rehabilitation, 8*(3), 48-59.

Cantu, R. C. (2001). Posttraumatic retrograde and anterograde amnesia: Pathophysiology and implications in grading and safe return to play. *Journal of Athletic Training, 36*(3), 244-248.

Echemendia, R. J., & Julian, L. J. (2001). Mild traumatic brain injury in sports: Neuropsychology's contribution to a developing field. *Neuropsychology Review, 11*(2), 69-86.

Lovell, M. R., Collins, M. W., Iverson, G. L., Johnston, K. M., & Bradley, J. P. (2004). Grade 1 or "ding" concussions in high school athletes. *American Journal of Sports Medicine, 32*(1), 47-54.

Giza, C. C., & Hovda, D. A. (2001). Neurometabolic cascade of concussion. *Journal of Athletic Training, 36*(3), 228-235.

Symonds, S. C. (1962). Concussion and its sequelae. *The Lancet, 1*(1), 1-5.

Collins, M. W., Iverson, G. L., Lovell, M. R., McKeag, D. B., Norwig, J., & Maroon, J. (2003). On-field predictors of neuropsychological and symptom deficit following sports-related concussion. *Clinical Journal of Sports Medicine, 13*(4), 222-229.

Mergner, T., Maurer, C., Peterka,R. (2003). A multisensory posture control model of human upright stance. In: *Progress of Brain Research*, vol 142. C. Prablanc, D. Pelisson and Y. Rosetti (eds), Elsevier Science, pp.189-201.

Lishman, W. A. (1988). Physiogenesis and psychogenesis in the post-concussional syndrome. *Biological Journal of Psychiatry, 153*, 460-469.

Ingelsoll, C. D., & Armstrong, C. W. (1992). The effect of closed-head injury on postural sway. *Medicine in Science, Sports & Exercise, 24*, 739-743.

Wober, C., Oder, W., Kollegger, H., Prayer, L., Baumgartner, C., & Wober-Bingol, C. (1993). Posturagraphic measurement of body sway in survivors of severe closed-head injury. *Archive of Physical Medical Rehabilitation, 74*, 1151-1156.

Geurts, A., Knoop, J., & van Limbeek, J. (1999). Is postural control associated with mental functioning is the persistent postconcussion syndrome? *Archive Physical Rehabilitation, 80*, 144-149.

Rieman, B. & Guskiewicz, K. (2002). Effect of mild head injury on postural stability as measured through clinical balance testing. *Journal of Athletic Training, 35*, 19-25.

Valovich, T., Periin, D., Gansneder, B. (2003). Repeat administration elicits a practice effect with the balance error scoring system but not with the standardized assessment of concussion in high school athletes. *Journal of Athletic Training, 38*(10), 51-56.

Peterson, C., Ferrara, M., Mrazik, M., Piland, S., Elliott, R. (2003). Evaluation of neuropsychological domain scores and postural stability following cerebral concussion in sport. *Clinical Journal of Sport Medicine, 13*(4), 230-237.

Cavanaugh, J., Guskiewicz, K., Stergiou, N. (2005a). A nonlinear dynamic approach for evaluating postural control: new directions for the management of sport-related cerebral concussion. *Sport Medicine, 35(11),* 935-950.

Cavanaugh, J., Guskiewicz, K., Giuliani, C., Marshall, S., Mercer, V., Stergiou, N. (2005b). Detecting altered postural control after cerebral concussion in athletes with normal postural stability. *British Journal of Sports Medicine, 39(11),* 805-811.

Cavanaugh, J.T., Guskiewicz, K.M., Giuliani, C., Marshall, S., Merser, V.S., Stergion, N. (2006). Recovery of postural control after cerebral concussion: new insights using approximate entropy. *Journal of Athletic Training, 41*(3), 305-313.

Kushner, D. (1998). Mild traumatic brain injury: Toward understanding manifestations and treatment. *Archive of Internal Medicine, 158*, 10-24.

Povlishock, J. T., Erb, D. E. , & Astruc, J. (1992). Axonal response to traumatic brain injury: reactive axonal change, deafferentation and neuroplasticity, *Journal of Neurotrauma, 9(*suppl.1), 189-200.

Heitger, M.H., Jones, R.D., Dalrymple-Alford, J.C., et al. (2006). Motor deficits and recovery during the first year following mild closed head injury. *Brain Injury, 20*(8), 807-824.

Geets,W., & Louette, N (1985). Early EEG in 300 cerebral concussions. *EEG and Clinical Neurophysiology, 14(4),* 333-338.

McClelland, R. J., Fenton, G. W. , Rutherford, W. (1994). The postconcussional syndrome revisited. *Journal of the Royal Society of Medicine, 87*, 508-510.

Tebano, T. M., Cameroni, M., Gallozzi ,G., Loizzo, A., Palazzino, G., Pessizi, G., & Ricci, G. F. (1988). EEG spectral analysis after minor head injury in man. *EEG and Clinical Neurophysiology, 70*, 185-189.

Montgomery, A., Fenton, G. W., McCLelland, R. J., MacFlyn, G., & Rutherford, W. H. (1991). The psychobiology of minor head injury. *Psychological Medicine, 21*, 375-384.

Pratar-Chand, R., Sinniah, M., & Salem, F. A. (1988). Cognitive evoked potential (P300): a metric for cerebral concussion. *Acta Neurologia Scandinavia, 78*, 185-189.

Watson, W. R., Fenton, R. J. McClelland, J., Lumbsden, J., Headley, M., & Rutherford, W. H. (1995). The post-concussional state: Neurophysiological aspects. *British Journal of Psychiatry, 167*, 514-521.

Thatcher, R. W., Walker, R. A., Gerson, I., & Geisler, F. H. (1989). EEG discriminant analyses of mild head injury. *EEG and Clinical Neurophysiology, 73*, 94-106.

Thornton, K. E. (1999). Exploratory investigation into mild brain injury and discriminant analysis with high frequency bands (32-64 Hz). *Brain Injury, 13(7)*, 477-488.

Slobounov, S., Sebastianelli, W., Simon, R. (2002). Neurophysiological and behavioral Concomitants of Mild Brain Injury in College Athletes. *Clinical Neurophysiology, 113*, 185-193.

Thompson, J., Sebastianelli, W., Slobounov, S. (2005). EEG and postural correlates of mild traumatic brain injury in athletes. *Neuroscience Letters, 377*, 158-163.

Slobounov, S., Sebastianelli, W., Moss, R. (2005). Alteration of posture-related cortical potentials in mild traumatic brain injury. *Neuroscience Letters, 383*, 251-255.

McClincy, M. P., Lovell, M. R., Pardini, J., Collins, M. W., & Spore, M. W. (2006). Recovery from sports concussion in high school and collegiate athletes. *Brain Injury, 20*(1), 33-39.

Guskiewicz, K. M., Riemann, B. L., Perrin, D. H., & Nashner, L. M. (1997). Alternative approaches to the assessment of mild head injury in athletes. *Medical Science and Sports Exercise, 29*(Suppl. 7), 5213-5221.

Lovell, M. R., & Collins, M. W. (1998). Neuropsychological assessment of the college football player. *Journal of Head Trauma Rehabilitation, 13*(2), 9-26.

Alves, W., Macciocchi, S. N., & Barth, J. T. (1993). Postconcussive symptoms after uncomplicated mild head injury. *Journal of Head Trauma Rehabilitation, 8*(3), 48-59.

Chelune, G. J., Naugle, R.I., Luders, H., Sedlak, J., and Awad, I.A. (1993). Individual change after epilepsy surgery: Practice effects and base-rate information. *Neuropsychology, 7*(1), 41-52.

Iverson, G. L. and Gaetz, M. (2004). Practical Considerations for Interpreting Change Following Brain Injury. In M. R. Lovell, Collins, M.W., Echemendia, R.J., and Barth, J.T. (Ed.), *Traumatic Brain Injury in Sports* (Vol. 1, pp. 323-356). New York: Taylor and Francis.

Mackin, R. S., Sabsevitz, D.S., Julian, L., Junco, R., Dwyer, M. and Echemendia, R.J. (1997). *Stability Coefficients and Practice Effects of Neuropsychological Tests in College Athletes: Preliminary Findings.* Paper presented at the Sports Related and Nervous System Injuries Orlando, Florida.

Barth, J. T., Alves, W.M., Ryan, T.V., Macciocchi, S.N., Rimel, R.W., Jane, J.A., and Nelson, W.E. (1989). Mild head injury in sports: Neuropsychological sequelae and recovery of function. In H. E. H. Levin, and A. Benton (Ed.), *Mild Head Injury* (pp. 257-275). New York, NY: Oxford University Press.

McCrea, M., Barr, W.B., Guskiewicz, K., Randolph, C.R., Marshall, S.W., Cantu, R., Onate J.A., and Kelly, J.P. (2005). Standard regression-based methods for measuring recovery after sport-related concussion. *Journal of the International Neuropsychological Society, 11*, 58-69.

Echemendia, R. J., and Cantu, R.C. (2004). Return to Play Following Cerebral Brain Injury. In M. R. Lovell, Collins, M.W., Echemendia, R.J., and Barth, J.T. (Ed.), *Traumatic Brain Injury in Sports* (Vol. 1, pp. 479-498). New York: Taylor and Francis.

Dikmen, S. S., Heaton, R.K., Grant, I., and Temkin, N.R. (1999). Test-retest reliability and practice effects of Expanded Halstead-Reitan Neuropsychological Test Battery. *Journal of the International Neuropsychological Society, 5*(4), 346-356.

Jacobson, N. S., and Truax., P. (1991). Clinical significance: A statistical approach to defining meaningful change in psychotherapy research. *Journal of Consulting and Clinical Psychology, 59*(1), 12-19.

Barr, W. B., and McCrea, M. (2001). Sensitivity and specificity of standardized neurocognitive testing immediately following sports concussion *Journal of the International Neuropsychological Society, 7*(6), 693-702.

Hinton-Bayre, A. D., Geffen, G.M., Geffen, L.B., McFarland, K.A., and Friis, P. (1999). Concussion in contact sports: Reliable change indices of impairment and recovery *Journal of Clinical and Experimental Neuropsychology, 21*(1), 70-86.

Iverson, G. L., Lovell, M.R., and Collins, M.W. (2003). Interpreting Change on ImPACT Following Sport Concussion. *Clinical Neuropsychologist, 17*(4), 460-467.

Brandt, J., and Benedict, R.H.B. (2001). *Hopkins Verbal Learning Test-Revised professional manual* Odessa, FL: Psychological Assessment Resources.

Trennery, M. R., Crosson, B., DeBoe, J., and Leber, W.R. (1989). *Stroop Neuropsychological Screening Test Manual.* Odessa, FL: Psychological Assessment Resources.

Reitan, R. M. (1958). Validity of the Trail-Making Test. *Perceptual and Motor Skills, 8*, 271-276.

Smith, A. (1982). *Symbol Digit Modalities Test Manual (Revised).* Los Angeles: Western Psychological Services.

Echemendia, R. J. (1997). Neuropsychological assessment of the college athlete: The Penn State concussion program. *Neuropsychology, 10*(2), 189-193.

Basso, M. R., Bornstein, R.A., and Lang, J.M. (1999). Practice effects on commonly used measures of executive function across twelve months. *Clinical Neuropsychologist, 13*(3), 283-292.

Bornstein, R. A., Baker, G.B., and Douglass, A.B. (1987). Short-term retest reliability of the Halstead-Reitan Battery in a normal sample. *The Journal of Nervous and Mental Disease, 175*(4), 229-232.

desRosiers, G. a. K., D. (1987). Cognitive assessment in closed head injury: Stability, validity, and parallel forms for two neuropsychological measures of recovery. *The International Journal of Clinical Neuropsychology, 9*(4), 162-173.

Matarazzo, J. D., Wiens, A.N., Matarazzo, R.G., and Goldstein, S.G. (1974). Psychometric and clinical test-retest reliability of the Halstead Impairment Index in a sample of healthy, young, normal men. *The Journal of Nervous and Mental Disease, 158*(1), 37-49.

Craddick, R.A., and Stern, M.R. (1963). Practice effects on the trail making test. *Perceptual and Motor Skills, 17*, 651-653.

Mitrushina, M., and Satz, P. (1991). Effects of repeated administration of a neuropsychological battery in the elderly. *Journal of Clinical Psychology, 47*(6), 790-801.

Dikmen, S. S., Reitan, R. M., and Temkin, N. R. (1983). Neuropsychological recovery in head injury. *Archives of Neurology, 40*(6), 333-338.

Dye, O. A. (1979). Effects of practice on trail making test performance. *Perceptual and Motor Skills, 48*, 296.

Macciocchi, S. N., Barth, J. T., Littlefield, L., and Cantu, R.C. (2001). Multiple concussions and neuropsychological functioning in collegiate football players. *Journal of Athletic Training, 36*(3), 303-306.

Macciocchi, S. N., Barth, J. T., Alves, W., Rimel, W. R., and Jane, J. A. (1996). Neuropsychological functioning and recovery after mild head injury in collegiate athletes. *Neurosurgery, 39*, 510-514.

Guskiewicz, K. M., Ross, S. E., and Marshall, S. W. (2001). Postural stability and neuropsychological deficits after concussion in collegiate athletes. *Journal of Athletic Training, 36*, 263-273.

Davidson, D. J., Zacks, R. T., and Williams, C. C. (2003). Stroop interference, practice, and aging. *Aging Neuropsychology and Cognition, 10*(2), 85-98.

Houx, P. J., Shepherd, J., Blauw, G-J., Murphy, M. B., Ford, I., Bollen, E. L., Buckley, B., Stott, D. J., Jukema, W., Hyland, M., Gaw, A., Norrie, J., Kamper, A. M., Perry, I. J., MacFarlane, P. W., Edo Meinders, A., Sweeney, B. J., Packard, C. J., Twomey, C., Cobbe, S. M., and Westendorp, R. G. (2002). Testing cognitive function in elderly populations: The PROSPER study. *Journal of Neurology, Neurosurgery, and Psychiatry, 73*(4), 385-389.

Stroop, J. R. (1935). Studies of interference in serial verbal reactions. *Journal of Experimental Psychology, 18*, 643-662.

Golden, C. G. (1978). A manual for clinical and experimental users: Stroop color and word test. Stoelting: Wood Dale, IL.

Connor, A., Franzen, M., and Sharp, B. (1988). Effects of practice and differential instructions on Stroop performance. *The International Journal of Clinical Neuropsychology, 10*(1), 4.

Hinton-Bayre, A. D., and Geffen, G. (2004). Australian rules football and rugby league. In M. R. Lovell, Collins, M. W., Echemendia, R. J., and Barth, J. T. (Ed.), *Traumatic Brain Injury in Sports*. New York: Taylor and Francis.

Franzen, M. D., Tishelman, A.C., Sharp, B.H., and Friedman, A.G. (1987). Test-retest reliability of the Stroop Word Color Test across two intervals. *Archives of Clinical Neuropsychology, 2*, 265-272.

Uchiyama, C. L., D'Elia, L. F., Dellinger, A. M., Selnes, O. A., Becker, J. T., Wesch, J. E., Chen, B.B., Satz, P., van Gorp, W., and Miller, E. N. (1994). Longitudinal comparison of alternate versions of the Symbol Digit Modalities Test: Issues of form comparability and moderating demographic variables. *Clinical Neuropsychologist, 8*(2), 209-218.

Bohnen, N., Twijnstra, A, and Jolles, J. (1992). Post-traumatic and emotional symptoms in different subgroups of patients with mild head injury. *Brain Injury, 6*(6), 481-487.

Erlanger, D., Feldman, D., Kutner, K., Kaushik, T., Kroger, H., Festa, J., Barth, J., Freeman, J., and Broshek, D. (2003). Development and validation of a web-based neuropsychological test protocol for sports-related *return to play* decision-making. *Archives of Clinical Neuropsychology, 18*, 293-316.

Reitan, R., Davison, L. A. (1974). *Clinical neuropsychology: Current status and applications*. Washington, DC: VH Winston.

Gronwall, D., & Wrightson, P. (1975). Cumulative effect of concussion. *Lancet, 2*(7943), 995-997.

Smith, A. (1973). *Symbol Digit Modalities Test. Manual*. Los Angeles: Western Psychological Services.

Macciocchi, S. N., Barth, J. T., Alves, W., Rimel, R. W., & Jane, J. A. (1996). Neuropsychological functioning and recovery after mild head injury in collegiate athletes. *Neurosurgery, 39*(3), 510-514.

Hinton-Bayre, A. D., Geffen, G. M., & McFarland, K. A. (1999). Sensitivity of neuropsychological tests to the acute effects of concussion in contact sport. In D. T. B. Murdoch, E. Ward (Ed.), *Brain Impairment and Rehabilitation: A National Perspective* (pp. 75-81). Brisbane: Academic Press.

Pellman, E. J., Lovell, M. R., Viano, D. C., Casson, I. R., & Tucker, A. M.. (2004). Concussion in professional football: Neuropsychological testing—Part 6. *Neurosurgery, 55,* 1290-1305.

Maroon, J. C., Lovell, M. R., Norwig, J., Podell, K., Powell, J. W., & Hartl, R. (2000). Cerebral concussion in athletes: evaluation and neuropsychological testing. *Neurosurgery, 47*(3), 659-669; discussion 669-672.

Podell, K. (2004). Computerized assessment of sports-related brain injury. In R. J. E. M. R. Lovell, J. T. Barth, M. W. Collins (Ed.), *Traumatic Brain Injury in Sports* (pp. 375-393). Lisse, The Netherlands: Swets & Zeitlinger.

Bailes, J. E., & Cantu, R. C. (2001). Head injury in athletes. *Neurosurgery, 48*(1), 26-45; discussion 45-26.

Powell, J. W., & Barber-Foss, K. D. (1999). Traumatic brain injury in high school athletes. *JAMA, 282*(10), 958-963.

Iverson, G. L., Lovell, M.R., Collins, M. W. (2005). Validity of ImPACT for measuring processing speed following sports-related concussion. *Journal of Clinical & Experimental Neuropsychology, 27*(6), 683-689.

McCrea, M., Kelly, J. P., Randolph, C., et al. (1998). Standardized assessment of concussion (SAC): On-site mental status evaluation of the athlete. *J Head Trauma Rehab, 13,* 27-35.

Aubrey M, C. R., Dvorak J, Graf-Bauman T, Johnston KM, Kelly J, Lovell MR, McCrory P, Meeuwisse WH, Schamasch P. (2002). Summary and agreement statement of the 1st international symposium on concussion in sport, Vienna 2001. *Clinical Journal of Sport Medicine, 12,* 6-11.

CHAPTER 20

TRAUMATIC BRAIN INJURIES IN CHILDREN

1. INTRODUCTION

Advances in pediatric neuroscience have significantly improved our understanding of the consequences of traumatic brain injuries (TBI) in recent years. However, the vast majority of questions involving the functioning and recovery of a human brain post-head injury remain unanswered. It is still a significant problem to find the threshold of damage in the child brain versus the adult brain as a result of TBI. We are still unclear about biological predispositions to residuals of head injury after a single episode of concussion in children. We still do not know the optimal period of time required for a full recovery after concussion. Overall, there are no validated criteria at this time in terms of return to sport participation by children suffering from TBI. These are just a few questions that many clinicians and researchers struggle to resolve in order to understand the nature of concussion in children. We hope that gaining more knowledge in these areas will allow us to influence this process at an earlier stage in an attempt to stop the process of brain tissue damage and to speed up its recovery.

It should be noted that much previous research on concussion has been at the stage of animal models. Moreover, most of our knowledge about concussive processes in a human brain involves the adult brain, which has completed its developmental cycle and does not offer much variation in terms of anatomical and functional plasticity. There is even less known about neurodynamics of concussive processes and recovery in children, whose young brain remains in a state of constant developmental changes. There is an enormous amount of variation, introduced into the picture of pediatric concussion that has to do with the brain's developmental phase at the time of the child's injury, its capacity for plasticity and adaptation to the injury, and other factors. This issue becomes even more complicated when we attempt to determine the effects of mild head injury on a still developing young brain.

There is some uncertainty regarding the definition and the effects of mild head injury (mild TBI) on an adult's brain and even more on a child's brain. Moreover, there are methodological issues regarding the outcome measures of post-injury recovery. This is due to the fact that the brain is still developing and, therefore, must simultaneously cope with overcoming the challenges of the cerebral insult and to meet the demands of normal development. For example, in case of an adult onset of head injury, the outcomes of recovery are measured based on already established neurocognitive abilities. However, in the case of a young brain, the outcome is determined by the recovery of established and newly acquired abilities.

While the extent of our knowledge of pediatric concussive processes is much less than the knowledge of this type of injury in adults, it is important to note that the rate of concussion among children and adolescents appears to be greater than that of adults. In fact, while researchers focus mainly on adults when investigating sport related concussions, overall, children and adolescent athletes represent the largest athlete group across many sports. The statistics are alarming, as millions of children are being injured in motor vehicle accidents and in sports, as well as other recreational activities. Nation-wide surveys show that children are also very likely to get injured as a result of falls involving bicycles, swings, playground structures, toddler walkers, stairs, etc. In fact, according to the Centers for Disease Control, <u>children ages six to fourteen years have a higher risk</u> of sustaining a head injury than any other age group.

Clearly, in many cases, concussion in both children and adults go unacknowledged and unrecognized by medical professionals, teachers and parents. The lack of knowledge about concussion is closely related to an unsafe attitude towards risky activities, including both organized and recreation sports. Yet, a large number of children who have sustained a concussion experience short and long-term effects that interfere with their everyday functioning. Despite our increased attention, as parents, teachers, and health providers, to children's academic performance and behavioral functioning, difficulties and changes in these areas are not generally attributed to possible post-concussive symptoms. Parents and teachers almost never miss difficulties and struggles that children experience at home and in school, such as behavioral and emotional disregulation, disinhibition, not following rules, inability to comply with authority or delayed speech and language development. Cognitive difficulties, such as slowed information processing, decreased sustained attention, impaired recall of academic material, and many others are also noticed, when a child's academic performance declines. However, the deficits from both categories are rarely recognized as post-concussive symptoms. Thus, it is vitally important to increase the awareness by health care professionals, parents and teachers of the effects of mild head injuries that may interrupt the developmental processes of a young brain.

Children frequently experience minor accidents while playing sports, riding bicycles, or simply rushing in the midst of excitement. Sometimes, what their parents, teachers or caregivers perceive as a minor bump on the head or a bad scrapes, constitutes a mild concussion that bears a multitude of subtle or more visible neuropsychological and neurobehavioral problems, which require immediate professional attention. Some mild head injuries involve very minor, if any, structural brain damage that can be easily observed by way of radiological tests. However, what often missed in such situations are functional deficits that significantly impair a child's functioning in school and at home.

Comprehensive neuropsychological examination, which has been accepted as a part of the overall assessment that includes other diagnostic techniques such as CT, MRI, EEG, and SPECT, in delineating any brain-behavior deficits secondary to brain injury, is capable of accurately detecting such functional deficits (NYSPA, 2005). The Social Security Administration has defined neuropsychological testing as the "administration of standardized tests that are reliable and valid with respect to assessing impairment in brain functioning" (SSA, 2002). The American Academy of Neurology has rated neuropsychological testing as "Established" with Class II evidence and a Type A recommendation (AAN, 1996). In fact, neuropsychological examination provides unique information about the injured brain that is not available from other tests and procedures, because "… The sensitivity of neuropsychological tests is such that they often reveal abnormality in the absence of positive findings on CT and MRI scans. Moreover, they can identify patterns of impairment that are not determinable through other procedures, leading to appropriate treatment recommendations." (NYSPA, 2005). Consequently, more accurate diagnostic results lead to more effective treatment plans to ameliorate child's suffering after a concussion.

2. EPIDEMIOLOGY OF PEDIATRIC TBI

Injuries sustained by high school athletes have resulted in 500,000 doctor visits, 30,000 hospitalizations and a total cost to the healthcare system of nearly two billion dollars per year (Adirim & Cheng, 2003). Injured athletes are forced to miss school days, which also can cost their parents hours of lost work productivity. In addition to the physical and financial cost of sport-related injuries, injured athletes may experience negative psychological consequences and develop psychological trauma, including mood disturbances and lowered self esteem (Smith et al., 1990). Pediatric closed head trauma deserves our utmost attention as clinicians, researchers, parents and teachers because it may affect our children's future health and development. It may negatively affect their cognitive, emotional and behavioral functioning in all settings, including home and school. Frequently, their treating pediatricians are the first professionals who realize the connection between deteriorating emotional, cognitive and behavioral functioning and a recent concussion. Sometimes, school teachers, nurses, and counselors make that connection and urge parents to seek medical attention.   Many such referrals involve pediatric neuropsychologists, who are experienced in identifying various aspects of brain-behavior relationship and in detecting any deficits in this profile. It should be noted that a large number of children are referred for pediatric neuropsychological examination by their pediatricians (Yeates et al., 1995).

Despite the urgency of this pressing health problem among children, it has been difficult to obtain accurate statistics regarding the incidence and prevalence of closed head injuries in the United States. Frequently, the existing local, regional, or national registries record only those instances of pediatric head injuries that require hospital admission.   Therefore, these records may omit many milder head injuries, such as bumps and scrapes resultant from minor slip and fall or bicycle accidents, which still inflict a deteriorating effect on child's growing brain.   Moreover, epidemiological studies frequently employ various definitions of the injury and data collection techniques. They also focus on varying cases and age groups. Therefore, Kraus (1995) pointed out that the rates of pediatric head injury vary widely among the nine published studies, averaging approximately 180 per 100,000 children per year in children under fifteen years of age.

In terms of the severity of head injuries, the United States National Coma Databank has determined that approximately 85% of all head injuries that require medical treatment are mild in nature; approximately 8% are moderately severe; and the remaining 6% are severe (Lauerssen et al., 1988). Similarly, the National Pediatric Trauma Registry reports that 76% of injuries are mild, 10% are moderate, and 15% are severe (Lescohier & DiScala, 1993). Thus, even though the rates of mild head injuries are likely to be substantially underestimated by the existing reporting guidelines, it is clear that these numbers are alarming. These "scary" statistics suggest that thousands of children and adolescents suffer from various post-concussive neurobehavioral and neurocognitive difficulties that greatly affect their academic achievement and everyday functioning.

3. WHO ARE AT HIGH RISK FOR TBI?

Several demographic factors play important role in the incidence of closed head injuries in children. It has become known that boys appear to be at considerably higher risk for closed head trauma than girls. Some studies state that the ratio of boys to girls rises from approximately 1.5:1 for preschool children to approximately 2:1 for school age children and adolescents (Kraus, 1995). The change appears to reflect a sharp increase in head injuries among males and a gradual decrease among females (Kraus et al., 1986).

Age is another significant factor in determining the risk for traumatic head injury in children. Numerous studies report that the rate of head injuries is relatively stable from birth to age five. In this age group, head injuries occur in about 160 per 100,000 children. After age five, the overall incidence gradually increases until early adolescence and then shows rapid growth, reaching a peak incidence of approximately 290 per 100,000 by age eighteen (Kraus et al., 1986). In addition to the gender and age variables, some studies indicate that the incidence rates may also vary as a function of family characteristics. In fact, it was found that open and effective communication, acceptance, adaptive coping skills, among others, play a moderating function not only in expediting recovery from head injury, but also in preventing the incidence of such trauma in children (Yeates et al., 1997).

Some studies have implicated family socioeconomic status. These findings are, however, inconsistent and vary from study to study. For instance, one study has shown that the incidence of brain injury among children was related to the median family income, as determined from census tract data (Kraus et al., 1990). Children's age and ethnicity, however, did not affect the relationship of TBI incidence and income. Another study found no relationship between parental education or income and the incidence of head injury (Klauber et al., 1986).

4. PEDIATRIC MORBIDITY AND MORTALITY

TBI is a leading cause of death among children and adolescents and results in substantial neurobehavioral morbidity for survivors. Kraus (1995) surveyed and concluded that approximately 40% to 50% of pediatric fatalities result from various traumas associated with brain injury. Many studies report the mortality rate among children less than fifteen years of age as anywhere from ten to twenty per 100,000 children (Annergers et al., 1980; Kraus, 1995). Most researchers agree that the mortality rate is higher among children with severe injuries than among those with mild injuries, although their statistics vary. Fletcher and colleagues (1995) indicate that approximately 50% of children brought to the emergency room with severe head injuries die. Kraus (1995) reports that the fatality rate among hospital admissions ranges from 12% to 62% for severe injuries, less than 4% for moderate injuries, and less than 1% for mild injuries among children and adolescents.

Despite inconsistency in the reporting of the mortality rates, research studies conclude that pediatric survivors of closed head trauma frequently experience adverse consequences. Again, children with severe injuries experience greater residual deficits and poorer outcome than those with milder head traumas. Although researchers utilize various measures of outcome of the closed head injury, they usually include the Glasgow Outcome Scale (Jennett & Bond, 1975), which differentiates five outcome categories, such as death, persistent vegetative state, severe disability, moderate disability, and "good recovery." Kraus (1995) demonstrated that 75% and 95% of children with closed head injuries displayed a "good recovery," 10% showed a moderate disability, 1-3% showed a severe disability, and less than 1% remained in a persistent vegetative state.

It is important to note that the measure of a "good recovery," using the Glasgow Outcome Scale, does not mean that the child will not experience neurobehavioral impairment or associated functional disabilities (Koelfen et al., 1997). Given the fact that mild TBI are likely to receive a much better outcome prediction than more severe head injuries, it would be a mistake to believe that mild concussions will not lead to any neurobehavioral or cognitive deficits. Bruce and Schut (1982) estimated that, with intensive care and rehabilitation, most of these children recover from closed head injury. However, about 50% of them continue to suffer from long-term neurological and cognitive deficits.

How can we detect these deficits to properly address them? Comprehensive neuropsychological examination by an experienced pediatric neuropsychologist produces a profile of strengths and weaknesses in all areas of brain functioning and delineates their relationship to the observed behavior.   Thus, the initial neuro-psychological assessment, consultations with treating doctors, parents and teachers, as well as follow-up neuropsychological evaluations are crucial in order to effectively treat and rehabilitate children with a head injury of any severity.

5. CLASSIFICATION OF PEDIATRIC TBI

It should be noted that an accurate classification of pediatric concussions is still problematic due to a lack of comprehensive research. Generally, there are three major types of brain injury that have been identified by clinicians.   Those are concussions, contusions, and lacerations. Concussion is a continuum of clinical syndromes, which may range in severity from brief amnesia to a prolonged coma following a head injury.   The term concussion, however, is frequently used to describe a mild head injury with rather transient disruption of neural functioning, such as disturbed orientation, short-term memory, equilibrium, speech and vision. This definition needs to be updated since numerous current research studies clearly indicate the possibility of long-term disabilities in children suffering from even a mild single episode of concussion. A contusion is a bruising or microscopic hemorrhage that occurs along the superficial levels of the brain. Lacerations are described as tears in the brain tissue that are often associated with penetrating or depressed skull fractures.

Glasgow Coma Scale (GCS) is one of the most commonly utilized measures of injury severity (Teasdale & Jennett, 1974). It ranges from three to fifteen. Most health care providers use the GCS with both the pediatric and the adult population. Simpson and Reilly (1982) have modified the GCS for use with children, scoring a child's best responses in motor, verbal and eye-open modalities. The total sum of these scores is then compared to the logical (not empirical) normal aggregate score for children of compatible ages. With the maximum total score being fourteen, children over the age of five years without any evidence of head trauma would be expected to obtain the highest score. Children from two to five years may score as high as thirteen; ages one to two up to twelve; six to twelve months up to eleven; and infants to six months up to a score of nine. Others have also attempted to adapt the GCS for use with the pediatric population (Fay et al., 1993).

While moderate and severe head injuries are defined more precisely, mild traumatic brain injuries (MTBI) are difficult to assess. Frequently, the definition of MTBI ranges from a bump on the head to a concussion. However, the importance of investigating the residual effects of mild head injury is crucial, as they are known to cause a large number of intracranial injuries (Schutzman & Greenes, 2001). In fact, Schutzman and Greenes (2001) have shown that the prevention of secondary injury associated with mild and moderate head trauma, in persons who initially appear to be at low risk, accounts for the largest reductions in head trauma mortality. Thus, without a definition of MTBI, the findings of various studies that look at residuals of MTBI are equivocal.

The definition of the Mild Traumatic Brain Injury, developed by the Mild Traumatic Brain Injury Committee of the Head Injury Interdisciplinary Special Interest Group of the American Congress of Rehabilitation Medicine (ACRM, 1993), states that such "injury involves a traumatically induced physiological disruption of brain function, as manifested by at least one of the following: (1) any period of loss of consciousness, (2) any loss of memory for events immediately before or after the accident, (3) any alteration in mental state at the time of the accident, and (4) focal neurological deficit(s) that may or may not be transient, but for which the severity of the injury does not exceed the following: (5) loss of consciousness of approximately 30 minutes or less; (6) after 30 minutes, an initial GCS of 13-15; and (7) post-traumatic amnesia not greater than 24 hours."

Traditionally, the GCS score has been used widely to assess neurological state of an injured brain. However, Post-Traumatic Amnesia

(PTA) is now viewed as a more accurate value of assessment and prediction of child's recovery, because it appears to be more sensitive to minor damages than GCS. Post-Traumatic Amnesia is measured from the very moment of injury to the time when the child is alert, oriented, and, most importantly, is able to continuously encode, retain and retrieve his memories of his moment-to-moment experiences. Thus, a child who received a GCS of fourteen or fifteen might be alert and even oriented, but will fail to record his moment-to-moment experiences as evidenced by his patchy and vague recollection of the events transpiring around him hours or days after the injury. The Children's Orientation and Amnesia Test (COAT; Ewing-Cobbs et al., 1990) was developed as a standardized assessment instrument for measuring both the presence and duration of PTA in children. In fact, this group of researchers has found that COAT effectively predicts memory functions for up to twelve months following a brain injury. Surveys have shown that when children enter the emergency room with mild head injury, they frequently present with normal neurological findings, normal mental status and no evidence of skull fracture (Levin, et al., 1982; Schutzman et al., 2001).

Despite negative findings of structural damage, mild head injuries in children may be accompanied by headaches, lethargy, irritability, withdrawal, and emotional lability (Begali, 1992; Boll, 1982). Moreover, there is a variation in the severity of MBTI, suggesting three subtypes: mild, moderate and severe concussion (Matz, 2003). The typical problems that follow MBTI include a constellation of somatic, cognitive and emotional symptoms, such as headache, dizziness, photosensitivity, sleep disturbance, decreased memory and attention, increased confusion, slowed thinking, emotional lability, irritability, and anger (Alexander, 1995). While these symptoms are more readily identified and reported by adults, children have a much harder time understanding their experiences and relating them to their caretakers. Instead, the sequence of behavioral and emotional difficulties arises along with cognitive problems, which are readily observed at home and in school. To conclude the description of types of traumatic head injuries, moderate and severe TBI are defined as follows. Moderate traumatic head injury is defined by a lowest post-resuscitation GCS score of nine to twelve, or thirteen to fifteen with a brain lesion on radiological images or a depressed skull fracture. Severe head injury is defined by a lowest post-resuscitation GCS score of three to eight (Williams et al., 1990).

6. TYPICAL SEQUELAE OF PEDIATRIC HEAD INJURY

6.1. Neurological Sequelae

Closed head injuries can be associated with a variety of neurological effects, which often depend on the nature and location of brain damage. Potential effects include paresis, peripheral neuropathy, movement disorders, endocrine disturbances, and seizures. Neuroimaging studies have indicated that severe traumatic brain injuries often result in a gradual and prolonged process of white matter degeneration, with associated cerebral atrophy and ventricular enlargement; in some cases, ventricular dilation is associated with hydrocephalus (Bigler, 1997).

There is a growing body of research and clinical findings that show that younger children appear to be especially vulnerable to early post-traumatic seizures, which typically occur within the first week after head injury. Yablon (1993), for example, has shown that such seizure activity occurs in approximately 10% of younger children. This rate seems to decline with the increasing age of a child (approximately 5% occurrence among older children). Other studies (McLean et al., 1995) indicate that, sometimes, early seizures involve focal status epilepticus, which might be associated with mass lesions. The data show that seizure activity often develops within the first two years after injury. Although the occurrence of post-traumatic seizures does not automatically place children at risk for later seizure disorder, seizures persist in about 2% of cases. Lastly, penetrating injuries or depressed skull fractures, which occur in approximately 10% of all head injuries, are associated with greater incidence of seizure activity.

Severe head traumas are not the only conditions that may result in significant neurocognitive and neurobehavioral residual deficits. A growing body of research shows that, although many pediatric patients with MTBI progress to full recovery, a large number of children sustain permanent neuronal damage and develop chronic, disabling symptoms over the course of weeks or even months post injury (Matz, 2003; Mazzola et al., 2002). Massagli and Jaffe (1994) indicate that some children with MTBI go on to develop headaches, tinnitus, fatigue, emotional lability, irritability that lasts for many days and weeks post injury. Korinthenberg and colleagues (2004) demonstrated that sixty-four out of ninety-eight children who sustained concussion showed abnormal EEG findings within twenty-four hours. However, after four to six weeks post trauma, twenty-four out of ninety-eight of these children continued to complain of post-traumatic headaches, fatigue, sleep disturbances, anxiety and affect lability. It should be mentioned that this post-traumatic symptomatology did not correlate with neurological or EEG findings that were observed immediately after the injury.

Thus, it is important not to overlook such symptoms in concussed children with negative radiological findings. In fact, Schutzman and Greenes (2001) indicate that, although radiographic evidence of intracranial injury is not uncommon among children with MTBI, mild head injury does not necessarily involve abnormalities in mental status at the initial exam in the emergency room, focal findings on neurological examination, or evidence of skull fracture. They posit that the majority of pediatric patients in their study sustained mild TBI, however, they saw a large number of intracranial injuries.

While most mild concussions do not seem to produce clearly defined anatomical correlates to subjective complaints, some studies find specific morphological changes in the brain tissue. For instance, Begali (1992) has determined that even mild concussions can produce minor damage to brain stem nuclei. Other studies present evidence that some concussions are associated with damage to the reticular activating system in the brain stem, which regulates arousal and attention, and cerebral hemispheres (Jane & Rimel, 1982; Levin et al., 1982; Pang, 1985).

6.2. Neuropsychological Sequelae

The research literature pertaining to the neuropsychological consequences of closed head injuries in children is extensive and has been the subject of several previous reviews (Arffa, 1998; Baron et al., 1995; Fletcher& Levin, 1988; Fletcher et al., 1995; Levin et al., 1995). Typically, the first mediating factor that is considered in projecting the nature and severity of neuropsychological sequelae of pediatric head injury is the severity of the injury. Moreover, unlike adult survivors of TBI, children may take much longer to recover from any type of brain injury. In fact, Beaumont (1983) and Luria (1963) have indicated that children may take up to five to six years to recover from severe TBI. Others delineated a shorter time line for recovery, suggesting that some children show significant improvement as early as two to three years post-injury (Barth & Macciocchi, 1985).

In addition, repeated concussions may lengthen the speed of recovery with each successive injury (Binder & Rattok, 1989). For example, Witol and Webbe (1994) demonstrated that soccer players who hit the ball repeatedly with their heads had been found to achieve lower scores on neuropsychological measures. Moreover, those children who played sports more frequently scored lower than their peers, especially on measures of attention and information processing (Abreau et al., 1990).

Overall, the research studies indicate that closed head injuries are frequently associated with deficits in the following areas of brain functioning: alertness and orientation, attention and memory, executive functions, language skills, nonverbal skills, cortical sensory and motor skills, academic achievement, adaptive functioning and behavioral adjustment. It has been well documented that these deficits are often manifested in poor school performance and the need for special educational services. Children with TBI were found to show severe learning deficits years after the injury, which were not initially apparent. In fact, according to NIH (1998), these deficits frequently go unrecognized in both the pediatric and adult populations. In addition, Johnson (1992) found that children with TBI continue to show memory and language deficits even after their successful return to school. The younger a child is at the onset of TBI, the more vulnerable he or she may be to persistent academic problems (Ewing-Cobbs et al., 2004).

Academic problems that persist for years after the head injury are very disturbing, because they reflect a significant compromise in functioning. Academic functioning in the school environment is the childhood equivalent of occupational functioning for adults. Therefore, neuropsychological examination can often greatly facilitate the child's re-assimilation to the school and assist in better planning of resources to help the child.

Despite the wealth of existing knowledge about the impact of mild head trauma on the neurocognitive and neurobehavioral functioning of children, many critical questions remain unanswered. Thus, this topic remains a controversial area. Some researchers argue that mild injury does not produce any noticeable cognitive or behavioral deficits, while others claim that even mild head trauma in children may produce acute declines in intellectual functioning (Tremont et al., 1999). Despite the controversy, research studies consistently indicate the presence of mild to moderate post MTBI residual sequelae that significantly interfere with every aspect of a child's functioning. For instance, children under the age of twelve with minor head injuries have been found to show behavioral difficulties such as low frustration tolerance and attention deficit, even four years post injury (Klonoff et al., 1977). On the other hand, the majority of findings suggest that global changes in IQ are very infrequent in MTBI, although some found ten- to fifteen-point differences among children with MTBI and non-neurologically injured youngsters (Tremont et al., 1999). Most studies reveal a post traumatic decline in language and attentional functions, which may persist for six months or longer after the injury. More specifically, children tend to struggle with language comprehension and verbal information processing (Begali, 1992). Frequently, these children are left to repeat a grade or placed in remedial classes due to information processing deficit that significantly interferes with their overall cognitive functioning (Alves & Jane, 1985); Boll, 1983; Ylvisaker, 1985).

Matz (2003) reviewed several studies and concluded that children with MTBI suffer an immediate negative impact on attention, short-term memory, information processing, as well as a decline in verbal memory and abstract reasoning. A decline in processing speed and perceptual organization are frequently seen in children with mild head injury. Tremont and colleagues (1999) found that children who were admitted with GCS scores of thirteen to fifteen showed the greatest decline in processing speed.

Some researchers have found a strong relationship between MTBI and hyperactivity in large samples of children (Bijur et al., 1990), suggesting a causal relationship between increased hyperactivity and inattention and head injury. However, it was also found that many children who displayed an increased rate of behavioral problems following MTBI, had behavioral problems prior to their head injury. Thus, this finding has supported the opposing hypothesis that children with more behavioral problems are at higher risk for accidents that result in TBI (Brown et al., 1981).

In some instances, emotional sequelae involve more severe disturbances. The data show that approximately 20% of pediatric survivors of MTBI were found to have a new psychiatric disorder within two years post-injury (Shaffer, 1995). Many researchers confirmed that the damage to central nervous system is the most powerful risk factor for later development of psychiatric disorders in children (Shaffer, 1995; Teeter & Semrud-Clikerman, 1997). The persistent deficits that are seen in MTBI are subtler and show greater variability than those deficits that are seen in moderate and severe TBI. The neuropsychological deficits seen in children with moderate TBI tend to resemble those seen in MTBI rather than in severe TBI. The findings of studies that looked at children with moderate TBI are, however, inconsistent, because many researchers placed these children in the same group with the MTBI (Ewing-Cobbs et al., 1989). Literature review by Asarnow and colleagues (1995) from 1971 to 1993 revealed that more than half of these studies did not include moderate TBI as a separate group to be compared to the mild and the severe TBI groups.

The pediatric survivors of severe TBI are frequently discharged from the hospitals with a prognosis for good recovery. However, this does not mean full recovery. More of these children sustain temporary to permanent physical and neuropsychological deficits than children with mild and moderate injuries (Jennett & Teasdale, 1981; Kraus, 1995). It has been found that 80% of children with severe TBI develop specific educational needs and require modified educational environment two years post-injury (Ewing-Cobbs et al., 1991). Those who experienced coma endure even greater impairment. In fact, the length of a coma is positively correlated with greater cognitive impairment and inability to return to school (Ruff et al., 1993). The extent of such neurocognitive deficits is great and involves naming, verbal fluency, writing, memory, attention, organization, and other functions (Ewing-Cobbs et al., 1986, 1991; Jaffe et al., 1985). Moreover, 61.9% of pediatric survivors of severe TBI were found to have a new psychiatric disorder within two years post-injury (Shaffer, 1995).

Table 1 illustrates the most common cognitive and neurobehavioral sequelae of mild head injury in children and adolescents. This neuropsychological profile reflects the most common changes in functions and may or may not be present in every case. Clinical practice and research have clearly demonstrated that specific functions remain stable, while others show an increase or decrease in frequency and severity in a concussed brain. However, the presence, persistence and severity of each deficit will vary from child to child, creating a unique profile of neuropsychological functioning layered over child's unique developmental dimensions.

Table 1. Typical neuropsychological profile of concussion

Functions IncreasedDecreased	Stable	
Irritability	Short-term memory	
Long-term memory		
Anger	Attention	Overlearned skills
Disinhibition	Concentration	Overall intelligence
Lethargy	Processing speed	
Dizziness	Comprehension	
Headache	Coordination	
Fatigue	Abstract reasoning	
	Emotional lability	Balance
Hyperactivity	Organization	
	Visuo-spatial skills	

7. RECOVERY

Traditionally, it was believed that children have a higher propensity for recovery following head injury than adults. Fortunately, a myth stating that early onset of brain injury is easily offset by neuronal plasticity has been debunked. In fact, some authors posit that earlier onset of injury is related to even more significant later deficits than later injury (Teeter, 1997). Perhaps, the basis of such controversies is the complexity of interplay among injury characteristics, environmental influences, and developmental factors, which is difficult to assess using current methodologies and, therefore, generates disagreement among the outcomes of various research projects. While many studies report short- and long-term negative effects of severe, moderate and mild brain injury in children, some studies find such effects only among children with severe injuries, but not among children with mild head traumas (Massagli et al., 1994; Tremont et al., 1999). The stated controversies, nevertheless, helped to establish a now well know fact that, although neuronal plasticity facilitates recovery within the first six months post injury, there is rapidly growing evidence of the devastating effects of head trauma on young children (Mazzola et al., 2002; Wellons et al., 2003).

When considering the importance of the positive findings of some of these studies, it is crucial to consider the personal and financial costs of pediatric head injury over a lifetime. These costs are tremendous and deserve our utmost attention. The truth is that even mild neuronal injury in children may produce a cascade of deficits that require long-term treatment. Compared to adults, children tend to suffer more diffuse than focal brain injuries due to differences in the biomechanical profile and tissue properties (Mazzola et al., 2002). In fact, a child's brain has a much higher water content and incomplete myelenization. Thus, it is improper to equate recovery of an adult brain relative to a child's brain from a mild concussion, as it disregards the physiological and neurochemical properties of a growing brain.

There are differences in the recovery process among various pediatric age groups. For instance, younger children show different patterns of recovery and their future learning abilities are affected because of their incomplete development (Brazelli et al., 1994; Johnson, 1992). It has been found that the earlier onset of injury is related to more significant later deficits than later injury (Teeter et al., 1997). The developing brain may be more vulnerable to the damage because of the rapid growth spurts that occur in the early stages. Kolb and Whishow (1990) suggest three critical age divisions that influence the prognosis of recovery in pediatric patients: (1) less than one year of age, (2) between one and five years of age, (3) more than five years of age.

Some of these differences in recovery account for the fact that structures that do not generally develop until later in life may be compromised by early damage, and this injury may not be obvious until a few years later (Rourke et al., 1983). Particularly vulnerable to this condition are tasks that involve the frontal lobe and association areas of the brain, which do not assume adult-like functions until twelve years of age or later (Teeter & Semrud-Clickeman, 1997). Frontal lobe generally involved in an ability to monitor behavior and to allow a person to change behavior according to situation. Association areas allow for the integration of information from various modalities, such as visual-motor, visual-spatial, auditory-visual, etc.

Head injury sustained by very young children was found to produce both receptive and expressive language deficits than injury sustained by toddlers (Ewing-Cobbs et al., 1989). As the children get older, their expressive language remains more susceptible to TBI than their receptive language. Recovery in motor and visual-spatial skills, which also relate to writing, was found to be lower in younger adolescents than in older adolescents (Thompson et al., 1994). Lastly, despite the significance of age of onset as a determining factor in the outcome of recovery, the depth and duration of the coma and the child's age appear to be the most important factors associated with the final outcome. Other prognostic indicators of the outcome include pupilary and optokinetic responses, intracranial pressure, extent of retrograde and post-traumatic amnesia, level of activity, and neurological findings (Levin, Benton & Grossman, 1982; Menkes & Batzdorf, 1985).

8. PREVENTION

Our current knowledge of pediatric head injury leads us to consider at least two levels of prevention. Naturally, the first level emphasizes our prerogative to protect children from such injuries. This task could be attained by providing children and caretakers with educational information regarding the use of bicycle helmets and motor vehicle safety. The devices, such as seat belts, air bags, and helmets, when used correctly, are known to lessen the impact and resultant primary brain injuries in children (Mazzola et al., 2002). Moreover, more aggressive steps to decrease alcohol-related motor vehicle accidents and to increase the number of programs geared towards child-abuse and neglect prevention are needed. In addition, Rivara (1995) discovered certain parental characteristics that are closely associated with pediatric TBI. Those predictors are parental alcohol abuse and perception of injury.

The second level of prevention encompasses the prevention of the damaging effects of recent head trauma, or secondary head injury effects. This initiative should occur immediately after the injury and should include not only physical, neurological and radiological evaluations, but also neuropsychological exams. Neuropsychological services have already become an integral part of the inpatient rehabilitation medical treatment plan. They are routinely utilized in cases of severe and moderate head injuries. However, given the known effects of mild head trauma on a child's brain, neuropsychological exams should be ordered as a preventive measure against future short and long-term residual effects of MTBI. As discussed above, multiple cognitive and neurobehavioral deficits arise from mild head injuries weeks and months after the trauma itself, causing an enormous multitude of problems in school and at home.

Thus, these mild concussions deserve to be taken more seriously and the children deserve to be helped.

Finally, in addition to the previously described prevention techniques, vigorous multimodal research of pediatric head injury will allow us to design much more effective treatment interventions focused on prevention of secondary injuries. There are many questions yet to be answered. In terms of future research, there is a need to further examine the neural substrates of the neuropsychological deficits that occur in childhood injury. Studies that capitalize on advances in neuroimaging to measure underlying neuropathology and to correlate these measures with neuropsychological functioning will enhance our understanding of the recovery outcomes. Moreover, the combination of neuropsychological measures and neuroimaging in research studies of children with mild head injuries could help to resolve controversies regarding the long-term consequences of such injuries.

CONCLUSION

It is critically important to consider a child's family, social environment, psychological predispositions, gender etc., as predictors of children at risk for TBI. Numerous family and environmental characteristics are closely related to both cognitive and behavioral functioning following closed head injury in children. We need to further understand which pre- and post-injury environmental factors modulate the brain injury related characteristics and how this process ultimately affects the child's treatment and recovery. Overall, longitudinal studies that observe children over a period of several years have to be conducted in order to fully understand the complex nature of the concussive processes, recovery of functions, and developmental aspects of a child's brain. In addition, given the increasing number of children participating in sports and the negative physical, financial, overall health and psychological well-being consequences experienced by injured athletes, injury surveillance studies are needed more than ever than before.

Acknowledgements

The earlier version of this chapter was elaborated upon by Dr. Rimma Danov and published in Slobounov & Sebastianelli (Eds.), "Foundation of sport-related brain injuries", Springer, 2006. The author especially appreciate Dr. Danov' current insights on pediatric concussion (personal communication, 2007).

REFERENCES

Adirim, T.A., Cheng, T.L. (2003). Overview of injuries in the young athletes. Sports Medicine, 33(1), 75-81.

Smith, A.M., Scott, S.G., Wiese, D.M. (1990). The psychological effects of sport injuries: coping. *Sports Medicine, 9(6)*, 352-369.

New York State Psychological Association. (2005). Neurodiagnostic Examination. *New York State Psychological Association: Statement.* 6 Executive Park Drive, Albany, New York 12203

Social Security Administration. (2002). Disability evaluation under Social Security. *SSA Publication Number 64-039.*

American Academy of Neurology: Therapeutics and Technology Subcommittee of the American Academy of Neurology. (1996). Assessment: Neuropsychological testing of adults. Consideration for neurologists. *Neurology, 47*, 592-599.

Yeates, K.O., Ris, M.D., & Taylor, H.G. (1995). Hospital referral patterns in pediatric neuropsychology. *Child Neuropsychology, 1*, 56-62.

Kraus, J.F. (1995). Epidemiological features of brain injury in children: Occurrence, children at risk, causes and manner of injury, severity, and outcomes. In S.H. Broman & M.E. Michel (Eds.), *Traumatic head injury in children* (pp. 22-39). New York: Oxford University Press.

Lauerssen, T. G., Klauber, M. R., & Marshall, L. F. (1988). Outcome from head injury related to patient's age: A longitudinal prospective study of adult and pediatric head injury. *Journal of Neurosurgery, 68,* 409-416.

Lescohier, I., & DiScala, C. (1993). Blunt trauma in children: Causes and outcomes of head versus intracranial injury. *Pediatrics, 91*, 721-725.

Kraus, J.F., Fife, D., Cox, P., Ramstein, K., & Conroy, C. (1986). Incidence, severity, and external causes of pediatric brain injury. *American Journal of Diseases of Children, 140*, 687-693.

Yeates, K. O., Taylor, H.G., Drotar, D., et al., (1997). Pre-injury family environment as a determinant of recovery from traumatic brain injuries in school-age children. *Journal of the International Neuropsychological Society, 3*, 617-630.

Kraus, J.F., Rock, A., & Hamyari, P. (1990). Brain injuries among infants, children, adolescents, and young adults. *American Journal of Diseases of Children, 144*, 684-691.

Klauber, M.R., Barrett-Connor, E., Hofstetter, C.R., & Micik, S.H. (1986). A population-based study of nonfatal childhood injuries. *Preventative Medicine, 15*, 139-149.

Annergers, J. F., Grabow, J.D., Kurland, L. T., & Laws, E. R. (1980). The incidence, causes, and secular trends of head trauma in Olmsted County, Minnesota, 1935-1974. *Neurology*, 30, 912-919.

Fletcher, J.M., Levin, H.S., & Butler, I.J. (1995). Neurobehavioral effects of brain injury in children: Hydrocephalus, traumatic brain injury, and cerebral palsy. In M.C. Roberts (Ed.), *Handbook of pediatric psychology* (2nd ed., pp.362-383). New York: Guilford Press.

Jennett, B., & Bond, M. (1975). Assessment of outcome after severe brain damage: A practical scale. *Lancet, i*, 480-484.

Koelfen, W., Freund, M., Dinter, D., Schmidt, B., Koenig, S., & Schultze, C. (1997). Long-term follow up of children with head injuries classified as "good recovery" using the Glasgow Outcome Scale: Neurological, neuropsychological, and magnetic resonance imaging results. *European Journal of Pediatrics, 156*, 230-235.

Bruce, D. A., & Schut, L. (1982). Concussion and contusion following pediatric head trauma. In: McLaurin R.L. (Ed.), *Pediatric Neurosurgery: surgery of the developing nervous system.*

Teasdale, G. & Jennett, B. (1974). Assessment of coma and impaired consciousness: A practical scale. *Lancet, ii*, 81-84.

Simpson, D., & Reilly P. (1982). Pediatric coma scale. *Lancet, 2*, 450.

Fay, G. C., Jaffe, K. M, Pollisar, N. L. et al. (1993). Mild pediatric traumatic brain injury: a cohort study. *Archives of Physical Medicine and Rehabilitation, 74*, 895-901.

Schutzman, S.A. & Greenes, D.S. (2001). Pediatric mild head trauma. *Annals of Emergency Medicine, 3*, 65-74.

American Congress of Rehabilitation Medicine. (1993). Definition of mild traumatic brain injury. *Journal of Head Trauma Rehabilitation, 8*, 86-87.

Ewing-Cobbs, L., Levin, H.S., Fletcher, J.M., Miner, M. E., & Eisenberg, H.M. (1990). The Children's Orientation and Amnesia Test: relationship to severity of acute head injury and to recovery of memory. *Neurosurgery, 27*, 683-691.

Levin, H. S., Benton, A. L., & Grossman, R. G. (1982). *Neurobehavioral consequences of closed head injury*. New York: Oxford University Press.

Begali, V. (1992). *Head injury in children and adolescents* (2 nd ed.). Brandon, VT: Clinical Psychology Publishing.

Boll, T. (1982). Behavioral Sequelae of head injury. In P. Cooper (Ed.), *Head injury* (pp. 363-377). Baltimore, MD: Williams & Wilkins.

Matz, P. G. (2003). Classification, diagnosis, and management of mild traumatic brain injury: A major problem presenting in a minor way. *Seminars in Neurosurgery, 14 (2)*, 125-130.

Alexander, M.P. (1995). Mild traumatic brain injury: Pathophysiology, natural history, and clinical management. *Neurology, 45*, 1253-1260.

Williams, D., Levin, H., & Eisenberg, H. (1990). Mild head injury classification. *Neurosurgery, 27 (3)*, 267-286.

Bigler, E.D. (1997). Brain Imaging and behavioral outcome in traumatic brain injury. In E.D. Bigler, E. Clark, & J.E. Farmer (Eds.), *Childhood traumatic brain injury: Diagnosis, assessment, and intervention* (pp.7-32). Austin, TX: PRO-ED.

Yablon, S. A. (1993). Post-traumatic seizures. *Archives of Physical Medicine and Rehabilitation, 74*, 983-1001.

McLean, D. E., Kaitz, E.S., Kennan, C.J., Dabney, K., Cawley, M.F., & Alexander, M.A. (1995). Medical and surgical complications of pediatric brain injury. *Journal of Head Trauma Rehabilitation, 10*, 1-12.

Mazzola, C. A. & Adelson, P.D. (2002). The ABCs of pediatric head trauma. *Seminars in Neurosurgery, 13 (1)*, 29-37.

Massagli, T.L & Jaffe, K.M. (1994). Pediatric traumatic brain injury: Prognosis and rehabilitation. *Pediatric Annals, Jan 23 (1)*, 29-36.

Korinthenberg, R., Schreck, J., Weser, J., & Lehmkuhl, G. (2004). Post-traumatic syndrome after minor head injury cannot be predicted by neurological investigations. *Brain & Development, 26*, 113-117.

Jane, J.A., & Rimel, R. W. (1982). Prognosis in head injury. *Clinical Neurosurgery, 29*, 346-352.

Pang, D. (1985). Pathophysiologic correlates of neurobehavioral syndromes following closed head injury. In M. Ylvisaker (Ed.), *Head injury rehabilitation: Children and adolescents* (pp.3-70). San Diego, CA: College-Hill Press.

Arffa, S. (1998). Traumatic brain injury. In C. E. Coffey & R. A. Brumback (Eds.), *Textbook of pediatric neuropsychiatry* (pp.1093-1140). Washington, DC: American Psychiatric Association.

Baron, I. S., Fennell, E. B., & Voeller, E. B. (1995). *Pediatric neuropsychology in the medical setting*. New York: Oxford University Press.

Fletcher, J.M., & Levin, H.S. (1988). Neurobehavioral effects of brain injury in children. In D.K. Routh (Ed.), *Handbook of pediatric psychology* (pp.258-295). New York: Guilford Press.

Fletcher, J.M., Levin, H.S., & Butler, I.J. (1995). Neurobehavioral effects of brain injury in children: Hydrocephalus, traumatic brain injury, and cerebral palsy. In M.C. Roberts (Ed.), *Handbook of pediatric psychology* (2nd ed., pp.362-383). New York: Guilford Press.

Levin, H.S., Ewing-Cobbs, L., & Eisenberg, H.M. (1995). Neurobehavioral outcome of pediatric closed-head injury. In S.H. Broman & M.E. Michel (Eds.), *Traumatic head injury in children* (pp. 70-94). New York: Oxford University Press.

Beaumont, J.G. (1983). *Introduction to neuropsychology.* New York: Guilford Press.

Luria, A.R., (1963). *Restoration of function after brain injury.* New York: Macmillan.

Barth, J. T., & Macciocchi, S. N. (1985). The Halstead-Reitan Neuropsychological Test Battery. In C. Newmark (Ed.), *Major psychological assessment techniques* (pp. 381-414). Boston: Allyn & Bacon.

Binder, L.M., & Rattok, J. (1989). Assessment of the postconcussive syndrome after mild head trauma. In M. D. Lezak (Ed.), *Assessment of the behavioral consequences of head trauma* (pp. 37-48). New York: Liss.

Witol, A., & Webbe, F. (1994). Neuropsychological deficits associated with soccer play. *Archives of Clinical Neuropsychology, 9*, 204-205.

Abreau, F., Templer, D. L., Schulyer, B.A., & Hutchinson, H. T. (1990). Neuropsychological assessment of soccer players. *Neuropsychology, 4*, 175-181.

NIH Consensus Development Panel (1998). *Rehabilitation of persons with traumatic brain injury.* NIH Consensus Statement, 1998 Oct 26-28, 16(1), 1-41.

Johnson, D. A. (1992). Head injured children and education: A need for greater delineation and understanding. *British Journal of Educational Psychology, 62*, 404-409.

Ewing-Cobbs, L., Barnes, M., Fletcher, J. M., et al. (2004). Modeling of longitudinal academic achievement scores after pediatric traumatic brain injury. *Developmental Neuropsychology, 25*, 107-133.

Tremont, G., Mittenberg, W., & Miller, L. J. (1999). Acute intellectual effects of pediatric head trauma. *Child Neuropsychology, 5*, 104-114.

Klonoff, H., Low, L.D., & Clark, C. (1977). Head injuries in children: A prospective five year follow-up. *Journal of Neurology, Neurosurgery and Psychiatry, 40*, 1211-1219.

Alves, W. M., & Jane, J. A. (1985). Mild brain injury: Damage and outcome. In D.Becker & J. T. Povlishock (Eds.), *Central nervous system trauma: Status report* (pp. 255-271). Bethesda, MD: National Institutes of Health.

Boll, T. (1983). Minor head injury in children- Out of sight but not out of mind. *Journal of Clinical Child Psychology, 12*, 74-80.

Ylvisaker, M. (Ed.). (1985). *Head injury rehabilitation: Children and adolescents.* San Diego: College Hill.

Bijur, P.E., Haslum, M., & Golding, J. (1990). Cognitive and behavioral sequelae of mild head injury in children. *Pediatrics, 86*, 337-344.

Brown, G., Chadwick, O., Shaffer, D., Rutter, M., & Traub, M. (1981). A prospective study of children with head injuries: III. Psychiatric sequelae. *Psychological Medicine, 11*, 63-78.

Shaffer, D. (1995). Behavioral sequelae of serious head injury in children and adolescents: The British studies. In S.H. Broman & M.E. Michel (Eds.), *Traumatic head injury in children* (pp.55-69). New York: Oxford University Press.

Teeter, P. A., & Semrud-Clikerman, M. (1997). *Child neuropsychology Assessment and interventions for neurodevelopmental disorders.* Boston: Allyn & Bacon.

Ewing-Cobbs, L., Fletcher, J. M., & Levin, H. S. (1989). Intellectual, motor, and language sequelae following closed head injury in infants and preschoolers. *Journal of Pediatric Psychology, 14*, 531-547.

Jennett, B., & Teasdale, G. (1981). *Management of head injuries.* Philadelphia: Davis.

Ewing-Cobbs, L., Iovino, I., Fletcher, J. M., Miner, M. E. & Levin, H. S. (1991). Academic achievement following traumatic brain injury in children and adolescents. *Journal of Clinical and Experimental Neuropsychology, 13*, 93.

Asarnow, R. F., Satz, P., Light, R., Zaucha, K., Lewis, R., & McCleary, C. (1995). The UCLA study of mild closed head injuries in children and adolescents. In S.H. Broman & M.E. Michel (Eds.), *Traumatic head injury in children* (pp.117-146). New York: Oxford University Press.

Ewing-Cobbs, L., Fletcher, J. M., & Levin, H. S. (1986). Neurobehavioral Sequelae following head injury in children: Educational implications. *Journal of Head Trauma Rehabilitation, 1*, 57-65.

Ruff, R. M., Marshall, L., Crouch I., Klauber, M. R., & Smith, E. A. (1993). Predictors of outcome following head trauma. *Brain Injury, 2*, 101-111.

Jaffe, K. M., Mastrilli, J., Molitor, C. B., & Valko, A. (1985). Physical rehabilitation. In M. Ylvisaker (Ed.), *Head injury rehabilitation: Children and adolescents* (pp. 167-195). San Diego, CA: College-Hill Press.

Wellons, J.C. & Tubbs, R.S. (2003). The management of pediatric traumatic brain injury. *Seminars in Neurosurgery, 14 (2)*, 111-118.

Brazelli, B., Colombo, N., Della Sala, S., & Spinnier, H. (1994). Spared and impaired cognitive abilities after bilateral frontal damage. *Cortex, 30*, 27-51.

Johnson, D. A. (1992). Head injured children and education: A need for greater delineation and understanding. *British Journal of Educational Psychology, 62*, 404-409.

Kolb, B., & Whishow, I. (1990). *Human Neuropsychology.* NY: Freeman.

Rourke, B. P., Bakker, D. J., Fisk, J. L., & Strang, J. D. (1983). *Child neuropsychology: An introduction to theory, research, and clinical practice.* New York: Guilford Press.

Ewing-Cobbs, L., Fletcher, J. M., & Levin, H. S. (1989). Intellectual, motor, and language sequelae following closed head injury in infants and preschoolers. *Journal of Pediatric Psychology, 14*, 531-547.

Thompson, N.M., Francis, D.J., Stuebing, K.K., Fletcher, J.M., Ewing-Cobbs, L., Miner, M.E., Levin, H.S., & Eisenberg, H. (1994). Motor, visual-spatial, and somatosensory skills after closed-head injury in children and adolescents: A study of change. *Neuropsychology, 8*, 333-342.

Levin, H. S., Benton, A. L., & Grossman, R. G. (1982). *Neurobehavioral consequences of closed head injury.* New York: Oxford University Press.

Menkes, J.H., & Batzdorf, U. (1985). *Textbook of Child Neurology.* Philadelphia: Lea & Febiger.

Rivara, F. P. (1995). Developmental and behavioral issues in childhood injury prevention. *Journal of Developmental and Behavioral Pediatrics, 16*, 362-370.

PART V: INJURY REHABILITATION

CHAPTER 21

INTEGRATED INJURY REHABILITATION

1. INTRODUCTION

UK Trade unions define rehabilitation as "any method by which people with sickness or injury (that interferes with their ability to work to their normal or full capacity) can be returned to work." (TUC, 2000) They specifically stress that no profession has a monopoly on rehabilitation. Accordingly, a multimodal approach is often the best for the sake of a patient. According to Nocon & Baldwin (1998), there is some consensus among clinicians regarding the overall goals of rehabilitation. This can be summarized as the following:

(a) The general objective of rehabilitation is to restore (to the maximum degree possible) function (physical and mental) and role participation (within the family, social network, or workforce). With regard to injury rehabilitation in athletes, the social network and/or workforce may be assumed as the athletic environment, such as teammates, coaches and athletic staff. Regaining mental function, for example, assumes restoring confidence in the injured body part's ability to meet the demands of practices and competitions . Specifically, it is important that rehabilitation is aimed at gaining confidence in situations where the injury occurred (Podlog & Eklund, 2007);

(b) Rehabilitation usually requires a combination of therapeutic, psychosocial, and work-related intervention that address the clinical problem as well as issues in the individual's physical and social environment. With regard to athletes recovering from injury, work-related interventions may assume not only re-acquisition of physical and psychological attributes (i.e. skill and confidence in the ability to perform rehabilitation exercise routines) but also regaining pre-injury sport-specific skills (e.g., execute multiple somersaults requiring strength, coordination and spatial orientation);

(c) Rehabilitation services need to: be responsive to the patient's needs and wishes; be goal-oriented; involve numerous agencies and disciplines; and be available when required. It should be added that the goals of rehabilitation services with special emphasis on a timely return to sport participation in athletics should include psychological interventions aimed at:

- Facilitation of the rehabilitation process;
- Maintenance of emotional equilibrium;
- Mobilization of existing coping resources;

- Enhancement of mental readiness for performance;
- Promotion of a sense of self-efficacy (adopted from Hail, 1993).

(d) Rehabilitation is a function of number of complementary services: it is not necessarily a separate service. However, it is an unfortunate reality that rehabilitation protocol is a primary concern of medical practitioners. It is still near practice when coaches, injured athletes, medical practitioners and sports psychologists combine their collective mind and effort to develop a mutually acceptable goal-oriented rehabilitation strategy. It is well known that goal setting links motivation to action (Locke & Latham, 1985). In rehabilitation, goal setting should parallel the treatment plan, establish an implicit statement and commitment by the athlete, and identify treatment as a collaborative process (Hail, 1993). This means again that "multiple services" but not a "separate service" are in charge of goal-setting and implementation of rehabilitation protocols. It is worth stressing a basic guide to goal-setting in rehabilitation, including but not limited to the following.

Goals in rehabilitation setting should be:

Specific and measurable (e.g., increase ROM in two weeks and reach ninety degrees of flexion in an injured joint);

Stated in *positive* **NOT** *negative* language (e.g., my teammates fully recovered from ACL damage in nine months, so I can do the same);

Challenging but *realistic* (i.e. overly difficult goals set up for failure; similarly, *lack of proper planning is planning to fail*);

Timetable for completion (i.e. gradual progression toward full recovery and timely and safe return to sport participation and competitions);

Integration of short, intermediate, and long-term goals (e.g., control for inflammation, followed by increased ROM, strength, overall fitness, enhanced functional resources … return to sport participation, full return to competition);

Outcome goals linked to the *process* goal (e.g., full return to competition provided that all physical and functional capabilities are restored via integrative rehabilitation);

Personalized and *internalized* (i.e., an injured athlete should embrace goals as self-set, not as something imposed from outside);

Monitored and *evaluated* (i.e., careful evaluation and quantification of progress allow for modification of treatment plan to prevent maladaptive responses. This also assumes flexibility versus rigidity in goal setting);

Sport goals linked to *life* goals (i.e., potential life benefits of experiencing injury and associated rehabilitation. An injury may have the potential to enrich the development of psychological interventions for sport injury rehabilitation, Udry, 1999);

Overall, with regard to the sports environment, the general objective of the rehabilitation of athletes suffering from injury is a timely return to sport participation. Specifically, rehabilitation protocols for injured athletes are aimed at the re-acquisition of pre-injury health status in terms of regaining: (a) physical capacities such as strength, flexibility, endurance; (b) physiological functions such as mobilization, aerobic/anaerobic conditioning, intermuscular coordination, overall fitness level; (c) psychological well-being, including emotional equilibrium, efficacy of beliefs, confidence etc; (d) psychosocial status (i.e., relatedness issues of social support and assistance from coaches, rehabilitation specialists and teammates (Podlog & Eklund, 2007).

The presence of any symptoms, both physical and functional, or signs of psychological trauma, as evidenced by fear of re-injury, avoidance behavior, experienced pain in any form should be considered as an index of under-recovery. This would require a comprehensive re-evaluation of the treatment protocol and objectives of rehabilitation by all involved agencies. Most importantly, it is now well-accepted among medical professionals caring for injured athletes that *"symptoms resolution may NOT be indicative of injury resolution per se"* (Dr. Sebastianelli, MD, Director of Athletic Medicine, Penn State University, personal communication, 2007).

Most often, return-to-sport participation criteria are defined by the presence and/or absence of symptoms of injury. Specifically, physical symptoms resolution (i.e., no evidence of residual tissue damage, restored anatomical integrity of joint, etc.) and functional symptoms resolution (i.e., ROM, strength, stamina) are among two major criteria of return-to-play. Regarding traumatic brain injury, athletes are allowed to return to play when common symptoms of concussion (i.e., headache, fatigue, light or sound sensitivity, etc) are resolved. However, the resolution of these symptoms does not necessarily indicate injury resolution per se. Residual dysfunctions and structural damage may still be present, but not observed due to numerous factors, including both extrinsic (e.g., lack of sensitivity of the assessment tools) and intrinsic (e.g., athlete's desire to quickly return to sport participation because of an "injured athlete is worthless" attitude). Another factor to consider: humans, in general, and athletes in particular, are able to compensate for mild or even severe physical and functional deficits because of redundancy in the human motor and cognitive systems. This in turn allows for the reallocation of existing resources such that undamaged pathways and functions are used to perform cognitive and motor tasks. This functional reserve and overall capability to accomplish testing protocol gives the appearance that an athlete has returned to pre-injury health status while in actuality the injury is still present and hidden from the observer. As a result, a premature return to sport participation based upon physical symptoms resolution may put athletes at high risk for recurrent injuries and the development of permanent psychological trauma. It is a growing concern among medical practitioners and coaches that athletes with an initial injury are prone to suffering from recurrent and more severe injuries. It is feasible to suggest that one of the major factors of recurrent injuries in athletes is the premature return to sport participation based upon the questionable assessment of symptoms resolution.

2. BASIC CONCEPTS OF REHABILITATION

Traditionally, the consequences of injury are usually described in terms of three essential concepts: (a) impairment; (b) disability; and (c) handicap.

"***Impairment*** refers to any loss or abnormality of psychological, physiological or anatomical structure of function." (World Heath Organization, 1980, p.27)

"***Disability*** refers to any restriction or lack (resulting from an impairment) of ability to perform an activity in the manner or within the range considered normal for a human being." World Heath Organization, 1980, p.29)

"***Handicap*** refers to a disadvantage for a given individual, resulting from an impairment or a disability that limits or prevents the fulfillment of a role that is normal (depending on age, sex and social and cultural factors) for the individual." (World Heath Organization, 1980, p.29)

The term *impairment* simply labels the effect of injury on the tissue and its function. The term *disability* assesses the impairment due to injury in terms of its effects on what would be considered a normal profile of activities for a fit person. The term *handicap* places disability within the context of that particular person's previous abilities, expectations and aspirations (Rose, et al., 2005). Although the model is not universally accepted, these terms aim at defining the progression of consequences of traumatic injury, rather than the current and stationary status of an injured person. Accordingly, these terms may imply the link between the initial injury and the eventual outcome, and most importantly, identify the most important targets for rehabilitation strategies. In other words, the evolution of a traumatic injury rather than its initial physical status at the site of the accident must be fully understood within the rehabilitation protocols.

2.1. Basic Principles of Rehabilitation

Rehabilitation is reactivation having the intermediate goals of a progressive increase in activity level and the restoration of function (Wadell,

2004). From this perspective, exercise has direct physiologic benefits, but that is really only the means to an end. It may help to focus on graded activity rather than progressive exercise, because that is the goal. Recovery and return to sport participation require change(s) in behavior, and injury symptoms resolution. The behavior of injured athletes, including avoidance behavior, is driven by motivation (i.e., the direction and intensity of efforts to fully recover and return to duties), beliefs, cognitive awareness (i.e., direction of thoughts) of the situation and emotionality (e.g., fear of re-injury and fear to be useless). That said, when properly designed and fully integrated into a comprehensive multimodal rehabilitation program, physical activity and exercise should be regarded as a heart of the rehabilitation protocol.

The principal role of the therapist in the rehabilitation is to design the combination tasks contingent with a patient's cognitive and motor functionality status that will facilitate the restoration of diminished functions. The rehabilitation therapist should have an opportunity to vary task demands, progressively increase task difficulty, and switch between tasks to constantly challenge the patient and maintain his or her optimal level of effort. Basic rehabilitation principles regardless of type of the injury and patients' individual differences should be considered as follows:

(a) The patient should be **actively engaged** in repetitive problem solving situations rather than passively repeating the required tasks. It is well-documented that active participation, along with repetitive problem solving situations, are the essential components of memory for spatial layout enhancement in cognitively disabled individuals (Antee et al., 1996);

(b) The patient should be **constantly challenged** both physically and mentally to enhance diminished functions and overall brain plasticity (Merzenich, NIH, 2004, personal communication). It should be noted that patients should be "optimally challenged" according to their capacities in order to maintain high levels of motivation. This principle is in line with Alexander Luria's famous assumption that practice and training are particular types of *repetition without repetition* (personal interaction with Dr. Luria, 1975);

(c) The patient should be involved in **functionally relevant** tasks. It is well-known, for example, that task-specific and/or goal-oriented practice promotes enhanced functional recovery by modulating cortical reorganization (personal communication, 2004, see also: Hallett, 2001) and is an important factor for long-term rehabilitation. Patients always vary in terms of symptoms and their resolution, thus, individuated approaches should be implemented depending upon what specific functions (i.e., sensory-motor, cognitive: memory, executive functions, information processing, balance, etc.) are diminished as a result of injury;

(d) The patient should practice tasks in a sequence from **simple to more complex**. In addition, modulation of task complexity should be accompanied by augmented and task-relevant feedback to enhance learning, selective attention and optimal motivation. Augmented feedback (in a way of superimposing on-line the quality/quantity of task accomplishment) is an essential component of directing selective attention during the re-acquisition of perceptual-motor skills (Newell, personal communication, 2007);

(e) The therapeutic intervention in terms of providing task parameters should be **adaptive** in order to meet the patient's demands and individualities. Appropriate adjustments to task complexity and available feedback have profound effects on learning (Newell, personal communication, 2007). Overall, an important feature of VR-based rehabilitation should be the adaptability of task parameters (simplicity/ complexity) and schedules based upon both the initial level of disability and progress upon recuperation;

(f) The gains and skills achieved during therapeutic intervention should be **transferable** to "real-life" tasks. Indeed, one of the major challenges facing clinicians in medical rehabilitation is identifying intervention methods that are effective, motivating and that transfer to the ability to operate in the "real" world (Rose et al., 2005);

(g) Finally, the patient should be involved in the treatment protocols, thereby stimulating the participant's improved ability to **compensate** for residual deficits and adopt strategies for the more effective allocation of his or her remaining attentional resources (Cicerone, 2002). In other words, both restoration and compensation for diminished function should be key features of VR rehabilitation protocols.

3. MODELS OF INTEGRATED REHABILITATION

3.1 Stages of Return to Sport Following Injury

As mentioned in the previous text, the major goal of injury rehabilitation in the athletic environment is return to sport participation. Clearly, there is a conceptual disparity between return to practices versus full return to sport participation and competition, that is often a confusing issue among coaches, athletes and medical practitioners. Some coaches strongly believe that if an athlete is ready to participate in practices he or she should be eligible to compete, unless he/she can't contribute to team success due to residual skill problems. In contrast, there is an opinion among coaches that being able to practice is not enough challenge for athletes to safely return to full participation, including competitions. The major challenge identified is the lack of psychological readiness to meet the challenges of competitions.

This is the major rational for a number of conceptual models dealing with return-to-sport following injury recovery, including Taylor & Taylor's *stages of return* model, Andersen's *biopsychosocial* model, and Ryan &

Deci's *self-determination* model. Although it has undergone little if any empirical evaluation (Podlog & Eklund, 2007), Taylor & Taylor's *return to sport* model deserves special consideration and is worth discussing in some detail.

Taylor & Taylor's (1997) model of return to sport is composed of five stages interactively involving both physical and psychological attributes. These stages, including the *"Initial return," "Recovery Confirmation," "Return of Physical and Technical Abilities," "High Intensity Training"* and *"Return to Competition"* in essence outline the progression of athletes' healing, restoration of diminished physical and functional disabilities and psychological recuperation underlying a safe return to sport participation.

During the "Initial return" and "Recovery Confirmation" stages of recovery, an athlete would have an opportunity to appraise the evolution of the injury based on a comprehensive evaluation of the healing process. Physical healing from injury is the major concern at these stages. During the initial healing process, the body releases specific hormones and chemicals to remove dead tissues, control for inflammation, limit swelling, create new arteries, and bring collagen for tissue regeneration (Flint, 2007).

The healing process within these two stages, as proposed by Taylor & Taylor's (1997) model, is reminiscent of two initial phases of injury (Flint, 2007) based on the body's reaction to trauma and the kinds of biological functions that occur in order to heal the damaged tissue. Specifically, there is an acute (*inflammatory*) phase followed by the fibroblastic (*proliferative*) phase of healing process. During the acute phase of injury, the body usually reacts to the trauma with vascular, chemical and cellular mediators in an attempt to control inflammation and speed up the healing process. Depending upon the injury, the inflammatory phase lasts from two to four days but may be contingent upon received therapeutic interventions. The signs of transition to the next, *fibroblastic*, phase of the healing process are reduction of swelling, redness, heat and pain, indicating the healing is progressing in the right direction. At this phase, along with reduced inflammation the rebuilding of new tissues begins and new blood vessels are built. Again, these are the positive signs of healing. Although on the one hand, the obvious signs of reduced inflammation and pain may manifest proper rehabilitation protocol, on the other hand, this is a "gray" area, when athletes may feel that they can return to sport and often frustrated at being held back by medical professionals. In these initial stages of injury healing, athletes should be educated about the healing process and the reality of pain. Injured athletes should clearly understand the current status (diagnosis) and future development (prognosis) of the injury. Moreover, they should be aware of their changing status from *athlete* to *injured athlete*.

The physical healing process at these phases should be matched with psychological skills and strategies to achieve both clinical and psychological goals in the following manner (adopted from Flint, 2007):

Acute Phase of Injury

<u>Clinical Goals</u>   <u>Psychological Goals</u>

- Reduce swelling   Increase knowledge of injury
- Reduce pain   Increase coping resources
- Protect from re-injury   Increase knowledge of support
                                       system
- Maintain existing ROM   Positive self-talk
- Educate about nutrition   Direct rational thought process
- Educate about healing

Fibroblastic Phase of Injury

<u>Clinical Goals</u>   <u>Psychological Strategies</u>

- Increase ROM   Relaxation, goal setting
- Increase flexibility   Rational Emotive Therapy
- Increase proprioception   Knowledge of setbacks
- Regain muscle stamina   Healing imagery
- Address scar tissue
- Maintain cardiovascular fitness
- Address other physical issues

The third stage of return to sport according to Taylor & Taylor's model is the *return of physical and technical abilities.* This stage assumes that an injured athlete successfully transits from the initial stage of healing and has partially re-acquired diminished functions and is ready to meet more challenging demands of rehabilitation. It should be noted that this intermediate stage of return to sport is usually long and perceived by athletes as a NO progress phase, resulting in frustration and setbacks. Setbacks can be physical, manifested in, for example, a return of swelling, pain, plateaus with no sign of increased ROM, strength or flexibility. Setbacks can be mental as well, as evidenced by athletes' negative self-talk, poor verbalization about the injury, apathy, generalized anxiety caused by uncertainty about the future evolution of the injury. It is critically important at this stage to facilitate an athlete's coping resource, maintain his/her adherence to rehabilitation protocol and increase social support.

The fourth stage of return to sport according to Taylor & Taylor's model is *"high intensity training,"* which marks the conclusion of the athlete's identification as *injured* or *rehabilitating*. Athletes at this stage should achieve or even exceed their pre-injury status in terms of physical skills and conditioning. The goal of continuous rehabilitation at this stage is to enter the final stage and be ready to compete. This stage of healing is reminiscent of the *"maturation or remodeling phase"* proposed by Flint (2007). Indeed, this is another cornerstone in the healing process since that weak and unruly matrix of new collagen experiences strong mechanical stressors that help to align the matrix fibers along these lines of stress. This realignment of collagen may take up to eighteen months to complete and is critical to the regaining of muscle strength. It is also the critical phase when the majority of sport-specific functions should be rehabilitated, as the athlete may now safely return to competition.

The functional healing process at these phases should be matched with psychological skills and strategies to achieve both clinical and psychological goals in the following manner (adopted from Flint, 2007):

Maturation Phase of Injury

<u>Functional Goals</u>   <u>Psychological Readiness</u>

- Increase strength   Increase confidence in body parts
- Increase power   Increase confidence in skill
- Increase sport specific functions
- Increase sport specific cardiovascular fitness
- Increase joint position sense
- Increase agility
- Address all psychological issues

Finally, the fifth stage of return to sport is *"return to competition."* That may undoubtedly be a source of both excitement and trepidation/anxiety because of uncertainty regarding the "success" of the rehabilitation efforts (Taylor & Taylor, 1997). It is a critical time for athletes to be advised that the achievement of pre-injury health status does not guarantee a successful return to competition. For example, the injured athlete's teammates who were injury free may have further progressed in their skills and competitive experience, thus an athlete who used to be "top star" on the team prior to injury might lose his or her status. This may create confusion and may lead to psychological trauma, if not anticipated and properly treated. A psychological consultation with athletes at this stage of healing may be beneficial in allowing the athletes to express their concerns and fears. If this is the case, it is critically important to redirect an athlete's irrational thoughts onto positive aspects and to re-establish emotional equilibrium.

The final stage of the healing process is complete when discharge parameters from medical professional are met. These include but are not limited to (adopted: from Flint, 2007):

- 90% strength (endurance and power) compared to opposite side;
- Full pain-free ROM;
- Full flexibility;
- Pain free;
- Full proprioception;
- Sport-specific cardiovascular fitness;
- Psychological issues addressed (or referred)

Athletes and coaches' education about the healing process and its stages is important in that this gives them an opportunity to fully appreciate the impact of injury on an athlete's entire life. It should be emphasized that meeting discharge parameters and returning to full participation is a process rather than a cornerstone. This is the evolution of formerly diminished function and structural deficits rather than an abrupt transition from "injured athlete" to "athlete."

Several critiques against Taylor & Taylor's *stage* model are outlined by Podlog & Eklund (2007) as follows:

- It is unclear the time frame of initiation and termination of stages, since no specific tests were introduced in this model. That said, it is important to note that huge individual differences among athletes, types of injury etc. along with many other uncontrollable variables make it impossible to propose explicit tests to clarify when and where each stage begins and ends. This comes to the second major critique of this stage model regarding individual differences among injured athletes;
- It is claimed that the model fails to consider individual differences with regard to injured athletes' ability to transit from one stage of healing to another. Also, it is not justified differential rate of healing among athletes underlying progression from one stage to another. This criticism may be applicable to any model of human behavior including stage model of return to sports. There is no generalization in terms of athletes' responses to injury, because no single injury alike in terms of mechanisms, symptoms resolution and its effect on "entire" athlete;
- It is claimed that no empirical research exists to support that there is five-stage healing process in athletes suffering from injury. Of course, athletes may or may not experience all five stages identified by Taylor & Taylor's mode, however, considering the fact of reminiscence of this model with "phase of healing" proposed by Flint (2007) it is feasible to suggest that five-stage generalized model makes a lot of sense;

- Finally, the point is raised about linearity of progression from "initial" stage to "return to competition" stage. For example, it may be the case that the athletes who do not receive recovery confirmation feedback may not transit onto the next stage of healing process. Clearly, it is NOT the purpose of generalized model to consider multiple confounding and concomitant factors influencing whole process. Model is not a conservative solution, but rather a viewpoint of evolution of event where every single case should be treated individually.

It is not surprisingly, that *stage model* of return to sport proposed by Taylor & Taylor (1997) has been subjected to numerous criticisms. Overall, it was argued the impossibility to clearly define time frame and specific physical and psychological properties of each stage under study. Moreover, well-controlled research is lacking to justify major predisposition of this model. Finally, type of injury (i.e. mild, moderate, severe, etc.), individual differences among injured athletes, initial fitness level, age, gender, social and cultural issues to stress just a few may be confounding factors influencing healing process. That said, stage model of return to sport may provide a solid foundation for future research and implementation in clinical practice dealing with injured athletes. Specifically, careful consideration of this model by coaches and medical practitioners may prevent numerous cases when athletes returned to sport prematurely solely based on physical symptoms resolution. Numerous psychosocial factors may induce enormous stress during rehabilitation or even upon return to sport that may modulate whole healing process of injury.

3.2. Biopsychological Model

The *biopsychosocial* model (Andersen, 2001; Brewer, 2001; Brewer, Andersen & van Raalte, 2002) is another well-recognized conceptualization examining the multimodal issues facing athletes returning to sport following injury. However, before discussing Andersen's model, suffice it to stress that in a similar manner, a *biopsychosocial* model examining low back pain and disability that is currently widely accepted in clinical community, was proposed back in 1977 by Engel, and modified by Waddel in 2004. The details of this model have been already discussed in the previous text (see Part 3, chapter 14). Interestingly, it is a model of human illness rather than of disease or pain, that examines the interactive effect of physical, psychological and social elements on a person suffering from pain and disability. Again, this model is not a conservative solution, but rather a notion of evolution of event where every single case and every single clinical element should be treated individually. On the other hand, this is an interactive model of healing & habilitation. There are four clinical elements in this model: (a) physical dysfunction; (b) beliefs and coping; (c) distress; and (d) social interaction. Those four elements influence pain and disability. In fact, a *biopsychosocial* model has been accepted as a primary concept of pain and disability (ISF: WHO, 2000). According to this model, participation *restriction*, activity *limitations* (due to personal factors) and *impairments* (e.g., body structure & functions) have to be examined at social (e.g. cultural), psychological (e.g., illness behavior, beliefs, coping strategies and emotions) and biological (e.g., neurophysiology, tissue damage) levels of analysis.

Now back to Andersen's proposal, which rightfully claims that returning an athlete to full activity is a complicated and multifaceted process including biological, psychological and social variables. The schematics of this model have been recently presented and described by Podlog & Eklund (2006; 2007). In essence, there are seven key components of this model, including:

- Characteristics of the injury;
- Sociodemographic variables;
- Biological variables;
- Psychological variables;
- Social/contextual variables;
- Intermediate biophysical outcomes;
- Sports injury rehabilitation outcome;

Indeed this model is self-explanatory, suggesting that every single variable may or may not play a role in the healing process. For example, as stressed by Brewer (2001), the characteristics of an injury (e.g., severity, site, history of injury, number of injuries etc.), interactively with sociodemographic factors, such as an athlete's age, gender and socioeconomic status, may influence biological and psychological variables. In turn, these multiple interactions may influence the "rehabilitation environment and, overall, the healing process. There are also multiple connections in this model underlying the effect of psychological factors (e.g., personality predisposition, cognition and affect) on biological rehabilitation outcome (e.g., increased ROM, strength, flexibility). Therefore, this model is labeled as a "heuristic unity framework" (Podlog & Eklund, 2007), giving coaches and medical practitioners important clues regarding an injured athlete's multimodal response to injury. An interesting proposition, though not supported by any empirical evidence, is that sociodemagraphic, biological, psychological, and social/contextual variables may *play an equally important role* in an athlete's recuperation on the way to full return to sport participation. Although this makes sense, it is indeed problematic as to how to implement this proposition in clinical practice, and most importantly, how to control the individual effect of these variables on the healing process. Another valuable critique was raised that the biopsychosocial model is not specifically designed for and does not address the issue of return-to-sport transition (Podlog & Eklund, 2007). It seems that the time has come that the empirical data-driven and comprehensive approach to injury rehabilitation incorporating Taylor & Taylor's *stage* and Andersen's biopsychosocial models should be elaborated. See also the schematics of the biopsychosocial model below (adopted from D. Pargman (Ed.), Psychological Bases of Sport injuries, Third edition, 2007, Chapter 7: Psychosocial considerations of the return to sport following injury (L. Podlog & R. Eklund, 2007)).

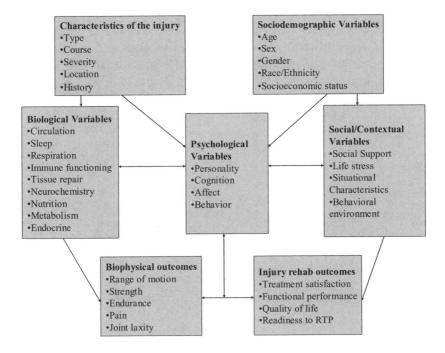

Figure 1. Brewer, R.W., Andersen, M.B., & Van Raatle, J.L. (2002). Psychological aspects of injury rehabilitation: Toward a biopsychosocial approach. In D.I. Mostofsky & L. Zaichkovsky (Eds), Medical and psychological aspects of sport and exercise (pp.41-54), Morgantown, WV: Fitness Information Technology.

3.3. Self-Determination Model

Athletes returning to sport participation following injury may perceive this critical transition from "injured athlete" to "athlete" as a *challenge* or as a *threat* (Taylor et al., 2003). A positive scenario when return to sport is perceived as a challenge stimulates an athlete's coping resources and further facilitates the healing process both physically and emotionally. At this stage of recovery, an athlete may exceed the pre-injury health status and in fact, may gain potential benefits in terms of enrichment of his/her psychological experience (Udry, 1999). Athletes may take advantage of the injury to re-evaluate their values and self-worth. In addition to love for the sport and excitement to be back in shape, the perception of the injury and awareness of its potential benefits may be an integral part of an athlete's maturation as a person. There are numerous examples when athletes who previously suffered from severe injuries were able not only to fully recover but also to reach a peak of excellence upon return to sport.

Another scenario is that when return to sport participation for some reason is perceived as a *threat*. A whole gamut of psychological (e.g., fear of re-injury, anxiety, chronic pain, etc) and behavioral responses (e.g., avoidance reaction, etc) may induce enormous stress, including a career ending outcome. Athletes' perception of the stressors of returning to sport participation is summarized in the following figure (Adopted from Podlog & Eklund, 2007).

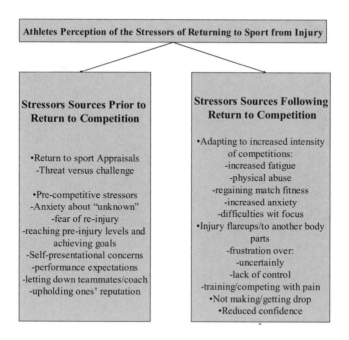

Figure 2. Athletes' Perception of the Stressors of Returning to Sport from Injury. (Adopted from Podlog & Eklund (2007). Psychological considerations of the return to sport following injury. In D. Pargman (Ed.), Psychological Bases of Sport Injuries, (pp.116). Morgantown, WV: Fitness Information Technology.)

In fact, it has been shown that coaches' perception of the stressors of returning to sport from injury is often similar to that expressed by the athletes (Padlog & Eklund, 2007). Although individual differences may play a significant role, there are at least three common concerns experienced by coaches and athletes, including but not limited to:

- Fear of re-injury;
- Fitness problems;
- Effect of physical limitation on the athlete's ability to perform.

Coaches' most common perceptions of the stressors of returning to sport following injury may be found in the following figure.

Coach Perception of the Stressors of Returning to Sport from Injury

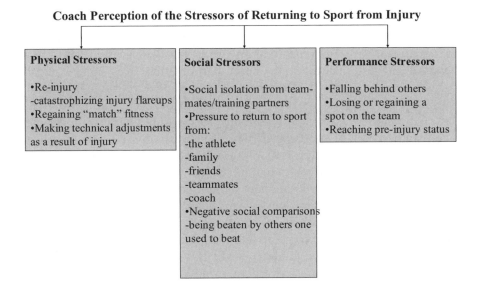

Physical Stressors	**Social Stressors**	**Performance Stressors**
•Re-injury -catastrophizing injury flareups •Regaining "match" fitness •Making technical adjustments as a result of injury	•Social isolation from team-mates/training partners •Pressure to return to sport from: -the athlete -family -friends -teammates -coach •Negative social comparisons -being beaten by others one used to beat	•Falling behind others •Losing or regaining a spot on the team •Reaching pre-injury status

Figure 3. Coaches' Perceptions of the Stressors of Returning to Sport from Injury. (Adopted from Podlog & Eklund (2007). Psychological considerations of the return to sport following injury. In D. Pargman (Ed.), Psychological Bases of Sport Injuries, (p.121). Morgantown, WV: Fitness Information Technology.)

Overall, coaches and athletes' experiences of returning to sport participation may indicate at least three psychosocial areas of concern and/or difficulties that may influence the development of "challenge" or "threat" scenarios. These include:

- Competence
- Autonomy
- Relatedness

In fact, these three psychological areas are the cornerstone for the self-determination theory (Rayn & Deci, 2000) that may facilitate or undermine heath, psychological well-being and motivation. The details of the self-determination theory emphasizing the effect of intrinsic (self) motivation on human behavior may be found elsewhere, but the implication of this conceptual framework for injured athletes is worth considering in some detail. The following text was developed based on Podlog & Eklund's (2007) review of self-determination theory as applied to return to sport from injury.

Competence: Injured athletes returning to sport participation may experience concerns about physical competency. Specifically, athletes may express doubt that they have achieved the pre-injury level of fitness and specific skills allowing them to make contribution to team. Competence and/or lack of competence may also be manifested in athletes' underestimation of their ability to accomplish the task. The competence issue may also be considered from the perspectives of cognitive and emotional responses to rehabilitation progress. Numerous reports in the literature suggest that return to sport participation may induce negative cognitive (e.g., negative thoughts) and emotional (e.g., fear, anger & frustration, generalized anxiety) responses. Moreover, competency-based fears and concerns associated with return to sport are the most prominent as a well-studied area of sports psychology. Interestingly, the fear of re-injury may be a source of concern not only at initial stages of injury recovery but also after return to full participation. The long-term reduction of self-efficacy may also be a prominent consequence of athletes' feeling of incompetence about future performance and the ability to accomplish personal and external expectations (Feltz, personal communication, 2007).

It is reported that for some athletes re-injury fears may persist for years after return to competition (Bianco et al., 1999). This may accompany the negative social comparisons regarding an athlete's post-injury performance in relation to his or her injury-free teammates (e.g., losing to fellow competitors one used to beat easily prior to injury). Another issue of concern is that athletes lose the flavor and competitive experience, which in turn may cause decline in confidence upon return to competition. A number of psychological intervention strategies can be recommended to assist athletes with competence issues in general, and to address the perceived physical and psychological insecurity, in particular. Among them are: a combination of muscle relaxation and healing imagery; desensitization; virtual reality and neurofeedback (see details in the following chapters of this book); goal-setting, etc. Interested readers may be advised to visit a paper by Podlog & Eklund (2007) underlining the details and reviewing the literature on competency-based fear responses in athletes recovering from injury.

Autonomy: The reality of athletics is that injured athletes often experience directly or indirectly an enormous amount of pressure to return to sport participation as soon as possible. Self-induced pressure to return to the sport prematurely is a major source of dependence (i.e. lack of autonomy). Elite athletes often hide their injury to demonstrate that they are O.K. and ready to play. Numerous anecdotal facts demonstrate that athletes remained in the field with fractures and/or competed with serious illnesses requiring surgery. There also outside pressures, coming from the surrounding athletic environment and dictated by "sports ethics." As discussed in the previous text, current practice in terms of return-to-play criteria is that an athlete may be cleared for sport participation based upon physical symptoms resolution. An injured athlete may be rehabilitated physically but not necessarily psychologically. It is recommended that injured athletes' psychological issues should be addressed or referred (see Flint, 2007). However it is conventional wisdom among athletes and coaches that "athletes are tough enough to deal with their mental problems on their own." As a matter of fact, this is an erroneous statement, pressing athletes prematurely return to the sport before the "injury" and/or psychological trauma is resolved. Not surprisingly, numerous observations suggest that a premature return to sport participation is the major cause of recurrent injuries with severe consequences affecting the overall well-being of athletes.

In addition to self-induced pressure, there are numerous social and cultural roots inducing pressure to prematurely return to duties in the contemporary sports environment. With the current pressure to "win at all cost," to "get bonuses" upon wining, "no pain no gain," "play with pain to get gain," "no less than 100% effort" and the overall attitude that "pain and injury are temporary but fame is forever," elite athletes are highly susceptible to injury and its associated doubt, worry and concerns upon return to sport. The loss of identity, as being an "injured athlete" but not "athlete," is another source of lack of autonomy while in rehabilitation. This pressurized environment is problematic for the normal healing process.

In fact, support for this notion of pressure is evidenced by a number of reports. For example, according to Bianco et al. (2001), athletes who reported pressure to return to sport (i.e., those less autonomous) indicated less favorable consequences upon the actual return to sport participation. These athletes experienced recurrent injuries, performance deterioration, lack of confidence and overall emotional dysfunctions. On the other hand, increased autonomy has been shown to be associated with more positive cognitive and emotional responses upon return to competition (Podlog & Eklund, 2006). This recent study clearly demonstrated that a higher level of self-determination (i.e., more intrinsic forms of motivation) has been correlated with decreased fear of re-injury, resulting in more positive appraisals and overall psychological outcomes. Overall, injured athletes may have different levels of autonomy in returning to practices and competitions, which may influence their self-perception of the consequences of return to sport. The lack of autonomy may be associated with perceived *threat*, whereas autonomy may be associated with perceived *challenge*, leading to joy and satisfaction at being "back on the race."

Relatedness: In previous chapters of this book, while interviewing collegiate coaches and athletes suffering from injuries, the question was raised: *"In your opinion, and in terms of psychological recovery, do you think that athletes recover better, or faster, from an injury if the injury is given more attention, or less attention?* This question is directly related to the issue of relatedness, since injured athletes may often feel a sense of alienation and social isolation from their friends, teammates and fellow competitors (Podog & Eklund, 2007). A number of reports reported concern that perceived lack of attention from coaches may induce the feeling of being cut off, further leading to diminished social interaction with teammates. This situation may create additional stressors in the recovery and the whole healing process. It used to be an "unspoken rule" in elite athletics that "…no place for injured athletes in training facility," since seeing athletes on crutches may induce a sense of insecurity among healthy teammates. They could feel that "I will be the next." More recently, this attitude has been gradually changing; now a policy has even been implemented that an injured athlete should be considered as a member of the team, requiring participation in practices and competitions. Here are several coaches' responses to the issue of *relatedness* with respect to injured athletes in collegiate athletics:

Question: In your opinion, and in terms of psychological recovery, do you think that injured athletes would recover better or faster if given more attention or less attention?

Coach Ganter (Penn State football): *I would have to say more attention is needed when you coach injured athletes. I can not picture an injury getting less attention, there is no motivation there, there is no we got to get you back, that is a tough one, but I would have to say more attention would help promote the quicker recovery. I think the better rehabilitation and more attention a guy would get and more encouragement would get them back quicker.*

Coach Battista (Penn State Ice Hockey): *All players are different. Most injured players would respond better to a coach or trainer who gave positive feedback and who pay more attention to them, "Looking better already, almost there, can't wait to have you back." Some however need to have it downplayed while others need to be told straight up that its not that bad, get over it! Thus, it is all about individual responses to injury. The tougher the player, the less attention he or she needed while recovery from injury.*

Here is an example of a completely opposite opinion regarding the relatedness issue from a professional coach's perspective:

Coach Josean (**Spain, soccer**). *If player injured, this is his own problem to get back to routine, otherwise I will someone who take care of the business. My job, as a coach is to take care of players' getting ready for the game. The player's job is to be healthy and play their best.*

While interviewing injured athletes a similar question was asked: *In your opinion, and in terms of psychological recovery, do you as an injured athlete think you would recover better or faster if given more attention or less attention?*

The injured athlete's opinion regarding the issue of relatedness can be summarized as the following: *In dealing with my injury it helped that the coach and medical staff gave me the attention that was needed. It helped me pull through the process better, but I wouldn't say faster. The reason for that is, because I think I got myself motivated come back faster since I wanted to be back so badly. Obviously, getting attention helps even more, because it gives you more motivation to do better and hopefully come back fully recovered. When coaches, medical staff, or even parents show attention to the injured it creates a sense of trust. Trust is a major issue that comes along with sport. If you don't have trust than what do you have? Makes you feel that you don't have a stable person to fall back on.*

Clearly, most athletes' desire to quick return to sport participation and make a contribution to team success, thus maintaining the sense of relatedness (connectedness) to coaches, teammates and training partners, may provide a good buffer against the feeling of alienation and isolation (Podlog & Eklund, 2007). In fact, there are numerous psychological studies indicating that social support and assistance from various sources may act as a "prophylaxis" against the isolation associated with injury and a return to sport participation. For example, it was reported by injured athletes that social support from coaches and physiotherapists was instrumental in preventing a premature return to sport and in reassuring athletes about the success of rehabilitation. Moreover, social support has also been reported as a helpful tool in controlling athletes' fear about overstressing the recovered body parts and helping injured athletes to set realistic expectations about the outcome of the healing process (Bianco, 2001).

There is indication in the literature (Carrol, 1998; sf: Podlog & Eklund, 2007), that social support can be improved by organizing specific training programs for injured athletes, including counseling the athletes on the difficulties dealing with return to sport and competitions and if necessary referring injured athletes to other professionals (e.g., sports psychologist). It is necessary also to keep athletes involved in sport to enable an easy transition from "injured athlete" to "athlete" status. Being involved in team activities, injured athletes may observe, learn and improve their skills and analyzing strategies from the sideline (Erlmer & Thomas, 1990). Taking the extra time to practice (including healing imagery) on a one-to-one basis with injured athletes not only can enable an easy transition back into sport but also can provide coaches with the opportunity to monitor an injured athlete's progress with the healing process. Thus, upon athletes' return to full participation coaches can efficiently introduce new skills gradually and without inducing injury, assess athletes' physical fitness and conditioning properly and give athletes skill-related feedback accordingly (Podlog & Eklund, 2007).

CONCLUSION

The reality of clinical practice in dealing with athletic injury and its rehabilitation is dictated by the ultimate goal of the return to pre-injury status as fast as possible. Most often, return-to-sport participation criteria are defined by the presence and/or absence of symptoms of injury. Specifically, physical symptoms resolution (i.e., no evidence of residual tissue damage, restored anatomical integrity of joint, etc.) and functional symptoms resolution (i.e., ROM, strength, stamina) are among two major criteria of return-to-play. However, symptoms resolution is not necessarily indicative of injury resolution. Humans in general, and athletes in particular, are able to compensate for mild or even severe physical and functional deficits because of redundancy in the human motor system. This in turn allows for reallocation of existing resources such that undamaged pathways and functions are used to perform cognitive and motor tasks. This functional reserve gives the appearance that an athlete has returned to pre-injury health status while in actuality the injury is still present and hidden from the observer. As a result, a premature return to sport participation based upon physical symptoms resolution may put athletes at high risk for recurrent injuries and the development of permanent psychological trauma.

Over the last decade or so, an integrative approach to the rehabilitation of injured athletes has been promoted. Specifically, by the collective efforts of sports psychologists and clinical practitioners, psychological and psychosocial interventions tend to be gradually incorporated into traditional rehabilitation protocols. Several models and concepts of the integrative rehabilitation of injured athletes have been developed. Taylor & Taylor's (1997) *stage* model of the return to sport participation, the *biopsychosocial* model of injury rehabilitation (Andersen, 2001) and the *self-determination* theory (Ryan & Deci, 2000) are among most popular theoretical constructs to-date. Of course, there are several limitations and inconsistencies present in these models (in fact, as in any existing model of human behavior). However, overall these models and assumptions surely may be considered as a solid foundation for future empirical research in sport and exercise psychology with a specific interest in injury rehabilitation.

In summary, although for many years there has been consensus that psychological substrates and interventions have an important role in sport injury rehabilitation (Flint, 2007), there is still a lack of well-controlled research regarding how these interventions should be effectively incorporated into clinical practice. For example, healing imagery, relaxation, desensitization, etc., have been shown to be important coping strategies when dealing with injury. However, how these strategies can complement and/or conflict with rehabilitation exercise routines is still an important scientific quest. Surely, these interactions (i.e., an integrative approach to injury rehabilitation) should be explored in order to achieve a maximal outcome allowing athletes' timely (NOT speedy) return to sport participation. Maybe it is the proper time to re-consider the trade union definition of rehabilitation, by adding: "...people with sickness or injury...can <u>timely and safely</u> return to duties... based upon injury rather than symptoms resolution."

REFERENCES

TUC Consultation document on rehabilitation: getting better at betting back. Trades Union Congress, London, 2000.

Nocon, A., Baldwin, S. *Trends in rehabilitation policy*. King's Fund. London, 1998.

Podlog, L., & Eklund, R. (2007). The psychosocial aspects of return to sport following serious injury: A review of the literature from a self-determination perspective. *Psychology of Sport and Exercise, 8*, 535-566.

Hail, J. *Psychology of Sport Injury*. Urbana-Champaign: Human Kinetics Publishers, 1993.

Locke, E.A., & Latham, G.P. (1985). The application of goal-setting to sports. Journal of Sport Psychology, 7, 205-222.

Udry, E. (1999). The paradox of injuries: Unexpected positive consequences. In D. Pargman (Ed). *Psychological bases of sport injuries* (2nd ed. pp.79-88). Morgantown, W.V: Fitness Information Technology.

Flint, F.A. (2007). Matching psychological strategies with physical rehabilitation: Integrated Rehabilitation. In D. Pargmam (ed). *Psychological bases of sport injuries*. pp.319-334. Morgantown: Fitness Information Technology.

World Health Organization. (1980). International classification of impairment, disabilities and handicaps: a manual of classification relating to the consequences of disease. Geneva: WHO.

Rose, D., Brooks B., Rizzo, A. (2005). Virtual reality in brain damage rehabilitation. *CyberPsychology and Behavior, 8*, 241-262

Waddell, G. *The back pain revolution*. Second Edition. Edinburgh: Churchill Liningstone, 2004.

Hallett, M. (2001). Plasticity of the motor cortex and recovery from stroke. *Brain Research Review, 36*, 169-174.

Cicerone, K.D. (2002). Remediation of 'working attention' in mild traumatic brain injury. *Brain Injury, 16*(3), 185-195.

Taylor, J & Taylor, S. Psychological approach to sports injury rehabilitation. Geithesburg, MD: Aspen, 1997.

Flint, F. (2007). Matching psychological strategies with physical rehabilitation: Integrated rehabilitation. In In D. Pargman (Ed). *Psychological bases of sport injuries* (2ⁿᵈ ed. pp.320-334). Morgantown, W.V: Fitness Information Technology.

Andersen, M.B. (2001). Returning to action and the prevention of future injury. In J. Crossman (Ed.), Coping with sports injuries: Psychological strategies for rehabilitation (pp. 162-173). Melbourne: Oxford University Press.

Podlog, L., & Eklund, R. (2006). A longitudinal investigation of competitive athletes' return to sport following serious injury. *Journal of Applied Sport Psychology.*

Brewer, B.M., Anderson, M.B., van Raatle, J.L. (2002). Psychological aspects of sport injury rehabilitation: Toward a biopsychosocial approach. In D. Mostofsky & L. Zaichkowsky (Eds.), *Medical Aspects of Sport and Exercise* (pp.41-54). Morgantown: WV: Fitness Information Techonology.

Engel, G.L. (1977). He need for a new medical model: a challenge for biomedicine. *Science, 196*, 129-136.

Brewer, B.W. (2001). Emotional adjustment to sport injury. In J. Crossman (Ed.), *Coping with Sport Injuries: Psychological Strategies for Rehabilitation* (pp.1-19). Melbourne.: Oxford University Press.

Taylor, J., Stone, K.R., Mullin, N.J., et al. Comprehensive sports injury management: From examination of injury to return to sport. Austin, Texas: Pro-ed, 2003.

Udry, E. (1999). Paradox of injuries: Unexpected positive consequences. In D. Pargman (ed.). Psychological bases of sport injuries, (2ⁿᵈ edition, pp. 79-88. Morgantown: WV: FIT.

Ryan, R., & Deci, E.L. (2000). Self-determination theory and the facilitation of intrinsic motivation, social development and well-being. *American Psychologist, 55*, 68-78.

Bianco, T., Malo, S., & Orlick, T. (1999). Sport injury and illness: Elite skiers describe their experience. *Research Quarterly for Exercise and Sport, 70*, 157-169.

Bianco, T., & Eklund, R. (2001). Conceptual consideration for social support research in sport and exercise settings: The instance of sport injury. Journal of *Sport and Exercise Psychology, 2*, 85-107.

Ermler K.L., Thomas, T. (1990). Intervention strategies for the alienating effect of injury. *Athletic Training, 25*, 269-271.

TUC 2000. Consultation document on rehabilitation: getting better at betting back. Trades Union Congress, London.

Nocon, A., Baldwin, S. *1998). Trends in rehabilitation policy. King's Fund. London.

Podlog, L., & Eklund, R. (2007). The psychosocial aspects of return to sport following serious injury: A review of the literature from a self-determination perspective. *Psychology of Sport and Exercise, 8*, 535-566.

Hail, J. (1993). Psychology of Sport Injury. Urbana-Champaign: Human Kinetics Publishers.

Locke, E.A., & Latham, G.P. (1985). The application of goal-setting to sports. Journal of Sport Psychology, 7, 205-222.

Udry, E. (1999). The paradox of injuries: Unexpected positive consequences. In D. Pargman (Ed). Psychological bases of sport injuries (2ⁿᵈ ed. pp.79-88). Morgantown, W.V: Fitness Information Technology.

Flint, F.A. (2007). Matching psychological strategies with physical rehabilitation: Integrated Rehabilitation. In D. Pargmam (ed). *Psychological bases of sport injuries.* pp.319-334.Morgantown: Fitness Information Technology.

Andersen, M.B. (2001). Returning to action and the prevention of future injury. In J. Crossman (Ed.), Coping with sports injuries: Psychological strategies for rehabilitation (pp. 162-173). Melbourne: Oxford University Press.

Engel, G.L. (1977). He need for a new medical model: a challenge for biomedicine. *Science, 196*, 129-136.

Waddell, G. (2004). The back pain revolution. Second Edition. Edinburgh: Churchill Liningstone.

Brewer, B.W. (2001). Emotional adjustment to sport injury. In J. Crossman (Ed.), Coping with Sport Injuries: Psychological Strategies for Rehabilitation (pp.1-19). Melbourne.: Oxford University Press.

Brewer, B..M., Anderson, M.B., van Raatle, J.L. (2002). Psychological aspects of sport injury rehabilitation: Toward a biopsychosocial approach. In D. Mostofsky & L. Zaichkowsky (Eds.), Medical Aspects of Sport and Exercise (pp.41-54). Morgantown: WV: Fitness Information Techonology.

Bianco, T., Malo, S., & Orlick, T. (1999). Sport injury and illness: Elite skiers describe their experience. *Research Quarterly for Exercise and Sport, 70*, 157-169.

Bianco, T., & Eklund, R. (2001). Conceptual consideration for social support research in sport and exercise settings: The instance of sport injury. Journal of *Sport and Exercise Psychology, 2*, 85-107.

Podlog, L., & Eklund, R. (2006). A longitudinal investigation of competitive athletes' return to sport following serious injury. *Journal of Applied Sport Psychology.*

Ermler K.L., Thomas, T. (1990). Intervention strategies for the alienating effect of injury. *Athletic Training, 25*, 269-271.

CHAPTER 22

EEG & NEUROFEEDBACK IN REHABILITATION

1. INTRODUCTION

One important turning point in the history of electrophysiology was the report by Galvani in 1791 that nerves contain an intrinsic form of electrical activity. Some sixty years later Du Bois-Reymond demonstrated that activity in a peripheral nerve was accompanied by recordable changes in the electrical potential of the nerve. With this discovery, the scientific community began to search for various factors that would be associated with this electrical activity. During this period one important theoretical question was the location of various forms of activity in the brain. Richard Carton, by studying rabbits and monkeys, was able to demonstrate a connection between an external sensory stimulation such as light and concomitant electrical activity in the brain. Specifically, he was able to show that electrodes on the scalp of these animals could reflect "feeble currents" associated with a variety of stimuli. This marked one of the initial demonstrations of the EEG with animals.

From the initial demonstration of the EEG with animals, it was some fifty-four years later that the technique was demonstrated in humans. In the 1920s Hans Berger was able to show potential differences between recording sites related to cortical processes. He named this electrical activity the "Elektrenkephalogramm." In his first set of papers, Berger sought to determine what factors were involved in the production of the EEG and was able to determine that EEG was related to activity within the brain and to rule out other physiological activity such as cerebral pulsations, cerebral blood flow, blood flow through scalp vessels, heart rate activity, muscle activity, eye movements and electrical properties of the skin (Berger, 1929). Berger took his studies beyond the physiological level and was one of the first to suggest that periodic fluctuations of the EEG might be related in humans to cognitive processes such as arousal, memory and consciousness. In determining the nature of the EEG Berger was initially surprised to discover that EEG changes were ones of quality rather than quantity. For example, as an individual moved from a relaxed state to one of stimulation and activity, Berger noted that the EEG did not increase in amplitude but rather changed in the quality of the wave forms. He initially identified these two different EEG wave forms as that of alpha activity and that of beta activity, with alpha being associated with cortical inactivity and beta with cortical activity.

2. PHYSIOLOGICAL BASIS

During the past century there was some debate as to the nature of the EEG. Although initially thought to result from summated action potential which fires in all or none fashion (cf., Adrian & Matthews, 1934), this has been shown not to be the case. For example, Li & Jasper (1953) were able to record EEG in cats even after neural action potentials were abolished using deep anesthesia. Current views suggest that the EEG originates in the depolarizations of the dendritic trees of pyramidal cells (Lutzenberger, Elbert, & Rockstroh, 1987; Lopes da Silva, 1991). Specifically, graded postsynaptic potentials of the cell body and dendrites of vertically orientated pyramidal cells in cortical layers three to five, give rise to the EEG recorded on the scalp. The ability to record at the scalp the relatively small voltage from these actions results from the fact that pyramidal cells tend to share a similar orientation and polarity and may be synchronously activated.

The electroencephalogram or EEG is typically recorded at the scalp surface near the ear, and represents the moment-to-moment electrical activity of the brain. The electroencephalogram or EEG is produced by the summation of synaptic currents that arise on the dendrites and cell bodies of billions of cortical pyramidal cells that are primarily located a few centimeters below the scalp surface. The synaptic currents involve neurotransmitter storage and release which are dependent on the integrity of the sodium/potassium and calcium ionic pumps located in the membranes of each neuron. Metabolic activity is the link between EEG/MEG and PET, SPECT and fMRI, which are measures of blood flow dynamics. Glucose regulation and restoration of ionic concentrations occur many milliseconds and seconds and minutes after electrical impulses and synaptic activity, and therefore, blood flow changes are secondary to the nearly instantaneous electrical activity and metabolic activities that give rise to the EEG at each moment of time (Thatcher & John, 1977).

2.1. Recording and Patterns of EEG Activity

To record the EEG, electrical signals of only a few microvolts must be detected on the scalp. This can be accomplished by amplifying the differential between two electrodes, at least one of which is placed on the scalp. Since the signal must be amplified almost one million times, care must be taken that the resulting signal is indeed actual EEG and not artifact. Where the electrodes are placed and how many are used depend on the purpose of the recording. Today, almost all EEG procedures use a variety of EEG helmets with up to 256 electrodes build into the helmet, although it is always possible to record EEG from only two electrodes. Those recording helmets that use 128 to 256 electrodes are generally referred to as dense array EEG recordings. If the spatial distribution of some aspect of the EEG is the research question, then multiple electrodes distributed over the scalp are required. Of course, one can record from many fewer electrodes depending upon the empirical questions that are being asked. For example, if one is only interested in EEG responses associated with movement, then one may chose to record from regions of the scalp lying above the motor areas of the brain.

Historically, the system of locating electrodes in EEG is referred to as the International 10-20 system (Jasper, 1958). The name 10-20 refers to the fact that electrodes in this system are placed at sites 10% and 20% from four anatomical landmarks. In the front, the nasion (the bridge of the nose) is used. In the rear of the head, the inion (the bump at the back of the head just above the neck) is used. The left and right landmarks are the preauricular points (depressions in front of the ears above the cheekbone). In this system, letters refer to areas of the brain: 0 = occipital, P = parietal, C = central, F = frontal, and T = temporal. Numerical subscripts indicate laterality (odd numbers left, even right) and degree of displacement from the midline (subscripted z). Thus, C_3 describes an electrode over the central region of the brain on the left side whereas C_z would refer to an electrode placed at the top of the scalp above the central area. With the development of dense array systems, the historical 10-20 system has been greatly expanded. See Figure 1 for details of the 10-20 system.

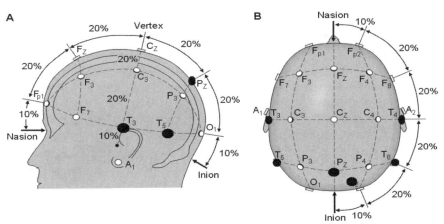

Figure 1. The position of the nineteen electrodes used for data acquisition located according to the 10-20 system. A1 and A2 are referring electrodes. The electrodes filled with black color are where the ten selected EEG features come from.
(adopted from: http://bulter.cc.tut.fi/~malmivuo/bem/bembook/13/13x/1202ax.htm, and modified by the author)

Two specific types of EEG recording are called monopolar and bipolar recordings. In order to understand this point, let us remember that EEG recordings reflect the difference in voltage between signals at two electrodes. What this means is that if the exact same cortical signal were present at two separate sites on which our electrodes were placed, then we would record a straight line reflecting no difference in activity between the two sites. Of course, this never happens, since there are always differences in activity between recording sites. In monopolar recordings the idea is to find a site that is not reflective of EEG activity per se to use as a reference site. Common sites used for this purpose are the ear (or ears), the mastoid, or even the nose. Other researchers have suggested that a useful reference to use is that of the average reference. This procedure basically takes a network of electrodes spaced across the scalp and mathematically averages these together. This mathematical average value is then used as the reference.

In bipolar recording, each electrode is located to record from an active site on the scalp. Thus, one could compare the difference in EEG activity between the right frontal with that of the left frontal areas. One might use such a procedure to infer whether the left or right hemisphere, for example, was more involved in a particular task. This type of procedure has traditionally been used in clinical settings to identify unusual pathological waveforms such as epileptic discharges.

The rhythmic variations of the EEG are continually present at the surface of the scalp from well before birth to death. In fact, the absence of EEG for twenty-four hours has been used as an indicator of "brain death." Additionally, EEG has been used to denote states of consciousness as found in sleep, epilepsy and brain pathology. As we will see in this book, EEG has also been used to denote brain trauma as found in stroke and concussion. The various frequencies and distributions of specific patterns of the EEG wax and wane, providing the brain researcher and clinician with a constant record of the changing patterns of electrical activity of the brain. Some aspects of the EEG may appear almost random while other fluctuations appear periodic. We have a variety of signal processing techniques to help us describe the EEG, but in general we use two basic parameters. These are amplitude and frequency. Some EEG patterns are extremely reliable and can be visually observed, as would have been required in the days before computer analysis. These patterns have been identified in their order of discovery by the Greek letters α (alpha), ß (beta), δ (delta), and so forth.

2.2. Types of EEG Activity

Alpha activity can be seen in about three-fourths of all individuals when they are awake and relaxed. Asking these individuals to relax and close their eyes will result in recurring periods of several seconds in which the EEG consists of relatively large, rhythmic waves of about 8-12 Hz. This is the *alpha rhythm*, the presence of which has been related to relaxation and the lack of active cognitive processes. If someone who displays alpha activity is asked to perform cognitive activity such as solving an arithmetic problem in his/her head, alpha activity will no longer be present in the EEG. This is referred to as alpha blocking. Typically, with cognitive activity the alpha rhythm is replaced by high frequency low amplitude EEG activity referred to as beta activity. Since Berger first discovered the alpha rhythm, a variety of studies have focused on its relationship to psychological processes, and the broad developments of the cognitive and affective neurosciences amplified this interest (see Shaw, 2003 for a review). Based on factor analysis of alpha activity some have suggested that alpha activity be divided into two or three separate frequency bands (Klimesch, 1999).

Beta activity occurs when one is alert. Traditionally, lower-voltage variations ranging from about 18 to 30 Hz have been referred to as beta, and higher frequency lower-voltage variations ranging from about 30 to 70 Hz or higher as gamma. Initial work suggested that gamma activity is related to the brain's ability to integrate a variety of stimuli into a coherent whole. For example, Catherine Tallon-Baudry and her colleagues (Tallon-Baudry, Bertrand, Delpuech, & Pernier, 1997) showed individuals pictures of a hidden Dalmatian dog that was difficult to see because of the black and white background. After training individuals to see the dog, there were differences in the gamma band response, suggesting differential responses to meaningful versus non-meaningful stimuli.

Additional patterns of spontaneous EEG activity include delta activity (0.5-4 Hz), theta activity (5-7 Hz), and lambda and K-complex waves and sleep spindles, which are not defined solely in terms of frequency. *Theta* activity refers to EEG activity in the 4-8 Hz range. Grey Walter (1953), who introduced the term theta rhythm, suggested that theta was seen at the cessation of a pleasurable activity. More recent research has theta associated with such processes as hypnagogic imagery, REM (rapid eye movement) sleep, problem solving, attention, and hypnosis. Source analysis of midline theta suggests that the anterior cingulate is involved in its generation (Luu & Tucker, 2003). Schacter (1977), in an early review of theta activity, suggested that there are actually two different types of theta activity: First there is theta activity associated with low levels of alertness as would be seen as one falls asleep. And second, there is theta activity associated with attention and active and efficient processing of cognitive and perceptual tasks. This is consistent with the suggestion of Vogel et al (1968) that there two types of behavioral inhibition, one associated with a gross inactivation of an entire excitatory process resulting in less active behavioral states, and one associated with selective inactivity as seen in over-learned processes.

Delta activity is low frequency (0.5-4Hz.) and has been traditionally associated with sleep in healthy humans as well as pathological conditions. The pathological conditions associated with delta have included cerebral infarct, contusion, local infection, tumor, epileptic foci and subdural hematoma. The basic idea is that these types of disorders influence the neural tissue, which in turn creates abnormal neural activity in the delta range by cutting off these tissues from major input sources. Although these observations were first seen with intracranial electrodes, more recent work has used MEG and EEG techniques. EEG delta activity is also the predominant frequency of human infants during the first two years of life.

2.3. Analysis and Quantification

Historically, EEG technicians in clinical settings underwent extensive training in order to be able to recognize the visual patterns of EEG related to sleep stages and neurological disorders. Some frequencies, such as the alpha rhythm, are easy to recognize, while the presence of other EEG frequencies are more difficult to detect. Since visual pattern recognition is subjective, EEG researchers sought quantitative procedures for describing EEG activity. With the advent of integrated computer chips, quantitative analysis of the EEG has become less of a practical problem. In order to do a quantitative analysis, it is first necessary to convert the continuous analog EEG signal into a digital form, which is accomplished by an analog to digital converter. Once the signal is represented as individual numbers in a time series, then these numbers can be manipulated mathematically. One of the first questions that must be determined is the sampling rate of the digital converter so that an accurate EEG record can be obtained. Based on a variety of engineering studies, the smallest sampling rate recommended is that of twice the highest frequency that one wishes to detect. Thus, if one wanted to study an EEG signal between 4 and 30 Hz, then one would have to record the EEG at a sampling rate of at least 60 Hz. However, most researchers sample at four to eight times the highest frequency under consideration, to ensure accurate detection of the EEG.

2.4. Frequency Analysis

One of the most common frequency analysis techniques is that of Fourier analysis. The technique is named after the French mathematician Fourier, who suggested that any given time series can be described as a corresponding sum of sine and cosine functions. Using this information, he described how to determine in the frequency domain the amplitude and phase information of a known temporal signal. One simple way of understanding this procedure is to imagine that one has a variety of templates which represent each frequency band under consideration. Thus, one could have an 8 Hz template, a 9 Hz template, a 10 Hz template and so forth. By simply placing each template on top of the signal, you could determine how closely the signal fit that template. This is basically the procedure that Fourier analysis uses. It takes an EEG signal in time and describes it in terms of how much of each frequency is represented in the signal. Thus, Fourier converts a time-based signal to a frequency-based signal. In the 1960s a mathematical algorithm was developed that speeds computations of this procedure; it is referred to as the Fast Fourier Transform (FFT) and is used by most computer programs today.

An analysis technique related to Fourier analysis is that of coherence analysis. Whereas Fourier analysis gives the frequency for a given electrode, coherence gives the covariance of this measure for a pair of electrodes. Thus, coherence tells you how the EEG signal at each of two electrodes is related. In simple terms, coherence reflects the manner in which two signals covary at a particular frequency. That is to say if the EEG at the right frontal electrode and the left frontal electrode both demonstrated a frequency of 8 Hz, then we would see greater coherence between the two electrodes than if they did not. In doing the coherence analysis, one can also obtain a measure of phase or synchrony. That is to say, we can determine if two signals of the same frequency have peaks and valleys at the same time. Using coherence, Thatcher and his colleagues have studied how the brains of children develop patterns of EEG activity in different areas as they mature, as well as EEG changes with brain damage, which he discusses in this volume.

2.5. American Academy of Neurology and Quantitative EEG

In 1997 the American Academy of Neurology officially acknowledged and supported the widespread use of "Digital EEG" in support of visual examination of EEG traces by a neurologist. In the same AAN position paper, qEEG was arbitrarily restricted or assigned the less worthy category "Experimental," as distinct from "Clinically Acceptable." This is important because the outdated, flawed and politically motivated 1997 ANN position opposing qEEG still holds sway in 2005 and it still influences insurance companies and it still restricts the availability of twenty-first century technology to people with serious clinical problems, including brain injury in athletes.

One is struck by the fact that the less worthy categories according to the AAN 1997 paper include many serious neurological and psychological conditions, such as traumatic brain injury, learning disabilities, language disorders, schizophrenia, depression, addition disorders, obsessive compulsive disorders, autism, bipolar disorders, etc. (Nuwer, 1997). One is also struck by the fact that AAN has not revised its 1997 position to more accurately represent the scientific literature and the scholarly rebuttal publications (Hughes and John, 1999; Hoffman et al, 1999 and Thatcher et al, 1999). Another remarkable fact is that the 1997 AAN assignment of qEEG to the "unworthy" category occurred without a proper review of the scientific literature and without any citations that rebutted the last twenty years of quantitative EEG studies. It is also remarkable that the AAN position paper supported visual examination of the EEG tracings as the "Gold Standard" for acceptance in courts and for third party reimbursement when it is well known that subjective visual examination of EEG traces is unreliable and inferior to quantitative analyses (Cooper et al, 1974; Woody, 1966; 1968; Majkowski et al, 1971; Volavka et al, 1971; Niedermeyer & Lopez Da Silva, 1995). The subjectivity and the lack of inter-rater and intra-rater reliability in the visual analysis of EEG tracings is explained in the primary textbook that neurologists study before taking an EEG examination:

> "There is simply no firm rule concerning the manner in which the reader's eyes and brain have to operate in this process. Every experienced electroencephalographer has his or her personal approach to EEG interpretation. This is also true for the manner in which the EEG report is written. Although standardization is an important goal in many areas of EEG technology, experienced electroencephalographers should not abandon a certain individualistic spirit...." (Niedermeyer & Lopes Da Silva, 1995, pp., 185-186)

As mentioned previously, in response to the AAN 1997 position paper, Hughes and John (1999) wrote a rebuttal that included 248 publications and systematically categorized and analyzed the consistency and high sensitivity of quantitative EEG studies in all of the areas that the AAN labeled as "experimental," and they also showed that the sensitivity and specificity of the AAN's alleged "clinically valid" categories was often lower than those of the category that the AAN labeled as "experimental." The Hughes and John (1999) rebuttal was the first paper to show that the AAN 1997 position paper was a sham, and Hughes and John's rebuttal was followed by two additional rebuttals that cited the scientific literature and pointed out misrepresentations and omissions in the 1997 AAN position paper (Hoffman et al, 1999; Thatcher et al, 1999). Nevertheless, the 1997 AAN position paper still holds sway in the minds of many neurologists and insurance companies in the year 2005, to the disadvantage of millions of people, including athletes who may have suffered a brain injury or those who had the misfortune of having a traumatic brain injury of any type.

Below is a partial list of organizations, in contrast to the AAN, that do support or certify by examination a Ph.D. and/or M.D. properly trained and experienced in EEG and qEEG, including the use of qEEG for the evaluation of mild to severe traumatic brain injury. The list below helps demonstrate that the AAN is not the relevant community of users of qEEG.

- American Medical EEG Society
- American Board of EEG and Clinical Neurophysiology
- American Psychological Association
- EEG and Clinical Neuroscience Society
- International Society for NeuroImaging in Psychiatry
- International Society for Brain Electrical Activity

- American Board of Certification in Quantitative Electroencephalography
- Biofeedback Certification Institute of America
- Association for Applied Psychophysiology and Biofeedback
- International Society for Neuronal Regulation
- Society for Applied Neuroscience

2.6. Simultaneous EEG and Quantitative EEG

Figure 2 below illustrates some of the features in a typical modern quantitative EEG analysis which can be activated rapidly by a few mouse clicks on a small home computer using free educational software or by using inexpensive FDA registered commercial qEEG software. The EEG traces are viewed and examined at the same time that quantitative analyses are displayed, so as to facilitate and extend analytical power.

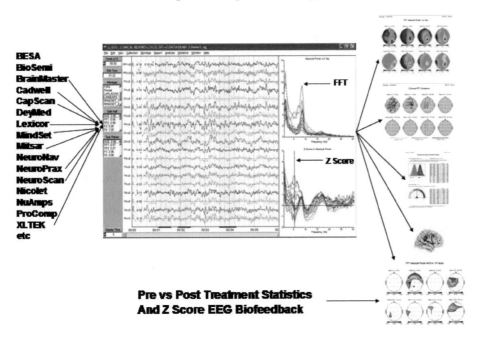

Figure 2. Example of qEEG analyses in which calibrated EEG digital data is imported, test re-test and split half reliabilities are computed, spectral analyses are performed (FFT) and compared to a normative database (e.g., Z Scores) and discriminant analyses and color topographic maps are produced and 3-dimensional source localization is measured and objective pre-treatment vs. post-treatment or pre-mediation vs. post-medication statistics are calculated within a few minutes using the same computer program.

Since 1929 when the human EEG was first measured (Berger, 1929), modern science has learned an enormous amount about the current sources of the EEG and the manner in which ensembles of synaptic generators are synchronously organized. It is known that short distance local generators are connected by white matter axons to other local generators that can be many centimeters distant. The interplay and coordination of short distance local generators with the longer distant white matter connections has been mathematically modeled and shown to be essential for our understanding of the genesis of the EEG (Nunez, 1981; 1995; Thatcher and John, 1977; Thatcher et al., 1986).

The first qEEG study was conducted by Hans Berger (1929) when he used the Fourier transform to spectrally analyze the EEG, because Dr. Berger recognized the importance of quantification and objectivity in the evaluation of the electroencephalogram (EEG). The relevance of quantitative EEG (qEEG) to the diagnosis and prognosis of various forms of brain injury stems directly from the quantitative EEG's ability to measure the consequences of brain damage both in terms of structural and functional deficiencies.

2.7. Time Domain Analysis

2.7.1 EEG Slow Potentials

If you were told that once you heard a tone a picture would follow a few seconds later, you would notice a slow negative potential being generated once the tone sounded. This slow negative potential, which is generally measured at the vertex, is the contingent negative variation, or CNV. The CNV is generated in the laboratory by presenting a first or warning stimulus, which signals that a second stimulus will follow in a specific time period. In most studies, the second stimulus signals cognitive or task processing. Walter et al., (1964) described the CNV as an expectancy measure, since the first stimulus suggests the second will follow.

Another form of event related potentials are very slow potentials which precede and accompany movement or other activities. If a person is asked to press a button as he or she wishes, it can be seen that as early as a second before movement begins, a recognizable EEG waveform starts to develop. A recording made with an electrode placed over the central areas of the cortex displays increasingly negative until, in the few milliseconds before a movement occurs, there is often a slight positive dip in the wave followed by a steep negative slope, which is terminated simultaneously with the beginning of the movement. The beginning of the movement is accompanied by a large positive deflection and a recovery to the original baseline. This complex of waveforms is not uniformly distributed. Technically, this slow increase in surface negativity is referred to as the *Bereitschaftspotential* (BP), or the *Readiness potential (RP)*. See Figure 3 below as an illustration.

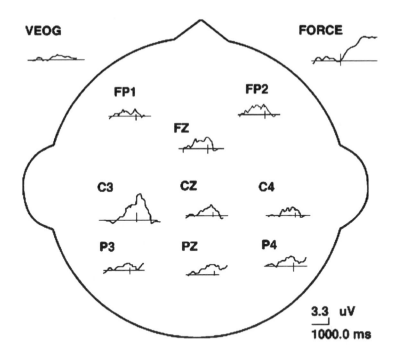

Figure 3. *Bereitschaftspotential* (BP) extracted from raw EEG prior to initiation of isometric force production task.

The readiness potential is maximal at the vertex and initially equal in amplitude over both hemispheres of the brain. One research paradigm is to signal to the subject which hand to use to make the movement. Prior to the movement, this potential begins to lateralize and becomes maximal over the motor cortex contralateral to the body part moved. Early speculation (e.g., Kutas and Donchin, 1980) suggested that this beginning of lateralization reflects the point in time at which the response side is determined (i.e., to move the left or right hand). Since the information contained within the RP includes non-motor processes as well as motor processes, researchers have suggested that by subtracting the response from one hemisphere from that of the opposite hemisphere, it would be possible to obtain a more pure measure of motoric preparation for a response. This measure has been referred to as the lateralized readiness potential (LRP) and has become an important tool in the study of the neural basis of human cognitive-motor processing. For example, the LRP has been shown to be related to preparations for differential rates of force development and that speeded tasks versus accuracy tasks show the largest LRPs (Ray et al., 2000).

To summarize, the development of this measure was based upon the assumption that the asymmetry of the RP could be used as an index for the preparation of specific motor acts. To eliminate any RP asymmetries that may contain activity lateralized with respect to nonmotoric processes, the LRP was calculated as the difference between recording sites contralateral and ipsilateral to the responding hand, averaged over left- and right-hand responses (see de Jong, Wierda, Mulder & Mulder, 1988; Gratton, Coles, Sirevaag, Eriksen & Donchin, 1985; 1988 for alternative ways to calculate the LRP). The LRP's special significance in cognitive and sensorimotor research stems from the fact that this component offers a continuous analog measure of the differential engagement of the left versus right hand associated with cued or uncued voluntary reactions (see Hackley & Miller, 1995 for a review of this work).

The growing popularity of the LRP is due to the fact that its neuroanatomical and functional correlates are better understood than those of most other endogenous event-related potentials. Surface and depth recording indicate that the LRP is mainly generated by the primary motor cortex. Moreover, the foreperiod LRP was found to be twice as large preceding complex movements (subjects were requested to press a sequence of three keys, using the index, ring and middle fingers) than preceding simple ones (only index finger keystroke was required). Also, it has been reported that lateralization tends to be larger preceding a short sequence (one press with the index finger) than preceding a longer sequence (three presses with the same finger). These and other studies support the hypothesis that lateralized preparatory activity in the motor cortex varies with specific properties of the planned movement.

It should be noted that the event related potentials, including evoked responses, readiness potentials, and CNVs, are generally much smaller in amplitude than spontaneous EEGs and are therefore often not discernible in the raw or untreated record. In order to examine ERPs, special recording and data treatment procedures are necessary.

3. QEEG CURRENT SOURCE LOCALIZATION

Figure 4 shows the axial, coronal and sagital views of the current sources of the qEEG in a TBI patient. This is just one of many examples in which the qEEG provides an inexpensive and accurate neuroimage of the focal source of abnormal EEG patterns in a patient who was hit by a blunt object in the right parietal region. In this figure, the focal location of the injury is clearly evident and is validated by the CT-scan results in which a right hemisphere epidural hematoma developed following the injury. The method of source localization, which is called Low Resolution Electromagnetic Tomography (LORETA) and was developed by Pascual-Margui et al (1994), is a well established and inexpensive (it is free) neuroimaging method based on qEEG that is also helpful in the evaluation of coup contra-coup patterns.

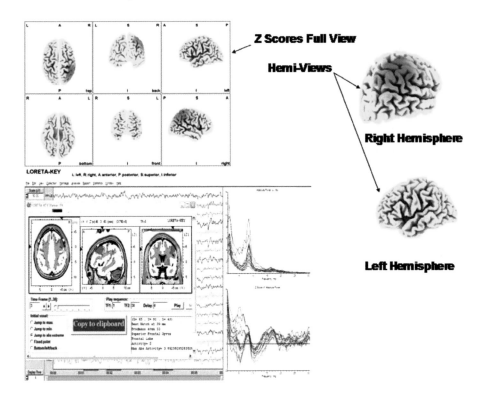

Figure 4. Example of the use of Low Resolution Electromagnetic Tomography (LORETA) to evaluate the effects of TBI involving a patient hit with a bat on the head near the right parietal lobe. The lower left panel is the digital EEG and qEEG that are simultaneously available for the evaluation of the EEG with the Key Institute LORETA control panel superimposed on the EEG. The upper and right panels are examples of the location of Z score deviations from normal which were confined to the right parietal and right central regions and are consistent with the location of impact.

The use of LORETA as a qEEG neuroimaging tool for the evaluation of mild brain injury (TBI), as a first important step toward the design of rehabilitation protocols, has also been published by Korn et al (2005). In this study the generators for abnormal rhythms in the mild TBI patients were closely related to the anatomical locations as measured by SPECT, thus providing additional concurrent validation of qEEG and TBI.

3.1. Dipole Models

Moving a magnet under a piece of paper covered with iron filings results in changing patterns of the filings as the magnet is moved. A similar procedure occurs in relation to electrical activity generated within the brain. Such a procedure is called dipole modeling. Using computers, one determines what type of pattern on the scalp would be produced by different generators in the brain. The pattern generated by the computer could then be compared to actual recorded EEG data. The computer can continue to move the dipole within the imagined brain until the theoretical pattern of EEG matches the actual pattern of EEG activity. Although dipole modeling offers one way of determining localization of activity, there are better methods for determining more exact localization of processes in the cortex, including structural techniques such as MEG, PET, and fMRI.

3.2. Other Brain Imaging Techniques

Magnetoencephalogram (MEG) uses a SQUID (Superconducting Quantum Interference Device) to detect the small magnetic field gradients that are produced when neurons are active, exiting and entering the surface of the head. MEG signals are similar to EEG ones but have one important advantage. This advantage stems from the fact that magnetic fields are not distorted when they pass through the cortex and the skull which makes localization of sources more accurate than with EEG. It should be noted that MEG is only sensitive to tangential activity, which limits it to activity located in the sulci or cortical folds. In order to make a measurement, an individual simply places his or her head within the sensing device, which typically contains a large array of sensors that do not require physical contact with the head. Since measuring magnetic fields using MEG is a complex process requiring liquid helium, which must be super-cooled twenty-four hours a day, this system is expensive both to acquire and to maintain.

Positron emission tomography (PET) systems measure variations in cerebral blood flow that are correlated with brain activity. It is through blood flow that the brain obtains oxygen and glucose from which it gets its energy. By measuring changes in blood flow in different brain areas, it is possible to infer which areas of the brain are more or less active during particular tasks. Blood flow using PET is measured by injecting a tracer (a radioactive isotope) into the blood stream which is recorded by the PET scanner (a gamma ray detector). The general procedure is to make a measurement during a control task, which is subtracted from the reading taken during an experimental task. Although it takes some time to make a PET reading, which reduces its value in terms of temporal resolution, it is able to determine specific areas of the brain that are active during different types of processing. Since PET can measure almost any molecule that can be radioactively labeled, it can be used to answer specific questions about perfusion, metabolism, and neurotransmitter turnover. Some of PET's main disadvantages include expense, the need for a cyclotron to create radioactive agents, the injection of radioactive tracers which limit the number of experimental sessions that can be run for a given individual, and limited temporal resolution.

Like PET, functional magnetic resonance imaging (fMRI) is based on the fact that blood flow increases in active areas of the cortex. However, it uses a different technology than PET in that in fMRI local magnetic fields are measured in relation to an external magnet. Specifically, hemoglobin, which carries oxygen in the bloodstream, has different magnetic properties before and after oxygen is absorbed. Thus, by measuring the ratio of hemoglobin with and without oxygen, the fMRI is able to map changes in cortical blood and infer neuronal activity. Although fMRI has the same temporal disadvantage as PET, it has a number of advantages, including better spatial resolution and the ability to do repeated images on one individual. Details regarding brain imaging technologies can be found elsewhere and are far beyond the scope of this chapter.

4. EEG BIOFEEDBACK/NEUROFEEDBACK

Electroencephalograhic (EEG) biofeedback, often referred to as *neurofeedback*, is an operant conditioning procedure whereby an individual modifies the amplitude, frequency or coherency of the neurophysiological dynamics of his or her own brain Roth et al, 1967; Fox & Rudell, 1968; Rosenfeld et al, 1969; Rosenfeld and Fox, 1971; Rosenfeld, 1990). Initially in the context of sleep research, Berry Sterman and associates conducted a series of studies investigating learned suppression of a previously rewarded cup-press response for food in cats (Roth, Sterman, & Clemente, 1967). During learned suppression of this response, the appearance of a particular EEG rhythm over the sensorimotor cortex emerged above non-rhythmic low voltage background activity. This rhythm was characterized by a frequency of 12–20 Hz, not unlike EEG sleep spindles, with a spectral peak around 12–14 Hz, and has been referred to as the "sensorimotor rhythm" (SMR). These EEG researches decided to further study this distinct rhythm directly, attempting to apply the operant conditioning method to see if cats could be trained to voluntarily produce SMR, by making a food reward contingent on SMR production. In fact, cats learned this feat of EEG self-regulation with apparent ease, and the behavior associated with SMR production was one of corporal immobility, with SMR bursts regularly preceded by a drop in muscle tone (Wyrwicka & Sterman, 1968; Sterman, Wyrwicka, & Roth, 1969).

More recently, fMRI studies in human subjects have shown that the SMR EEG pattern is clearly associated with an increase in metabolic activity in the striatum of the basal ganglia nuclear complex (Birbaumer, 2005). Further, examining fMRI changes in children with ADHD who improved significantly in cognitive tests after SMR neuro-feedback training, Lavesque & Beauregard (2005) have observed a specific and significant increase in metabolic activity in the striatum. Collectively, these findings support the notion that the state changes underlying the SMR are associated with functional changes in the striatum.

It was also suggested that SMR activity reflects synchronized thalamo-cortical oscillations resulting from decreased background muscle and reflex tone and suppressed movement in a context of directed attention. Accordingly, it was concluded that these oscillations, associated with reduced motor tone, movement intention, and the resulting proprioceptive input to ventrobasal thalamus, project strong afferent volleys to cortical target neurons. These volleys result in a cascade of LTP-enhanced motor alterations which are stabilized and consolidated over time. This "unconscious" learning process is further stabilized through arousal reduction and related EEG oscillations following reward in the form of PRS. Such a conclusion supports the proposition that SMR neurofeedback efficacy is based on potentiation of inhibitory mechanisms in sensorimotor pathways (Sterman, 1996). This enhancement appears to be progressive and sustained, affecting function beyond the neurofeedback context. That said, it is important to note that the exact physiological foundations of SMR are not well understood, however, the practical ability of humans and animals to directly modify their scalp recorded EEG through feedback is well established (Fox and Rudell, 1968; Rosenfeld et al, 1969; Hetzler et al, 1977; Sterman, 1996).

It should be mentioned that a different neurofeedback approach, based on the measurement of "slow cortical potential" (SCR) shifts (between positive and negative polarity) has also proved successful in the treatment of epilepsy (e.g., Rockstroh et al, 1993; Kotchoubey et al., 2001). Conceptually, SCP and SMR neurofeedback share the same goal of reducing cortical excitability. As negative slow potentials are reflective of lowered excitation thresholds (through depolarization) in the apical dendrites of cortical pyramidal neurons, while positive slow potentials represent raised excitation thresholds (for a review, see Birbaumer, 1997), SCP training in epileptics is aimed at enabling the patient to voluntarily produce cortical inhibition (i.e., positive SCPs), and thus interrupt seizure onset. Interestingly, Birbaumer (2005) also reports that both the SMR pattern in the EEG and learned increases in positive SCPs were associated with increased metabolic activity in the striatum of the basal ganglia, suggesting a convergent effect of SMR and SCP training (Birbaumer, 2005).

An emerging and promising treatment approach is the use of quantitative EEG technology and EEG biofeedback training for the treatment of mild to moderate TBI. One of the earliest EEG biofeedback studies was by Ayers (1987), who used alpha qEEG training in 250 head injury cases and demonstrated a return to pre-morbid functioning in a significant number of cases. Peniston et al (1993) reported improved symptomology using EEG biofeedback in Vietnam veterans with combat related post-traumatic disorders. Trudeau et al (1998) reported high discriminant accuracy of qEEG for the evaluation of combat veterans with a history of blast injury. More recently Hoffman et al (1995), in a biofeedback study of fourteen TBI patients, reported that approximately 60% of mild TBI patients showed improvement in self reported symptoms and/or in cognitive performance as measured by the

MicroCog assessment test after forty sessions of qEEG biofeedback. Hoffman et al (1995) also found statistically significant normalization of the qEEG in those patients who showed clinical improvement. Subsequent studies by Hoffman et al (1996a; 1996b) confirmed and extended these findings by showing significant improvement within five to ten sessions. A similar finding of qEEG normalization following EEG biofeedback was reported by Tinius and Tinius (2001) and Bounias et al (2001; 2002). Ham and Packard (1996) evaluated EEG biofeedback in forty patients with posttraumatic headache and reported that 53% showed at least moderate improvement in headaches; 80% reported moderate improvement in the ability to relax and cope with pain and 93% found biofeedback helpful to some degree. Thornton and Carmody (2005) reported success in using EEG biofeedback for attention deficit disorders in children with a history of TBI. An excellent review of the qEEG biofeedback literature for the treatment of TBI is in Duff (2004). Further elaborations of neurofeedback therapy may be implemented within the scope of the concept of brain-computer interface (BCI). An example of a possible algorithm for BCI is shown in Figure 5 below.

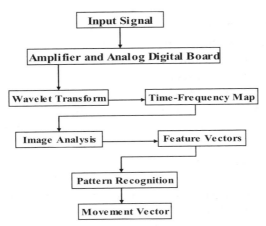

Figure 5. BCI algorithm proposed by Slobounov et al., (1997) for control of diminished arm and knee motion.

Overall, the skilled practice of clinical neurofeedback requires a solid understanding of the neurophysiology underlying EEG oscillation, operant learning principles and mechanisms, as well as an in-depth appreciation of the ins and outs of the various hardware/software equipment options open to the practitioner. It is suggested that the best clinical practice includes the systematic mapping of quantitative multi-electrode EEG measures against a normative database before and after treatment to guide the choice of treatment strategy and document progress towards EEG normalization (Sterman & Egner, 2007).

CONCLUSION

The purpose of this chapter is to overview the electrical activity of the brain as measured by the EEG. EEG activity is described in terms of frequency bands including *alpha, beta, delta, theta* and *gamma*. Although EEG is generally reduced following head trauma, EEG delta has been shown to be particularly sensitive to trauma and pathology. In addition to measuring ongoing EEG activity, researchers have also examined time-locked EEG segments in relation to particular stimuli. Cognitive as opposed to sensory evoked potentials have been shown to be more influenced by cortical trauma. Finally, various EEG measures, including slow wave potentials, event-related desynchronization, and lateral readiness potential, were illustrated by describing studies that have focused on motor related activities.

The EEG biofeedback is a treatment regimen that marries the basic science of qEEG and TBI with a cost effective method of symptom amelioration. The fact that the effects of mild TBI can be detected with two to five electrodes emphasizes the practical and cost efficient aspect of this technology in the evaluation of athletes. For example, blue tooth technology and amplifiers inside a football helmet may potentially almost instantly evaluate the neurological status of an athlete with a head injury and thus can be used to ameliorate the effects of brain injury as well as to understand the long term consequences and rates of recovery from TBI.

On a final note, *neurofeedback* therapy to-date has been greatly aided by advances in computer technology and software. However, the ease in development of new software programs for feedback functionality and displays, together with the entry into the field of a diverse group of professionals and "semi-professionals," has led to an unfortunate lack of consensus on methodology and standards of practice. In turn this has contributed to considerable reluctance by the academic and medical communities to endorse the field of neurofeedback.

Acknowledgements

Some sections of this chapter were elaborated upon by Dr. Robert Thatcher and published in Slobounov & Sebastianelli (Eds.), "Foundation of sport-related brain injuries", Springer, 2006. The author especially appreciates Dr. Thatcher's insights on EEG correlates of concussion (personal communication, 2007).

REFERENCES

Berger, H. (1929). Uber das Elektrenkephalogramm des Menschen. Translated and reprinted in Pierre Gloor, Hans Berger on the electroencephalogram of man. *Electroencephalography and clinical neurophysiology (Supp. 28) 1969, Amsterdam:Elsevier.*

Adrian, E., & Matthews, B. (1934). Berger rhythm: Potential changes from the occipital loves of man. *Brain, 57*, 355-385.

Li. C., & Jasper, H. (1953). Microelectrode studies of the electrical activity of the cerebral cortex in the cat. *Journal of Physiology, 121*, 117-140.

Lutzenberger, W., Elbert, T., & Rockstroh, B. (1987). A brief tutorial on the implications of volume conduction for the interpretation of the EEG. *Journal of Psychophysiology, 1*, 81-89.

Lopes da Silva, F. (1991). Neural mechanisms underlying brain waves: From neural membranes to networks. *Electroencephalography and Clinical Neurophysiology, 79*, 81-93.

Thatcher, R.W., & John, E.R. (1997). *Functional Neuroscience, Vol. I*, New Jersey: Erlbaum Associates.

Jasper, H.H. (1958). The ten-twenty electrode system of the International Federation. *EEG Clinical Neurophysiology, 10*, 371-375.

Shaw, J. (2003). *The brain's alpha rhythms and the mind.* Amsterdam: Elsevier.

Klimesch, W. (1999). EEG alpha and theta oscillations reflect cognitive and memory performance: A review and analysis. *Brain Research Reviews, 29*, 169-195.

Tallon-Baudry, C., Bertrand, O., Delpuech, C., & Pernier, J. (1997). Oscillatory gamma-band (30-70 Hz) activity induced by a visual search task in humans. *Journal of Neuroscience, 17*, 722-734.

Walter, W.G. (1953). *The living brain.* New York: W.W. Norton.

Luu, P. & Tucker, D. (2003). Self-regulation and the executive functions: Electrophsiological clues. In A. Zani & A. Proverbio (eds.), *The Cognitive Electrophysiology of Mind and Brain.* New York: Academic Press.

Schacter, D.L. (1977). EEG theta waves and psychological phenomena: A review and analysis. *Biological Psychology, 5*, 47-82.

Vogel, W., Broverman, D.M., & Klaiber, E.L. (1968). EEG and mental abilities. *Electroencephalography and Clinical Neurophysiology, 24*, 166-175.

Nuwer, M.R. (1997). Assessment of digital EEG, quantitative EEG and EEG brain mapping report of the American Academy of Neurology and the American Clinical Neurophysiology Society. *Neurology, 49*, 277-292.

Hughes, J.R., & John, E.R. (1999). Conventional and quantitative electroencephalography in psychiatry. *Neuropsychiatry, 11(2)*, 190-208.

Hoffman, D.A., Lubar, J.F., Thatcher, R.W., Sterman, B.M., Rosenfeld, P.J., Striefel, S., Trudeau, D., and Stockdale, S. (1999). Limitation of the American Academy of Neurology and American Clinical Neurophysiology Society Paper on QEEG. *Journal of Neuropsychiatry. and Clinical Neurosciences, 11(3)*, 401-407.

Thatcher, R.W., Moore, N., John, E.R., Duffy, F., Hughes, J.R. and Krieger, M. (1999). QEEG and traumatic brain injury: Rebuttal of the American Academy of Neurology 1997 Report by the EEG and Clinical Neuroscience Society. *Clinical Electroencephalograph, 30(3)*, 94-98.

Cooper, R., Osselton, J.W. and Shaw, J.G. (1974). *EEG Technology.* Butterworth & Co, London.

Woody, R.H. (1966). Intra-judge Reliability in Clinical EEG. *Journal of Clinical Psychology, 22*, 150-161.

Woody, R.H. (1968). Inter-judge Reliability in Clinical EEG. *Journal of Clinical Psychology, 24*, 151-161.

Volavaka, J., Matousek, M., Roubicek, J., Feldstein, S., Brezinova, N., Prior, P. I., Scott, D.F. and Synek, V. (1971). The reliability of visual EEG assessment. *Electroencephalography & Clinical Neurophysiology, 31*, 294-302.

Majkowski, J., Horyd, W., Kicinska, M., Narebski, J., Goscinski, I., and Darwaj, B. (1971). Reliability of electroencephalography. *Polish Medical Journal, 10*, 1223-1230.

Niedermeyer, E. & Ds Silva, F.L.(1995). *Electroencephalography: Basic principles, clinical applications and related fields*. Baltimore: Williams & Wilkins.

Berger, H. (1929). Uber das Electrenkephalogramm des Menschen. *Archiv. Fur. Psychiatrie und Neverkrankheiten*, 87, 527-570.

Nunez, P. *Electrical Fields of the Brain*, Oxford University Press, Cambridge, 1981.

Nunez, P. *Neocortical dynamics and human EEG rhythms*, Oxford University Press, New York, 1995.

Thatcher, R.W., Krause, P., & Hrybyk, M. (1986). Corticocortical associations and EEG coherence: a two compartmental model. *Electroencephalography and Clinical Neurophysiology, 64*, 123-143.

Walter, W., Cooper, V., Aldridge, W. C., McCallum, W., & Winter, A. (1964). Contingent negative variation: an electrical sign of sensorimotor association and expectancy in the human brain. *Nature, 203*, 380-384.

Kutas, M. & Donchin, E. (1980). Preparation to respond as manifested by movement related brain potentials. *Brain Research, 202*, 95-115.

Ray, W., Slobounov, S., Mordkoff, J., Johnston, J., & Simon, R. (2000). Rate of force development and the lateralized readiness potential. *Psychophysiology, 37*, 757-765.

de Jong, R., Weirda, M., Mulder, G., & Mulder, I. (1988). Use of partial stumulus information in response processing. *Journal of Experimental psychology: Human perception and performance, 14*, 682-692.

Gratton, G., Coles, M., Sirevaag, E., Eriksen, C. & Donchen, E. (1988). Pre- and poststimulus activation of response channels: A psychophysiological analysis. *Journal of Experimental Psychology: Human perception and performance, 14*, 331-344.

Pascual-Marqui, R.D., Michel, C.M. & Lehmann, D. (1994). Low resolution electromagnetic tomography: A new method for localizing electrical activity in the brain. *International Journal of Psychophysiology, 18*, 49-65.

Korn, A., Golan, H., Melamed, I., Pascual-Marqui, R., Friedman, A. (2005). Focal cortical dysfunction and blood-brain barrier disruption in patients with Postconcussion syndrome. *Journal of Clinical Neurophysiology, 22(1)*, 1-9.

Roth, S.R., Sterman, M.B., Clemente, C.C. (1967). Comparison of EEG correlates of reinforcement, internal inhibition, and sleep. *Electroencephalography and Clinical Neurophysiology, 23*, 509-520.

Fox, S.S. Rudell, A.P. (1968). Operant controlled neural event: formal and systematic approach to electrical codifing of behavior in brain. *Science, 162*, 1299-1302.

Rosenfeld, J.P., Rudell, A.P. Fox, S.S. (1969). Operant control of neural events in humans. *Science, 165*, 821-823.

Rosenfeld, J.P. & Fox, S.S. (19710. Operant control of a brain potential evoked by a behavior. *Physiology and Behavior, 7*, 489-494.

Rosenfeld, J.P. (1990). Applied psychophysiology and bioffedback of event-related potentials (Brain Waves): Historical perspective, review, future directions. *Biofeedback and Self-Regulation, 15(2)*, 99-119.

Wyrwicka, W., Sterman, M.B. (1968). Instrumental conditioning of sensorimotor cortex EEG spindles in the waking cat. *Physiological Behavior, 3*, 703-707.

Sterman, M.B., Wyrwicka, W., Roth, S.R. (1969). Electrophysiological correlates and neural substrates of alimentary behaviour in the cat. *Annals of N.Y. Academy of Science, 157*, 723-739.

Birbaumer, N. (2005). Breaking the silence: Brain-computer interfaces in paralysis. *Proceedings of the Annual Conference, International Society For Neuronal Regulation, 13*, 2.

Levesque J, Beauregard M. Effect of neurofeedback training on the neural substrates of selective attention in children with attention-deficit/hyperactivity disorder: A functional magnetic resonance imaging study. *Submitted* (2005).

Sterman, M.B. (1996). Physiological origins & functional correlates of EEG rhythmic activities - implications for self-regulation. *Biofeedback and Self Regulation, 21*, 3-33.

Rockstroh, B., Elbert, T., Birbaumer, N., et al. (1993). Cortical self-regulation in patients with epilepsies. *Epilepsy Research, 14*, 63-72.

Kotchoubey, B., Strehl, U., Uhlmann, C., et al. (2001). Modification of slow cortical potentials in patients with refractory epilepsy: a controlled outcome study. *Epilepsia, 42*, 406-416.

Birbaumer., N. (1997). Slow cortical potentials: their origin, meaning, and clinical use. In: *Brain and behaviour – past, present and future*. Boxtel GJM, von Böcker KBE (Eds.), pp.25-39. University Press, Tilborg.

Ayers, M.E. (1987). Electroencephalographic neurofeedback and closed head injury of 250 individuals. In: *National Head Injury Syllabus*. Head Injury Foundation, Washington, DC, pp. 380-392.

Peniston, E.G., Marrianan, D.A., Deming, W.A. (1993). EEG alpha-theta brainwave synchronization in Vietnam theater veterans with combat-related post-traumatic stress disorder and alcohol abuse. *Advances in Medical Psychotherapy, 6*, 37-50.

Trudeau, D.L., Anderson, J., Hansen, L.M., Shagalov, D.N., Schmoller, J., Nugent, S. and Barton, S. (1998). Findings of mild traumatic brain injury in combat veterans with PTSD and a history of blast concussion. *Journal of Neuropsychiatry and Clinical Neuroscience, 10(3)*, 308-313.

Hoffman, D.A., Stockdale, S., Hicks, L., et al. (1995). Diagnosis and treatment of head injury. *Journal of Neurotherapy, 1(1)*, 14-21.

Hoffman, D.A., Stockdale, S., Van Egren, L., et al. (1996a). Symptom changes in the treatment of mild traumatic brain injury using EEG neurofeedback. *Clinical Electroencephalography (Abstract). 27(3)*, 164.

Hoffman, D.A., Stockdale, S., Van Egren, L., et al. (1996b). EEG neurofeedback in the treatment of mild traumatic brain injury. *Clinical Electroencephalography (Abstract), 27(2)*, 6.

Tinius, T. P., & Tinius, K. A. (2001). Changes after EEG biofeedback and cognitive retraining in adults with mild traumatic brain injury and attention deficit disorder. *Journal of Neurotherapy, 4(2)*, 27-44.

Bounias, M., Laibow, R. E., Bonaly, A., & Stubblebine, A. N. (2001). EEG-neurobiofeedback treatment of patients with brain injury: Part 1: Typological classification of clinical syndromes. *Journal of Neurotherapy, 5(4)*, 23-44.

Bounias, M., Laibow, R. E., Stubbelbine, A. N., Sandground, H., & Bonaly, A. (2002). EEG-neurobiofeedback treatment of patients with brain injury Part 4: Duration of treatments as a function of both the initial load of clinical symptoms and the rate of rehabilitation. *Journal of Neurotherapy, 6(1)*, 23-38.

Thornton, K. and Carmody, D.P. (2005). Electroencephalogram biofeedback for reading disability and traumatic brain injury. *Child Adolescence Psychiatry, 14(1)*, 137-62.

Ham, L.P., Packard, R.C. (1996). A retrospective, follow-up study of biofeedback-assisted relaxation therapy in patients with posttraumatic headache. *Biofeedback and Self Regulation, 21(2)*, 93-104.

Duff, J. (2004). The usefulness of quantitative EEG (QEEG) and neurotherapy in the assessment and treatment of post-concussion syndrome. *Clinical EEG in Neuroscience, 35(4)*, 198-209.

Sterman, B., & Egner, T. (2007). Foundation and practice of clinical neurofeedback. *Unpublished paper.*

CHAPTER 23

VIRTUAL REALITY IN INJURY REHABILITATION

1. INTRODUCTION

Virtual reality (VR) in a general sense can be described as a *computer technology* which allows us to create a detailed three dimensional representation of particular real life or imaginary situations, which can be examined and manipulated and within which one can move around (Rose, 1996). Its most vital characteristic is "presence," i.e., the feeling of being immersed in the computer generated world. This created world should be interactive in a sense that an individual may be able to navigate through and manipulate its components (i.e., objects, visual scenes, etc). Within the virtual world created by the computer every response that the user makes has a consequence to which she/he must adapt in terms of both mental processing and overt behavior. In addition, the computer generated *virtual environment* (VE) enables us to temporarily isolate a person from his/her normal sensory and motor experiences and/or environment and substitute for it an artificial environment built by the VR programmers.

The visual, tactile and auditory aspects of the computer generated virtual environment can be delivered to the users separately or collectively via visual display 3D units, 3D stereo sound systems, pressure and heat emitting devices (e.g., data gloves, body suit). These realistic looking and/or altered sensory experiences in turn are dependent upon the user's movement within the VR environment, movement which is captured by various interface devices such as a joy stick, a wireless computer mouse and other motion tracking sensors. For example, if the user in the virtual building looks to the left or right, the host VR computer will detect the head motion (i.e., direction of gaze) and alter accordingly the visual images relayed to the user via the visual display unit. Similarly, if the user "touches" the objects in the VR world, the host computer will detect this action and deliver the sensation of touch (i.e. tactile pressure) via a human-computer interface (i.e. data glove). Over the past year, the potential uses of VR have extended far beyond computer games. Recently, VR technology has been applied to the training of surgeons, pilots, drivers, fire fighters, war fighters, athletes. Virtual environments avoid the danger, expense, and problems with control and monitoring users' behavioral responses. Similarly, VR technologies are widely used by engineers, architects, and designers to create and visualize desired future structures, buildings, landscapes, etc. Finally, VR technologies are widely used now in the medical rehabilitation of diminished motor and cognitive functions.

There are different VR hardware configurations currently available for clinical practices. The important principle for VR application is that the hardware solution should be suitable for injured patients, cost-effective and able to accommodate the demands for the assessment and rehabilitation of diminished functions. It is also of major importance that the latency of updates in the VR environment remains very low, such that the users do not perceive a delay in update to the virtual environment when moving their head. The lightweight Head Mounted Display (HMD) model V8 from Virtual Research is most popular hardware used in medical rehabilitation. This HMD uses an active matrix LCD with true (630x3x480 pixel resolution and provides bright, vibrant color and CRT quality images with a field of view of sixty degrees diagonally. It should be noted, however, there is a possibility for motion sickness in some patients (i.e., stroke, TBI, etc.), which may be a serious limiting factor for VR application for clinical purposes. Specifically, VR users wearing the HMD may experience *"cybersickness,"* which is a form of motion sickness with reported symptoms including nausea, vomiting, eyestrain, disorientation, ataxia, and vertigo. *Cybersickness* is believed to be a consequence of incongruity between different sensory modalities. This is thought to happen when there is a conflict between the perception of visual, proprioceptive, vestibular and auditory stimuli. The *cybersickness* may happen when sensory information from the VR environment is mismatched with the user's feeling and/or expectations based upon his or her "real world" sensory experiences. Therefore, alternative commercially available VR systems suitable for these patients should be explored (e.g., CAVE, S-Medallion and one-wall VR system).

Overall, VR has recently emerged as a promising method in various domains of therapy, offering the potential to achieve significant successes in assessment, treatment and improved outcome, thereby increasing its efficacy. Continuing advances in VR technologies along with cost reductions have stimulated both the research and development of VR systems aimed at the psychological, physical/behavioral and emotional rehabilitation of injured individuals. A growing number of diverse occupations that currently use the immersive and interactive properties of VR include drivers, parachutists, fire-fighters, soldiers, divers, surgeons etc. There is an obvious potential use of VR for the rehabilitation of individuals suffering from traumatic and/or neurological injuries. Virtual reality enables the clinician to place a patient in a variety of precisely controlled computer-simulated environments that are entirely safe, and which are intentionally programmed to avoid the patient being at a disadvantage as a consequence of sensory or motor impairments. By the creation of this "level playing field" it is possible to observe how a patient behaves – how active, how interested, how distractible, he or she is during the rehabilitation protocols. VR may allow the patient to be exposed to a sensory world of a complexity that it is impossible to deliver in any other way (Rose, 1996). That said, it is important to note that those involved in the development of medical VR to be of use to therapists and medical professionals, should familiarize themselves with the clinicians' terminology and some details of the problems (i.e., restricted mobility, cognitive deficits, etc.) they seek to challenge.

2. VIRTUAL REALITY IN MEDICAL REHABILITATION

2.1. Basic Concepts of Rehabilitation

There still a lack of conceptually solid, well-controlled and ecologically valid rehabilitation programs (including for the initial assessment of disability) for injured patients. Virtual Reality (VR) has the potential to overcome these challenges. As mentioned in the previous section of this book (see Part 5, Chapter 21), traditionally, the consequences of tissue damage including damage to the brain are usually described in terms of three essential concepts: (a) impairment; (b) disability, and (c) handicap.

"*Impairment* refers to any loss or abnormality of psychological, physiological or anatomical structure of function." (World Heath Organization, 1980, p.27)

"*Disability* refers to any restriction or lack (resulting from an impairment) of ability to perform an activity in the manner or within the range considered normal for a human being." World Heath Organization, 1980, p.29)

"*Handicap* refers to a disadvantage for a given individual, resulting from impairment or a disability that limits or prevents the fulfillment of a role that is normal (depending on age, sex and social and cultural factors) for the individual." (World Heath Organization, 1980, p.29)

From the perspective of *impairment*, referring primarily to compromised anatomical structures and physiology, there is little direct evidence that VR technologies can be used to improve the situation. However, VR may potentially provide a powerful tool to increase the capability of patients to interact with an enriched virtual environment that can be created contingent upon whatever motor abilities the patient has. Specifically, clinicians agree that this environmental interaction is vital to the injury rehabilitation process. Stimulating the activation of impaired functions by manipulations of sensory and/or motor stimuli in a VR environment may facilitate recuperation. This has been documented in numerous recent studies within the conceptual framework of "*neuroplastisity*" (Hallett, 2001).

The more obvious efficacy is the VR application for the rehabilitation of *disability*, which is commonly observed as a failure to perform a normal activity in a normal manner. It has been shown that there is great potential to re-train an individual with disabilities in a VR environment, since training in real life situations is most often impractical due to a patient's sensory, motor or cognitive disabilities. For example, in the treatment of patients with sensory loss, certain aspects or categories of sensory stimuli can be accentuated at the program level to offset sensory deprivation. Also, the movement within the training situation can be precisely tuned and adjusted to whatever motor capabilities the patient has. In addition, the provision of well-controlled feedback within the framework of "augmented reality" (e.g., patient-friendly visual and auditory information in the form of a bar graph or sound effects) superimposed upon virtual images and scenes, may significantly speed up the re-training process. For example, linguistic labels may be superimposed upon computer-generated realistic looking objects to improve associative memory in patients suffering from *agnosia*. Certainly, to warrant the feasibility of VR rehabilitation of disabilities, it is important to demonstrate that what is learned in a VR environment can be safely transferred to real life situations. For example, the improved spatial memory of patients in a VR environment should be demonstrated in their ability to orient themselves in real life environments (e.g., to find their way to the store, bank, etc., just a few blocks from home).

As defined in the previous text, the term *handicap* refers to the disadvantage for a given individual resulting from impairment or disability. There are numerous examples of technological devices developed within the framework of a "human-computer interface" to overcome severe (including permanent) disabilities. For example, there are devices for utilizing the eye movements of paralyzed patients to operate computer keyboards, to control a wheelchair and to communicate with the surrounding environment. It is possible that ultimately one can imagine a situation in which disabled individuals might be able to carry out tasks in their real world environment by operating within a linked virtual version of it (Rose, 1996). Thus, the combined resources of robotics and VR technologies may empower the disabled individual with permanent physical and functional deficits to an extent undreamed of a few years ago.

2.2. VR Rehabilitation Principles

The principal role of the therapist in the rehabilitation is to design a combination of VR-based tasks contingent with a patient's cognitive and motor functionality status that will facilitate the restoration of diminished functions. The rehabilitation therapist via an *Executive Controller* (i.e., a combination of available interface devices and specifically designed computer modules/VR software) should have an opportunity to vary task demands, progressively increase task difficulty, switch between tasks to constantly challenge the patient and maintain his or her optimal level of effort. It is important to stress that patients should always be challenged by the VR environment and task complexity to avoid adaptation leading to the saturation of rehabilitation progress.

Various degrees of "enriched environments" have been previously designed and created via VR including: *Navigation Maze, Virtual House, Interactive Kitchen, Virtual Store, Virtual Room, Virtual Person*, etc. These were specifically developed for the assessment of diminished/disabled functions in patients with various sensory-motor and cognitive disabilities (e.g., stroke, TBI). Additional features may one or more of task-related augmented feedback, sequential order of task presentation and modulation of task difficulty according to the principles of learning and re-acquisition of cognitive-motor skills. A set of VR-based tasks and treatment procedures focused on the rehabilitation of specific diminished functions (i.e., memory, attention, spatial awareness, balance, etc.) should be designed and implemented based on the following principles, which are similar to the general principles of integrated rehabilitation:

(a) The patient should be *actively engaged* in repetitive problem solving situations rather than passively repeating the required tasks. It is well-documented that active participation, along with repetitive problem solving situations, is the essential component of memory for spatial layout enhancement in cognitively disabled individuals (Antee et al., 1996). The advantages of VR technologies should be considered. For example, creating a "computer game environment" where patients would be actively involved in problem-solving situations created by 3D graphics;

(b) The patient should be *constantly challenged* both physically and mentally to enhance diminished functions and overall brain plasticity (Merzenich, NIH, 2004, personal communication). It should be noted that patients should be "optimally challenged" according to their capacities in order to maintain high levels of motivation. This principle is in line with Alexander Luria's famous assumption that practice and training are particular types of *repetition without repetition* (personal interaction with Dr. Luria, 1975). The advantages of VR technologies should be considered. For example, manipulating the degree of difficulty of task performance along with on-line control of results outcome. Thus, the level of challenge and the patient's current capabilities (i.e., current range of motion, strength, coordination, etc.) can be properly regulated at the VR program level;

(c) The patient should be involved in *functionally relevant* tasks. It is well-known, for example, that task-specific and/or goal-oriented practice promotes enhanced functional recovery by modulating cortical reorganization (Hallett, 2001) and is an important factor for long-term rehabilitation. Patients always vary in terms of symptoms and their resolution, thus, individuated approaches should be implemented depending upon what specific functions (i.e., sensory-motor, cognitive, memory, executive functions, information processing, balance, etc.) are diminished as a result of injury. For example, if most of a patient's symptoms are within the cognitive domain, task-specific memory/attention/decision-making assessment and rehabilitation modules should be the targets of rehabilitation. It is important that at the end of rehabilitation an athlete should acquire the pre-injury level of performing the sport-related task (e.g., catching the ball, etc). Therefore, functional, and/or goal-oriented tasks (i.e., VR visualization, stereo and interactive systems) rather than simple force training or repetitive movement should be implemented within VR rehabilitation protocols;

(d) The patient should practice tasks in a sequence from *simple to more complex.* In addition, modulation of task complexity should be accompanied by augmented and task-relevant feedback to enhance learning, selective attention and optimal motivation. Augmented feedback (i.e., superimposing on-line the quality/quantity of task accomplishment) is an essential component of directing selective attention during the re-acquisition of perceptual-motor skills (Newell, personal communication, 2006). For example, the motor skills needed to be re-learned in the VR rehabilitation of the upper extremity can be classified according to difficulty level and defined within the scope of a hierarchical structure (e.g., reach-to-grasp can be decomposed into transport, orientation, pre-shape and actual grasp components). In fact, assessment of task complexity is not a trivial task but rather an individually dependent phenomenon requiring formal evaluation;

(e) The therapeutic intervention in terms of providing task parameters should be *adaptive* in order to meet the patient's demands and individual characterization. Appropriate adjustments to task complexity and available feedback have profound effects on learning (Newell, personal communication, 2006). Overall, an important feature of VR-based rehabilitation should be the adaptability of task parameters (simplicity/complexity) and schedules based upon both the initial level of disability and progress on recuperation. Overall, adaptive versus maladaptive response to VR rehabilitation protocols should be carefully evaluated to achieve maximal rehabilitation outcomes;

(f) The gain and skills achieved during therapeutic intervention should be *transferable* to "real-life" tasks. Indeed, one of the major challenges facing clinicians in medical rehabilitation is identifying intervention methods that are effective, motivating and that transfer to the ability to operate in the "real" world (Rose et al., 2005). Most of the clinically accepted traditional assessment and rehabilitation tests are "ecologically invalid." For example, a neurological patient with brain trauma can be asymptomatic based upon conventional neuropsychological memory testing. However, the same subject may found it extremely difficult to remember the pathway from his room to the bathroom, although it is only few turns in the hospital corridor;

(g) Finally, the patient should be involved in the treatment protocols stimulating his or her improved ability to compensate for residual deficits and to adopt strategies for the more effective allocation of his or her remaining attentional resources (Cicerone, 2002). In other words, both restoration of and compensation for diminished function should be key features of VR rehabilitation protocols. Overall, the most important key features of a VR-based rehabilitation are (a) gradual progression; (b) adaptability of task parameters; (c) control for initial performance level and progress; (d) patient's positive attitude; (e) motivation and challenge.

2.3. VR for Restoring Diminished Mobility of Selective Body Parts

Virtual reality has emerged as a promising technology in various domains of therapy and medical rehabilitation. Continuing advances in VR technology allowing for the creation of realistic looking custom environments along with patients' capability to actively interact with these environments, have supported the development of user-friendly, cost effective and assessable VR systems. A vast variety of clinical and basic science research is currently addressed using the advances in VR technologies. Among them, the restoration of diminished mobility as a result of injury or trauma (i.e., range of motion, ROM) is a growing target of VR application in rehabilitation medicine. Just a few examples of VR application for restoring diminished function are outlined below (see also Figure 1).

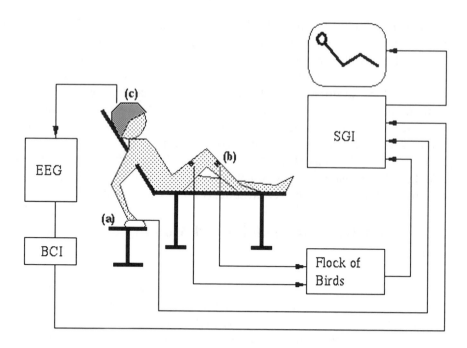

Figure 1. General schematics of VR application for restoring the knee range of motion.

The subject receives on-line feedback about current range of motion during rehabilitation sessions. Thus, the progress with rehabilitation and subjects confidence with its efficacy may be considerable increased, as in fact was documented in the author's clinical research. See also Figure 2 below for illustration. A patient's range of knee motion following injury can be visualized using computer graphics allowing to assess and to control the current range of motion of injured knee; short-term (prescribed) goal as well as the maximal desired range of motion after injury recovery. As can be seen from Figures 1 & 2, the range of motion was accurately assessed by motion tracking system (e.g. Flock of Birds). Online feedback to the patient was achieved by graphic programs.

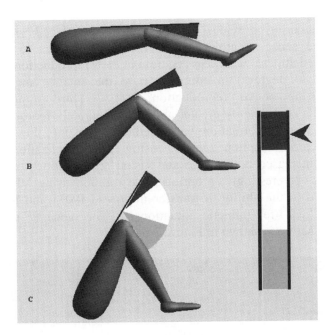

Figure 2. An example of 3D images of knee joint: where (a) current range of motion of injured knee; (b) short-term (prescribed) goal; and (c) maximal range of motion after injury recovery.

There are few studies directly suggesting the feasibility of VR in the rehabilitation of motor control and learning in patients suffering from various traumatic injuries. In our own work, the applicability of VR incorporated with motion tracking technologies for improving visual-kinesthetic integration and joint position sense in athletes after shoulder surgery has been documented (Slobounov et al., 1998; 1999). The ability to accurately assess and enhance a person's joint position sense is an inexact science at best. Accordingly, at this time a new method of evaluating and improving joint position sense using VR graphics incorporated with modern motion tracking technology (e.g., "Flock of Birds' Six Degrees of Freedom System") was examined in this study. Patients suffering from shoulder injuries due to athletic trauma (n=6) and healthy subjects (n=12) were evaluated for their ability to determine shoulder joint position in two different tasks. The first task was an active reproduction of a passively placed angle. The second task was a visual reproduction of a passively placed angle. In addition, a training protocol was added to the shoulder experiment to determine the effectiveness of proprioceptive training in conjunction with 3D visualization techniques. The primary findings from this pioneering VR research in a clinical setting were: First, a significant statistical difference (p<0.05) in the level of shoulder joint position sense seen in injured subjects as compared to healthy. This was not surprising and confirmed numerous previous studies indicating reduced sensitivity and proprioceptive sense in athletes with traumatic injuries to the joint(s). Second, injured athletes were less accurate in their reproduction of the larger angles (e.g., 80 degrees) in comparison to the smaller angles (e.g., 40 degrees) in the active reproduction task. Third, injured athletes demonstrated a significantly greater ability to accurately reproduce range of motion when visual information was added during testing. Finally and most importantly, proprioception training using 3D visualization techniques significantly increased both active and visual reproductions of a passively placed angle. Interestingly, proprioceptive training using VR technology was beneficial to the ability to transfer improved ROM and joint position sense into "real-life" athletic situations. The general view of the experimental set-up is shown in Figure 3 below.

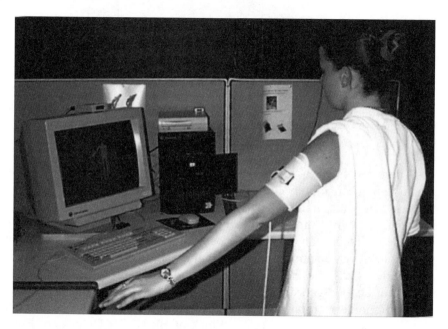

Figure 3. Joint position sense and range of shoulder motion can be improved by integrating visual information with kinesthetic sense.

There are also studies indicating the feasibility of VR for other clinical purposes. For example, a PC-based Virtual Reality System has been developed for the rehabilitation of hand movement in stroke patients. This system consists of two hand-input devices, a *CyberGlove* and force feedback, which collectively allow the patient to interact with the rehabilitation exercises. The properties of the exercises are highly task specific and targeted to recover specific functions, including range of motion, speed, fractionation of grasping tasks and strength. Several weeks of training using this system allowed the patients to significantly improve the desired movement properties. Also, VR incorporated with motion tracking technologies can be used to properly track the arm movement trajectory, as can be seen from the Figure 4 below:

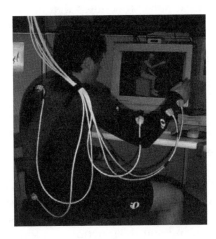

Figure 4. Motion tracking technology (FOB system) incorporated with computer graphics may help patient to visualize arm movement trajectory in real time aimed at restoring the reach abilities in a clinical setting.

Thus, it is concluded that VR technology along with a properly designed rehabilitation protocol may be useful in augmenting rehabilitation of the upper limb in patients suffering from stroke. It is reasonable to suggest that the VR technology may also be beneficial for other patients suffering from traumatic injuries to the limbs, which has resulted in temporary movement dysfunctions.

In fact, a series of valuable research has been conducted in Dr. Rizzo's laboratory at USC that examined the feasibility of VR in skill learning, including spatial rotation, depth perception, 3D field dependency, static and dynamic manual target tracking in 3D space. The numerous published reports of this group have clearly demonstrated training improvement in various subject populations, including young and elderly subjects, ADD patients, traumatic injury patients, etc. Indeed, the VR environment and proper types of 3D interaction scenarios may soon be invaluable for rehabilitation purposes across the lifespan. It is important to note that conventional physical and occupational therapy methods, although focused on muscle strength, balance and ROM, usually tend to be tedious, monotonous and provide limited opportunity for systematic grading, assessment of current performance and improvement. Most importantly, the gain and improvement achieved via conventional rehabilitation protocols may or may not be "transferrable" to real world tasks. For example, the patient may improve his or her range of motion in the shoulder joint by performing ROM tests in the training room. However, this patient may not be able to extend the arm above the head to catch the ball despite "recovered ROM." Indeed, one of the major challenges facing clinical practitioners today is identifying treatment procedures that are cost effective, motivating and transferable to the "real world." VR technologies may meet these challenges and soon will be a dominant approach in rehabilitating patients with restricted mobility.

2.4. Feasibility of VR for Assessment of Balance

The first step in any successful rehabilitation protocol is the accurate and comprehensive assessment of the patient's current status and diminished abilities. As was mentioned in the previous text (see Part 1), human posture and postural stability are a product of an extremely complex dynamical system that relies heavily on the integration of input from visual, proprioceptive, vestibular, etc., multimodal sensory sources. In fact, balance problem is the most prominent dysfunction resulting from traumatic injuries to the low extremities and to the brain (see Part 1 of this text for details). The feasibility of VR for the assessment of balance in neurological patients was clearly demonstrated in Dr. Keshner's laboratory at Temple University (see Figure 5 below).

Figure 5. Neurological patient is tested for balance using VR facility, force platform and motion tracking technologies.

Similarly, VR can be successfully used for the assessment of athletes suffering from balance abnormalities as a result of traumatic brain injury. In a series of recent studies we used the virtual room experimental paradigm to examine visual-kinesthetic sense in student-athletes recovering from concussion. In the following paragraphs some details of this research are described.

The VR-generated 3D stereo visual field consisted of a 6x8 foot one wall screen with rear projection, an Electrohome Marquis 8500 projector with full color stereo workstation field (2034x768 pixels) at 200 Hz, a dual Xenon processor PC with a Nvidia Quadro 4900 XGL graphics card, and StereoGraphics Inc. glasses. The VR system was synchronized with the AMTI force platform and a Flock of Birds motion analysis system. The subject was wearing a vest (approximate weight = 400 grams), with the attached Flock of Birds motion tracking sensors. The Flock of Birds sensors were placed on fixed anatomical hallmarks in order to measure the trunk kinematics. Specifically, the attached sensors approximated the position of the second metatarsophalangeal joints bilaterally on the rotation centers of the shoulders (specifically, the second metacarpophalangeal joins the spinoid process of C7 and L4-L5 vertebrae). An additional Flock of Birds sensor was located on the subject's head to interact with visual field motion. The visual field motion consisted of a realistic looking "moving room." An overview of the experimental set-up is shown in the Figure 6 below. In fact, both children and adults may be tested using various options of VR configurations.

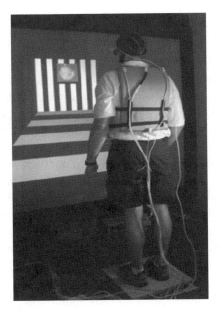

Figure 6. Penn State VR set-up to examine the presence of *"egomotion"* and visual-kinesthetic integration (Dr. Slobonov' Lab at Lash Football Building, PSU).

The subjects were tested pre-season and re-tested several times after they suffered from sport-related concussion. The subjects were instructed to produce whole body motion predominantly at the ankle joints in synchrony with the computer graphics-generated "moving room" visual scenes (30s in duration). There were three testing conditions in the major experimental session that differ from each other in terms of the frequency and rotational aspects of "room" motion. The initial condition (0) is the baseline measure with no variation in the visual field. The remaining two conditions are: 1) whole room forward-backward oscillations within 18 cm displacement at 0.2 Hz; and, 2) whole room lateral "roll" between 10-30 degrees at 0.2 Hz. Our pilot experiments indicate that these "moving room" conditions maximally induce responsive self-motion in normal controls and greatly destabilize posture as a result of concussion.

The center of pressure data (30s trial duration x 150 Hz sample rate) derived from the force platform (i.e., the area of the center of pressure, triggered by a signal from the VR host computer) were computed by CAS software. The center of pressure data derived from the force platform were computed by CAS software while the center of pressure time series was calculated using the following approximations:

$$CPx = (-My+Fx)/Fz \qquad\qquad (1)$$
$$CPy = (-Mx+Fy)/Fz \qquad\qquad (2)$$

Where: Mx and My are the moments about the x and y axes respectively and Fx, Fy, and Fz are the medio-lateral (M-L), anterior-posterior (A-P) and vertical ground reaction forces, respectively.

A specially developed MATLAB (The Mathworks, Natick, Mass., USA) program was used to estimate whole body kinematic data, obtained from the Flock of Birds motion tracking sensors. Specifically, the time series of the torso motion along X and Y axes, and the coherence values between the quantities of moving room and postural responses as reflected in the center of pressure dynamics and whole body kinematics were assessed using a specially developed m-code in MATLAB 6.5. Only the magnitude of coherence was calculated based on values relative to baseline to reduce the inter-subject variability of absolute coherence. It should be noted that coherence is a measure of the linear dependency (or coupling) of two signals at a specific frequency. The auto-spectra for each signal were calculated by using Welsh's averaged periodogram method. Coherence was calculated based on the cross-spectra f_{xy} and auto-spectra f_{xx}, f_{yy} with the spectra estimated from segments of data and the coherence R_{xy} estimated from the combined spectra:

$$R_{xy}(\lambda)|2=|f_{xy}(\lambda)|^2/(f_{xx}(\lambda)f_{yy}(\lambda)) \qquad\qquad (3);$$

The significance of coherence was also calculated. That is the confidence limit for zero coherence at the α %, and L is the number of disjoint segments: sig $(\alpha) = 1-(1-\alpha)^{1/(L-1)}$. In addition, a continuous wavelet transform (CWT) was performed to track the dynamics of coupling between the subject's body motion and visual scenes oscillation over the entire trial duration (30 s). The CWT is able to resolve both time and scale (frequency) events better than the short Fourier transform (STFT). In mathematics and signal processing, the continuous wavelet transform (CWT) of a function f is defined by:

$$\gamma(\tau, s) = \int_{-\infty}^{+\infty} f(t) \frac{1}{\sqrt{|s|}} \overline{\psi \left(\frac{t - \tau}{s} \right)} dt \qquad (4);$$

Where τ represents translation, s represents scale, which is related to frequency, and ψ is the mother wavelet. $\overline{z}$ is the complex conjugate of z. In this study the mother wavelet is a complex morlet wavelet, as it has both good time and frequency accuracy.

There are several findings of interest from this study. *First*, all subjects were clinically asymptomatic at day ten of testing after concussion episodes and were cleared for sport participation on the basis of neurological and neuropsychological assessments as well as clinical symptoms resolution. *Second*, there was overall a strong coherence between the center of pressure (CP) and Flock of Birds (FoB) data (on average 0.85, p<0.001). Therefore, in the following analyses either CP or FoB time series data are used to examine the coherency of postural responses with visual scene motion achieved by VR, as an indication of visual-kinesthetic integration/ disintegration. *Third*, baseline testing revealed that all subjects adjusted their postural movement to the direction and frequency of the virtual room motion. This was evidenced by high coherence values for both A-P (r=0.82, p<0.001) and Roll (r=0.85, p<0.001) VR conditions. *Fourth*, MANOVA revealed the main effect of testing on the CP/FB data when the subject was viewing the moving room in both the A-P (F [3, 55]=13.53, P<.01) and lateral (Roll), (F [3,55]=14.36, P<0.05) directions. The details of partial recovery of coherence can be seen from Figure 7.

A

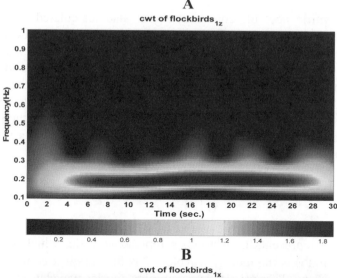

B

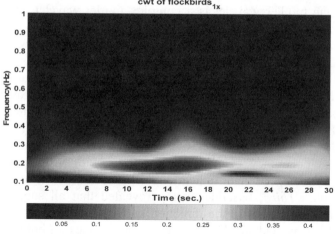

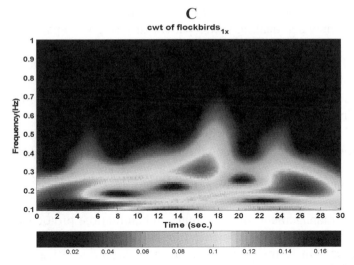

Figure 7. Time-frequency decomposition of the coherence values between motion of the VR room at .2 Hz and postural compensatory adjustments achieved by wavelet transforms. The plots are representing 2D color maps of evolution of coherence over the trial duration (30s) while subjects were viewing VR motion at .2 Hz. Where: (A) – baseline testing; (B & C) samples of impaired visual-kinesthetic coupling at 10 days post-injury. The, abnormal coupling at certain time within the trial duration is clearly revealed by Wavelet analysis.

Clearly, no other behavioral and neuropsychological signs of abnormality were observed in concussed individuals on day ten post-injury, but the signs of sensory-motor disintegration revealed by VR were still present. Indeed, a combination of various assessment methods and tools should be used by clinicians in order to make an accurate decision in terms of return to play and to identify athletes at risk for recurrent concussions. There is an ongoing challenge within our multi-disciplinary research group at Penn State University to predict athletes at risk for future concussions using modern VR technologies.

2.5. VR Application for Pain Management

The primary objectives of treating patients who experience pain behavior include but are not limited to: pain management, minimizing disability, improving quality of life, and most importantly, preventing progression of the disease and illness. Of these objectives, the most important is pain control, because pain has a strong association with disability and quality of life. Pain is often assumed to be a physical phenomenon, and to require just physical methods of control and management. However, as shown in previous chapters of this text (see Part 3), other factors such as personality, environment, perception and awareness of pain overall define the patient's pain experience and behavior, and are critical for proper pain management protocols. Accordingly, clinical practitioners when dealing with a patient suffering from pain should:

- Be aware of intervention strategies that will promote patients' active involvement in pain management programs;
- Understand the effects of conditioned learning on the establishment or maintenance of pain behavior;
- Understand the ways in which pain behavior is manifested;
- Know the principles of cognitive-behavioral pain management;
- Understand how psychological approaches are affected by the therapist-patient relationship (Strong et al., 2002).

Pain management during injury rehabilitation protocols traditionally includes various classes of medication and non-traditional pain management techniques. In terms of medication, a wide variety of anti-inflammatory drugs, both non-steroidal and steroidal, are currently available. In addition, opioid drugs, which are morphine-like, are commonly considered along with anti-depressant, anti-seizure and anti-anxiety medications. Specifically, the most commonly used pain treatments are based on the symptoms-modifying drugs. However, it should be noted that the use of drugs for the control of pain has significant drawbacks, since these drugs may cause strong adverse reactions and often drug dependency. Clearly, there is no single drug that is both completely effective and safe in treating patients suffering from pain. Accordingly, the significant risks for prolonged drug use have required the development of non-pharmacological treatment of patients experiencing pain behavior.

Among non-pharmacological pain treatment protocols are: heat, cold, electrical stimulation, massage therapy, light therapy, splints and orthoses, diet, acupuncture, weight loss, psychological treatment and physical exercise. Specifically, of all the non-pharmacological pain treatment protocols, exercise has been proven as one of the most efficient with the greatest beneficial effects. For example, exercise can reduce joint pain, increase range of motion (ROM), increase strength, reduce stress on the joint, improve overall biomechanics and prevent disability. Over the centuries, the effects of exercise on both the physical and psychological well-being of patients suffering from pain behavior have been documented by scientific and anecdotal evidence. The details of these non-pharmacological treatment protocols could be found elsewhere and are far beyond the scope of this chapter.

More recently, it has been recognized that virtual reality (VR) has the potential to improve the quality of life of patients for whom traditional (i.e., pharmacological) and other conventional therapeutic modalities are insufficient. Various VR applications have been developed with a specific focus on treatment of phobias (i.e., fear of pain), eating disorders, post-traumatic stress disorders, and pain management. It is an important advantage of VR for pain management that this allows various degrees of distraction during clinical treatments. Another advantage of VR application for pain management is that this allows the patients to become actively involved and participatory in a computer-generated virtual world. This creates a treatment environment where the surrounding environment changes in a natural way and responds to the patients' actions. Overall, VR therapy makes the whole treatment protocol more enjoyable and tolerable. This in turn creates a motivational climate for patients' willingness to participate and appreciate the rehabilitation outcome.

The feeling of immersion is one of the important prerequisites of VR systems that is achieved by combining real-time computer graphics, physics-based models, body motion tracking devices, 3D stereo visual displays, 3D sound systems and haptic sensations. By combining these properties, the VR could afford the situations where the patients are exposed to relevant emotional experiences (i.e., fear of burn) as "distractors." This allows the patients to more readily tolerate painful and/or unpleasant medical treatments. Recent VR research by Hoffman et al. (2001) has clearly indicated that this in fact has strong feasibility to control pain within multiple treatment protocols. Several researchers have also demonstrated the applicability of VR for dental pain control (Hoffman et al., 2001), for control of burn pain (Hoffman et al., 2000) and as a distraction intervention for women receiving chemotherapy (Schneider et al., 2004). Also, the beneficial effect of distraction using VR technology during lumbar punctures in adolescents with cancer has been reported by Sander, Wint et al., (2002).

One specific example of the VR system developed to distract patients from pain during wound care is a "virtual kitchen," where patients could open drawers, pick up pots, touch other objects and see 3D realistic looking computer generated objects (Hoffman et al., 2000). Another example is the use of VR games to distract patients from pain during their physical therapy sessions (Kline-Schoder, 2005). This system uses a force feedback steering wheel to apply torque to the patient as he or she navigates a virtual car via a virtual city. The overall principle of VR games is that the patient is encouraged to move his/her arms, as appropriate for the therapy chosen by the therapist. There is a commercially available game where the patient (as a soccer goalie) tries to deflect virtual soccer balls that are headed to the goalie. Specifically, the patient should move his or her arm to intersect the motion of the virtual ball in 3D stereo environment. If the patient's arm reaches the proper location at the right time in 3D space, the ball will be deflected accordingly. If not, a goal will be scored. Interestingly, the path and speed of the virtual balls can be pre-programmed and modified by the therapist depending upon the patient's condition so as to provide sufficient challenge and motivation. Overall, clearly VR game situations may distract patients from experienced pain while performing physical rehabilitation exercise aimed at improving coordination, strength, range of motion, along with the possibility to maintain patients' motivation and enjoyment.

2.6. VR for Cognitive Rehabilitation

The virtual reality (VR) application can provide an *enriched task environment* and a highly motivational context allowing interaction with a computer generated 3D world. In the following text, the focus will be on the feasibility of VR for assessment and rehabilitation of patients suffering from cognitive impairments, predominantly traumatic brain injuries (TBI) patients. A set of goal-oriented VR-based *computer graphics tasks* (assessment modules) have been recently developed, providing ecologically valid assessments of the amount of impairment in injured patients within both the cognitive and motor domains. Recently, numerous research laboratories have been involved in the development of concrete computer graphics *rehabilitation programs* (rehabilitation modules). This work is usually accomplished within the general framework of *neuroplasticity* and the basic principles of skill acquisition/learning, enabling the stimulation of neurological and functional recovery after injuries.

It should be noted that dramatic advances have been made in the application of VR in brain damage rehabilitation (see Rose et al., 2005, for details). Specifically, executive control function, memory impairment, attention deficit and spatial awareness are just a few cognitive impairments observed as result of traumatic injuries (predominantly traumatic brain injuries). It is a current challenge of Head Rehab, LLC. to fully develop a generic VR prototype aimed at the comprehensive assessment, and if necessary, rehabilitation of cognitive dysfunctions as a result of traumatic brain injury. Considerable progress has been made in terms of both the conceptualization and actual implementation of Head Rehab's proprietary VR prototype. Conceptually, current research clearly indicates the necessity to implement a multimodal perspective to fully appreciate the degree of functional deficits in patients suffering from cognitive abnormalities. Empirical evidence suggests that behavioral, sensory-motor, neuropsychological and neural mechanisms are interactively affected by brain injury; this may be evidenced via advanced methodologies. The need for a multimodal assessment of cognitive deficits was justified in the author's research laboratory and by Head Rehab, LLC via the application of different assessment modules addressing: (a) **memory**, including spatial memory, *recall* and *recognition* substrates; and (b) **attention**, including sustained and focus/selective awareness, etc.

2.6.1. Assessment of Spatial Memory: Virtual Hospital Corridor (VHC)

Every day spatial memory problems in patients with traumatic brain injury (**TBI**) became apparent while they are in the hospital. Commonly, brain injured patients experience memory problems; specifically, in assessing their spatial location with respect to some external objects in the surrounding environment. For example, if the patient left the hospital room, he/she would have difficulty in finding his/her way back, not remembering whether to turn *left* or *right,* move *forward* and/or *backward* in the hospital corridor. Even with some clues, these patients also experienced issues assessing the direction of their path as they moved. Accordingly, a virtual hospital corridor (**VHC**) was developed by the Head Rehab, LLC, (2007, proprietary information). It is reminiscent of a real hospital corridor, ultimately consisting of various degrees of pathway complexity (number of turns, complexity of the visual environment and number of objects along the way). The patient is required to navigate through the corridor using a 3D mouse or other interactive devices. An overview of the virtual hospital is shown in Figure 8 below:

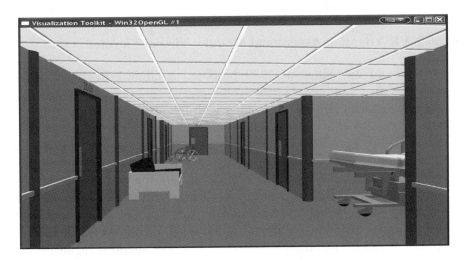

Figure 8. Virtual Hospital overview.

It has been shown that navigation via *virtual hospital* is a trivial task for normal controls, but is it extremely challenging task for patients with spatial memory impairment. It should be noted that this aspect of memory impairment is commonly overlooked when traditional memory assessment tools are used. The experimental set-up to assess the feasibility of VR spatial memory assessment module is shown in Figure 9 below.

Figure 9. VR experimental set-up to assess spatial memory in patients suffering from TBI.

2.6.2. VR Attention Assessment Module: Virtual Elevator

Sohlberg & Matter in 1989 proposed a clinical model of attention processes that outlined hierarchically organized levels of attention. The author's personal experience dealing with brain injured patients on a daily basis is clearly in support of this clinical model of attention. Specifically, these patients, especially in the acute stage of injuries, most often are unable to sustain attention even for a short period of time. Their distractibility level increases significantly, affecting both everyday life and academic/learning activities. The rate of recovery of this cognition function is influenced by numerous factors, such as the initial impact, degree of structural damage and/or initial functional deficits, although it is not well-documented in clinical research. Overall, the conceptual model of attention includes:

Focused attention – basic ability to respond to specific external stimuli

Sustained attention – commonly termed "concentration," referring to the maintenance of consistent behavioral responses during continuous and repetitive activity

Selective attention – commonly termed "freedom from distractibility"

Alternating attention – flexibility, allowing for shifts in the focus of attention

Divided attention – when two or more behavioral responses are required to activate simultaneously

A Virtual Reality proprietary prototype of Test of Everyday Attention (TEA) using the Virtual Elevator model has been recently developed by Head Rehab, LLC. Sustained attention, similar to the "Elevator Counting" test (Andrew et al., 2001) has been developed. The patient is "situated" in the Virtual Elevator (VE) moving up (from floor 1 to floor 12) and down (from floor 12 to floor 1). There are visual separations that patient should count in order to identify the *floor indicator* upon arrival (stop). An overview of the virtual scene is shown in Figure 10 below:

Figure 10. Virtual elevator overview.

This VR module has been tested using neurologically normal controls and brain injured patients. A cursory analysis of the results clearly indicates considerable difference between the subjects, indicating the deficits in sustained attention in neurological patients. As observed, the neurological patient accumulated six errors during performance of the task. In contrast, the normal control subject made only one error during the entire experimental session. Overall, Head Rehab's VR Elevator prototype has proven to be a highly sensitive tool for assessing the cognitive deficits (e.g., problems with sustained attention) in neurological patients.

2.6.3 Spatial Awareness: Virtual Person

VR rehabilitation of a patient's spatial awareness can be carried out in the context of "executive dysfunctions" using "active search" and "visual-constructional" VR-based task paradigms. Specifically, a *Whole Body Spatial Sequencing* design has been created by Head Rehab, LLC (proprietary information) to address this issue. The conceptual framework for this design has been developed based on the notion of *egocentric* memory (spatial knowledge relative to the observer) and *allocentric* memory (spatial knowledge relative to cues independent of the observer). It should be noted that there are three dimensions of spatial abilities – spatial relations and orientation; visualization; and kinesthetic imagery (ability to determine the spatial position of an object in relation to oneself) – all of which are selectively sensitive towards cognitive impairment as a result of trauma. Using the default *Virtual Person* (VP) position and user friendly VR software, injured patients may be involved creating sequences of frames representing different movement patterns, such as stepping over obstacles, sitting and standing, reaching movement, throwing a ball, etc. In this version of the Head Rehab LLC software, patients can create a variety of real-life daily activities. An example of frames to be created by patients is shown in Figure 11 below:

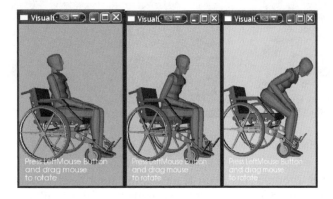

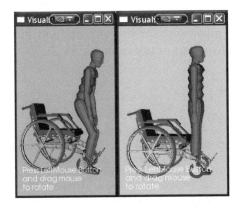

Figure 11. Sequences of bosy postures associted with sitting and standing task.

It should be noted that Head Rehab, LLC. software allows to "rotate", "change the view of observation" using various interactive devices (i.e. computer mouse, joysticks etc.) and animate created frames to show the created movement patterns.

CONCLUSION

Virtual Reality is an interactive, computer generated 3D environment that simulates the real world and provokes in the user a sensation of immersion. Numerous recent studies have examined the use of VR in skill, including spatial rotation, depth perception, 3D field dependency, static and dynamic manual target tracking and visual field-specific reaction time along the human life-span. More recently, VR has been used for the evaluation of the daily skills of patients, including those with vestibular dysfunctions, and those undergoing memory and motor rehabilitation (see Rose, 1996, for review). It is expected that the capabilities of VR will be especially evident in the neurological rehabilitation of injured athletes, where cognitive, emotional and behavioral problems often interact with physical impairments to reduce the overall level of functionality and moving ability. Finally, one of the major challenges facing clinicians in medical rehabilitation is identifying intervention strategies and techniques that are cost-effective, ecologically valid, motivating and transferable to functioning in the "real" world. Virtual reality, indeed, has the capacity to address these challenges.

Acknowledgements

Images and information included in section *2.6. "VR for Cognitive Rehabilitation"* have been provided by Head Rehab, LLC. I also acknowledge special contribution of Elena Slobounov, who has been developing VR applications for this project and has been instrumental in providing technical support of my research for the last 20 years.

REFERENCES

Rose, D. (1996). Virtual reality in rehabilitation following traumatic brain injury. Proc. 1st. Conference on disability, virtual reality & Associated Technologies. Maidenhead, UK.

World Health Organization. (1980). International classification of impairment, disabilities and handicaps: a manual of classification relating to the consequences of disease. Geneva: WHO.

Hallett, M. (2001). Plasticity of the motor cortex and recovery from stroke. *Brain Research Review, 36*, 169-174.

Atnee, E.A., Briiksm, B.M., Rose, F.D., Andrws, T.K., Leadbetter, A.G., Clifford, B.R. Memory processes and virtual environment: I can't remember what was there, but I can remember how I got there: Implications for people with disabilities", Proc. 1st Euro. Conf. Disability, Virtual Reality & Assoc. Tech, Maidenhead, UK, 1996.

Rose, D., Brooks B., Rizzo, A. (2005). Virtual reality in brain damage rehabilitation. *CyberPsychology & Behavior, 8*, 241-262.

Cicerone, K.D. (2002). Remediation of 'working attention' in mild traumatic brain injury. *Brain Injury, 16*(3), 185-195, 2002.

Slobounov, S., Kraemer, W., Sebastianelli, W., Simon, R., Poole, S. (1998). The efficacy of modern motion tracking and computer graphics technologies in a clinical setting. *Journal of Sport Rehabilitation, 7(1)*, 20-32.

Slobounov, S., Poole, S., Simon, R., Slobounova, E., Bush, J., Sebastianelli, W., Kraemer, W. (1999). The efficacy of modern technology to improve healthy and injured shoulder joint position sense. *Journal of Sport Rehabilitation, 8(1)*, 10-23.

Strong, J., Unruh, A, Wright, A., Baxter, D. *Pain: A textbook for therapists.* Churchill & Livingstone, 2002.

Hoffman, H.G., Patterson, D.R., Carrougher, G.L., Sharar, S.R. (2001). Effectiveness of virtual reality-based pain control with multiple treatments. Clinical Journal of Pain, 17, 229-235.

Hoffman, H.G., Garcia-Palacios, A., Patterson, D.R., et al. (2001). The effectiveness of virtual reality for dental pain control: a case study. CyberPsychology & Behavior, 1 (4), 527-535.

Schneider, S.M., Prince-Paul, M., Allen, M.J., Silverman, P., Talaba, D. (2004). Virtual reality as a distraction intervention for women receiving chemotherapy, *Oncology Nursing Forum, 31*, 81-88.

Sander Wint, S., Eshelman, D., Steele, J., Guzzetta, C.E. (2002). Effects of distraction using virtual reality glasses during lumbar punctures in adolescents with cancer. *Oncology Nursing Forum, 29*, E8-E15.

Hoffman, H.G., Doctor, J.N., Patterson, D.R. et al. (2000). Virtual reality as an adjunctive pain control during burn wound care in adolescent patients. Pain, 85, 305-309.

Sohlberg, M.M., & Mateer, C.A. (1989). *Introduction to cognitive rehabilitation: Theory and practice*, New York: Guilford Press.

Andrew, J., Bate, J.L., Mathias, J., & Crawford, R. (2001). Performance on the Test of Everyday Attention and Standard Tests of Attention following Severe Traumatic Brain Injury. *Clinical Neuropsychology, 15*(3), 405-422.

INDEX

A

ACL 12, 66, 204, 205, 244, 300
Accommodation in Training 25
Acute Injury 3, 322, 476
Acute traumatic injury 276, 279
Adaptation to training 25, 31
Adjustment to injury 151
Affective response to injury 150, 364
Anticipatory Posture (APA) 52
American Academy of Neurology (ANN) 499, 500
Amnesia 400-403, 417
    anterograde amnesia 406, 437
    post-traumatic amnesia 404
    retrograde amnesia 406
Amygdala 278

B

Basketball 13, 162, 202
Baseball 162, 202
Balance Training 65, 66, 526
    Stability 47, 47, 418
Balance Risk factor 43, 418
Block Composition 37
Biopsychosocial Model
    Pain 327
    Rehabilitation 479
Body weight 115
Bracing behavior 300, 302, 303
Brain imaging 54, 279
Brain plasticity 473

C

Catastrophic Injury 5
Cheerleading 169, 188, 221
Chronic Injury 3, 4, 198

CNS 81, 82, 243
Cognitive Testing 424, 433, 434, 436
Cognitive Responses to Injury 148, 149, 214, 364, 372
Cognitive 415
    Cognitive Tests 432-438
Concussion 59, 134, 199, 377
    Concussion Mechanisms 380
    Biomechanics 381
    Classification 399, 452
    Concussion signs & Symptoms 415
    Second Impact Syndrome 390
    Cumulative Effect 391
Collegiate coaches 119-142
Coaching Styles 113, 178
Cyclic Sports 231
    Cumulative effect 55, 416

D

Dancing 220
Disability 320, 321, 388, 472

E

Emotional trauma 251
Electroencephalography (EEG) 83, 84, 91, 419, 495, 496, 504, 505
    Bereitschaftspotential (BP) 502, 503
    Contingent Negative Variation (CNV) 502
    Neurofeedback 507-509
    EEG Types 469
        International 10/20 495
        qEEG 504

Lateralized Readiness
Potential (LRP) 504
Low resolution
Electromagnetic Tomography
(LORETA) 283, 504, 505
Movement-related cortical
potentials, MRCP 503
ERP 283
Endurance 226, 235
Epidemiology 4, 13, 362, 449

F

Fatigue 77
    Self-reports 77, 86
    Peripheral 79
    Central 81
Fatigue & Injury 85
Fatigue & Nutrition 103
    Central Fatigue
    Hypothesis 82
Fear 269, 288
    Quotes 270, 271
    Neurobiology 273
    Animal Models 274
    Human Models 276
    Perception 279, 280, 282
Fear of Injury 134, 289, 291,
298-304
    High order fear 273
    Fear conditioning 274,
275
Female Athletes 147, 296
Fencing 14, 122
Field Hockey 18
Figure skating 219
Fitness & Injury 217
Flexibility 225, 234, 478
Football 11, 19, 121, 171, 182,
204, 224, 249
Force platform 48

G

Gender effect 105, 357, 368
Glasgow coma score 453
Golf 14
Gymnastics 14, 121, 218
Grading Scales TBI 400-404

H

Handicap 472
Human Models of Fear 276

I

Ice Hockey 17, 122
Identical Elements Theory 28
    ImPACT, TBI 434, 435
Individualization in Training
29
Injury Traumatic 61, 343, 346
Injury Classification 6
Interview 119, 161-194
    Methods 152
    Types 153
    Skill 154
    Rapport 156
    Non-Verbal 157

K

Kinesiophobia 289, 292, 305
    Tampa Scale (TSK) 293-295
Knowledge-Attitude 116, 117

L

Lacrosse 17, 18, 166, 206
Low back Injury 7, 9
Loss of Consciousness (LOC)
417

M

Major Injury 5, 334
Magnetic Resonance Imaging
(MRI) 7
    fMRI 83, 84, 91, 277, 283
Minor/mild Injury 4, 322, 335
Moderate Injury 5, 322
Magnitoencephalogram (MEG)
506
Metabolic cascade TBI 382-386
Mild traumatic brain injury
(MTBI) 56
Multiple concussions 412

N

NCAA 11, 115
Neurological Injury 7, 8
Neuropsychology 432, 456
Nutrition 97, 98
    Vitamin 102
    Protein 98
    Carbohydrates 9
    Antioxidants 99
    Creatine 101
    Supplements 102
Neurochemical Cascade TBI
386-388

O

Observation Formats 298
Olympic Games 6
Overload in Training 25
Overuse Injury 202, 207

P

Pain 311, 531
    Quotes 312, 313
    Hyperalgesia 311, 318
    Allodynia 311, 312, 318

    Nociception 312, 314,
    320
Pain Pathway 315
    Types 317
    Gate Control Theory
    322
Pain Modulation 322
Pediatric TBI 448
    Pediatric Sequelae 455-459
    Neuropathology Pediatric
    TBI 455
PET 506
Periodization 29, 30, 33, 232
Perceived Risk 289
Post-concussive symptoms 416,
420
Posture 45-47
    Posture Control (models)
    51
    Cortical Control 53
    Posture Control
    Abnormal 56, 89
Prefrontal Cortex (PFC) 279
Prevention of Injuries 366, 461
Psychological Trauma 10, 245,
306, 332, 357
    Concept 245
    Components 250
    Predispositions 254
    Model 264
    Age Effect 358

Q

Q EEG 504, 505
    Q Current source localization
    505

R

Recovery pediatric MTBI 459
Rehabilitation 300, 469, 516, 534

Goals 470
Types of Goals 470
Principles 472
Models 474, 481
VR Rehab 521
Cognitive Rehab 534-539
Return-to-play 289, 408
Stages 474
Risk of TBI 450
Running 15, 174, 207
Rowing 15
Rugby 19, 28

S

Severe Injury 3, 334
Skeletal/muscle Injury 61, 343, 346
Skiing 211
Soft Tissue Injury 7
Sport Disabling Injury 5,
Soccer 18, 177, 194
Softball 13
Star Excursion Balance Test 90
Stress-Injury 147
Strength 227, 233
Swimming 16, 209
Symptoms TBI 422, 423

T

Tennis 16, 197, 198
Training Effects 34, 35
Track & Field 179
Training Specificity 25, 26
TMS 82, 84, 91
Traumatic brain injury (TBI) 20, 25, 248
Computerized testing 434-436
Pediatric evaluation 459
TBI evaluation 415
TBI grading scales 400-406
Trailmaking Tests 428, 429

V

Virtual reality (VR) 49, 66, 516
Volleyball 16, 124, 210
VTC Balance 57

W

Water Polo 19, 210
WHO 480
Wrestling 19, 300